ENVELOPING ALGEBRAS

North-Holland Mathematical Library

Board of Advisory Editors:

M. Artin, H. Bass, J. Eells, W. Feit, P. J. Freyd, F. W. Gehring, H. Halberstam, L. V. Hörmander, M. Kac, J. H. B. Kemperman, H. A. Lauwerier, W. A. J. Luxemburg, F. P. Peterson, I. M. Singer, and A. C. Zaanen

VOLUME 14

NORTH-HOLLAND PUBLISHING COMPANY
AMSTERDAM · NEW YORK · OXFORD

ENVELOPING ALGEBRAS

JACQUES DIXMIER

University of Paris VI

1977

NORTH-HOLLAND PUBLISHING COMPANY

AMSTERDAM · NEW YORK · OXFORD

North-Holland ISBN for the series: 0 7204 2450 X
North-Holland ISBN for this volume: 0 7204 0430 4

A translation of:

ALGEBRES ENVELOPPANTES
© BORDAS (Gauthier-Villars), Paris, 1974

Translated by:

Minerva Translations, Ltd., London

Published by:

NORTH-HOLLAND PUBLISHING COMPANY
AMSTERDAM · NEW YORK · OXFORD

Sole distributors for the U.S.A. and Canada:
ELSEVIER NORTH-HOLLAND, INC.
52 Vanderbilt Avenue
New York, N.Y. 10017

Library of Congress Cataloging in Publication Data

Dixmier, Jacques.
Enveloping algebras
(North-Holland mathematical library; v. 14)
Translation of Algèbres enveloppantes.
Bibliography: p. 359
Includes index.
1. Universal enveloping algebras. 2. Lie algebras.
3. Representations of algebras. 4. Ideals (Algebra)
I. Title.
QA252.3.D5713 512'.2 76-2622 ISBN 0-444-11077-1

PRINTED IN THE GERMAN DEMOCRATIC REPUBLIC

PREFACE

If G is a locally compact group, an important problem consists in determining the continuous representations of G in Hilbert spaces or even in more general topological vector spaces. Let us assume that G is a real Lie group. Let $\mathfrak{g}$ be the complex Lie algebra of G. The study of the finite-dimensional representations of G is almost entirely equivalent to that of the finite-dimensional representations of $\mathfrak{g}$. For infinite-dimensional representations there again exist relationships, which are clearly more delicate, between representations of G and representations of $\mathfrak{g}$; we shall return to this later. In any case, it seems reasonable to consider the representations of $\mathfrak{g}$.

The study of the representations of $\mathfrak{g}$ can be transformed into a problem of associative algebra by passage to the universal enveloping algebra $U(\mathfrak{g})$ of $\mathfrak{g}$. From Poincaré—Birkhoff—Witt's theorem, $\mathfrak{g}$ is embedded in $U(\mathfrak{g})$, and, if $(x_1, \ldots, x_n)$ is a basis for $\mathfrak{g}$, the monomials $x_1^{\nu_1} \cdots x_n^{\nu_n}$ (where $\nu_1, \ldots, \nu_n$ are integers $\geqq 0$) form a basis for the vector space $U(\mathfrak{g})$. Thus, although it is of infinite dimension, the algebra $U(\mathfrak{g})$ is completely open to calculation. For example, let us take for $\mathfrak{g}$ the 3-dimensional Heisenberg algebra $\mathfrak{h}$ having a basis (x, y, z) such that $[x, y] = z$ and $[x, z] = [y, z] = 0$. Then $U(\mathfrak{h})$ has the basis $(x^m y^n z^p)$, where $m, n, p \in \mathbf{N}$; we have

$$(x^m y^n z^p)\,(x^{m'} y^{n'} z^{p'}) = x^m y^n x^{m'} y^{n'} z^{p+p'}$$

and we must transform $y^n x^{m'}$. Now $y^n x^{m'} = y^{n-1} xyx^{m'-1} - y^{n-1} x^{m'-1} z$, and by recurrence we arrive at the multiplication table without difficulty.

Let us now return to the general case. Every representation π of $\mathfrak{g}$ may be extended in a unique way to a representation π' of $U(\mathfrak{g})$, and $\pi \mapsto \pi'$ is a bijection between the representations of $\mathfrak{g}$ and of $U(\mathfrak{g})$ which preserves equivalence, irreducibility, etc. Passing from $\mathfrak{g}$ to $U(\mathfrak{g})$ has the disadvantage of introducing an infinite-dimensional algebra, but allows us to use associative methods (maximal left ideals, localization, etc.).

Let us now seek the irreducible representations of $\mathfrak{h}$, or of $U(\mathfrak{h})$. In such a representation π, the endomorphism $\pi(z)$ must be scalar. If $\pi(z) = 0$, we have to choose $\pi(x)$ and $\pi(y)$ permutable, and the representation must be of dimension 1; we thus obtain all linear forms over $\mathfrak{h}$ which are zero in z. Let us assume that $\pi(z)$ is a scalar $\alpha \neq 0$. We must choose endomorphisms $\pi(x)$ and $\pi(y)$ such that

$$\pi(x)\,\pi(y) - \pi(y)\,\pi(x) = \alpha \cdot 1,$$

in such a way that π is irreducible. A well-known solution involves taking the following endomorphisms in the vector space $\mathbf{C}[X]$:

$$(\pi_\alpha(x)f)\,(X) = -\alpha X f(X), \qquad (\pi_\alpha(y)f)\,(X) = \frac{\mathrm{d}f}{\mathrm{d}X}.$$

But a deeper study reveals the existence of an enormous number of irreducible representations of $\mathfrak{h}$, even for α $(\neq 0)$ fixed. It seems that these representations defy classification. A similar phenomenon exists for $\mathfrak{g} = \mathfrak{sl}(2)$, and most certainly for all non-commutative Lie algebras.

On the other hand, let us return to a real Lie group G and its complex Lie algebra $\mathfrak{g}$. Let π be a unitary representation of G in a Hilbert space H. We construct a representation of $\mathfrak{g}$ associated with π in the following way: Let H_∞ be the set of vectors of H which are indefinitely differentiable for π. Then $\mathfrak{g}$ operates naturally in H_∞, whence we have a representation π_∞ of $\mathfrak{g}$ (or of $U(\mathfrak{g})$) in H_∞. Unfortunately, if π is (topologically) irreducible, π_∞ is not at all algebraically irreducible and may even contain (as, for example, in the case of the Heisenberg group) an infinite number of irreducible representations which are pairwise non-equivalent.

The passage from G to $\mathfrak{g}$ for infinite-dimensional representations does not, therefore, appear in an encouraging form at first sight.

But let A be an associative algebra. Let $A^\wedge$ be the set of classes of irreducible representations of A. For all $\pi \in A^\wedge$, Ker π is called a primitive ideal of A. Let Prim(A) be the set of primitive ideals of A. The mapping $\pi \mapsto \operatorname{Ker} \pi$ of $A^\wedge$ into Prim(A) is surjective, but not in general injective. Even if $A^\wedge$ is very large, Prim(A) can be of reasonable size. N. Jacobson has equipped it with a topology and termed it the structural space of A.

Let us now return to $A = U(\mathfrak{h})$. While $A^\wedge$ is very large, Prim(A) is completely calculable. It is the disjoint union of two subsets M_1 and M_1, where M_1 is the set of kernels of one-dimensional representations of A, and M_2 the set of kernels of the π_α (where α is a non-zero scalar).

On the other hand, the link between $G^\wedge$ (the set of classes of irreducible unitary representations of G) and Prim $U(\mathfrak{g})$ seems to be close. Let π be an irreducible unitary representation of G. Then, we repeat, π_∞ is not in general irreducible, but it is probable that Ker π_∞ is a primitive ideal of $U(\mathfrak{g})$; in any case, this has been demonstrated for G semi-simple and for G solvable. Let us assume that G is simply connected and nilpotent. Then the passage from π to Ker π_∞ defines a bijection from $G^\wedge$ onto the "Hermitean" subset of Prim $U(\mathfrak{g})$ [i.e., the set of primitive ideals of $U(\mathfrak{g})$ which are invariant under the canonical involution of $U(\mathfrak{g})$]. When G is not nilpotent, the relationship between $G^\wedge$ and Prim $U(\mathfrak{g})$ is more complicated and very little understood. Nevertheless, a close study of Prim $U(\mathfrak{g})$ would be useful for the study of $G^\wedge$, and will certainly be simpler.

Having claimed that we had to study the primitive ideals of $U(\mathfrak{g})$ and not its irreducible representations, we should now qualify this assertion. As a matter of fact, by imposing supplementary conditions on the representations under consideration, we obtain families of irreducible representations which we may hope to classify. Up to now, however, such families have only been usefully defined for G semi-simple. For example, we shall study certain representations of $\mathfrak{g}$ which are linked with a maximal compact subgroup of G.

Enveloping algebras are interesting for another reason. Several attempts have been made to extend the methods of algebraic geometry to the non-commutative case. It is not very clear which is the right category of algebras to consider. Enveloping algebras are certainly insufficient, but in certain respects they figure among the simplest non-commutative algebras. They therefore constitute a good starting point to study "non-commutative algebraic geometry". For this and for aesthetic reasons, we shall not take **C** but a commutative field of characteristic 0 as the base field (little can be done in the case where the characteristic >0). Apart from this, there is a more pressing reason: even if we are chiefly concerned with the case of **C**, certain methods require extensions of fields. (As a matter of fact, a small number of recent arguments even require taking a ring as base: this is scarcely reassuring for the future!)

The logical sequence of chapters is as follows:

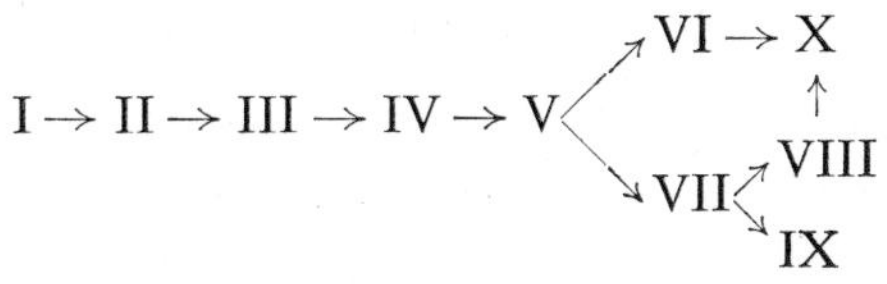

The study of enveloping algebras depends of course on a fairly detailed knowledge of Lie algebras. On the other hand, certain properties of Lie algebras may conveniently be established by the use of enveloping algebras (cf. 2.5, 7.2 and 7.5); in this book, we exploit this possibility. But the reader may then feel trapped in a vicious circle. For this reason, we establish in Chapter 1, using standard methods, those properties of Lie algebras which will be necessary in the rest of the book. (We pass fairly rapidly over the first few proofs, as this book ought not to be considered as an introduction to Lie algebras.) However, since the properties of root systems obviously do not depend on the theory of Lie algebras, we have restricted ourselves to recalling in the appendix, without proof, those properties which are indispensable.

Chapter 2 introduces the protagonists: the enveloping algebras. To study their primitive ideals, we need some information about their general two-sided ideals: this is the aim of Chapter 3. Chapter 4 deals with one of the links (but not the only one) between enveloping algebras and commutative algebras, i.e., the centres of enveloping algebras, of their quotients and of their rings of fractions.

Mainly by virtue of the concept of induced representations and its variants, we are able to construct simple representations of Lie algebras and hence the primitive ideals of enveloping algebras. This notion is studied in Chapter 5.

We now have the principal tools at hand. In Chapter 6, we determine all primitive ideals of $U(\mathfrak{g})$ when $\mathfrak{g}$ is solvable and the base field is algebraically closed. To do this, we use the orbital method, introduced by A. A. Kirillov for the case of nilpotent Lie groups.

Chapters 7, 8 and 9 deal with the case where $\mathfrak{g}$ is semi-simple. In Chapters 7 and 9, we study some particular simple representations, linked with the choice of a Cartan subalgebra (Chapter 7), or with a symmetric decomposition (Chapter 9). In Chapter 8, we determine, among other things, the minimal primitive ideals of $U(\mathfrak{g})$ for an algebraically closed base field. These chapters leave much to be desired: on the one hand, many of the problems remain to be solved; on the other hand, we have not been able to establish certain important results here because they depend on non-algebraic methods. One may hope that this situation will soon be remedied.

Chapter 10 is based on the whole of Chapters 1 to 8. If $\mathfrak{g}$ is any Lie algebra over an algebraically closed field, it is probable that the orbital method, in a suitable form, can again be applied. In any case, we succeed in constructing a large family of primitive ideals of $U(\mathfrak{g})$ linked with the "regular" linear forms on $\mathfrak{g}$.

Each chapter closes with "supplementary remarks". Here we specify certain bibliographical questions, we give counter-examples, and we indicate, without proof but with references, additional matter which is occasionally important. Such matter is not used in the text. The open problems (which are legion!) are collected at the end of the book.

Naturally we use a lot of general results of algebra, for which the reader is referred to N. Bourbaki, *Algèbre* (quoted as AL), and *Algèbre commutative* (quoted as AC). From time to time, we make use of the theory of algebraic groups, for which the reader is referred to C. Chevalley, *Theorie des groupes de Lie*, Chap. 2 (quoted as CH) and Chap. 3 (quoted as CH′), and A. Borel, *Linear algebraic groups* (quoted as BO). Various references are collected in the Appendix.

The bibliography contains, with a few exceptions, only essentially algebraic works. Certain ideas used in the book originated, however, in studies devoted to Lie groups (for example the publications of G. W. Mackey on induced representations); I regret that I was unable to quote them all.

The results which follow have, for several years, been the subject of seminars in Paris. My thanks are due to the lecturers, notably P. Bernat, N. Conze, M. Duflo, M. Raïs, G. Renault, R. Rentschler, and M. Vergne. Their lectures have very often served as the basis of my text. P. Bernat, M. Duflo, P. Gabriel, J. Lepowsky, S. Mauro and M. Vergne have constructively criticized the manuscript of this book.

It is impossible to conclude this introduction without at least quoting the names of I. M. Gelfand, Harish-Chandra, B. Kostant who founded the theory or contributed some of its most important advances.

PREFACE TO THE ENGLISH EDITION

Only a few minor changes have been made to the French edition: (1) the bibliography has been updated; (2) some of the problems which have since been solved have been replaced by fresh ones; (3) sections 7.6.3, 7.6.6, 7.6.10, (resp. 7.6.9) have been improved following suggestions by W. Borho (resp. J. Lepowsky); (4) a few minor additions and corrections have been made.

The curves shown on p. XIV have their origin in the study of $U(\mathfrak{sl}(3))$. They are due to Professor W. Borho, who kindly authorized me to reproduce them.

J. D.

CONTENTS

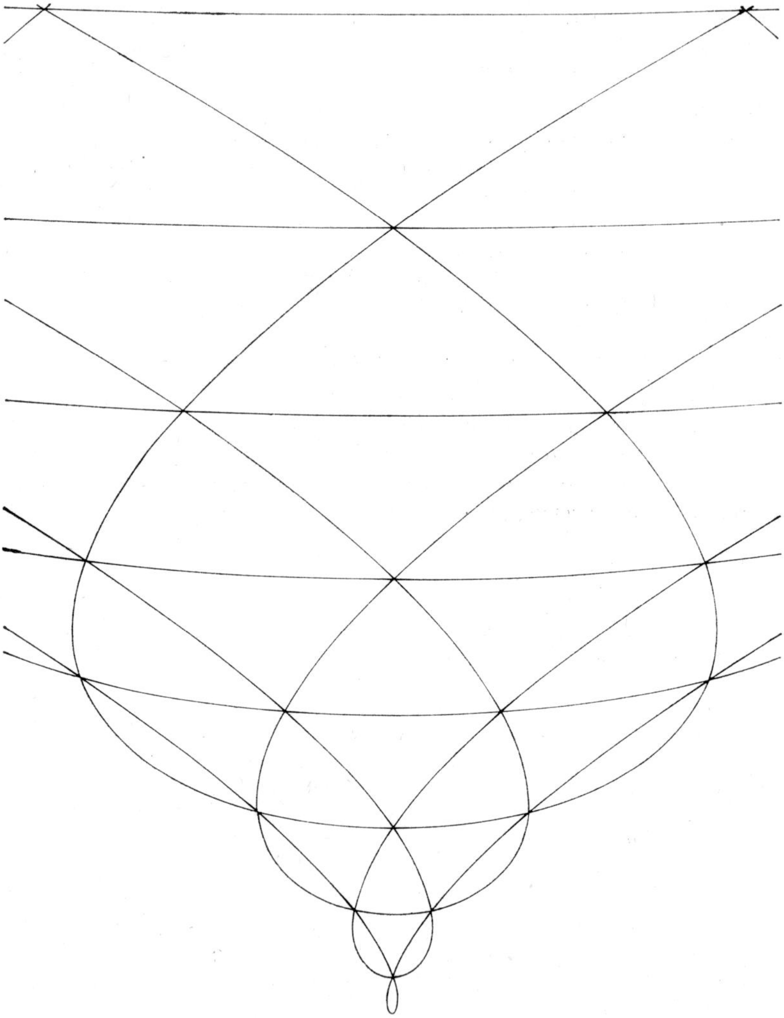

NOTATION

We first present the following very frequently used notation:

$$\begin{array}{ccc} U(\mathfrak{g}) \to K(\mathfrak{g}) & & S(\mathfrak{g}) \\ \uparrow \qquad\quad \uparrow & & \uparrow \\ Z(\mathfrak{g}) \to C(\mathfrak{g}) & & Y(\mathfrak{g}) \end{array}$$

$U(\mathfrak{g})$: the enveloping algebra of $\mathfrak{g}$;
$Z(\mathfrak{g})$: the centre of $U(\mathfrak{g})$;
$K(\mathfrak{g})$: the field of fractions of $U(\mathfrak{g})$;
$C(\mathfrak{g})$: the centre of $K(\mathfrak{g})$;
$S(\mathfrak{g})$: the symmetric algebra of $\mathfrak{g}$;
$Y(\mathfrak{g})$: the set of the elements of $S(\mathfrak{g})$ which vanish under the representation deduced from the adjoint representation.

$\mathfrak{gl}(V)$: 1.1.2.
$\mathbf{M}_n(k)$: 1.1.2.
$\mathfrak{gl}(n, k)$: 1.1.2.
E_{ij}: 1.1.2.
$\mathrm{Aut}(\mathfrak{g})$: 1.1.3.
$[\mathfrak{a}, \mathfrak{b}]$: 1.1.4.
$\mathrm{tr}(x)$: 1.1.5.
$\mathfrak{sl}(V)$: 1.1.5.
$\mathfrak{sl}(n, k)$: 1.1.5.
$\mathrm{ad}_{\mathfrak{g}}x$, $\mathrm{ad}\, x$: 1.1.10.
$\mathrm{Aut}_e(\mathfrak{g})$: 1.1.14.
$[V]$: 1.2.2.
$\oplus\, \varrho_i$: 1.2.3.
$\mathfrak{g}^\wedge$:1.2.5.
$\mathcal{IH}\,(V)$: 1.2.6.
V_ξ: 1.2.8, 1.2.13.
$\mathrm{mtp}(\xi, V)$: 1.2.8.
$V^{\mathfrak{g}}$: 1.2.10.
V^λ: 1.2.13.
$\varrho \otimes \varrho'$: 1.2.14.
$\otimes^p \varrho$: 1.2.14.
$S^p\varrho$: 1.2.14.
$\Lambda^p\varrho$: 1.2.14.
ϱ^*: 1.2.16.
$\mathscr{C}^i\mathfrak{g}$: 1.3.1.
$\mathscr{D}^i\mathfrak{g}$: 1.3.2.
$\mathfrak{g}^0(x)$: 1.9.6.
$\mathfrak{g}^*(x)$: 1.9.6.
$R(\mathfrak{g}, \mathfrak{h})$: 1.10.1, 1.10.22.
h_λ: 1.10.2.
$\langle\cdot,\cdot\rangle$: 1.10.3.
H_α: 1.10.4.
s_α: 1.10.9.
$W(\mathfrak{g}, \mathfrak{h})$: 1.10.10.
B_f: 1.11.1.
$\mathfrak{a}^f$: 1.11.1.

Unless otherwise indicated, k denotes a commutative field with characteristic 0, and $\mathfrak{g}$ denotes a Lie algebra over k, which from Section 1.3 on is of finite dimension. Although this may occasionally appear irksome, each additional hypothesis concerning $\mathfrak{g}$ or k will be explicitly stated in every case.

All vector spaces and all algebras have k as their base field, unless otherwise indicated. The symbols $\otimes$ *and* Hom *denote tensor products and spaces of homomorphisms taken over k; linear independence is to be understood as relative to k; etc.*

Topological notions refer to the Zariski topology.

CHAPTER 1

LIE ALGEBRAS

1.1. General remarks

1.1.1. A *Lie algebra* is a vector space $\mathfrak{g}$ together with a multiplication (usually termed a bracket and denoted by $(x,y) \mapsto [x,y]$) such that:

(1) $[x,y]$ depends linearly on x and y;

(2) $[x,x] = 0$ for all $x \in \mathfrak{g}$;

(3) $[x,[y,z]] + [y,[z,x]] + [z,[x,y]] = 0$ for all $x,y,z \in \mathfrak{g}$.

Properties (1) and (2) imply that $[y,x] = -[x,y]$ for all $x,y \in \mathfrak{g}$. Two elements x and y of $\mathfrak{g}$ are said to *commute* if $[x,y] = 0$. $\mathfrak{g}$ is said to be *commutative* if any two elements of $\mathfrak{g}$ are permutable.

1.1.2. Let A be an algebra. (Henceforth the word "algebra" alone will refer to an "associative algebra".) For $x,y \in A$, write $[x,y] = xy - yx$. Then A is a Lie algebra, called the *underlying* Lie Algebra of A. Let V be a vector space, and End(V) the algebra of endomorphisms of V. The underlying Lie algebra of End(V) is denoted by $\mathfrak{gl}(V)$. Let $\mathbf{M}_n(k)$ be the algebra of matrices with n rows and n columns with elements in k. The underlying Lie algebra of $\mathbf{M}_n(k)$ is denoted by $\mathfrak{gl}(n,k)$. By $(E_{ij})_{1\leq i,j\leq n}$ we denote the canonical base of $\mathfrak{gl}(n,k)$. (E_{ij} is the matrix (α_{kl}) such that $\alpha_{kl} = 0$ for $(k,l) \neq (i,j)$ and $\alpha_{ij} = 1$.)

1.1.3. Let $\mathfrak{g},\mathfrak{g}'$ be Lie algebras. A *homomorphism* of $\mathfrak{g}$ into $\mathfrak{g}'$ is a linear mapping φ of $\mathfrak{g}$ into $\mathfrak{g}'$ such that $\varphi([x,y]) = [\varphi(x),\varphi(y)]$ for any $x,y \in \mathfrak{g}$. *Isomorphisms* and *automorphisms* of Lie algebras are defined in the obvious way. The group of automorphisms of $\mathfrak{g}$ is denoted by $\mathrm{Aut}(\mathfrak{g})$.

1.1.4. If $\mathfrak{a},\mathfrak{b}$ are vector subspaces of $\mathfrak{g}$, we denote by $[\mathfrak{a},\mathfrak{b}]$ the set of linear combinations of elements of the form $[a,b]$, where $a \in \mathfrak{a}$ and $b \in \mathfrak{b}$. A *Lie subalgebra* (or an *ideal*) of $\mathfrak{g}$ is a vector subspace $\mathfrak{g}'$ of $\mathfrak{g}$ such that $[\mathfrak{g}',\mathfrak{g}'] \subset \mathfrak{g}'$ (or $[\mathfrak{g},\mathfrak{g}'] \subset \mathfrak{g}'$). If $\mathfrak{g}'$ is an ideal of $\mathfrak{g}$, the bracket in $\mathfrak{g}$ defines by passage to the quotient a bracket in the vector space $\mathfrak{g}/\mathfrak{g}'$ which makes $\mathfrak{g}/\mathfrak{g}'$ a Lie algebra, termed the *quotient Lie algebra* of $\mathfrak{g}$ by $\mathfrak{g}'$.

1.1.5. Let V be a finite-dimensional vector space. The set of the $x \in \mathfrak{gl}(V)$ whose trace $\mathrm{tr}(x)$ is zero is an ideal of $\mathfrak{gl}(V)$ denoted by $\mathfrak{sl}(V)$. The set of the $x \in \mathfrak{gl}(n,k)$ such that $\mathrm{tr}(x) = 0$ is an ideal of $\mathfrak{gl}(n,k)$ denoted by $\mathfrak{sl}(n,k)$.

A square matrix (α_{ij}) is termed *diagonal* (resp. *lower triangular, strictly lower triangular, upper triangular, strictly upper triangular*) if $\alpha_{ij} = 0$ for $i \neq j$ $(i < j,\ i \leqq j,\ i > j,\ i \geqq j)$. The set of diagonal (lower diagonal, etc.) matrices is a Lie subalgebra of $\mathfrak{gl}(n,k)$.

1.1.6. Let $\mathfrak{g}_1,\mathfrak{g}_2$ be Lie algebras. Let us provide the vector space $\mathfrak{g}_1 \times \mathfrak{g}_2$ with the bracket

$$[(x_1,x_2),(y_1,y_2)] = ([x_1,y_1],[x_2,y_2]) \quad \text{for} \quad x_1,y_1 \in \mathfrak{g}_1,\ x_2,y_2 \in \mathfrak{g}_2.$$

Then $\mathfrak{g}_2 \times \mathfrak{g}_2$ is a Lie algebra, termed the *product Lie algebra* of $\mathfrak{g}_1$ and $\mathfrak{g}_2$. $\mathfrak{g}_1$ and $\mathfrak{g}_2$ are ideals in $\mathfrak{g}_1 \times \mathfrak{g}_2$ with sum $\mathfrak{g}_1 \times \mathfrak{g}_2$ and such that $\mathfrak{g}_1 \cap \mathfrak{g}_2 = 0$. Conversely, let $\mathfrak{h}$ and $\mathfrak{k}$ be ideals of $\mathfrak{g}$ such that $\mathfrak{h} + \mathfrak{k} = \mathfrak{g}$ and $\mathfrak{h} \cap \mathfrak{k} = 0$; the mapping $(h,k) \mapsto h + k$ of $\mathfrak{h} \times \mathfrak{k}$ into $\mathfrak{g}$ is a Lie algebra isomorphism, by means of which we can identify $\mathfrak{g}$ and $\mathfrak{h} \times \mathfrak{k}$. There are analogous definitions and results for the product of a finite number of Lie algebras.

1.1.7. Let us provide $\mathfrak{g}$ with the bracket $(x,y) \mapsto -[x,y]$. Then $\mathfrak{g}$ will be a new Lie algebra, termed *opposite* to $\mathfrak{g}$.

1.1.8. Let k' be an extension of k. On the k'-vector space $\mathfrak{g} \otimes k'$ there exists one and only one multiplication extending that of $\mathfrak{g}$ and making $\mathfrak{g} \otimes k'$ a Lie algebra over k'; this Lie algebra is said to be *deduced from* $\mathfrak{g}$ *by extension of the field of scalars from* k *to* k'.

1.1.9. Let P and Q be subsets of $\mathfrak{g}$. The set of elements of Q which commute with all elements of P is termed the *centralizer* of P in Q. If $Q = \mathfrak{g}$, this centralizer is a Lie subalgebra of $\mathfrak{g}$. The centralizer of $\mathfrak{g}$ in $\mathfrak{g}$ is an ideal of $\mathfrak{g}$ called the *centre* of $\mathfrak{g}$. The centralizer in $\mathfrak{g}$ of an element x of $\mathfrak{g}$ is denoted by $\mathfrak{g}^x$.

Let $\mathfrak{h}$ be a Lie subalgebra of $\mathfrak{g}$. The set $\mathfrak{n}$ of the $x \in \mathfrak{g}$ such that $[x,\mathfrak{h}] \subset \mathfrak{h}$ is termed the *normalizer* of $\mathfrak{h}$ in $\mathfrak{g}$. It is a Lie subalgebra of $\mathfrak{g}$, and $\mathfrak{h}$ is an ideal of $\mathfrak{n}$.

1.1.10. A linear mapping D of $\mathfrak{g}$ into $\mathfrak{g}$ such that $D([x,y]) = [Dx,y] + [x,Dy]$ for all $x,y \in \mathfrak{g}$ is termed a *derivation* of $\mathfrak{g}$. The set $\mathfrak{d}$ of derivations of $\mathfrak{g}$ is a Lie subalgebra of $\mathfrak{gl}(\mathfrak{g})$.

For all $x \in \mathfrak{g}$, we denote by $\mathrm{ad}_{\mathfrak{g}} x$, or by $\mathrm{ad}\, x$, the mapping $y \mapsto [x,y]$ of $\mathfrak{g}$ into $\mathfrak{g}$. The mapping $x \mapsto \mathrm{ad}\, x$ is a homomorphism of $\mathfrak{g}$ into $\mathfrak{d}$. The derivations of $\mathfrak{g}$ of the form $\mathrm{ad}\, x$ are termed the *inner derivations* of $\mathfrak{g}$.

1.1.11. A vector subspace of $\mathfrak{g}$ which is stable under every derivation of $\mathfrak{g}$ is termed a *characteristic ideal* of $\mathfrak{g}$. If $\mathfrak{a}$ and $\mathfrak{b}$ are characteristic ideals of $\mathfrak{g}$, then $[\mathfrak{a},\mathfrak{b}]$ is a characteristic ideal of $\mathfrak{g}$.

1.1.12. If D is a nilpotent derivation of $\mathfrak{g}$, then $\exp D$ is an automorphism of $\mathfrak{g}$. For, if $x,y \in \mathfrak{g}$, we have

$$\begin{aligned}[(\exp D)x, (\exp D)y] &= \Big[\sum_{i \geq 0} \frac{1}{i!} D^i x, \sum_{j \geq 0} \frac{1}{j!} D^j y\Big] \\ &= \sum_{n \geq 0} \sum_{i+j=n} \frac{1}{i!\,j!} [D^i x, D^j y] \\ &= \sum_{n \geq 0} \frac{1}{n!} D^n([x,y]) = (\exp D)([x,y]).\end{aligned}$$

1.1.13. Let $\mathfrak{a}$ and $\mathfrak{b}$ be Lie algebras, and $b \mapsto D_b$ a homomorphism of $\mathfrak{b}$ into the Lie algebra of derivations of $\mathfrak{a}$. We define a multiplication in the vector space $\mathfrak{b} \times \mathfrak{a}$ by setting

$$[(b,a), (b',a')] = ([b,b'], [a,a'] + D_b a' - D_{b'} a) \quad \text{for} \quad a,a' \in \mathfrak{a},\ b,b' \in \mathfrak{b}.$$

It can be shown that $\mathfrak{b} \times \mathfrak{a}$ is a Lie algebra $\mathfrak{g}$, called the *semi-direct product of* $\mathfrak{b}$ *by* $\mathfrak{a}$ (corresponding to the homomorphism $b \mapsto D_b$). If we identify $\mathfrak{a}$ and $\mathfrak{b}$ canonically with vector subspaces of $\mathfrak{g}$, then $\mathfrak{a}$ is an ideal of $\mathfrak{g}$ and $\mathfrak{b}$ is a Lie subalgebra of $\mathfrak{g}$.

1.1.14. If $\mathfrak{g}$ is finite-dimensional, the group $\mathrm{Aut}\, \mathfrak{g}$ is an algebraic group in $\mathfrak{g}$ whose Lie algebra is the set of derivations of $\mathfrak{g}$ (CH, p. 179). Let

$\mathfrak{d}' = \mathrm{ad}(\mathfrak{g})$, which is an ideal of $\mathfrak{d}$; let $\mathfrak{a}$ be the smallest Lie algebra which is algebraic in $\mathfrak{g}$ and contains $\mathfrak{d}'$ (CH, p. 173); we then have $\mathfrak{d}' \subset \mathfrak{a} \subset \mathfrak{d}$. The irreducible algebraic group $\mathscr{A}$ in $\mathfrak{g}$ whose Lie algebra is $\mathfrak{a}$ (CH, p 129 and 156) is termed the *adjoint algebraic group* of $\mathfrak{g}$. Carrying over the structure, each automorphism of $\mathfrak{g}$ transforms $\mathfrak{d}'$ into $\mathfrak{d}'$, and hence $\mathfrak{a}$ into $\mathfrak{a}$ and $\mathscr{A}$ into $\mathscr{A}$. Thus $\mathscr{A}$ is a distinguished subgroup of Aut $\mathfrak{g}$. If $\mathfrak{a} = \mathfrak{d}'$, we simply say that $\mathscr{A}$ is the *adjoint group* of $\mathfrak{g}$.

Let x be an element of $\mathfrak{g}$ such that ad x is nilpotent. Then $k \cdot (\mathrm{ad}\, x)$ is an algebraic Lie subalgebra of $\mathfrak{d}'$ and hence of $\mathfrak{a}$, and the corresponding irreducible algebraic subgroup $\mathscr{G}$ of $\mathscr{A}$ is $\exp(k\, \mathrm{ad}\, x)$ (CH, p. 159). We denote by $\mathrm{Aut}_e(\mathfrak{g})$ the subgroup of $\mathscr{A}$ generated by the exp ad x, where $x \in \mathfrak{g}$ and ad x is nilpotent; the elements of $\mathrm{Aut}_e(\mathfrak{g})$ are termed the *elementary automorphisms* of $\mathfrak{g}$.

1.2. Representations

1.2.1. Let V be a vector space. A homomorphism ϱ of $\mathfrak{g}$ in $\mathfrak{gl}(V)$ is termed a *representation of* $\mathfrak{g}$ *in* V. The dimension of V is termed the *dimension* of ϱ and is denoted by $\dim \varrho$; we say that V is the *space of* ϱ. We often set $\varrho(x) = x_V$, and $\varrho(x)(v) = x_V v = xv$ for $x \in \mathfrak{g}$, $v \in V$; and we say that V is a $\mathfrak{g}$-*module*. If $\varrho(x) = 0$ for all $x \in \mathfrak{g}$, then V is said to be a *trivial* $\mathfrak{g}$-module.

If ϱ' is a representation of $\mathfrak{g}$ in V', a linear mapping u of V into V' such that $\varrho'(x)u = u\varrho(x)$ for all $x \in \mathfrak{g}$ is termed a $\mathfrak{g}$-*homomorphism* of V into V'.

The mapping $x \mapsto \mathrm{ad}\, x$ is a representation of $\mathfrak{g}$ in $\mathfrak{g}$ termed the *adjoint representation*.

1.2.2. Let ϱ_1 and ϱ_2 be representations of $\mathfrak{g}$ in V_1 and V_2. We say that ϱ_1 and ϱ_2 are *equivalent* if there exists a bijective $\mathfrak{g}$-homomorphism of V_1 into V_2. This is an equivalence relation, whence the notion of a *class of representations* and a *class of* $\mathfrak{g}$-*modules*. We denote by $[V]$ the class of a $\mathfrak{g}$-module V. To simplify matters, we sometimes use "representation" instead of "class of representations" and "$\mathfrak{g}$-module" instead of "class of $\mathfrak{g}$-modules".

1.2.3. Let $(V_i)_{i \in I}$ be a family of vector spaces. For all $i \in I$, let ϱ_i be a representation of $\mathfrak{g}$ in V_i. For all $x \in \mathfrak{g}$, let $\varrho(x)$ be the endomorphism $\oplus_{i \in I}\varrho_i(x)$ of $V = \oplus_{i \in I} V_i$. Then ϱ is a representation of $\mathfrak{g}$ in V, called the *direct sum of the* ϱ_i, and denoted by $\oplus_{i \in I}\varrho_i$ (or simply by $\varrho_1 \oplus \cdots \oplus \varrho_n$ if $I = \{1, \ldots, n\}$). The $\mathfrak{g}$-module V is also called the *direct sum of the* $\mathfrak{g}$-*modules* V_i. If all the ϱ_i are equivalent to the same representation σ, corresponding to a $\mathfrak{g}$-module W, ϱ is said to be a *multiple of* σ and V is said to be a *multiple of* W.

1.2.4. Let V be a vector space and ϱ a representation of $\mathfrak{g}$ in V. A vector subspace W of V is said to be *stable* under ϱ if $\varrho(x)(W) \subset W$ for all $x \in \mathfrak{g}$. In this case, the mapping $x \mapsto \varrho(x)|W$ is a representation of $\mathfrak{g}$ in W, called a *subrepresentation of* ϱ. If $\sigma(x)$ denotes the endomorphism of V/W which is deduced from $\varrho(x)$ by passing to the quotient, the mapping $x \mapsto \sigma(x)$ is a representation of $\mathfrak{g}$ in V/W, called a *quotient representation* of ϱ. The $\mathfrak{g}$-modules W and V/W are termed a *sub-*$\mathfrak{g}$*-module* and a *quotient* $\mathfrak{g}$*-module*, respectively, of the $\mathfrak{g}$-module V. A quotient $\mathfrak{g}$-module of a sub-$\mathfrak{g}$-module of V, or a sub-$\mathfrak{g}$-module of a quotient $\mathfrak{g}$-module of V, which amounts to the same, is termed a $\mathfrak{g}$*-subquotient* of V; the corresponding representation is termed a *subquotient representation* of ϱ.

1.2.5. A representation of $\mathfrak{g}$ in V is said to be *simple*, or *irreducible*, if $V \neq 0$ and the only stable vector subspaces of V are 0 and V. The corresponding $\mathfrak{g}$-module is also said to be *simple*. The set of classes of *finite-dimensional* simple $\mathfrak{g}$-modules of $\mathfrak{g}$ (or the set of classes of finite-dimensional simple representations of $\mathfrak{g}$) is denoted by $\mathfrak{g}^\wedge$.

1.2.6. Let ϱ be a representation of $\mathfrak{g}$ in V. A series $(V_0, V_1, \ldots, V_n)$ of sub-$\mathfrak{g}$-modules of V such that $V = V_0 \supset V_1 \supset \cdots \supset V_n = 0$ is termed a *composition series* of ϱ (or of the $\mathfrak{g}$-module V). A composition series $(V_0, V_1, \ldots, V_n)$ such that the $\mathfrak{g}$-modules $V_i V_{i+1}$ $(0 \leq i < n)$ are simple is termed a *Jordan—Hölder series*. (A Jordan—Hölder series exists if $\dim V < +\infty$, but not necessarily if $\dim V = +\infty$.) From the theory of operator groups, if $(V_0, \ldots, V_n)$ and $(V'_0, \ldots, V'_p)$ are two Jordan—Hölder series of V, then $p = n$ and a permutation σ of $\{0, 1, \ldots, n-1\}$ exists such that the $\mathfrak{g}$-modules V_i/V_{i+1} and $V'_{\sigma(i)}/V'_{\sigma(i)+1}$ are isomorphic for $0 \leq i < n$.

Let us assume the existence of a Jordan—Hölder series $(V_0, V_1, \ldots, V_n)$ of V. Let W be a simple subquotient of V. There exists $i \in \{0, \ldots, n-1\}$ such that W is isomorphic to V_i/V_{i+1}; the number of such integers i is termed the *multiplicity* of W in V; this multiplicity is independent of the choice of the Jordan—Hölder series. The set of classes of simple subquotients of V is denoted by $\mathcal{J}\mathcal{H}(V)$.

1.2.7. Let ϱ be a representation of $\mathfrak{g}$ in V. From the theory of operator groups, the following conditions are equivalent:

(a) ϱ is a direct sum of simple representations;

(b) there exists a family $(V_i)_{i \in I}$ of simple sub-$\mathfrak{g}$-modules of V such that $V = \sum_{i \in I} V_i$;

(c) for every sub-$\mathfrak{g}$-module V' of V there exists a sub-$\mathfrak{g}$-module V'' of V such that $V = V' \oplus V''$ (AL VIII, p. 32).

If these conditions are satisfied, ϱ is said to be *semi-simple* or *completely reducible.* The $\mathfrak{g}$-module V is also said to be *semi-simple.* If V is semi-simple, the $\mathfrak{g}$-subquotients of V are semi-simple.

1.2.8. Let V be a $\mathfrak{g}$-module, and S a simple $\mathfrak{g}$-module. The sum V_S of sub-$\mathfrak{g}$-modules of V which are isomorphic to S is a sub-$\mathfrak{g}$-module termed the *isotypic component of type S of V.* Every simple sub-$\mathfrak{g}$-module of V_S is isomorphic to S. The sum of the V_S for variable S is direct. Every $\mathfrak{g}$-endomorphism of V leaves the V_S stable. If V is semi-simple, every sub-$\mathfrak{g}$-module of V is the sum of its intersections with the V_S (AL VIII, p. 33—34).

Let V be a $\mathfrak{g}$-module, $\mathfrak{h}$ a Lie subalgebra of $\mathfrak{g}$ and ξ a simple $\mathfrak{h}$-module. The isotypic component of type ξ of the $\mathfrak{h}$-module V is denoted by V_ξ. Let us decompose V_ξ into the form $\oplus_{\lambda \in \Lambda} W_\lambda$, where each W_λ is a simple $\mathfrak{h}$-module which is isomorphic to ξ; then Card Λ is independent of the choice of the decomposition of V_ξ (AL VIII, p. 34), is termed the *multiplicity of ξ in V*, and is denoted by $\mathrm{mtp}(\xi, V)$. (For $\mathfrak{h} = \mathfrak{g}$ and V the sum of a finite number of simple modules, this terminology is consistent with 1.2.6.) We use a similar terminology when speaking of representations instead of modules.

1.2.9. Let V be a finite-dimensional vector space and u an endomorphism of V. Then u is said to be *diagonalizable* if there exists a basis B for V such that the matrix of u with repect to B is diagonal.

The following conditions are equivalent:

(i) there exists a basis $(e_1, \ldots, e_n)$ for V such that the matrix of u with respect to $(e_1, \ldots, e_n)$ is upper triangular;

(ii) there exists a basis $(e'_1, \ldots, e'_n)$ for V such that the matrix of u with respect to $(e'_1, \ldots, e'_n)$ is lower triangular;

(iii) the eigenvalues of u in an algebraically closed extension of k belong to k.

Then u is said to be triangularizable. (We can define strictly triangularizable endomorphisms in a obvious way, but these are merely nilpotent endomorphisms.)

A set M of endomorphisms of V is said to be *diagonalizable* if there exists a basis B of V such that, for every $u \in M$, the matrix of u with respect to B is diagonal. We define *triangularizable* and *strictly triangularizable* sets in a similar fashion.

A representation ϱ of $\mathfrak{g}$ in V is termed diagonalizable (triangularizable, strictly triangularizable) if $\varrho(\mathfrak{g})$ is diagonalizable (triangularizable, strictly triangularizable). We also say that the $\mathfrak{g}$-module V is diagonalizable (triangularizable, strictly triangularizable). To say that V is triangularizable amounts to saying that its simple subquotients are one-dimensional. To say that V is strictly triangularizable amounts to saying that its simple subquotients are trivial and one-dimensional. The $\mathfrak{g}$-subquotients of a triangularizable (strictly triangularizable) $\mathfrak{g}$-module are triangularizable (strictly triangularizable).

1.2.10. Let ϱ be a representation of $\mathfrak{g}$ in V. An element v of V is termed an *invariant* of ϱ, or of the $\mathfrak{g}$-module V, if $\varrho(\mathfrak{g})v = 0$. We denote the set of invariants of V by $V^{\mathfrak{g}}$. If ϱ is semi-simple, V is the direct sum of $V^{\mathfrak{g}}$ and the vector subspace generated by $\varrho(\mathfrak{g})(V)$ (indeed, this is obvious if ϱ is simple).

1.2.11. Let $0 \to V' \to V \to V'' \to 0$ be an exact sequence of $\mathfrak{g}$-module homomorphisms. We immediately deduce the exact sequence

$$0 \to V'^{\mathfrak{g}} \to V^{\mathfrak{g}} \to V''^{\mathfrak{g}}.$$

If, moreover, the $\mathfrak{g}$-module V is semi-simple, we even have the exact sequence

$$0 \to V'^{\mathfrak{g}} \to V^{\mathfrak{g}} \to V''^{\mathfrak{g}} \to 0.$$

In fact, V can be identified with $V' \oplus W$, where W is a sub-$\mathfrak{g}$-module of V; and the homomorphism $V \to V''$ defines an isomorphism of W onto V'' and hence of $W^{\mathfrak{g}}$ onto $V''^{\mathfrak{g}}$.

1.2.12. A one-dimensional representation of $\mathfrak{g}$ can be identified with a linear form λ on $\mathfrak{g}$ such that $\lambda([\mathfrak{g},\mathfrak{g}]) = 0$.

1.2.13. Let ϱ be a representation of $\mathfrak{g}$ in V, $\mathfrak{g}'$ a Lie subalgebra of $\mathfrak{g}$, and $\lambda \in \mathfrak{g}'^*$.

The set of the $v \in V$ such that $\varrho(x)v = \lambda(x)v$ for all $x \in \mathfrak{g}'$ is denoted by $V_{\lambda,\varrho}$, or simply by V_λ. The set V_λ is a vector subspace of V which is stable under $\varrho|\mathfrak{g}'$. If $V_\lambda \neq 0$, then $\lambda([\mathfrak{g}',\mathfrak{g}']) = 0$. As a matter of fact, the notation V_λ is a special case of that of 1.2.8).

The set of the $v \in V$ such that, for all $x \in \mathfrak{g}'$, $(\varrho(x) - \lambda(x))^n v = 0$ for n sufficiently large is denoted by $V^{\lambda,\varrho}$, or simply by V^λ. The set V^λ is a vector subspace of V, and $V^\lambda \supset V_\lambda$. For all $x \in \mathfrak{g}'$, V^λ is contained in the nilspace of $\varrho(x) - \lambda(x)$; hence the sum of the V^λ is direct and *a fortiori* the sum of the V_λ is direct.

1.2.14. Let ϱ and ϱ' be representations of $\mathfrak{g}$ in V and V', respectively. It can be shown that the mapping $x \mapsto \varrho(x) \otimes 1 + 1 \otimes \varrho'(x)$ of $\mathfrak{g}$ into $\mathrm{End}(V \otimes V')$ is a representation of $\mathfrak{g}$ in $V \otimes V'$, called the *tensor product of* ϱ *and* ϱ', and denoted by $\varrho \otimes \varrho'$. We define the tensor product of a finite number of representations in a similar way. In particular, for every integer $p \geqq 0$ we have the pth *tensor power of* ϱ, denoted by $\otimes^p \varrho$, which operates in $\otimes^p V$. The direct sum τ of the $\otimes^p \varrho$ for $p = 0, 1, \ldots$ is a representation of $\mathfrak{g}$ in the tensor algebra T of V. For all $x \in \mathfrak{g}$, $\tau(x)$ is the unique derivation of T which extends $\varrho(x)$. On passing to the quotient, this derivation defines a derivation $\sigma(x)$ (or $\varepsilon(x)$) of the symmetric algebra S (or the exterior algebra E) of V, and σ (or ε) is a representation of $\mathfrak{g}$ in S (or E) under which the homogeneous components S^p (or E^p) of S (or E) are stable (AL III, p. 128). The subrepresentation of σ defined by S^p is called the *symmetric* p^{th} *power of* ϱ and is denoted by $S^p\varrho$. The *exterior* p^{th} *power of* ϱ is similarly defined and denoted by $\Lambda^p\varrho$.

1.2.15. Let V and V' be $\mathfrak{g}$-modules, and $M = \mathrm{Hom}(V, V')$. For all $x \in \mathfrak{g}$ and $u \in M$, let $x_M u$ be the element $x_{V'}u - ux_V$ of M. It can be shown that M thus has a $\mathfrak{g}$-module structure and that $M^{\mathfrak{g}} = \mathrm{Hom}_{\mathfrak{g}}(V, V')$.

1.2.16. In particular, let us take $V' = k$, together with the null representation of $\mathfrak{g}$ in k. Then $M = V^*$, and, for all $x \in \mathfrak{g}$, we have $x_{V^*} = -{}^t x_V$. The $\mathfrak{g}$-module V^* is called the *dual* $\mathfrak{g}$*-module* of the $\mathfrak{g}$-module V; the representation $x \mapsto x_{V^*}$ is said to be the *dual* of the representation $x \mapsto x_V$. If ϱ is the representation corresponding to V, the dual representation is denoted by ϱ^*. The symmetric algebra $S(V^*)$ of V^* (which can be identified with the algebra of polynomial functions over V) thus has a $\mathfrak{g}$-module structure, from 1.2.14.

The dual representation of the adjoint representation is termed the *coadjoint representation.*

1.2.17. Let V and V' be finite-dimensional $\mathfrak{g}$-modules. The canonical vector space

$$\mathrm{Hom}(V, V') \to V^* \otimes V'$$

isomorphism is a $\mathfrak{g}$-module isomorphism.

1.2.18. If V and V' are triangularizable (or strictly triangularizable) finite-dimensional $\mathfrak{g}$-modules, it is clear that $V \otimes V'$, $\otimes^p V$, $S^p V$, $\Lambda^p V$,

$\mathrm{Hom}(V,V')$ and V^* are triangularizable (or strictly triangularizable) $\mathfrak{g}$-modules (1.2.9). We shall obtain a similar result for semi-simple $\mathfrak{g}$-modules later (1.7.8).

1.2.19. Let ϱ be a representation of $\mathfrak{g}$ in V and k' an extension of k. There is one and only one representation ϱ' of $\mathfrak{g}' = \mathfrak{g} \otimes k'$ in $V' = V \otimes k'$ such that $\varrho'(x) = \varrho(x) \otimes 1$ for all $x \in \mathfrak{g}$. We say that ϱ' is *deduced from* ϱ *by extension of the field of scalars from* k *to* k' and denote it by $\varrho_{k'}$.

(a) If ϱ' is simple, then ϱ is clearly simple.

(b) Let us assume that ϱ' is semi-simple and let us prove that ϱ is semi-simple. Let W be a sub-$\mathfrak{g}$-module of V. Then $W \otimes k'$ is a sub-$\mathfrak{g}$-module of V', and hence there is a projection p of V' onto $W \otimes k'$ which is a $\mathfrak{g}'$-homomorphism. Let h be a k-linear mapping of k' onto k such that $h(1) = 1$. The mapping $x \mapsto (1 \otimes h)(p(x \otimes 1))$ of V into $W \otimes k = W$ is a $\mathfrak{g}$-homomorphism which reduces to the identity mapping on W. This shows that ϱ is semi-simple.

(c) We now assume that ϱ is simple; in general ϱ' will not be simple. We say that ϱ is *absolutely simple* if ϱ remains simple for every extension of the base field (cf. 2.6.5). Nevertheless, we have the following result: if ϱ is simple, and k' is a finite extension of k, then ϱ' is the direct sum of a finite number of simple representations. Indeed, from (b) we can assume that k' is Galois over k. Let Γ be the Galois group of k' over k. Let T be a maximal element of the set of sub-$\mathfrak{g}'$-modules of V' whose intersection with V is 0. If T' is a sub-$\mathfrak{g}'$-module of V' strictly containing T, then $T' \cap V \neq 0$, and so $T' \cap V = V$ and $T' = V'$. Hence the $\mathfrak{g}'$-module V'/T is simple. There exists a sub-$\mathfrak{g}$-module W of V such that

$$\bigcap_{\gamma \in \Gamma} (1 \otimes \gamma)(T) = W \otimes k.$$

We have

$$W \subset T \cap V = 0,$$

hence

$$\bigcap_{\gamma \in \Gamma} (1 \otimes \gamma)(T) = 0.$$

Thus V' is isomorphic to a sub-$\mathfrak{g}'$-module of $\oplus_{\gamma \in \Gamma} V'/(1 \otimes \gamma)(T)$, which proves our assertion.

(d) If ϱ is semi-simple, and k' is a finite extension of k, then ϱ' is semi-simple from (c).

If ϱ is semi-simple and finite-dimensional, then ϱ' is semi-simple (for any k'). For, we can assume that ϱ is simple. On the other hand, from (b), we can assume that k' is Galois over k. Since $\dim \varrho' < +\infty$, V' has a simple sub-$\mathfrak{g}'$-module W'. Let Γ be the Galois group of k' over k. There exists a sub-$\mathfrak{g}$-module W of V such that

$$\sum_{\gamma\in\Gamma} (1 \otimes \gamma)(W') = W \otimes k'.$$

Then $W = V$ since V is simple, and hence V' is the sum of simple sub-$\mathfrak{g}'$-modules.

We shall see (2.6.9) that, if ϱ is semi-simple and $\mathfrak{g}$ is finite-dimensional then ϱ' is semi-simple.

1.2.20. Let ϱ be a finite-dimensional representation of $\mathfrak{g}$. For $x,y \in \mathfrak{g}$, we write $b(x,y) = \mathrm{tr}(\varrho(x)\varrho(y))$. Then b is a symmetric bilinear form on $\mathfrak{g}$, and is said to be *associated with* ϱ. If $x,y,z \in \mathfrak{g}$, we have

$$b([x,y],z) - b(x,[y,z]) = \mathrm{tr}(\varrho(x)\varrho(y)\varrho(z) - \varrho(y)\varrho(x)\varrho(z) - \varrho(x)\varrho(y)\varrho(z) + \varrho(x)\varrho(z)\varrho(y)) = 0,$$

hence

(1) $$b([x,y],z) = b(x,[y,z]).$$

Let $\mathfrak{a}$ be an ideal of $\mathfrak{g}$, and $\mathfrak{a}^\perp$ its orthogonal subspace with respect to b. Then

$$b([\mathfrak{a}^\perp,\mathfrak{g}],\mathfrak{a}) = b(\mathfrak{a}^\perp,[\mathfrak{g},\mathfrak{a}]) \subset b(\mathfrak{a}^\perp,\mathfrak{a}) = 0,$$

hence

$$[\mathfrak{a}^\perp,\mathfrak{g}] \subset \mathfrak{a}^\perp,$$

so that $\mathfrak{a}^\perp$ is an ideal of $\mathfrak{g}$.

1.2.21. If $\mathfrak{g}$ is finite-dimensional, the bilinear form associated with the adjoint representation of $\mathfrak{g}$ is called the *Killing form* of $\mathfrak{g}$. When we mention orthogonality in $\mathfrak{g}$, without further qualification, we are referring to the Killing form.

Let K be the Killing form of $\mathfrak{g}$. For all $x \in \mathfrak{g}$, let f_x be the element of $\mathfrak{g}^*$ defined by $f_x(y) = K(x,y)$ for all $y \in \mathfrak{g}$. It follows from (1) that the mapping $x \mapsto f_x$ is a $\mathfrak{g}$-homomorphism of $\mathfrak{g}$ (equipped with the adjoint representation) into $\mathfrak{g}^*$ (equipped with the coadjoint representation). This homomorphism φ is called the *Killing homomorphism*. The homomorphism of the algebra $S(\mathfrak{g})$ into the algebra $S(\mathfrak{g}^*)$ which extends φ is again a $\mathfrak{g}$-homomorphism (for the representations in $S(\mathfrak{g})$ and $S(\mathfrak{g}^*)$ deduced from the adjoint representation), and is also called a Killing homomorphism.

1.2.22. Throughout the rest of this book, the term "Lie algebra" will be used to denote "finite-dimensional Lie algebra". On the other hand, the representations which we study will often be infinite-dimensional.

1.3. Solvable and nilpotent Lie algebras

1.3.1. We write

$$\mathscr{C}^1\mathfrak{g} = \mathfrak{g}, \quad \mathscr{C}^2\mathfrak{g} = [\mathfrak{g},\mathfrak{g}], \quad \ldots, \quad \mathscr{C}^{i+1}\mathfrak{g} = [\mathfrak{g},\mathscr{C}^i\mathfrak{g}], \quad \ldots$$

By recurrence, we thus define a decreasing series of characteristic ideals, called the *descending central series* of $\mathfrak{g}$.

1.3.2. We write

$$\mathscr{D}^0\mathfrak{g} = \mathfrak{g}, \quad \mathscr{D}^1\mathfrak{g} = [\mathfrak{g},\mathfrak{g}], \quad \ldots, \quad \mathscr{D}^{i+1}\mathfrak{g} = [\mathscr{D}^i\mathfrak{g}, \mathscr{D}^i\mathfrak{g}], \quad \ldots$$

We thus define a decreasing series of characteristic ideals, called the *derived series* of $\mathfrak{g}$. We have $\mathscr{D}^i\mathfrak{g} \subset \mathscr{C}^{i+1}\mathfrak{g}$ for all i. If $\mathfrak{a}$ is a vector subspace of $\mathfrak{g}$ containing $[\mathfrak{g},\mathfrak{g}]$ then $\mathfrak{a}$ is an ideal of $\mathfrak{g}$ and the Lie algebra $\mathfrak{g}/\mathfrak{a}$ is commutative.

1.3.3. If $f: \mathfrak{g} \to \mathfrak{h}$ is a surjective Lie algebra homomorphism, then $f(\mathscr{C}^i\mathfrak{g}) = \mathscr{C}^i\mathfrak{h}, f(\mathscr{D}^i\mathfrak{g}) = \mathscr{D}^i\mathfrak{h}$ for all i, as can be seen by recurrence on i.

1.3.4. If k' is an extension of k, then

$$\mathscr{C}^i(\mathfrak{g} \otimes k') = (\mathscr{C}^i\mathfrak{g}) \otimes k', \qquad \mathscr{D}^i(\mathfrak{g} \otimes k') = (\mathscr{D}^i\mathfrak{g}) \otimes k' \quad \text{for all } i.$$

1.3.5. The following conditions are equivalent:

(i) there is an integer k such that $\mathscr{C}^k\mathfrak{g} = 0$;

(ii) there is an integer k such that $[x_1, [x_2, [\ldots, [x_{k-1}, x_k], \ldots,]]] = 0$ for all $x_1, \ldots, x_k \in \mathfrak{g}$;

(iii) there is a decreasing series $(\mathfrak{g}_1, \mathfrak{g}_2, \ldots, \mathfrak{g}_n)$ of ideals of $\mathfrak{g}$ such that $\mathfrak{g}_1 = \mathfrak{g}, \mathfrak{g}_n = 0$, and $[\mathfrak{g},\mathfrak{g}_i] \subset \mathfrak{g}_{i+1}$ for $i < n$.

In fact, it is obvious that (i) $\Leftrightarrow$ (ii) $\Rightarrow$ (iii); and if condition (iii) is satisfied, we can see by induction on i that $\mathscr{C}^i\mathfrak{g} \subset \mathfrak{g}_i$.

1.3.6. A Lie algebra is said to be *nilpotent* if it satisfies the equivalent conditions of 1.3.5. From condition (i) of 1.3.5, the centre of a non-null nilpotent Lie algebra is non-null.

1.3.7. The following conditions are equivalent:

(i) there is an integer k such that $\mathscr{D}^k\mathfrak{g} = 0$;

(ii) there is a decreasing series $(\mathfrak{g}_0, \mathfrak{g}_1, \ldots, \mathfrak{g}_n)$ of ideals of $\mathfrak{g}$ such that $\mathfrak{g}_0 = \mathfrak{g}, \mathfrak{g}_n = 0$, and $[\mathfrak{g}_i,\mathfrak{g}_i] \subset \mathfrak{g}_{i+1}$ for $i < n$.

In fact, it is obvious that (i) $\Rightarrow$ (ii); and if condition (ii) is satisfied, we can see by induction on i that $\mathscr{D}^i\mathfrak{g} \subset \mathfrak{g}_i$.

1.3.8. A Lie algebra is said to be *solvable* if it satisfies the equivalent conditions of 1.3.7. Every nilpotent Lie algebra is solvable.

1.3.9. A subalgebra and a quotient algebra of a nilpotent (or solvable) Lie algebra are nilpotent (or solvable). The product algebra of two nilpotent (or solvable) Lie algebras is nilpotent (or solvable). Let $\mathfrak{a}$ be an ideal of $\mathfrak{g}$; if $\mathfrak{a}$ and $\mathfrak{g}/\mathfrak{a}$ are solvable, then $\mathfrak{g}$ is solvable. Let k' be an extension of k; then $\mathfrak{g}$ is nilpotent (or solvable) if and only if $\mathfrak{g} \otimes k'$ is nilpotent (or solvable).

1.3.10. The following conditions are equivalent:

(i) $\mathfrak{g}$ is solvable;

(ii) there is a decreasing series $(\mathfrak{g}_0, \mathfrak{g}_1, \ldots, \mathfrak{g}_n)$ of Lie subalgebras of $\mathfrak{g}$ such that $\mathfrak{g}_0 = \mathfrak{g}$, $\mathfrak{g}_n = 0$, and such that, for $i < n$, $\mathfrak{g}_{i+1}$ is an ideal of codimension 1 in $\mathfrak{g}_i$.

Indeed, (i) $\Rightarrow$ (ii) from 1.3.2, and (ii) $\Rightarrow$ (i) from 1.3.9.

1.3.11. LEMMA. *Let V be a finite-dimensional vector space, ϱ a representation of $\mathfrak{g}$ in V, $\mathfrak{a}$ an ideal of $\mathfrak{g}$, and $\lambda \in \mathfrak{a}^*$. Then V_λ (cf. 1.2.13) is stable under $\varrho(\mathfrak{g})$.*

Let v_0 be a non-zero element of V_λ, and $x \in \mathfrak{g}$; we prove that $\varrho(x)v_0 \in V_\lambda$. We set $v_i = \varrho(x)^i v_0$ for $i = 1, 2, \ldots$ Let j be the largest integer such that $v_0, v_1, \ldots, v_j$ are linearly independent, and let

$$V' = kv_0 + kv_1 + \cdots + kv_j.$$

Then $v_{j+1} \in V'$, and hence

$$\varrho(x)(V') = kv_1 + kv_2 + \cdots + kv_{j+1} \subset V'.$$

We show that, for all $y \in \mathfrak{a}$ and for every integer $i \geqq 0$, we have

$$\varrho(y)v_i \in \lambda(y)v_i + kv_{i-1} + kv_{i-2} + \cdots + kv_0. \tag{1}$$

This is obvious for $i = 0$; and, if it is true for some integer i, then, since $[x,y] \in \mathfrak{a}$,

$$\begin{aligned}\varrho(y)v_{i+1} &= \varrho(y)\varrho(x)v_i = \varrho(x)\varrho(y)v_i + \varrho([y,x])v_i \\ &\in \varrho(x)(\lambda(y)v_i + kv_{i-1} + \cdots + kv_0) + kv_i + \cdots + kv_0 \\ &\subset \lambda(y)v_{i+1} + kv_i + \cdots + kv_0.\end{aligned}$$

This proves (1). We deduce that V' is stable under $\varrho(\mathfrak{a})$ and that

$$\operatorname{tr}(\varrho(y) \mid V') = (j+1)\lambda(y).$$

As $[y,x] \in \mathfrak{a}$, we see that

$$(j+1)\lambda([y,x]) = \operatorname{tr}((\varrho(y) \mid V')(\varrho(x) \mid V') - (\varrho(x) \mid V')(\varrho(y) \mid V')) = 0.$$

Then

$$\begin{aligned}\varrho(y)\varrho(x)v_0 &= \varrho(x)\varrho(y)v_0 + \varrho([y,x])v_0 \\ &= \varrho(x)\lambda(y)v_0 + \lambda([y,x])v_0 = \lambda(y)\varrho(x)v_0,\end{aligned}$$

whence $\varrho(x)v_0 \in V_\lambda$.

1.3.12. THEOREM ($\mathfrak{g}$ solvable). *Let V be a finite-dimensional vector space, and ϱ a representation of $\mathfrak{g}$ in V. If for all $x \in \mathfrak{g}$, $\varrho(x)$ is triangularizable (as it is when k is algebraically closed,) then ϱ is triangularizable.*

This is obvious when $\mathfrak{g} = 0$. Let us assume that $\dim \mathfrak{g} = n > 0$, and that the theorem is true for all dimensions $<n$. There exists in $\mathfrak{g}$ an ideal $\mathfrak{a}$ of codimension 1. From the induction hypothesis, and if $V \neq 0$, there exists $v \in V - \{0\}$ such that $\varrho(y)v = \lambda(y)v$ for all $y \in \mathfrak{a}$, with $\lambda(y) \in k$. Clearly, $\lambda \in \mathfrak{a}^*$. Then $V_\lambda \neq 0$ since $v \in V_\lambda$. Let $x \in \mathfrak{g}$ be such that $\mathfrak{g} = \mathfrak{a} \oplus kx$. Then $\varrho(x)(V_\lambda) \subset V_\lambda$ (1.3.11). Since $\varrho(x)$ is triangularizable, there is a non-zero eigenvector w for $\varrho(x)|V_\lambda$. Hence w is an eigenvector for $\varrho(kx) + \varrho(\mathfrak{a}) = \varrho(\mathfrak{g})$. Given this, the theorem follows by induction on $\dim V$.

1.3.13. COROLLARY (k algebraically closed, $\mathfrak{g}$ solvable). *Every finite-dimensional simple representation of $\mathfrak{g}$ has dimension* 1.

1.3.14. We say that $\mathfrak{g}$ is *completely solvable* if the adjoint representation of $\mathfrak{g}$ is triangularizable, that is, if there is a decreasing sequence of ideals of $\mathfrak{g}$ with dimensions $\dim \mathfrak{g}$, $\dim \mathfrak{g} - 1$, $\dim \mathfrak{g} - 2, \ldots, 0$. Then

$$\mathfrak{g} \text{ nilpotent} \quad \Rightarrow \quad \mathfrak{g} \text{ completely solvable} \quad \Rightarrow \quad \mathfrak{g} \text{ solvable},$$

from 1.3.5 (iii) and 1.3.10. If k is algebraically closed, a solvable Lie algebra is completely solvable (1.3.12).

1.3.15. THEOREM. *The following conditions are equivalent:*

(i) $\mathfrak{g}$ *is nilpotent*;

(ii) *for all* $x \in \mathfrak{g}$, $\operatorname{ad} x$ *is nilpotent.*

(i) $\Rightarrow$ (ii). This follows from condition 1.3.5 (ii).

(ii) $\Rightarrow$ (i). Assume that $\operatorname{ad} x$ is nilpotent for all $x \in \mathfrak{g}$. Let $\mathfrak{g}'$ be a solvable Lie subalgebra of maximal dimension of $\mathfrak{g}$; suppose that $\mathfrak{g}' \neq \mathfrak{g}$. Consider the adjoint representation of $\mathfrak{g}$, and let ϱ be its restriction to $\mathfrak{g}'$. Then $\mathfrak{g}'$ is stable under $\varrho(\mathfrak{g}')$. Let σ be the quotient representation of ϱ in $\mathfrak{g}/\mathfrak{g}'$, and φ the canonical mapping of $\mathfrak{g}$ onto $\mathfrak{g}/\mathfrak{g}'$. Since $\sigma(x)$ is nilpotent for all $x \in \mathfrak{g}'$, there exists $y \in \mathfrak{g}$ such that $\varphi(y) \neq 0$ and $\sigma(\mathfrak{g}')\varphi(y) = 0$ (1.3.12). Then $y \notin \mathfrak{g}'$ and $[\mathfrak{g}',y] \subset \mathfrak{g}'$. Consequently, $\mathfrak{g}' + ky$ is a Lie subalgebra of $\mathfrak{g}$, and $[\mathfrak{g}' + ky, \mathfrak{g}' + ky] \subset \mathfrak{g}'$, so that $\mathfrak{g}' + ky$ is solvable. This is a contradiction, and so $\mathfrak{g}' = \mathfrak{g}$. Let $n = \dim \mathfrak{g}$. By applying 1.3.12 to the adjoint representation of $\mathfrak{g}$, we see that $(\operatorname{ad} x_1)(\operatorname{ad} x_2) \cdots (\operatorname{ad} x_n) = 0$ for all $x_1, \ldots, x_n \in \mathfrak{g}$, and so $\mathfrak{g}$ is nilpotent.

1.3.16. Lemma. *Let V be a cector space, $\mathfrak{g} = \mathfrak{gl}(V)$, and $x \in \mathfrak{g}$. If x is nilpotent, then $\operatorname{ad}_{\mathfrak{g}} x$ is nilpotent.*

Let n be an integer such that $x^n = 0$, and $f = \operatorname{ad}_{\mathfrak{g}} x$. If $y \in \mathfrak{g}$, then $f^m(y)$ is the sum of terms of the form $\pm x^i y x^j$, where $i + j = m$. Hence $f^{2n}(y) = 0$.

1.3.17. Proposition. *Let V be a finite-dimensional vector space, and let ϱ be a representation of $\mathfrak{g}$ in V such that every element of $\varrho(\mathfrak{g})$ is nilpotent. Then:*

(i) *ϱ is strictly triangularizable.*
(ii) *The Lie algebra $\varrho(\mathfrak{g})$ is nilpotent.*

For all $x \in \varrho(\mathfrak{g})$, $\operatorname{ad}_{\varrho(\varrho)} x$ is the restriction to $\varrho(\mathfrak{g})$ of a nilpotent endomorphism (1.3.16), and hence is nilpotent. Consequently, $\varrho(\mathfrak{g})$ is nilpotent (1.3.15). To prove (i), we may assume that ϱ is injective by passage to the quotient; as $\mathfrak{g}$ is then nilpotent, (i) follows from 1.3.12.

1.3.18. Lemma. *Let V be a finite-dimensional vector space, let u and v be endomorphisms of V, and let p be an integer such that $(\operatorname{ad} u)^p v = 0$. Then, for all $\xi \in k$, the nilspace of $u - \xi$ is stable under v.*

By replacing u by $u - \xi$, we may assume that $\xi = 0$. Let N be the nilspace of u. We argue by induction on p, the case $p = 0$ being trivial. Let $w = [u,v]$. Then $(\operatorname{ad} u)^{p-1} w = 0$, and so $w(N) \subset N$. On the other hand, for all integers $q \geqq 0$ we have

$$[u^q, v] = \sum_{i+j=q-1} u^i w u^j. \tag{1}$$

Let n be an integer such that $u^n | N = 0$, and take $q \geqq 2n$. Then (1) proves that $u^q v | N = 0$, whence $v(N) \subset N$.

1.3.19. Theorem ($\mathfrak{g}$ nilpotent). *Let V be a finite-dimensional vector space, and ϱ a representation of $\mathfrak{g}$ in V. We assume that for all $x \in \mathfrak{g}$, $\varrho(x)$ is triangularizable. Then:*

(i) $V = \bigoplus_{\lambda \in \mathfrak{g}^*} V^\lambda$.
(ii) *Each V^λ is stable under ϱ.*
(iii) *For all $\lambda \in \mathfrak{g}^*$, the set of the $(\varrho(x) - \lambda(x)) | V^\lambda$ is strictly triangularizable.*

We prove the theorem by induction on $\dim V$, the case $\dim V = 0$ being trivial.

First case: for all $x \in \mathfrak{g}$, $\varrho(x)$ has a single eigenvalue $\lambda(x)$.

From 1.3.12, there exists a basis $(v_1, \ldots, v_n)$ for V with respect to which the matrix of $\varrho(x) - \lambda(x)$ is strictly lower triangular for all $x \in \mathfrak{g}$. Since $\varrho(x)v_n = \lambda(x)v_n$ for all $x \in \mathfrak{g}$, we see that $\lambda\ \mathfrak{g}^*$, and we have $V = V^\lambda$. Let $\mu \in \mathfrak{g}^*$, and assume that $V^\mu \neq 0$. If $\mu \neq \lambda$, there exists $x \in \mathfrak{g}$ such that $\mu(x) \neq \lambda(x)$. But $\varrho(x)$ has the eigenvalue $\mu(x)$, which is a contradiction. Hence $V^\mu = 0$ for $\mu \neq \lambda$ and the theorem has been proved for the first case.

Second case: there exists $x \in \mathfrak{g}$ such that $\varrho(x)$ has at least two distinct eigenvalues.

For all $\alpha \in k$, let N_α be the nilspace of $\varrho(x) - \alpha$. Then $V = \oplus_{\alpha \in k} N_\alpha$ and $\dim N_\alpha < \dim V$ for all $\alpha \in k$. The N_α are stable under $\varrho(\mathfrak{g})$ (1.3.18). We can thus apply the induction hypothesis to the subrepresentations of ϱ defined by the N_α. We are thus led to prove that, if the theorem is true for the representations ϱ_1 and ϱ_2, it is true for $\varrho_1 \oplus \varrho_2$, which is easy.

1.3.20. PROPOSITION ($\mathfrak{g}$ solvable). *Let K be the Killing form of $\mathfrak{g}$. Then* $K(\mathfrak{g},[\mathfrak{g},\mathfrak{g}]) = 0$.

Let us assume that k is algebraically closed. There exists a basis for $\mathfrak{g}$ such that, for all $x \in \mathfrak{g}$, the matrix of $\operatorname{ad} x$ with respect to this basis is lower triangular (1.3.12). If $x,y,z \in \mathfrak{g}$, the matrix of $\operatorname{ad}[y,z] = [\operatorname{ad} y, \operatorname{ad} z]$, and hence the matrix of $\operatorname{ad} x \cdot \operatorname{ad}[y,z]$, is then strictly lower triangular. Consequently

$$K(x,[y,z]) = \operatorname{tr}(\operatorname{ad} x \cdot \operatorname{ad}[y,z]) = 0,$$

whence $K(\mathfrak{g},[\mathfrak{g},\mathfrak{g}]) = 0$. In the general case, let $\bar{k}$ be an algebraic closure of k. Then $\mathfrak{g}' = \mathfrak{g} \otimes \bar{k}$ is solvable, and its Killing form K' is the extension of K, hence

$$K(\mathfrak{g},[\mathfrak{g},\mathfrak{g}]) \subset K'(\mathfrak{g}',[\mathfrak{g}',\mathfrak{g}']) = 0.$$

1.3.21. LEMMA. *Let $n \in \mathbf{N}$, $\mathfrak{g} = \mathfrak{gl}(n,k)$, $\lambda_1, \ldots, \lambda_n \in k$, let*

$$v = \lambda_1 E_{11} + \cdots + \lambda_n E_{nn} \in \mathfrak{g},$$

let f be a $\mathbf{Q}$-linear mapping of k into k, and let

$$v' = f(\lambda_1)E_{11} + \cdots + f(\lambda_n)E_{nn}.$$

Then:

(i) $(\operatorname{ad}_{\mathfrak{g}} v)E_{ij} = (\lambda_i - \lambda_j)E_{ij}$.

(ii) *The endomorphism v' is a polynomial in v.*

(iii) *The endomorphism $\operatorname{ad}_{\mathfrak{g}} v'$ is a polynomial in $\operatorname{ad}_{\mathfrak{g}} v$.*

The first assertion follows from an easy calculation. There is a polynomial $p \in k[X]$ such that $p(\lambda) = f(\lambda)$ for all $\lambda \in \{\lambda_1, \ldots, \lambda_n\}$, and hence $p(v) = v'$. Then

$$(\mathrm{ad}_{\mathfrak{g}} v')E_{ij} = (f(\lambda_i) - f(\lambda_j))E_{ij} = f(\lambda_i - \lambda_j)E_{ij},$$

and so (iii) follows from (ii).

1.3.22. LEMMA. *Let V be a finite-dimensional vector space, u an endomorphism of V, and $u = v + w$ its Jordan decomposition* (*with* $[v,w] = 0$, *v semi-simple, w nilpotent*) (CH, p. 72). *Let* $\mathfrak{g} = \mathfrak{gl}(V)$. *Then* $\mathrm{ad}_{\mathfrak{g}} u = \mathrm{ad}_{\mathfrak{g}} v + \mathrm{ad}_{\mathfrak{g}} w$ *is the Jordan decomposition of* $\mathrm{ad}_{\mathfrak{g}} u$.

By extension of the field of scalars, we may assume that k is algebraically closed. Then v is diagonalizable and hence $\mathrm{ad}_{\mathfrak{g}} v$ is diagonalizable (1.3.21). On the other hand, $\mathrm{ad}_{\mathfrak{g}} w$ is nilpotent (1.3.16), and $[\mathrm{ad}_{\mathfrak{g}} v, \mathrm{ad}_{\mathfrak{g}} w] = \mathrm{ad}_{\mathfrak{g}}[v,w] = 0$.

1.3.23. LEMMA. *Let V be a finite-dimensional vector space, and $\mathfrak{k}$ a Lie subalgebra of* $\mathfrak{gl}(V)$. *If* $\mathrm{tr}(xy) = 0$ *for all* $x,y \in \mathfrak{k}$, *then every element of* $[\mathfrak{k},\mathfrak{k}]$ *is nilpotent.*

We may assume that k is algebraically closed. Let $u \in [\mathfrak{k},\mathfrak{k}]$, and let $u = v + w$ be its Jordan decomposition. There exist $a_1, b_1, \ldots, a_r, b_r \in \mathfrak{k}$ such that $u = [a_1,b_1] + \cdots + [a_r,b_r]$. Let $(e_1, \ldots, e_n)$ be a basis for V such that $v = \lambda_1 E_{11} + \cdots + \lambda_n E_{nn}$ [we identify $\mathfrak{gl}(V)$ with $\mathfrak{gl}(n,k)$]. Let f be a $\mathbf{Q}$-linear mapping of k into k, and set $v' = f(\lambda_1)E_{11} + \cdots + f(\lambda_n)E_{nn}$. As v' is a polynomial in v (1.3.21) and hence commutes with w, $v'w$ is nilpotent and thus has trace zero. Consequently

$$f(\lambda_1)\lambda_1 + \cdots + f(\lambda_n)\lambda_n = \mathrm{tr}(v'v) = \mathrm{tr}(v'u)$$

$$= \sum_i \mathrm{tr}(v'a_ib_i - v'b_ia_i) = \sum_i \mathrm{tr}([v',a_i]b_i).$$

We set $\mathfrak{gl}(V) = \mathfrak{g}$. Then $\mathrm{ad}_{\mathfrak{g}} v$ is a polynomial in $\mathrm{ad}_{\mathfrak{g}} u$ (1.3.22), and $\mathrm{ad}_{\mathfrak{g}} v'$ is a polynomial in $\mathrm{ad}_{\mathfrak{g}} v$ (1.3.21), hence $[v',a_i] = (\mathrm{ad}_{\mathfrak{g}} v')a_i \in \mathfrak{k}$. From the assumption in the lemma, $\mathrm{tr}([v',a_i]b_i) = 0$. Hence

$$f(\lambda_1)\lambda_1 + \cdots + f(\lambda_n)\lambda_n = 0.$$

Let $(\mu_j)_{j \in J}$ be a basis for the vector space k over $\mathbf{Q}$. For all $\xi \in k$, we denote by ξ_j its jth coordinate ($j \in J$). Let us take f to be the mapping $\xi \mapsto \xi_j$. Then $\lambda_{1j}\lambda_1 + \cdots + \lambda_{nj}\lambda_n = 0$, whence, on taking the jth coordinate of the first member, $\lambda_{1j}^2 + \cdots + \lambda_{nj}^2 = 0$. We deduce that $\lambda_{ij} = 0$ for all i and j, whence $v = 0$. Hence $u = w$ is nilpotent.

1.3.24. THEOREM. *Let K be the Killing form of $\mathfrak{g}$. If*

$$K([\mathfrak{g},\mathfrak{g}],[\mathfrak{g},\mathfrak{g}]) = 0,$$

then $\mathfrak{g}$ is solvable.

Let $\mathfrak{k} = \operatorname{ad} \mathfrak{g}$. If $x,y \in [\mathfrak{k},\mathfrak{k}]$, we have $x = \operatorname{ad} u$, $y = \operatorname{ad} v$, with $u,v \in [\mathfrak{g},\mathfrak{g}]$, hence $\operatorname{tr}(xy) = K(u,v) = 0$. Thus every element of $\mathscr{D}^2\mathfrak{k}$ is nilpotent (1.3.23), so $\mathscr{D}^2\mathfrak{k}$ is nilpotent (1.3.17 (ii)), hence $\mathfrak{k}$ is solvable. The kernel $\mathfrak{c}$ of the adjoint representation of $\mathfrak{g}$ is the centre of $\mathfrak{g}$, and $\mathfrak{g}/\mathfrak{c}$ is solvable from the foregoing. Hence $\mathfrak{g}$ is solvable (1.3.9).

1.4. The radical. The largest nilpotent ideal

1.4.1. PROPOSITION. *Among the solvable ideals of $\mathfrak{g}$, there exists one which contains all the others.*

Let $\mathfrak{a}$ be a solvable ideal of $\mathfrak{g}$ of maximal dimension. Let $\mathfrak{b}$ be an arbitrary solvable ideal. Then $\mathfrak{a} + \mathfrak{b}$ is an ideal, and $(\mathfrak{a} + \mathfrak{b})/\mathfrak{b}$, which is isomorphic to a quotient of $\mathfrak{a}$, is solvable. Thus $\mathfrak{a} + \mathfrak{b}$ is solvable, whence $\mathfrak{a} + \mathfrak{b} = \mathfrak{a}$ and $\mathfrak{b} \subset \mathfrak{a}$.

1.4.2. The ideal of 1.4.1 is termed the *radical* of $\mathfrak{g}$.

1.4.3. PROPOSITION. *The radical of $\mathfrak{g}$ is the smallest ideal $\mathfrak{a}$ of $\mathfrak{g}$ such that $\mathfrak{g}/\mathfrak{a}$ has radical null.*

Let $\mathfrak{r}$ be the radical of $\mathfrak{g}$, φ the canonical homomorphism of $\mathfrak{g}$ onto $\mathfrak{g}/\mathfrak{r}$, $\mathfrak{r}'$ the radical ot $\mathfrak{g}/\mathfrak{r}$, and $\mathfrak{q} = \varphi^{-1}(\mathfrak{r}')$. Then $\mathfrak{q}/\mathfrak{r}$ is solvable, hence $\mathfrak{q}$ is solvable, whence $\mathfrak{q} \subset \mathfrak{r}$ and $\mathfrak{r}' = 0$. Let $\mathfrak{a}$ be an ideal of $\mathfrak{g}$ such that $\mathfrak{g}/\mathfrak{a}$ has radical null. Let ψ be the canonical homomorphism of $\mathfrak{g}$ onto $\mathfrak{g}/\mathfrak{a}$. Then $\psi(\mathfrak{r})$ is a solvable ideal of $\mathfrak{g}/\mathfrak{a}$ hence $\psi(\mathfrak{r}) = 0$ and $\mathfrak{r} \subset \mathfrak{a}$.

1.4.4. PROPOSITION. *Let $\mathfrak{g}_1,\mathfrak{g}_2$ be Lie algebras, and $\mathfrak{r}_1,\mathfrak{r}_2$ their radicals. Then the radical of $\mathfrak{g}_1 \times \mathfrak{g}_2$ is $\mathfrak{r}_1 \times \mathfrak{r}_2$.*

The proof is easy.

1.4.5. LEMMA. *Let $\mathfrak{a}$ be an ideal of $\mathfrak{g}$, V a finite-dimensional vector space, ϱ a simple representation of $\mathfrak{g}$ in V such that each element of $\varrho(\mathfrak{a})$ is nilpotent. Then $\varrho(\mathfrak{a}) = 0$.*

Let W be the set of those $v \in V$ such that $\varrho(\mathfrak{a})v = 0$. Then $W \neq 0$ (1.3.17(i)) and $\varrho(\mathfrak{g})(W) \subset W$ (1.3.11), hence $W = V$.

1.4.6. Lemma. *Let $\mathfrak{a}$ be an ideal of $\mathfrak{g}$, V a finite-dimensional vector space, ϱ a representation of $\mathfrak{g}$ in V, and $(V_0, V_1, \ldots, V_n)$ a Jordan—Hölder series of the $\mathfrak{g}$-module V. The following conditions are equivalent:*

(i) *for all $x \in \mathfrak{a}$, $\varrho(x)$ is nilpotent;*

(ii) *for all $x \in \mathfrak{a}$, we have*

$$\varrho(x)(V_0) \subset V_1, \quad \varrho(x)(V_1) \subset V_2, \ldots, \qquad \varrho(x)(V_{n-1}) \subset V_n.$$

(ii) $\Rightarrow$ (i). Obvious.

(i) $\Rightarrow$ (ii). This follows from 1.4.5.

1.4.7. Proposition. *Let V be a finite-dimensional vector space, ϱ a representation of $\mathfrak{g}$ in V, and b the bilinear form associated with ϱ. Then:*

(i) *The ideals $\mathfrak{a}$ of $\mathfrak{g}$ such that $\varrho(x)$ is nilpotent for all $x \in \mathfrak{a}$ are all contained in one of them, $\mathfrak{n}$ say.*

(ii) *Let $(V_0, V_1, \ldots, V_n)$ be a Jordan—Hölder series of the $\mathfrak{g}$-module V, and ϱ_i the representation of $\mathfrak{g}$ in V_i/V_{i+1} deduced from ϱ. Then*

$$\mathfrak{n} = \operatorname{Ker} \varrho_0 \cap \cdots \cap \operatorname{Ker} \varrho_{n-1}.$$

(iii) *$\mathfrak{n}$ is orthogonal to $\mathfrak{g}$ with respect to b.*

Let $\mathfrak{a}$ be an ideal of $\mathfrak{g}$. From 1.4.6, the following conditions are equivalent:

(a) for all $x \in \mathfrak{a}$, $\varrho(x)$ is nilpotent;

(b) $\mathfrak{a} \subset \operatorname{Ker} \varrho_0 \cap \cdots \cap \operatorname{Ker} \varrho_{n-1}$.

This proves (i) and (ii) at the same time. If $x \in \mathfrak{n}$ and $y \in \mathfrak{g}$, $\varrho(x)\varrho(y)$ is nilpotent, and thus $b(x,y) = 0$, whence (iii).

1.4.8. The ideal $\mathfrak{n}$ of 1.4.7 is termed the *largest nilpotency ideal of ϱ.*

1.4.9. Proposition. *Let $\mathfrak{n}$ be the largest nilpotency ideal of the adjoint representation of $\mathfrak{g}$. Then $\mathfrak{n}$ is the largest nilpotent ideal of $\mathfrak{g}$. It is orthogonal to $\mathfrak{g}$ with respect to the Killing form.*

Let $\mathfrak{a}$ be an ideal of $\mathfrak{g}$. The following conditions are equivalent:

(i) $\mathfrak{a}$ is nilpotent;

(ii) for all $x \in \mathfrak{a}$, $\operatorname{ad}_{\mathfrak{g}} x$ is nilpotent.

For, if $x \in \mathfrak{a}$ is such that $(\operatorname{ad}_{\mathfrak{a}} x)^n = 0$, then $(\operatorname{ad}_{\mathfrak{g}} x)^{n+1} = 0$; thus (i) $\Rightarrow$ (ii) has been proved and (ii) $\Rightarrow$ (i) follows from 1.3.15. Given this, 1.4.9 follows from 1.4.7.

1.5. Semi-simple Lie algebras

1.5.1. LEMMA. *Let $\mathfrak{g}'$ be an ideal of $\mathfrak{g}$, and K and K' the Killing forms of $\mathfrak{g}$ and $\mathfrak{g}'$. Then $K' = K|\mathfrak{g}' \times \mathfrak{g}'$.*

If $x,y \in \mathfrak{g}'$, then $\mathrm{ad}_\mathfrak{g} x$ and $\mathrm{ad}_\mathfrak{g} y$ map $\mathfrak{g}$ into $\mathfrak{g}'$, hence

$$K(x,y) = \mathrm{tr}(\mathrm{ad}_\mathfrak{g} x \; \mathrm{ad}_\mathfrak{g} y) = \mathrm{tr}(\mathrm{ad}_{\mathfrak{g}'} x \; \mathrm{ad}_{\mathfrak{g}'} y) = K'(x,y).$$

1.5.2. THEOREM. *The following conditions are equivalent:*
(i) *the radical of $\mathfrak{g}$ is null;*
(ii) *every commutative ideal of $\mathfrak{g}$ is null;*
(iii) *the Killing form of $\mathfrak{g}$ is non-degenerate.*

not (i) $\Rightarrow$ not (ii). Let us assume that the radical $\mathfrak{r}$ of $\mathfrak{g}$ is not null. Let $\mathscr{D}^i\mathfrak{r}$ be the last non-null ideal of the derived series of $\mathfrak{r}$; it is commutative. If $x \in \mathfrak{g}$, then $\mathrm{ad}_\mathfrak{g} x|\mathfrak{r}$ is a derivation of $\mathfrak{r}$, and hence leaves $\mathscr{D}^i\mathfrak{r}$ stable. Thus, $\mathscr{D}^i\mathfrak{r}$ is a non-null commutative ideal of $\mathfrak{g}$.

not (ii) $\Rightarrow$ not (iii). This follows from 1.4.9.

not (iii) $\Rightarrow$ not (i). Let $\mathfrak{a}$ be orthogonal to $\mathfrak{g}$ with respect to the Killing form, and let us assume that $\mathfrak{a} \neq 0$. Then $\mathfrak{a}$ is an ideal of $\mathfrak{g}$ (1.2.20). Its Killing form is zero (1.5.1), hence $\mathfrak{a}$ is solvable (1.3.24). Hence the radical of $\mathfrak{g}$ is not null.

1.5.3. A Lie algebra which satisfies the conditions of 1.5.2. is termed *semi-simple.*

1.5.4. If $\mathfrak{g}$ is semi-simple, its centre is null (1.5.2 (ii)), hence the adjoint representation of $\mathfrak{g}$ is injective.

1.5.5. PROPOSITION. *Let $\mathfrak{g}_1, \ldots, \mathfrak{g}_n$ be Lie algebras. Then $\mathfrak{g}_1 \times \cdots \times \mathfrak{g}_n$ is semi-simple if and only if $\mathfrak{g}_1, \ldots, \mathfrak{g}_n$ are semi-simple.*

This follows from 1.4.4.

1.5.6. PROPOSITION. *Let $\mathfrak{r}$ be the radical of $\mathfrak{g}$, and k' an extension of k. Then:*
(i) *The radical of $\mathfrak{g} \otimes k'$ is $\mathfrak{r} \otimes k'$.*
(ii) *$\mathfrak{g}$ is semi-simple if and only if $\mathfrak{g} \otimes k'$ is semi-simple.*

The assertion (ii) follows from 1.5.2 (iii). The ideal $\mathfrak{r} \otimes k'$ of $\mathfrak{g} \otimes k'$ is solvable and $\mathfrak{g} \otimes k'/\mathfrak{r} \otimes k' = (\mathfrak{g}/\mathfrak{r}) \otimes k'$ is semi-simple from (ii), whence (i).

1.5.7. LEMMA. *Let $\mathfrak{a}$ be an ideal of $\mathfrak{g}$, K the Killing form of $\mathfrak{g}$, and $\mathfrak{b}$ the orthogonal subspace of $\mathfrak{a}$ with respect to K. Assume that $\mathfrak{a}$ does not contain* *a non-null solvable ideal of* $\mathfrak{g}$. *Then* $\mathfrak{b}$ *is an ideal of* $\mathfrak{g}$ *which is complementary to* $\mathfrak{a}$, *so that* $\mathfrak{g} = \mathfrak{a} \times \mathfrak{b}$.

As $\mathfrak{b}$ is an ideal of $\mathfrak{g}$, $\mathfrak{a} \cap \mathfrak{b}$ is an ideal of $\mathfrak{g}$. Its Killing form is zero (1.5.1), hence $\mathfrak{a} \cap \mathfrak{b}$ is solvable (1.3.24), and so $\mathfrak{a} \cap \mathfrak{b} = 0$ by assumption. But $\dim \mathfrak{a} + \dim \mathfrak{b} \geqq \dim \mathfrak{g}$, and hence $\mathfrak{b}$ is complementary to $\mathfrak{a}$.

1.5.8. PROPOSITION ($\mathfrak{g}$ semi-simple). *Let* $\mathfrak{a}$ *be an ideal of* $\mathfrak{g}$, *and K the Killing form of* $\mathfrak{g}$.

(i) *The Lie algebras* $\mathfrak{a}$ *and* $\mathfrak{g}/\mathfrak{a}$ *are semi-simple.*

(ii) *Let* $\mathfrak{b}$ *be the orthogonal subspace of* $\mathfrak{a}$ *with respect to K. Then* $\mathfrak{b}$ *is an ideal of* $\mathfrak{g}$ *which is complementary to* $\mathfrak{a}$, *so that* $\mathfrak{g} = \mathfrak{a} \times \mathfrak{b}$.

The assertion (ii) follows from 1.5.7. Thus $\mathfrak{a}$ and $\mathfrak{b}$ are semi-simple (1.5.5), and hence $\mathfrak{g}/\mathfrak{a}$, which is isomorphic to $\mathfrak{b}$, is semi-simple.

1.5.9. PROPOSITION ($\mathfrak{g}$ semi-simple).

(i) $\mathfrak{g} = [\mathfrak{g},\mathfrak{g}]$.

(ii) *Every derivation of* $\mathfrak{g}$ *is an inner derivation.*

The algebra $\mathfrak{g}/[\mathfrak{g},\mathfrak{g}]$ is commutative and semi-simple and hence null, whence (i). Let D be a derivation of $\mathfrak{g}$. Let $\mathfrak{h} = k \times \mathfrak{g}$ be the semi-direct product of k by $\mathfrak{g}$ corresponding to the homomorphism $\lambda \to \lambda D$ of k into the Lie algebra of derivations of $\mathfrak{g}$. From 1.5.7, $\mathfrak{h}$ is the product Lie algebra of $\mathfrak{g}$ and an ideal. This ideal is of the form $k \cdot (1,x_0)$, where $x_0 \in \mathfrak{g}$. For all $x \in \mathfrak{g}$, we have

$$0 = [(1,x_0),(0,x)] = (0, Dx + [x_0,x]),$$

whence

$$Dx = -[x_0,x] \qquad \text{and} \qquad D = \operatorname{ad}(-x_0).$$

1.5.10. From 1.5.9 (ii), we can speak of the adjoint group of $\mathfrak{g}$ for $\mathfrak{g}$ semi-simple, and it is the neutral irreducible component of $\operatorname{Aut}(\mathfrak{g})$.

1.5.11. $\mathfrak{g}$ is said to be *simple* if $\dim \mathfrak{g} > 1$ and the only ideals of $\mathfrak{g}$ are 0 and $\mathfrak{g}$.

1.5.12. THEOREM. *A Lie algebra is semi-simple if and only if it is a product of simple Lie algebras.*

(i) Let us assume that $\mathfrak{g}$ is simple. Let $\mathfrak{a}$ be a commutative ideal of $\mathfrak{g}$. If $\mathfrak{a} \neq 0$, then $\mathfrak{a} = \mathfrak{g}$; since $\dim \mathfrak{g} > 1$, there are vector subspaces of $\mathfrak{g}$ which are distinct from 0 and $\mathfrak{g}$ and are ideals of $\mathfrak{g}$, which is a contradiction. Hence $\mathfrak{a} = 0$ and $\mathfrak{g}$ is semi-simple. It follows that, if $\mathfrak{g}_1, \ldots \mathfrak{g}_n$ are simple, then $\mathfrak{g}_1 \times \cdots \times \mathfrak{g}_n$ is semi-simple (1.5.5).

(ii) Let us assume that $\mathfrak{g}$ is semi-simple. By induction on dim $\mathfrak{g}$ we show that $\mathfrak{g}$ is a product of simple Lie algebras. This is obvious if the only ideals of $\mathfrak{g}$ are 0 and $\mathfrak{g}$. Otherwise, let $\mathfrak{a}$ be an ideal of $\mathfrak{g}$ distinct from 0 and $\mathfrak{g}$. Then $\mathfrak{g} = \mathfrak{a} \times \mathfrak{b}$ for $\mathfrak{a}$ and $\mathfrak{b}$ semi-simple (1.5.8), and it is sufficient to apply the induction hypothesis.

1.5.13. Proposition. *Let $\mathfrak{a}_1, \ldots, \mathfrak{a}_n$ be simple Lie algebras and $\mathfrak{g} = \mathfrak{a}_1 \times \cdots \times \mathfrak{a}_n$. Then the ideals of $\mathfrak{g}$ are the products of some of the $\mathfrak{a}_i$. In particular, the $\mathfrak{a}_i$ are the non-null minimal ideals of $\mathfrak{g}$.*

Every product of some of the $\mathfrak{a}_i$ is an ideal of $\mathfrak{g}$.

Let $\mathfrak{a}$ be an ideal of $\mathfrak{g}$. We may assume that $\mathfrak{a} \cap \mathfrak{a}_i \neq 0$ for $i = 1, \ldots, p$ and that $\mathfrak{a} \cap \mathfrak{a}_i = 0$ for $i = p+1, \ldots, n$. Then $\mathfrak{a} \cap \mathfrak{a}_i = \mathfrak{a}_i$ for $i \leqq p$, hence $\mathfrak{a} \supset \mathfrak{a}_1 \times \cdots \times \mathfrak{a}_p$. Then $[\mathfrak{a}, \mathfrak{a}_j] = 0$ for $j > p$, and so $\mathfrak{a}$ commutes with $\mathfrak{a}_{p+1} \times \cdots \times \mathfrak{a}_n$; consequently, $\mathfrak{a} \cap (\mathfrak{a}_{p+1} \times \cdots \times \mathfrak{a}_n)$, which is contained in the centre of the semi-simple Lie algebra $\mathfrak{a}_{p+1} \times \cdots \times \mathfrak{a}_n$, is null. Hence $\mathfrak{a} = \mathfrak{a}_1 \times \cdots \times \mathfrak{a}_p$.

1.5.14. If $\mathfrak{g}$ is semi-simple, the Killing homomorphisms $\mathfrak{g} \to \mathfrak{g}^*$ and $S(\mathfrak{g}) \to S(\mathfrak{g}^*)$ are isomorphisms (1.5.2 (iii)), under which we sometimes identify $\mathfrak{g}^*$ with $\mathfrak{g}$ and $S(\mathfrak{g}^*)$ with $S(\mathfrak{g})$.

1.6. Semi-simplicity of representations

1.6.1. Lemma ($\mathfrak{g}$ semi-simple). *Let V be a finite-dimensional vector space, ϱ a simple injective representation of $\mathfrak{g}$ in V, and b the bilinear form associated with ϱ. Then:*

(i) *b is non-degenerate.*

(ii) *Let $(x_1, \ldots, x_n)$ be a basis for $\mathfrak{g}$, and $(y_1, \ldots, y_n)$ the dual basis relative to b. If k is algebraically closed, then*

$$\varrho(x_1)\varrho(y_1) + \cdots + \varrho(x_n)\varrho(y_n) = \frac{\dim \mathfrak{g}}{\dim V} 1 .$$

Let us identify $\mathfrak{g}$ with $\varrho(\mathfrak{g})$ by means of ϱ. Let $\mathfrak{a}$ be the orthogonal subspace of $\mathfrak{g}$ with respect to b; it is an ideal of $\mathfrak{g}$ (1.2.20). We have $\mathrm{tr}(xy) = 0$ for $x, y \in \mathfrak{a}$, hence every element of $[\mathfrak{a}, \mathfrak{a}]$ is nilpotent (1.3.23), and so $[\mathfrak{a}, \mathfrak{a}]$ is nilpotent. Thus $\mathfrak{a}$ is solvable and consequently null since $\mathfrak{g}$ is semi-simple. We have thus proved (i).

To prove (ii), let $x \in \mathfrak{g}$, and set

$$[x, x_i] = \sum_j \lambda_{ij} x_j, \qquad [x, y_i] = \sum_j \mu_{ij} y_j .$$

Then

$$(1) \qquad \lambda_{ij} = b([x,x_i],y_j) = -b(x_i,[x,y_j]) = -\mu_{ji}.$$

Hence

$$\begin{aligned}[x,x_1y_1 + \cdots + x_ny_n] &= \sum_i [x,x_i]y_i + \sum_i x_i[x,y_i] \\ &= \sum_{i,j} \lambda_{ij}x_jy_i + \sum_{i,j} \mu_{ij}x_iy_j = 0.\end{aligned}$$

Assume that k is algebraically closed. As ϱ is simple, $x_1y_1 + \cdots + x_ny_n$ is a scalar λ (Schur's lemma). Hence

$$(\dim V)\lambda = \mathrm{tr}(x_1y_1 + \cdots + x_ny_n) = b(x_1,y_1) + \cdots + b(x_n,y_n) = \dim \mathfrak{g}.$$

1.6.2. LEMMA ($\mathfrak{g}$ semi-simple). *Let V be a finite-dimensional vector space, ϱ a representation of $\mathfrak{g}$ in V, and f a linear mapping of $\mathfrak{g}$ into V. The following conditions are equivalent:*

(i) $f([x,y]) = \varrho(x)f(y) - \varrho(y)f(x)$ *for any* $x,y \in \mathfrak{g}$;

(ii) *There exists* $v \in V$ *such that* $f(x) = \varrho(x)v$ *for all* $x \in \mathfrak{g}$.

If condition (ii) is satisfied, then

$$\begin{aligned}\varrho(x)f(y) - \varrho(y)f(x) &= \varrho(x)\varrho(y)v - \varrho(y)\varrho(x)v \\ &= \varrho([x,y])v = f([x,y]).\end{aligned}$$

We now assume (i) and prove (ii). The problem can be easily reduced to the case where k is algebraically closed.

(a) We assume that ϱ is simple and injective, and use the notation of Lemma 1.6.1 and its proof. We set

$$v = \lambda^{-1}(\varrho(x_1)f(y_1) + \cdots + \varrho(x_n)f(y_n)).$$

Then, for all $x \in \mathfrak{g}$,

$$\begin{aligned}\lambda(\varrho(x)v - f(x)) &= \varrho(x)(\varrho(x_1)f(y_1) + \cdots + \varrho(x_n)f(y_n)) \\ &\quad -(\varrho(x_1)\varrho(y_1)f(x) + \cdots + \varrho(x_n)\varrho(y_n)f(x)) \\ &= \sum_i ([\varrho(x),\varrho(x_i)]f(y_i) + \varrho(x_i)f([x,y_i])) \\ &= \sum_{i,j} \lambda_{ij}\varrho(x_j)f(y_i) + \sum_{i,j} \mu_{ij}\varrho(x_i)f(y_j) = 0.\end{aligned}$$

(b) We assume that ϱ is simple. Let $\mathfrak{a} = \mathrm{Ker}\,\varrho$. Then $\mathfrak{g} = \mathfrak{a} \times \mathfrak{b}$, with $\mathfrak{a}$ and $\mathfrak{b}$ semi-simple. From (a) there exists $v \in V$ such that $f(x) = \varrho(x)v$ for all $x \in \mathfrak{b}$. Since $\varrho(\mathfrak{a}) = 0$, $f([\mathfrak{a},\mathfrak{a}]) = 0$ by assumption (i). But $\mathfrak{a} = [\mathfrak{a},\mathfrak{a}]$ (1.5.9), hence $f(x) = 0 = \varrho(x)v$ for all $x \in \mathfrak{a}$.

(c) We consider the general case by induction on dim V. If ϱ is not simple, let W be a vector subspace of V which is stable under ϱ and such that $W \neq 0$ and $W \neq V$. Let φ be the canonical mapping of V onto V/W. From the induction hypothesis, there exists $v \in V$ such that $(\varphi \circ f)(x) = \varphi(\varrho(x)v)$ for all $x \in \mathfrak{g}$. We set $f'(x) = f(x) - \varrho(x)v \in W$. Then

$$f'([x,y]) = \varrho(x)f'(y) - \varrho(y)f'(x) \quad \text{for } x,y \in \mathfrak{g}.$$

From the induction hypothesis, there exists $v' \in W$ such that $f'(x) = \varrho(x)v'$ for all $x \in \mathfrak{g}$. Hence $f(x) = \varrho(x)(v + v')$ for all $x \in \mathfrak{g}$.

1.6.3. THEOREM ($\mathfrak{g}$ semi-simple). *Let V be a finite-dimensional vector space, and ϱ a representation of $\mathfrak{g}$ in V. Then ϱ is semi-simple.*

Let U be a vector subspace of V which is stable under ϱ, φ the canonical mapping of V onto V/U, and τ the quotient representation of ϱ in V/U. We set $L = \operatorname{Hom}(V/U,V)$ and $M = \operatorname{Hom}(V/U,U) \subset L$. For $x \in \mathfrak{g}$ and $l \in L$, we define $\lambda(x)l \in L$ by

$$(\lambda(x)l)(w) = \varrho(x)lw - l\tau(x)w \quad \text{for all } w \in V/U.$$

Then λ is a representation of $\mathfrak{g}$ in L (1.2.15). If $l \in M$, then $\lambda(x)l \in M$, and we can therefore consider the subrepresentation μ of λ defined by M. Let $l_0 \in L$ such that $\varphi \circ l_0 = \mathrm{id}_{V/U}$. Let f be the mapping of $\mathfrak{g}$ into L defined by $f(x) = \lambda(x)l_0$ for all $x \in \mathfrak{g}$. If $w \in V/U$, then

$$\begin{aligned}\varphi((\lambda(x)l_0)(w)) &= \varphi\varrho(x)l_0w - \varphi l_0\tau(x)w \\ &= \tau(x)\varphi l_0w - \varphi l_0\tau(x)w = \tau(x)w - \tau(x)w = 0,\end{aligned}$$

hence $f(x) \in M$. On the other hand,

$$f([x,y]) = \mu(x)f(y) - \mu(y)f(x)$$

from the implication (ii) $\Rightarrow$ (i) of 1.6.2. From the implication (i) $\Rightarrow$ (ii) of 1.6.2, there exists $m_0 \in M$ such that $f(x) = \mu(x)m_0$ for all $x \in \mathfrak{g}$. Then

$$\lambda(x)(l_0 - m_0) = f(x) - f(x) = 0 \quad \text{for all } x \in \mathfrak{g},$$

hence

$$n_0 = l_0 - m_0 \in \operatorname{Hom}_{\mathfrak{g}}(V/U,V).$$

Now $\varphi \circ n_0 = \varphi \circ l_0 = \mathrm{id}_{V/U}$, so $n_0(V/U)$ is complementary to U in V and stable under ϱ.

1.6.4. COROLLARY ($\mathfrak{g}$ semi-simple). *Let $\mathfrak{a}$ be a commutative Lie algebra, and ϱ a finite-dimensional representation of $\mathfrak{g} \times \mathfrak{a}$. The following conditions* *are equivalent:*

(i) ϱ *is semi-simple;*

(ii) *for all* $a \in \mathfrak{a}$, $\varrho(a)$ *is semi-simple.*

We may assume that k is algebraically closed (1.2.19).

(i) $\Rightarrow$ (ii). We can assume that ϱ is simple. Then $\varrho(a)$ is scalar for all $a \in \mathfrak{a}$ from Schur's lemma.

(ii) $\Rightarrow$ (i). Let V be the space of ϱ. Then $V = \bigoplus_{\lambda \in \mathfrak{a}^*} V^\lambda$. Each V^λ is stable under $\varrho(\mathfrak{g})$. If condition (ii) is satisfied, $\varrho(a)|V^\lambda$ is scalar for all $a \in \mathfrak{a}$. Hence ϱ is semi-simple since $\varrho|\mathfrak{g}$ is semi-simple (1.6.3).

1.6.5. Proposition. *Let V be a finite-dimensional vector space, and $\mathfrak{g}$ a semi-simple Lie subalgebra of $\mathfrak{gl}(V)$. Then $\mathfrak{g}$ contains the semi-simple and nilpotent components of its elements.*

We may assume that k is algebraically closed. Let $\mathscr{V}$ be the set of vector subspaces of V which are stable under $\mathfrak{g}$. For all $W \in \mathscr{V}$, let $\mathfrak{g}_W$ be the set of those $x \in \mathfrak{gl}(V)$ such that $x(W) \subset W$ and $\mathrm{tr}(x|W) = 0$. It is a Lie subalgebra of $\mathfrak{gl}(V)$ containing $\mathfrak{g}$ because $\mathfrak{g} = [\mathfrak{g},\mathfrak{g}]$. Let $\mathfrak{n}$ be the normalizer of $\mathfrak{g}$ in $\mathfrak{gl}(V)$. Let $\mathfrak{g}_* = \mathfrak{n} \cap (\bigcap_{W \in \mathscr{V}} \mathfrak{g}_W)$. Let $x \in \mathfrak{g}_*$, and let s and n be its semi- simple and nilpotent components. Then $s^\sim = \mathrm{ad}_{\mathfrak{gl}(V)} s$ and $n^\sim = \mathrm{ad}_{\mathfrak{gl}(V)} n$ are the semi-simple and nilpotent components of $x^\sim = \mathrm{ad}_{\mathfrak{gl}(V)} x$ (1.3.22). On the other hand, s and n ($s^\sim$ and $n^\sim$) are polynomials in x ($x^\sim$), so $s \in \mathfrak{g}_*$ and $n \in \mathfrak{g}_*$. Finally we prove that $g_* = \mathfrak{g}$. Since $\mathfrak{g}$ is a semi-simple ideal of $\mathfrak{g}_*$, there is an ideal $\mathfrak{a}$ of $\mathfrak{g}_*$ such that $\mathfrak{g}_* = \mathfrak{g} \times \mathfrak{a}$ (1.5.7). Let $a \in \mathfrak{a}$, and let W be a minimal element of $\mathscr{V} - \{0\}$. Then $a|W$ is scalar from Schur's lemma, and $\mathrm{tr}(a|W) = 0$, hence $a|W = 0$. Now V is the sum of the minimal elements of $\mathscr{V} - \{0\}$ (1.6.3), hence $a = 0$ and $\mathfrak{a} = 0$.

1.6.6. Corollary. *With the notation of* 1.6.5, *an element x of $\mathfrak{g}$ is a semi-simple (or nilpotent) endomorphism of V if and only if* $\mathrm{ad}_\mathfrak{g} x$ *is semi-simple (or nilpotent).*

Let s,n be the semi-simple and nilpotent components of x. From 1.3.22 and 1.6.5, $\mathrm{ad}_\mathfrak{g} s$ and $\mathrm{ad}_\mathfrak{g} n$ are the semi-simple and nilpotent components of $\mathrm{ad}_\mathfrak{g} x$. Since the adjoint representation of $\mathfrak{g}$ is injective,

$$x = s \;\Leftrightarrow\; \mathrm{ad}_\mathfrak{g} x = \mathrm{ad}_\mathfrak{g} s, \qquad x = n \;\Leftrightarrow\; \mathrm{ad}_\mathfrak{g} x = \mathrm{ad}_\mathfrak{g} n,$$

and the corollary is proved.

1.6.7. Corollary ($\mathfrak{g}$ semi-simple). *Let $x \in \mathfrak{g}$. The following conditions are equivalent:*

(i) $\mathrm{ad}_{\mathfrak{g}} x$ *is semi-simple (nilpotent);*

(ii) *there is a finite-dimensional injective representation* ϱ *of* $\mathfrak{g}$ *such that* $\varrho(x)$ *is semi-simple (nilpotent);*

(iii) *for every finite-dimensional representation* ϱ' *of* $\mathfrak{g}$, $\varrho'(x)$ *is semi-simple (nilpotent).*

(iii) $\Rightarrow$ (ii). This is obvious.

(ii) $\Rightarrow$ (i). This follows from 1.6.6.

(i) $\Rightarrow$ (iii). Let us assume that (i) is true and let ϱ' be a finite-dimensional representation of $\mathfrak{g}$; let $\mathfrak{g}' = \mathfrak{g}/\mathrm{Ker}\,\varrho'$, ϱ'' be the representation of $\mathfrak{g}'$ deduced from ϱ' by passage to the quotient, and y the canonical image of x in $\mathfrak{g}'$; then $\mathrm{ad}_{\mathfrak{g}'} y$ is semi-simple (nilpotent), hence $\varrho''(y)$ is semi-simple (nilpotent) from 1.6.6.

1.6.8. We assume that $\mathfrak{g}$ is semi-simple. An element x of $\mathfrak{g}$ which satisfies the conditions of 1.6.7 is called *semi-simple (nilpotent)*. From 1.6.5, each element z of $\mathfrak{g}$ may be written uniquely as $z = s + n$, with s semi-simple, n nilpotent, and $[s,n] = 0$. We say that s (n) is the *semi-simple (nilpotent) component* of z. If $\mathfrak{g}'$ is a semi-simple Lie subalgebra of $\mathfrak{g}$ containing z, then $s \in \mathfrak{g}'$ and $n \in \mathfrak{g}'$ (e.g., from 1.6.5).

1.6.9. THEOREM. *Let* $\mathfrak{r}$ *be the radical of* $\mathfrak{g}$. *There exists a Lie subalgebra* $\mathfrak{s}$ *of* $\mathfrak{g}$ *such that* $\mathfrak{g} = s \oplus \mathfrak{r}$.

(The algebra s, being isomorphic to $\mathfrak{g}/\mathfrak{r}$, is thus semi-simple.)

We prove this by induction on $\dim \mathfrak{r}$. If there is an ideal $\mathfrak{a}$ of $\mathfrak{g}$ such that $0 \subset \mathfrak{a} \subset \mathfrak{r}$, $\mathfrak{a} \neq 0$, $\mathfrak{a} \neq \mathfrak{r}$, the induction hypothesis applied to $\mathfrak{g}/\mathfrak{a}$, whose radical is $\mathfrak{r}/\mathfrak{a}$, proves that there is a Lie subalgebra $\mathfrak{b}$ of $\mathfrak{g}$ such that $\mathfrak{b} \cap \mathfrak{r} = \mathfrak{a}$, $\mathfrak{b} + \mathfrak{r} = \mathfrak{g}$. Thus $\mathfrak{b}/\mathfrak{a}$ is isomorphic with $\mathfrak{g}/\mathfrak{r}$, hence semi-simple, and $\mathfrak{a}$ is a solvable ideal of $\mathfrak{b}$, hence $\mathfrak{a}$ is the radical of $\mathfrak{b}$. Since $\mathfrak{a} \neq \mathfrak{r}$, the induction hypothesis applied to $\mathfrak{b}$ proves that there is a Lie subalgebra $\mathfrak{s}$ of $\mathfrak{b}$ such that $\mathfrak{b} = \mathfrak{s} \oplus (\mathfrak{b} \cap \mathfrak{r})$. Then

$$\mathfrak{s} \cap \mathfrak{r} \subset \mathfrak{s} \cap \mathfrak{b} \cap \mathfrak{r} = 0$$

and

$$\mathfrak{s} + \mathfrak{r} = \mathfrak{s} + (\mathfrak{b} \cap \mathfrak{r}) + \mathfrak{r} = \mathfrak{b} + \mathfrak{r} = \mathfrak{g}.$$

Henceforth we assume that $\mathfrak{r} \neq 0$ and that the only ideals of $\mathfrak{g}$ contained in $\mathfrak{r}$ are 0 and $\mathfrak{r}$. In particular, $[\mathfrak{r},\mathfrak{r}] = 0$, and the centre $\mathfrak{c}$ of $\mathfrak{g}$ is equal to 0 or $\mathfrak{r}$. If $\mathfrak{c} = \mathfrak{r}$, the adjoint representation ϱ of $\mathfrak{g}$ defines a representation of $\mathfrak{g}/\mathfrak{r}$ (which is semi-simple) in $\mathfrak{g}$; from 1.6.3, there is a vector subspace $\mathfrak{s}$ of $\mathfrak{g}$ which is complementary to $\mathfrak{r}$ and stable under ϱ; hence $\mathfrak{s}$ is an ideal of $\mathfrak{g}$ and the theorem is proved. We henceforth assume that $\mathfrak{c} = 0$.

Let M be the set of the $u \in \text{End}(\mathfrak{g})$ such that $u(\mathfrak{g}) \subset \mathfrak{r}$ and that $u|\mathfrak{r}$ is a homothety [whose ratio will be denoted by $\lambda(u)$]. Let N be the set of the $u \in M$ such that $\lambda(u) = 0$; it is a vector subspace of codimension 1 in M. Let $P = \text{ad}_\mathfrak{g}\mathfrak{r}$; then $P \subset N$ because $[\mathfrak{r},\mathfrak{r}] = 0$. Let σ be the representation of $\mathfrak{g}$ in $\text{End}(\mathfrak{g})$ deduced from ϱ (1.2.15); we recall that $\sigma(x)u = [\text{ad}_\mathfrak{g}x,u]$ for $x \in \mathfrak{g}$ and $u \in \text{End}(\mathfrak{g})$. Then $\sigma(\mathfrak{g})M \subset N$ and $\sigma(\mathfrak{g})P \subset P$. For $x \in \mathfrak{r}$, $y \in \mathfrak{g}$ and $u \in M$, we have

$$(\sigma(x)u)(g) = [x,u(y)] - u([x,y]) = -\lambda(u)[x,y]$$

since $[\mathfrak{r},\mathfrak{r}] = 0$; hence

$$\sigma(x)u = -\lambda(u)\, \text{ad}_\mathfrak{g}x, \tag{1}$$

so that $\sigma(\mathfrak{r})M \subset P$. The representation σ' of $\mathfrak{g}$ in M/P deduced from σ is therefore null over $\mathfrak{r}$ and consequently semi-simple (1.6.3). We have

$$\sigma'(\mathfrak{g})(M/P) \subset N/P,$$

hence there exists $u_0 \in M$ such that $\lambda(u_0) = -1$ and $\sigma(\mathfrak{g})u_0 \subset P$. Since $\mathfrak{c} = 0$, there exists, for all $x \in \mathfrak{g}$, a unique element $\psi(x) \in \mathfrak{r}$ such that $\sigma(x)u_0 = \text{ad}_\mathfrak{g}\psi(x)$. The mapping ψ of $\mathfrak{g}$ into $\mathfrak{r}$ is linear, and $\psi(x) = x$ for $x \in \mathfrak{r}$ from (1). Hence $\mathfrak{s} = \text{Ker}\,\psi$ is complementary to $\mathfrak{r}$ in $\mathfrak{g}$. Since $\mathfrak{s}$ is the set of the $x \in \mathfrak{g}$ such that $\sigma(x)u_0 = 0$, $\mathfrak{s}$ is a Lie subalgebra of $\mathfrak{g}$.

1.7. Reductive Lie algebras

1.7.1. PROPOSITION. *Let $\mathfrak{r}$ be the radical of $\mathfrak{g}$. Let $\mathfrak{a}_1$ be the intersection of the kernels of the finite-dimensional simple representations of $\mathfrak{g}$. Let $\mathfrak{a}_2$ be the intersection of the largest nilpotency ideals of the finite-dimensional representations of $\mathfrak{g}$. Then:*

(i) $\mathfrak{a}_1 = \mathfrak{a}_2 = [\mathfrak{g},\mathfrak{g}] \cap \mathfrak{r} = [\mathfrak{g},\mathfrak{r}]$.

(ii) *The ideal $\mathfrak{a}_1$ is nilpotent.*

(iii) *In particular, if $\mathfrak{g}$ is solvable, $[\mathfrak{g},\mathfrak{g}]$ is nilpotent.*

From 1.4.7, we have $\mathfrak{a}_1 = \mathfrak{a}_2$. The ideal $\mathfrak{a}_2$ is contained in the largest nilpotency ideal of the adjoint representation, and hence is nilpotent and contained in $\mathfrak{r}$. If λ is a linear form on $\mathfrak{g}$ such that $\lambda([\mathfrak{g},\mathfrak{g}]) = 0$, then λ is a simple representation of $\mathfrak{g}$, hence $\mathfrak{a}_1 \subset \text{Ker}\,\lambda$; this proves that $\mathfrak{a}_1 \subset [\mathfrak{g},\mathfrak{g}]$. Thus $\mathfrak{a}_1 \subset [\mathfrak{g},\mathfrak{g}] \cap \mathfrak{r}$. Let $\mathfrak{s}$ be a Lie subalgebra of $\mathfrak{g}$ such that $\mathfrak{g} = \mathfrak{s} \oplus \mathfrak{r}$ (1.6.9). We have $[\mathfrak{s},\mathfrak{s}] = \mathfrak{s}$ (1.5.9), hence

$$[\mathfrak{g},\mathfrak{g}] = \mathfrak{s} + [\mathfrak{s},\mathfrak{r}] + [\mathfrak{r},\mathfrak{r}] = \mathfrak{s} \oplus [\mathfrak{g},\mathfrak{r}],$$

whence $[\mathfrak{g},\mathfrak{g}] \cap \mathfrak{r} = [\mathfrak{g},\mathfrak{r}]$. Finally let us prove that $[\mathfrak{g},\mathfrak{r}] \subset \mathfrak{a}_2$. Let ϱ be a finite-dimensional representation of $\mathfrak{g}$; we must prove that $\varrho(x)$ is nilpotent for all $x \in [\mathfrak{g},\mathfrak{r}]$; taking 1.5.6 (i) into account, we may assume k to be algebraically closed. Then let σ be a finite-dimensional simple representation of $\mathfrak{g}$ in a space V. There exists $\lambda \in \mathfrak{r}^*$ such that $V_\lambda \neq 0$ (1.3.12). From 1.3.11, V_λ is stable under $\sigma(\mathfrak{g})$, and hence equal to V. Then $\sigma(\mathfrak{r})$ consists of scalar endomorphisms, and hence $\sigma([\mathfrak{g},\mathfrak{r}]) = 0$. This proves that $[\mathfrak{g},\mathfrak{r}] \subset \mathfrak{a}_1 = \mathfrak{a}_2$.

1.7.2. The ideal $\mathfrak{a}_1$ of 1.7.1 is termed the *nilpotent radical* of $\mathfrak{g}$. If $\mathfrak{g}$ is solvable, then $\mathfrak{a}_1 = [\mathfrak{g},\mathfrak{g}]$.

1.7.3. PROPOSITION. *Let $\mathfrak{r}$ be the radical of $\mathfrak{g}$, and $\mathfrak{t}$ its nilpotent radical. The following conditions are equivalent:*

(i) *the adjoint representation of $\mathfrak{g}$ is semi-simple;*

(ii) $\mathfrak{g}$ *is the product of a semi-simple Lie algebra and a commutative Lie algebra;*

(iii) *there exists a finite-dimensional representation of $\mathfrak{g}$ such that the associated bilinear form is non-degenerate;*

(iv) *there exists a finite-dimensional semi-simple injective representation of $\mathfrak{g}$;*

(v) $\mathfrak{t} = 0$;

(vi) $\mathfrak{r}$ *is the centre of* $\mathfrak{g}$.

(i) ⇒ (ii). Let us assume that the adjoint representation of $\mathfrak{g}$ is semi-simple. Then $\mathfrak{g}$ is the direct sum of minimal non-null ideals $\mathfrak{a}_1, \ldots, \mathfrak{a}_n$, and hence $\mathfrak{g} = \mathfrak{a}_1 \times \cdots \times \mathfrak{a}_n$. For all i, $\mathfrak{a}_i$ has only the ideals 0 and $\mathfrak{a}_i$, and hence $\mathfrak{a}_i$ is simple or one-dimensional. Thus $\mathfrak{g}$ is the product of a semi-simple Lie algebra and a commutative Lie algebra.

(ii) ⇒ (iii). Let us assume that $\mathfrak{g} = \mathfrak{g}_1 \times \mathfrak{g}_2$, with $\mathfrak{g}_1$ semi-simple and $\mathfrak{g}_2$ commutative. Then the adjoint representation ϱ_1 of $\mathfrak{g}_1$ has an associated bilinear form which is non-degenerate. Clearly, $\mathfrak{g}_2$ has a finite-dimensional representation ϱ_2 whose associated bilinear form is non-degenerate. We now identify ϱ_1 and ϱ_2 with representations of $\mathfrak{g}$ by virtue of the canonical homomorphisms $\mathfrak{g} \to \mathfrak{g}_1$ and $\mathfrak{g} \to \mathfrak{g}_2$. Then $\varrho_1 \oplus \varrho_2$ is a finite-dimensional representation of $\mathfrak{g}$ whose associated form is non-degenerate.

(iii) ⇒ (iv). This follows from 1.4.7 (iii).

(iv) ⇒ (v). Obvious.

(v) ⇒ (vi). If $\mathfrak{t} = 0$, we have $[\mathfrak{g},\mathfrak{r}] = 0$, and hence $\mathfrak{r}$ is contained in the centre $\mathfrak{c}$ of $\mathfrak{g}$. On the other hand, $\mathfrak{c} \subset \mathfrak{r}$.

(vi) ⇒ (i). If $\mathfrak{r}$ is equal to the centre $\mathfrak{c}$ of $\mathfrak{g}$, the adjoint representation of $\mathfrak{g}$, which has the kernel $\mathfrak{c}$, defines a representation of $\mathfrak{g}/\mathfrak{r}$ which is semi-simple from 1.6.3.

1.7.4. If $\mathfrak{g}$ satisfies the conditions of 1.7.3, we say that $\mathfrak{g}$ is *reductive*. Because of criterion (ii) of 1.7.3, the properties of reductive Lie algebras can almost always be trivially deduced from the corresponding properties of semi-simple Lie algebras, and we shall use them without formally stating them in this more general framework.

1.7.5. Let $\mathfrak{h}$ be a Lie subalgebra of $\mathfrak{g}$. We say that $\mathfrak{h}$ is *reductive in* $\mathfrak{g}$ if the representation $x \mapsto \mathrm{ad}_{\mathfrak{g}} x$ of $\mathfrak{h}$ is semi-simple. Then the subrepresentation $x \mapsto \mathrm{ad}_{\mathfrak{h}} x$ of $\mathfrak{h}$ is semi-simple and hence $\mathfrak{h}$ is reductive.

1.7.6. PROPOSITION. ($\mathfrak{g}$ semi-simple). *Let K be the Killing form of $\mathfrak{g}$ and $\mathfrak{m}$ a Lie subalgebra of $\mathfrak{g}$ satisfying the following conditions:*

(a) *$K|\mathfrak{m} \times \mathfrak{m}$ is non-degenerate;*

(b) *if $x \in \mathfrak{m}$, the semi-simple and nilpotent components of x relative to $\mathfrak{g}$ belong to $\mathfrak{m}$.*

Then $\mathfrak{m}$ is reductive in $\mathfrak{g}$.

From 1.7.3 (iii), $\mathfrak{m}$ is reductive. Let $\mathfrak{c}$ be the centre of $\mathfrak{m}$, $x \in \mathfrak{c}$, and s and n be the semi-simple and nilpotent components of x. We have $n \in \mathfrak{m}$. Since $\mathrm{ad}_{\mathfrak{g}} n$ is a polynomial without constant term in $\mathrm{ad}_{\mathfrak{g}} x$ (AL VIII, p. 108), we have $n \in \mathfrak{c}$. If $y \in \mathfrak{m}$, $\mathrm{ad}_{\mathfrak{g}} n \cdot \mathrm{ad}_{\mathfrak{g}} y$ is nilpotent, hence $K(n,y) = 0$, and thus, from condition (a), $n = 0$. Hence $\mathrm{ad}_{\mathfrak{g}} x$ is semi-simple, which proves that $\mathfrak{m}$ is reductive in $\mathfrak{g}$ (1.6.4).

1.7.7. PROPOSITION ($\mathfrak{g}$ semi-simple). *Let $\mathfrak{a}$ be a Lie subalgebra of $\mathfrak{g}$ which is reductive in $\mathfrak{g}$, $\mathfrak{m}$ the centralizer of $\mathfrak{a}$ in $\mathfrak{g}$, and K the Killing form of $\mathfrak{g}$.*

(i) *The restriction of K to $\mathfrak{m}$ is non-degenerate.*

(ii) *If $x \in \mathfrak{m}$, the semi-simple and nilpotent components of x relative to $\mathfrak{g}$ belong to $\mathfrak{m}$.*

(iii) *The algebra $\mathfrak{m}$ is reductive in $\mathfrak{g}$.*

(iv) *$\mathfrak{g} = \mathfrak{m} \oplus [\mathfrak{a},\mathfrak{g}]$, and $[\mathfrak{a},\mathfrak{g}]$ is the orthogonal subspace of $\mathfrak{m}$.*

Since $\mathfrak{a}$ is reductive in $\mathfrak{g}$, we have $\mathfrak{g} = \mathfrak{m} \oplus [\mathfrak{a},\mathfrak{g}]$ (1.2.10). If $x \in \mathfrak{a}$, $y \in \mathfrak{m}$, and $z \in \mathfrak{g}$, then $[x,y] = 0$, hence

$$K([z,x],y) = K(z[x,y]) = 0,$$

and hence $\mathfrak{m}$ and $[\mathfrak{g},\mathfrak{a}]$ are orthogonal. This proves (i) and (iv), (ii) is obvious, and (iii) follows from (i), (ii) and 1.7.6.

1.7.8. PROPOSITION. *Let* ϱ_1, ϱ_2 *be finite-dimensional semi-simple representations of* $\mathfrak{g}$. *Then* $\varrho_1 \otimes \varrho_2$ *is semi-simple.*

We may assume k to be algebraically closed and ϱ_1, ϱ_2 to be simple. Passing to the quotient by the nilpotent radical of $\mathfrak{g}$, which is contained in $\operatorname{Ker} \varrho_1 \cap \operatorname{Ker} \varrho_2$, we may assume $\mathfrak{g}$ to be reductive. Let $\mathfrak{g} = \mathfrak{a} \times \mathfrak{c}$, where $\mathfrak{c}$ is the centre of $\mathfrak{g}$ and $\mathfrak{a}$ is semi-simple. Then $\varrho_1(\mathfrak{c})$, which commutes with $\varrho_1(\mathfrak{g})$, consists of scalar endomorphisms; the same applies to $\varrho_2(\mathfrak{c})$ and consequently to $(\varrho_1 \otimes \varrho_2)(\mathfrak{c})$. Now $(\varrho_1 \otimes \varrho_2)|\mathfrak{a}$ is semi-simple (1.6.3).

1.7.9. PROPOSITION. *Let* $\mathfrak{h}$ *be a Lie subalgebra of* $\mathfrak{g}$ *which is reductive in* $\mathfrak{g}$, *and* ϱ *a representation of* $\mathfrak{g}$ *in* V.

(i) *Let* W *be the sum of the finite-dimensional simple sub-*$\mathfrak{h}$*-modules of* V. *Then* W *is a sub-*$\mathfrak{g}$*-module of* V.

(ii) *If* ϱ *is semi-simple and finite-dimensional then* $\varrho|\mathfrak{h}$ *is semi-simple.*

Let W_0 be a finite-dimensional simple sub-$\mathfrak{h}$-module of V. We shall consider $\mathfrak{g}$ as an $\mathfrak{h}$-module by virtue of the representation $x \mapsto \operatorname{ad}_{\mathfrak{g}} x$ of $\mathfrak{h}$ in $\mathfrak{g}$. Then $\mathfrak{g} \otimes W_0$ is a semi-simple $\mathfrak{h}$-module (1.7.8). Let θ be the linear mapping of $\mathfrak{g} \otimes W_0$ into V such that $\theta(x \otimes w) = xw$ for $x \in \mathfrak{g}$ and $w \in W_0$. This is an $\mathfrak{h}$-homomorphism, for if $y \in \mathfrak{h}$, we have

(1) $\theta([y,x] \otimes w + x \otimes y \cdot w) = [y,x] \cdot w + x \cdot y \cdot w = y \cdot x \cdot w = y \cdot \theta(x \otimes w)$.

Hence $\theta(\mathfrak{g} \otimes W_0)$ is a finite-dimensional semi-simple sub-$\mathfrak{h}$-module of V, whence $\mathfrak{g} \cdot W_0 = \theta(\mathfrak{g} \otimes W_0) \subset W$. This proves (i). To prove (ii), we may assume ϱ to be simple, and then, from (i), $W = V$.

1.7.10. PROPOSITION. *Let* V *be a finite-dimensional semi-simple* $\mathfrak{g}$*-module, and* S *the symmetric algebra of* V *which has a* $\mathfrak{g}$*-module structure in a natural way. The algebra* $S^{\mathfrak{g}}$ *of invariant elements of* S *is of finite type.*

For all $\delta \in \mathfrak{g}^\wedge$, let S_δ be the sum of sub-$\mathfrak{g}$-modules of S of class δ (cf. 1.2.8). We have $S^{\mathfrak{g}} = S_{\delta_0}$, where δ_0 is the one-dimensional null representation of $\mathfrak{g}$. From 1.7.8, $S = \bigoplus_{\delta \in \mathfrak{g}^\wedge} S_\delta$; let $p \mapsto p^\natural$ be the projection of S onto S_{δ_0} defined by this decomposition. If $s \in S_{\delta_0}$, the mapping $p \mapsto ps$ of S into S is a $\mathfrak{g}$-homomorphism, hence $S_\delta s \subset S_\delta$ for all $\delta \in \mathfrak{g}^\wedge$, and consequently $(sp)^\natural = sp^\natural$ for all $p \in S$.

Let $\overline{S}$ be the ideal of the elements of S without constant term. Let I be the ideal of S generated by $S^{\mathfrak{g}} \cap \overline{S}$, and let $(s_1, \ldots, s_p)$ be a generating system of the ideal I. We may assume that $s_1, \ldots, s_p$ belong to $S^{\mathfrak{g}} \cap \overline{S}$, and are homogeneous. Let S_1 be the subalgebra of $S^{\mathfrak{g}}$ generated by the

s_i and 1. Let us show that $S^{\mathfrak{g}} = S_1$. In order to do so, we shall prove that every homogeneous element s of $S^{\mathfrak{g}}$ is in S_1, reasoning by induction on the degree $n > 0$ of s. Since $s \in I$, we have $s = \sum_{i=1}^{p} s_i s_i'$, the s_i' being elements of S which we may assume to be homogeneous with deg s_i' = deg s − deg $s_i < n$. Then

$$s = s^{\natural} = \sum_{i=1}^{p} (s_i s_i')^{\natural} = \sum_{i=1}^{p} s_i s_i'^{\natural}.$$

The $s_i'^{\natural}$ are elements of $S^{\mathfrak{g}}$ which are homogeneous of degree $<n$; they are therefore in S_1 by the induction hypothesis.

1.7.11. Proposition. *Let* $\mathfrak{a}$ *be a commutative Lie algebra,* $\mathfrak{b}$ *a semi-simple Lie algebra,* $\mathfrak{g} = \mathfrak{a} \times \mathfrak{b}$, $a \in \mathfrak{a}$, $b \in \mathfrak{b}$, *and* $x = a + b \in \mathfrak{g}$.

(a) *The following conditions are equivalent:*

(i) b *is semi-simple in* $\mathfrak{b}$*;*

(ii) *there exists a finite-dimensional semi-simple injective representation* ϱ *of* $\mathfrak{g}$ *such that* $\varrho(x)$ *is semi-simple;*

(iii) *for every finite-dimensional semi-simple representation* ϱ' *of* $\mathfrak{g}$, $\varrho'(x)$ *is semi-simple.*

(b) *The following conditions are equivalent:*

(i) $a = 0$ *and* b *is nilpotent in* $\mathfrak{b}$*;*

(ii) *there exists a finite-dimensional semi-simple injective representation* ϱ *of* $\mathfrak{g}$ *such that* $\varrho(x)$ *is nilpotent;*

(iii) *for every finite-dimensional semi-simple representation* ϱ' *of* $\mathfrak{g}$, $\varrho'(x)$ *is nilpotent.*

Since $\mathfrak{g}$ has a finite-dimensional semi-simple injective representation, the implications (a) (iii) ⇒ (ii) and (b) (iii) ⇒ (ii) are obvious. Since $\varrho'(a)$ is semi-simple for every finite-dimensional semi-simple representation ϱ' of $\mathfrak{g}$ (1.6.4), we have (a) (ii) ⇒ (i) ⇒ (iii). The assertion (b) (i) ⇒ (iii) is obvious. Finally, let ϱ be a finite-dimensional semi-simple injective representation of $\mathfrak{g}$ such that $\varrho(x)$ is nilpotent, and let us prove that $a = 0$ (whence (b) (ii) ⇒ (i)). We may assume k to be algebraically closed. Let $\varrho = \varrho_1 \oplus \cdots \oplus \varrho_p$, where the ϱ_i are simple, and let s and n be the semi-simple and nilpotent components of b in $\mathfrak{b}$. Then $\varrho(n)$ is nilpotent, $\varrho(s + a)$ is semi-simple, $\varrho(n)$ and $\varrho(s + a)$ commute and their sum $\varrho(x)$ is nilpotent, hence $\varrho(s + a) = 0$. For $i = 1, \ldots, p$, $\varrho_i(a)$ is scalar and $\varrho_i(s)$ has trace zero, and hence $\varrho_i(a) = \varrho_i(s) = 0$. Consequently, $\varrho(a) = 0$ and $a = 0$.

1.7.12. In a reductive Lie algebra $\mathfrak{g}$, an element x is said to be *semi-simple* (or *nilpotent*) if it satisfies the conditions (a) (or (b)) of 1.7.11. This generalizes 1.6.8. As in 1.6.8, we define the *semi-simple and nilpotent components* of any element of $\mathfrak{g}$.

1.7.13. PROPOSITION ($\mathfrak{g}$ reductive). *Let $\mathfrak{h}$ be a Lie subalgebra of $\mathfrak{g}$ which is reductive in $\mathfrak{g}$, $x \in \mathfrak{h}$, and y (or z) the semi-simple (or nilpotent) component of x in $\mathfrak{h}$. Then y (or z) is the semi-simple (or nilpotent) component of x in $\mathfrak{g}$.*

Let ϱ be a finite-dimensional semi-simple representation of $\mathfrak{g}$. Then $\varrho|\mathfrak{h}$ is semi-simple (1.7.9 (ii)), and therefore $\varrho(y)$ is semi-simple and $\varrho(z)$ is nilpotent. Hence y is semi-simple in $\mathfrak{g}$ and z is nilpotent in $\mathfrak{g}$.

1.8. Representations of $\mathfrak{sl}(2,k)$

1.8.1. The elements

$$e = \begin{pmatrix} 0 & 1 \\ 0 & 0 \end{pmatrix}, \quad f = \begin{pmatrix} 0 & 0 \\ 1 & 0 \end{pmatrix}, \quad h = \begin{pmatrix} 1 & 0 \\ 0 & -1 \end{pmatrix}$$

of $\mathfrak{sl}(2,k)$ form a basis for $\mathfrak{sl}(2,k)$. We have

$$[h,e] = 2e, \qquad [h,f] = -2f, \qquad [e,f] = h.$$

Let r be an integer $\geqq 0$. Let ϱ_r be the linear mapping of $\mathfrak{sl}(2,k)$ into $\mathbf{M}_{r+1}(k)$ such that

$$\varrho_r(h) = \begin{pmatrix} r & 0 & 0 & \cdots & 0 \\ 0 & r-2 & 0 & \cdots & 0 \\ 0 & 0 & r-4 & \cdots & 0 \\ \cdot & \cdot & \cdots & \cdots & \cdot \\ 0 & 0 & 0 & \cdots & -r \end{pmatrix},$$

$$\varrho_r(f) = \begin{pmatrix} 0 & 0 & 0 & \cdots & \cdot & 0 \\ 1 & 0 & 0 & \cdots & \cdot & 0 \\ 0 & 1 & 0 & \cdots & \cdot & 0 \\ \cdot & \cdot & \cdot & \cdots & \cdot & \cdot \\ 0 & 0 & 0 & \cdots & 1 & 0 \end{pmatrix},$$

$$\varrho_r(e) = \begin{pmatrix} 0 & \mu_1 & 0 & \cdots & 0 \\ 0 & 0 & \mu_2 & \cdots & 0 \\ 0 & 0 & 0 & \cdots & 0 \\ \cdot & \cdot & \cdot & \cdots & \mu_r \\ 0 & 0 & 0 & \cdots & 0 \end{pmatrix},$$

where $\mu_i = i(r - i + 1)$. It is easy to prove that ϱ_r is a representation of $\mathfrak{sl}(2,k)$ in k^{r+1}.

1.8.2. PROPOSITION. *The representation ϱ_r is simple.*

Let W be a vector subspace of k^{r+1} which is non-null and stable under ϱ_r. Let $w \in W - \{0\}$. Let $(e_1, \ldots, e_{r+1})$ be the canonical basis of k^{r+1}. For a suitable choice of n, $\varrho_r(f)^n w$ is a non-zero multiple of e_{r+1}. Hence $e_{r+1} \in W$. By transforming e_{r+1} by powers of $\varrho_r(e)$, it can be seen that $e_i \in W$ for all i, and hence $W = V$.

1.8.3. LEMMA. *Let ϱ be a representation of $\mathfrak{sl}(2,k)$ in V, let $\mu \in k$, and let v_0 be an element of V such that $\varrho(h)v_0 = \mu v_0$.*

(i) $\varrho(h)\varrho(e)v_0 = (\mu + 2)\varrho(e)v_0$ *and* $\varrho(h)\varrho(f)v_0 = (\mu - 2)\varrho(f)v_0$.

(ii) *Let* $v_i = \varrho(f)^i v_0$ *for* $i = 0, 1, 2, \ldots$ *Let us assume that* $\varrho(e)v_0 = 0$. *Then* $\varrho(e)v_i = i(\mu - i + 1)v_{i-1}$ *for* $i > 0$ *(we set* $v_{-1} = 0$*).*

We have

$$\varrho(h)\varrho(e)v_0 = \varrho(e)\varrho(h)v_0 + 2\varrho(e)v_0 = (\mu + 2)\varrho(e)v_0,$$
$$\varrho(h)\varrho(f)v_0 = \varrho(f)\varrho(h)v_0 - 2\varrho(f)v_0 = (\mu - 2)\varrho(f)v_0.$$

The assertion (ii) is obvious for $i = 0$. If it is true for i, then

$$\begin{aligned}\varrho(e)v_{i+1} &= \varrho(e)\varrho(f)v_i = \varrho(f)\varrho(e)v_i + \varrho(h)v_i \\ &= i(\mu - i + 1)\varrho(f)v_{i-1} + (\mu - 2i)v_i = (i + 1)(\mu - i)v_i.\end{aligned}$$

1.8.4. THEOREM. *Let ϱ be an $(r + 1)$-dimensional simple representation of $\mathfrak{sl}(2,k)$. Then ϱ is equivalent to ϱ_r.*

Let V be the space of ϱ. In parts (a), (b) and (c) of the proof, we assume k to be algebraically closed.

(a) There exists a $v \in V - \{0\}$ which is an eigenvector of $\varrho(h)$. Let $\varrho(h)v = \lambda v$. From 1.8.3 (i), we have

$$\varrho(h)\varrho(e)^i v = (\lambda + 2i)\varrho(e)^i v.$$

As $\varrho(h)$ has only a finite number of eigenvalues, there exists i_0 such that $\varrho(e)^{i_0}v \neq 0$, $\varrho(e)^{i_0+1}v = 0$. Setting $\varrho(e)^{i_0}v = v_0$, $\lambda + 2i_0 = \mu$, we have

$$\varrho(h)v_0 = \mu v_0, \qquad \varrho(e)v_0 = 0.$$

(b) Setting $\varrho(f)^i v_0 = v_i$, we have $\varrho(h)v_i = (\mu - 2i)v_i$. Let v_s be the last non-zero v_i. Then

$$\varrho(h)v_0 = \mu v_0, \qquad \varrho(h)v_1 = (\mu - 2)v_1, \qquad \ldots, \qquad \varrho(h)v_s = (\mu - 2s)v_s$$

(hence the sequence $(v_0, \ldots, v_s)$ is free), and

$$\varrho(f)v_0 = v_1, \quad \varrho(f)v_1 = v_2, \quad \ldots, \quad \varrho(f)v_{s-1} = v_s, \quad \varrho(f)v_s = 0$$

Finally, $\varrho(e)v_i = i(\mu - i + 1)v_{i-1}$ for all $i \geqq 0$ (1.8.3). It can be seen that $kv_0 + \cdots + kv_s$ is stable under ϱ, and therefore equal to V. Consequently, $s = r$.

(c) We have $(r+1)(\mu - r)v_r = \varrho(e)v_{r+1} = \varrho(e) \cdot 0 = 0$, and hence $\mu = r$. We thus establish that ϱ is equivalent to ϱ_r.

(d) Let $\bar{k}$ be an algebraic closure of k. From the foregoing, the eigenvalues of $\varrho(h)$ in $\bar{k}$ are rational integers. Hence $\varrho(h)$ has a non-zero eigenvector in V, and we can repeat the preceding proof *verbatim*.

1.8.5. COROLLARY. *Every finite-dimensional representation of* $\mathfrak{sl}(2,k)$ *is equivalent to a representation of the form*

$$\varrho_{r_1} \oplus \varrho_{r_2} \oplus \cdots \oplus \varrho_{r_p}.$$

This follows from 1.6.3 and 1.8.4. For, from 1.7.3 (iv), $\mathfrak{sl}(2,k)$ is semisimple.

1.9. Cartan subalgebras

1.9.1. A nilpotent Lie subalgebra of $\mathfrak{g}$ which is equal to its normalizer is termed a *Cartan subalgebra* of $\mathfrak{g}$.

1.9.2. PROPOSITION. *Let* $\mathfrak{h}$ *be a Lie subalgebra of* $\mathfrak{g}$, *and* k' *an extension of* k. *Then* $\mathfrak{h}$ *is a Cartan subalgebra of* $\mathfrak{g}$ *if and only if* $\mathfrak{h} \otimes k'$ *is a Cartan subalgebra of* $\mathfrak{g} \otimes k'$.

This is obvious.

1.9.3. THEOREM. *Let* $\mathfrak{h}$ *be a nilpotent Lie subalgebra of* $\mathfrak{g}$. *We consider the representation* $x \mapsto \mathrm{ad}_{\mathfrak{g}} x$ *of* $\mathfrak{h}$ *in* $\mathfrak{g}$, *whence, we have for all* $\lambda \in \mathfrak{h}^*$, *a vector subspace* $\mathfrak{g}^\lambda$ *of* $\mathfrak{g}$ (1.2.13). *We assume that, for all* $x \in \mathfrak{h}$, $\mathrm{ad}\, x$ *is triangularizable. Then:*

(i) $\mathfrak{g} = \oplus_{\lambda \in \mathfrak{h}^*} \mathfrak{g}^\lambda$.
(ii) $[\mathfrak{g}^\lambda, \mathfrak{g}^\mu] \subset \mathfrak{g}^{\lambda+\mu}$; *in particular*, $[\mathfrak{g}^0, \mathfrak{g}^\mu] \subset \mathfrak{g}^\mu$.
(iii) *The set* $\mathfrak{g}^0$ *is a Lie subalgebra of* $\mathfrak{g}$ *containing* $\mathfrak{h}$.
(iv) $\mathfrak{h}$ *is a Cartan subalgebra of* $\mathfrak{g}$ *if and only if* $\mathfrak{g}^0 = \mathfrak{h}$.
(v) *If* $\mathfrak{g}^0$ *is nilpotent, then* $\mathfrak{g}^0$ *is a Cartan subalgebra of* $\mathfrak{g}$.

(i) This follows from 1.3.19.

(ii) Let $x \in \mathfrak{h}$, $y \in \mathfrak{g}^\lambda$ and $z \in \mathfrak{g}^\mu$. We set $\lambda(x) = \alpha$ and $\mu(x) = \beta$. Then

$$(\text{ad } x - (\alpha + \beta))[y,z] = [(\text{ad } x - \alpha)y,z] + [y,(\text{ad } x - \beta)z],$$

and hence by recurrence

$$(\text{ad } x - (\alpha + \beta))^n[y,z] = \sum_{p=0}^{n} \frac{n!}{p!(n-p)!} [(\text{ad } x - \alpha)^p y, (\text{ad } x - \beta)^{n-p} z].$$

For n sufficiently large, all terms on the right-hand side are zero. Hence $[y,z] \in \mathfrak{g}^{\lambda+\mu}$, whence (ii).

(iii) We have $[\mathfrak{g}^0,\mathfrak{g}^0] \subset \mathfrak{g}^0$ from (ii), and hence $\mathfrak{g}^0$ is a Lie subalgebra of $\mathfrak{g}$. Since $\mathfrak{h}$ is nilpotent, we have $\mathfrak{h} \subset \mathfrak{g}^0$.

(iv) Let $\mathfrak{n}$ be the normalizer of $\mathfrak{h}$ in $\mathfrak{g}$. If $x \in \mathfrak{h}$, then $(\text{ad}_\mathfrak{g} x)(\mathfrak{n}) \subset \mathfrak{h}$, and hence $\text{ad}_\mathfrak{g} x|\mathfrak{n}$ is nilpotent. Consequently, $\mathfrak{h} \subset \mathfrak{n} \subset \mathfrak{g}^0$. If $\mathfrak{g}^0 = \mathfrak{h}$, we have $\mathfrak{n} = \mathfrak{h}$ and $\mathfrak{h}$ is a Cartan subalgebra of $\mathfrak{g}$. Let us assume that $\mathfrak{g}^0 \neq \mathfrak{h}$. Let ϱ be the representation $x \mapsto \text{ad}_{\mathfrak{g}^0} x$ of $\mathfrak{h}$, and let σ be the quotient representation of ϱ in $\mathfrak{g}^0/\mathfrak{h}$. By applying 1.3.17 (i) to σ, we see that $\mathfrak{n} \neq \mathfrak{h}$; hence $\mathfrak{h}$ is not a Cartan subalgebra.

(v) Let us assume $\mathfrak{g}^0$ to be nilpotent. We now apply the foregoing with $\mathfrak{h}$ replaced by $\mathfrak{g}^0$, and we denote the new subspace $\mathfrak{g}^0$ by $\mathfrak{k}$. Then $\mathfrak{g}^0 \subset \mathfrak{k}$. But the inclusion $\mathfrak{h} \subset \mathfrak{g}^0$ implies that $\mathfrak{g}^0 \supset \mathfrak{k}$. Thus $\mathfrak{g}^0 = \mathfrak{k}$, and, from (iv), $\mathfrak{g}^0$ is a Cartan subalgebra.

1.9.4. COROLLARY. *Let $\mathfrak{h}$ be a Cartan subalgebra of $\mathfrak{g}$. Then $\mathfrak{h}$ is a maximal nilpotent Lie subalgebra of $\mathfrak{g}$.*

We may assume k to be algebraically closed. Let $\mathfrak{h}'$ be a nilpotent Lie subalgebra of $\mathfrak{g}$ containing $\mathfrak{h}$. With the notation of 1.9.3, we have $\mathfrak{h}' \subset \mathfrak{g}^0 = \mathfrak{h}$, and hence $\mathfrak{h}' = \mathfrak{h}$.

1.9.5. PROPOSITION. *Under the assumptions of* 1.9.3, *let K be the Killing form of $\mathfrak{g}$. Then:*

(i) $K(\mathfrak{g}^\lambda,\mathfrak{g}^\mu) = 0$ *if* $\mu \neq -\lambda$.

(ii) $K(\mathfrak{h}, \mathfrak{g}^\lambda) = 0$ *if* $\lambda \neq 0$.

(iii) *If* $x,y \in \mathfrak{h}$, *then*

$$K(x,y) = \sum_{\alpha \in \mathfrak{h}^*} (\dim \mathfrak{g}^\alpha)\alpha(x)\alpha(y).$$

Let $x \in \mathfrak{g}^\lambda$ and $y \in \mathfrak{g}^\mu$. Then $(\text{ad } x \text{ ad } y)(\mathfrak{g}^\nu) \subset \mathfrak{g}^{\nu+(\lambda+\mu)}$. If $\lambda + \mu \neq 0$, then $\text{ad } x \text{ ad } y$ is nilpotent, and therefore $K(x,y) = 0$. This proves (i), whence, in particular, (ii). Let $x,y \in \mathfrak{h}$. Then, with respect to a suitable basis,

$(\operatorname{ad} x \operatorname{ad} y)|\mathfrak{g}^{\alpha}$ has a lower triangular matrix, with $\alpha(x)\alpha(y)$ as its only eigenvalue (1.3.19 (iii)), whence (iii).

1.9.6. Let $x \in \mathfrak{g}$. We denote the nilspace of ad x, i.e., $\bigcup_{n \geqq 0} \operatorname{Ker} (\operatorname{ad} x)^n$, by $\mathfrak{g}^0(x)$. We denote the intersection of the images of the $(\operatorname{ad} x)^n$ by $\mathfrak{g}^*(x)$. We know that $\mathfrak{g} = \mathfrak{g}^0(x) \oplus \mathfrak{g}^*(x)$; this decomposition is termed the *Fitting decomposition of* $\mathfrak{g}$ defined by x. We have

$$[\mathfrak{g}^0(x), \mathfrak{g}^0(x)] \subset \mathfrak{g}^0(x), \qquad [\mathfrak{g}^0(x), \mathfrak{g}^*(x)] \subset \mathfrak{g}^*(x);$$

for, we may assume k to be algebraically closed, and our assertion then follows from 1.9.3 applied with $\mathfrak{h} = kx$.

1.9.7. PROPOSITION. *Under the assumptions of* 1.9.3, *we assume* $\mathfrak{g}$ *to be semi-simple. Then* $\mathfrak{g}^0$ *is reductive in* $\mathfrak{g}$.

From 1.9.5 (i), the restriction of the Killing form of $\mathfrak{g}$ to $\mathfrak{g}^0$ is non-degenerate. Let $x,x' \in \mathfrak{g}$, let s and s' be their semi-simple components, and let n and n' be their nilpotent components. If $x' \in \mathfrak{g}^0(x)$, then $[s,x'] = 0$, hence $[s,s'] = 0$, and so $s' \in \mathfrak{g}^0(x)$. This proves that, if $x' \in \mathfrak{g}^0$, then $s' \in \mathfrak{g}^0$, and hence $n' \in \mathfrak{g}^0$. It is now sufficient to apply 1.7.6.

1.9.8. For all $x \in \mathfrak{g}$, let us consider the characteristic polynomial of $\operatorname{ad}_{\mathfrak{g}} x$:

$$\det (T - \operatorname{ad} x) = T^n + a_{n-1}(x)\, T^{n-1} + a_{n-2}(x)\, T^{n-2} + \cdots,$$

where T is an indeterminate. The a_i are polynomial functions over $\mathfrak{g}$. If $x \in \mathfrak{g}$, dim $\mathfrak{g}^0(x)$ is the smallest integer p such that $a_p(x) \neq 0$. Let l be the smallest integer such that a_l is not identically zero. l is said to be the *rank* of $\mathfrak{g}$. An element x of $\mathfrak{g}$ is termed *generic* if $a_1(x) \neq 0$. If k' is an extension of k, the rank of $\mathfrak{g} \otimes k'$ is equal to l.

1.9.9. THEOREM. *Let x be a generic element of* $\mathfrak{g}$ *and* $\mathfrak{k}$ *the nilspace of* ad x. *Then* $\mathfrak{k}$ *is the only Cartan subalgebra if* $\mathfrak{g}$ *containing x.*

If $y \in \mathfrak{k}$, then $\mathfrak{g}^0(x)$ and $\mathfrak{g}^*(x)$ are stable under $\operatorname{ad}_{\mathfrak{g}} y$ (1.9.6). Let S (or R) be the set of the $y \in \mathfrak{k}$ such that $\operatorname{ad}_{\mathfrak{g}} y|\mathfrak{g}^*(x)$ is bijective (or such that $\operatorname{ad}_{\mathfrak{k}} y$ is not nilpotent). Then R and S are open in $\mathfrak{k}$, and $x \in S$. If $R \neq \emptyset$, there exists $y \in R \cap S$, and then $\dim \mathfrak{g}^0(y) < \dim \mathfrak{k}$, which contradicts the assumption that x is generic. Thus, for all $y \in \mathfrak{k}$, $\operatorname{ad}_{\mathfrak{k}} y$ is nilpotent and hence $\mathfrak{k}$ is nilpotent. From 1.9.3 (v) applied with $\mathfrak{h} = kx$, $\mathfrak{k}$ is a Cartan subalgebra. Finally, if $\mathfrak{k}_1$ is a Cartan subalgebra containing x, then $\mathfrak{k}_1 \subset \mathfrak{k}$ and hence, from 1.9.4, $\mathfrak{k}_1 = \mathfrak{k}$.

1.9.10. Theorem 1.9.9 proves the *existence* of Cartan subalgebras.

A Cartan subalgebra $\mathfrak{h}$ of $\mathfrak{g}$ is said to be *splitting* if, for all $x \in \mathfrak{h}$, $\mathrm{ad}_{\mathfrak{g}} x$ is triangularizable. If $\mathfrak{h}$ is a splitting Cartan subalgebra of $\mathfrak{g}$, the non-zero elements λ of $\mathfrak{h}$ such that $\mathfrak{g}^{\lambda} \neq 0$ are termed the *roots* of $\mathfrak{g}$ relative to $\mathfrak{h}$.

1.9.11. THEOREM (k algebraically closed). *Let $\mathfrak{h}$ and $\mathfrak{k}$ be Cartan subalgebras of $\mathfrak{g}$. There exists $\alpha \in \mathrm{Aut}_e(\mathfrak{g})$* (*cf.* 1.1.14) *such that $\alpha(\mathfrak{h}) = \mathfrak{k}$.*

Let $\lambda_1, \ldots, \lambda_n$ be the roots of $\mathfrak{g}$ relative to $\mathfrak{h}$, pairwise distinct. Then $\mathfrak{g} = \mathfrak{h} \oplus \mathfrak{g}^{\lambda_1} \oplus \cdots \oplus \mathfrak{g}^{\lambda_n}$. If $x \in \mathfrak{g}^{\lambda_i}$, then $\mathrm{ad}_{\mathfrak{g}} x$ is nilpotent since $(\mathrm{ad}_{\mathfrak{g}} x)(\mathfrak{g}^{\lambda_i}) \subset \mathfrak{g}^{\lambda + \lambda_i}$; hence $\exp \mathrm{ad}_{\mathfrak{g}} x$ is a well-defined element of $\mathrm{Aut}_e(\mathfrak{g})$. Let f be the mapping of the vector space $\mathfrak{h} \times \mathfrak{g}^{\lambda_1} \times \cdots \times \mathfrak{g}^{\lambda_n}$ into the vector space $\mathfrak{g}$ defined by

$$f(h, x_1, \ldots, x_n) = (\exp \mathrm{ad}\, x_1) \cdots (\exp \mathrm{ad}\, x_n) h.$$

If we set $\dim \mathfrak{g} = p$, then $(\mathrm{ad}_{\mathfrak{g}} x_i)^p = 0$ for all i, and consequently

$$f(h, x_1, \ldots, x_n) = \left(\sum_{f=0}^{p-1} \frac{1}{j!} (\mathrm{ad}\, x_1)^j \right) \cdots \left(\sum_{j=0}^{p-1} \frac{1}{j!} (\mathrm{ad}\, x_n)^j \right) h,$$

so that f is polynomial.

Let $\mathfrak{h}'$ be the set of the $x \in \mathfrak{h}$ such that $\lambda_1(x) \neq 0, \ldots, \lambda_n(x) \neq 0$. Let $h_0 \in \mathfrak{h}'$ and let T be the linear mapping which is tangent to f at $(h_0, 0, \ldots, 0)$. If $h \in \mathfrak{h}$ and $x \in \mathfrak{g}^{\lambda_1}$, we have

$$f(h_0 + h, 0, 0, \ldots, 0) = h_0 + h,$$

$$f(h_0, x, 0, \ldots, 0) = (\exp \mathrm{ad}\, x)\, h_0 = h_0 + [x, h_0] + \frac{1}{2!} [x, [x, h_0]] + \cdots$$

and hence

$$T(h, 0, 0, \ldots, 0) = h, \qquad T(0, x, 0, \ldots, 0) = [x, h_0].$$

Consequently,

$$T(\mathfrak{h} \times 0 \times \cdots \times 0) = \mathfrak{h},$$

and

$$T(0 \times \mathfrak{g}^{\lambda_1} \times 0 \times \cdots \times 0) = -[h_0, \mathfrak{g}^{\lambda_1}] = \mathfrak{g}^{\lambda_1}$$

since $h_0 \in \mathfrak{h}'$.

Similarly, $T(0 \times 0 \times \mathfrak{g}^{\lambda_2} \times \cdots \times 0) = \mathfrak{g}^{\lambda_2}$, etc. The above proves that T is surjective.

Since $\mathfrak{h}'$ is open and non-empty, we conclude that $f(\mathfrak{h}' \times \mathfrak{g}^{\lambda_1} \times \cdots \times \mathfrak{g}^{\lambda_n}$ contains an open non-empty subset A of $\mathfrak{g}$ (BO, pp. 39, 75). *A fortiori*, $\mathrm{Aut}_e(\mathfrak{g})\mathfrak{h}' \supset A$. We shall define $\mathfrak{k}'$ relative to $\mathfrak{k}$ as $\mathfrak{h}'$ was defined relative to $\mathfrak{h}$. Then $\mathrm{Aut}_e(\mathfrak{g})\mathfrak{k}$ contains an open non-empty subset B of $\mathfrak{g}$. Since

$A \cap B \neq \emptyset$, there exist $s,t \in \mathrm{Aut}_e(\mathfrak{g}), h_1 \in \mathfrak{h}'$ and $k_1 \in \mathfrak{k}'$ such that $sh_1 = tk_1$, or $(t^{-1}s)h = k_1$. Now $\mathfrak{h}$ and $\mathfrak{k}$ are nilspaces of ad h_1 and ad k_1 respectively, hence $(t^{-1}s)(\mathfrak{h}) = \mathfrak{k}$.

1.9.12. COROLLARY. *Let* $\mathfrak{h}$ *be a Cartan subalgebra of* $\mathfrak{g}$. *Then* $\dim \mathfrak{h}$ *is the rank of* $\mathfrak{g}$, *and the set of generic elements of* $\mathfrak{g}$ *belonging to* $\mathfrak{h}$ *is open and non-empty in* $\mathfrak{h}$.

We may assume k to be algebraically closed. The corollary then follows from 1.9.9 and 1.9.11.

1.9.13. PROPOSITION. *Let* $\mathfrak{g}'$ *be a Lie subalgebra of* $\mathfrak{g}$. *Every element of* $\mathfrak{g}'$ *which is generic in* $\mathfrak{g}$ *is also generic in* $\mathfrak{g}'$.

For all $x \in \mathfrak{g}'$, let $u(x) = \mathrm{ad}_{\mathfrak{g}} x$, let $u_1(x)$ and $u_2(x)$ be the endomorphisms of $\mathfrak{g}'$ and $\mathfrak{g}/\mathfrak{g}'$ deduced from $u(x)$ by restriction and by passage to the quotient respectively; and let $d(x)$, $d_1(x)$ and $d_2(x)$ be the dimensions of the nilspaces of $u(x)$, $u_1(x)$ and $u_2(x)$. Let

$$d_1 = \inf_{x \in \mathfrak{g}'} d_1(x), \qquad d_2 = \inf_{x \in \mathfrak{g}'} d_2(x).$$

The set A of the $x \in \mathfrak{g}'$ such that $d_1(x) = d_1$ and $d_2(x) = d_2$ is open and non-empty in $\mathfrak{g}'$. On the other hand, $d(x) = d_1(x) + d_2(x)$ for all $x \in \mathfrak{g}'$. Thus, if $x \in \mathfrak{g}'$ is generic in $\mathfrak{g}$, we have $x \in A$.

1.10. The system of roots of a split semi-simple Lie algebra

1.10.1. A pair $(\mathfrak{g},\mathfrak{h})$ where $\mathfrak{g}$ is a semi-simple Lie algebra and $\mathfrak{h}$ a splitting Cartan subalgebra of $\mathfrak{g}$ is termed a *split semi-simple Lie algebra*. The set of roots of $\mathfrak{g}$ relative to $\mathfrak{h}$ is denoted by $R(\mathfrak{g},\mathfrak{h})$.

1.10.2. THEOREM. *Let* $(\mathfrak{g},\mathfrak{h})$ *be a split semi-simple Lie algebra, let* $R = R(\mathfrak{g},\mathfrak{h})$, *and let* K *be the Killing form of* $\mathfrak{g}$. *Then:*

(i) $\mathfrak{g} = \mathfrak{h} \oplus (\oplus_{\alpha \in R} \mathfrak{g}^{\alpha})$, *and* $\dim \mathfrak{g}^{\alpha} = 1$ *for all* $\alpha \in R$.

(ii) *The Lie algebra* $\mathfrak{h}$ *is commutative. If* $h \in \mathfrak{h}$ *and* $x \in \mathfrak{g}^{\alpha}$, *then* $[h,x] = \alpha(h)x$. *If* $\alpha,\beta \in R$, *then* $[\mathfrak{g}^{\alpha},\mathfrak{g}^{\beta}] \subset \mathfrak{g}^{\alpha+\beta}$. *If* $\alpha \in R$, *then* $-\alpha \in R$, *and* $\mathfrak{h}_{\alpha} = (\mathfrak{g}^{\alpha},\mathfrak{g}^{-\alpha}]$ *is a one-dimensional vector subspace of* $\mathfrak{h}$; *it contains one and only one element* H_{α} *such that* $\alpha(H_{\alpha}) = 2$.

(iii) *If* $\alpha,\beta \in \mathfrak{h}^*$ *and* $\alpha + \beta \neq 0$, *then* $\mathfrak{g}^{\alpha}$ *and* $\mathfrak{g}^{\beta}$ *are orthogonal with respect to* K. *The restriction of* K *to* $\mathfrak{g}^{\alpha} \times \mathfrak{g}^{-\alpha}$ *(in particular, to* $\mathfrak{h} \times \mathfrak{h}$*) is non-degenerate. If* $x,y \in \mathfrak{h}$, *then* $K(x,y) = \sum_{\alpha \in R} \alpha(x)\alpha(y)$.

(iv) *The elements of R generate $\mathfrak{h}^*$.*

(v) *If $\alpha \in R$, the vector subspace $S_\alpha = \mathfrak{h}_\alpha + \mathfrak{g}^\alpha + \mathfrak{g}^{-\alpha}$ is a Lie subalgebra of $\mathfrak{g}$. If $X_\alpha \in \mathfrak{g}^\alpha - \{0\}$, there exists one and only one $X_{-\alpha} \in \mathfrak{g}^{-\alpha}$ such that $[X_\alpha, X_{-\alpha}] = H_\alpha$; let φ be the linear mapping of $\mathfrak{sl}(2,k)$ into $\mathfrak{g}$ such that $\varphi(e) = X_\alpha$, $\varphi(f) = X_{-\alpha}$ and $\varphi(h) = H_\alpha$ (with the notation of* 1.8.1); *then φ is an isomorphism of the Lie algebra $\mathfrak{sl}(2,k)$ onto the Lie algebra $\mathfrak{s}_\alpha$.*

(a) We already know that $\mathfrak{g} = \mathfrak{h} \oplus (\oplus_{\alpha \in R} \mathfrak{g}^\alpha)$ and that $[\mathfrak{g}^\alpha, \mathfrak{g}^\beta] \subset \mathfrak{g}^{\alpha+\beta}$.

(b) The assertions of (iii), except for the last one, follow from 1.9.5 and from the fact that K is non-degenerate. They imply that, if $\alpha \in R$, then $-\alpha \in R$.

(c) Let $x \in \mathfrak{h}$ such that $\alpha(x) = 0$ for all $\alpha \in R$. Then $K(x,y) = 0$ for all $y \in \mathfrak{h}$ (1.9.5 (iii)), and hence $x = 0$ from (b). This proves (iv).

(d) Let $h \in \mathfrak{h}$, and let d and n be the semi-simple and nilpotent components of $\mathrm{ad}_\mathfrak{g} h$, respectively. From 1.3.19 (iii), we have $dx = \alpha(h)x$ for $x \in \mathfrak{g}^\alpha$. We shall now show that $d([z,z']) = [dz,z'] + [z,dz']$ for all values of $z,z' \in \mathfrak{g}$. It is sufficient to prove this for $z \in \mathfrak{g}^\alpha$ and $z' \in \mathfrak{g}^{\alpha'}$; then $[z,z'] \in \mathfrak{g}^{\alpha+\alpha'}$, so that

$$\begin{aligned} d([z,z']) &= (\alpha + \alpha')(h)[z,z'] \\ &= [\alpha(h)z,z'] + [z,\alpha'(h)z'] = [dz,z'] + [z,dz']. \end{aligned}$$

From 1.5.9 (ii), there exists $u \in \mathfrak{g}$ such that $d = \mathrm{ad}_\mathfrak{g} u$. For all $h' \in \mathfrak{h}$, we have $0 = d(h') = [u,h']$, and hence u belongs to the normalizer of $\mathfrak{h}$, i.e. to $\mathfrak{h}$. Then $n = \mathrm{ad}_\mathfrak{g}(h - u)$, and $h - u \in \mathfrak{h}$; since n is nilpotent, we have $\alpha(h - u) = 0$ for all $\alpha \in R$, and hence, from (c), $h - u = 0$. Thus, for $x \in \mathfrak{g}^\alpha$, we have $[h,x] = [u,x] = dx = \alpha(h)x$. In particular, $\mathfrak{h}$ is commutative.

(e) Since $K|\mathfrak{h} \times \mathfrak{h}$ is non-degenerate, there exists, for all $\lambda \in \mathfrak{h}^*$ one and only one $h_\lambda \in \mathfrak{h}$ such that $\lambda(h) = K(h_\lambda,h)$ for all $h \in \mathfrak{h}$. If $x \in \mathfrak{g}^\alpha$ and $y \in \mathfrak{g}^{-\alpha}$, then, for all $h \in \mathfrak{h}$,

$$\begin{aligned} K(h,[x,y]) &= K([h,x],y) = K(\alpha(h)x,y) = \alpha(h)K(x,y) \\ &= K(h_\alpha,h)K(x,y) = K(h,K(x,y)h_\alpha), \end{aligned}$$

and hence

$$(1) \qquad [x,y] = K(x,y)h_\alpha.$$

Since the restriction of K to $\mathfrak{g}^\alpha \times \mathfrak{g}^{-\alpha}$ is non-degenerate, we deduce that, for all $\alpha \in R$,

$$(2) \qquad \mathfrak{h}_\alpha = [\mathfrak{g}^\alpha, \mathfrak{g}^{-\alpha}] = kh_\alpha.$$

(f) Let $\alpha \in R$. We choose $x \in \mathfrak{g}^\alpha$, $y \in \mathfrak{g}^{-\alpha}$ such that $K(x,y) = 1$, so that $[x,y] = h_\alpha$. If $\alpha(h_\alpha) = 0$, we have $[h_\alpha,x] = [h_\alpha,y] = 0$, and therefore $\mathfrak{g}' = kh_\alpha + kx + ky$ is a nilpotent subalgebra of $\mathfrak{g}$. We now apply 1.3.12 to the representation $z \mapsto \mathrm{ad}_\mathfrak{g} z$ of $\mathfrak{g}'$; if $z \in [\mathfrak{g}',\mathfrak{g}']$, it can be seen that the eigenvalues of $\mathrm{ad}_\mathfrak{g} z$ in an algebraically closed extension of k are all zero; since $h_\alpha \in [\mathfrak{g}',\mathfrak{g}']$, we have $\beta(h_\alpha) = 0$ for all $\beta \in R$, which is impossible. Hence $\alpha(h_\alpha) \neq 0$. Consequently, there exists one and only one $H_\alpha \in \mathfrak{h}_\alpha$ such that $\alpha(H_\alpha) = 2$. The proof of (ii) is complete.

(g) Let $\alpha \in R$, and let X_α be a non-zero element of $\mathfrak{g}^\alpha$. There exists an $X_{-\alpha} \in \mathfrak{g}^{-\alpha}$ such that $K(X_\alpha,X_{-\alpha}) \neq 0$. Then $[X_\alpha,X_{-\alpha}]$ is a non-zero element of $\mathfrak{h}_\alpha$, and by a suitable choice of $X_{-\alpha}$ we have $[X_\alpha,X_{-\alpha}] = H_\alpha$. Since

$$[H_\alpha,X_\alpha] = \alpha(H_\alpha)X_\alpha = 2X_\alpha,$$

and
$$[H_\alpha,X_{-\alpha}] = -\alpha(H_\alpha)X_{-\alpha} = -2X_{-\alpha},$$

the linear mapping φ of $\mathfrak{sl}(2,k)$ into $\mathfrak{g}$ such that $\varphi(e) = X_\alpha$, $\varphi(f) = X_{-\alpha}$, $\varphi(h) = H_\alpha$ is an isomorphism of $\mathfrak{sl}(2,k)$ onto $kX_\alpha + kX_{-\alpha} + \mathfrak{h}_\alpha$.

(h) Let $\alpha \in R$, and let us assume that $\dim \mathfrak{g}^\alpha > 1$. Let y be a non-zero element of $\mathfrak{g}^{-\alpha}$. There exists a non-zero element X_α of $\mathfrak{g}^\alpha$ such that $K(X_\alpha,y) = 0$. Let $X_{-\alpha} \in \mathfrak{g}^{-\alpha}$ such that $[X_\alpha,X_{-\alpha}] = H_\alpha$; let us define φ as in (g), and let ϱ be the representation $z \mapsto \mathrm{ad}_\mathfrak{g}\varphi(z)$ of $\mathfrak{sl}(2,k)$. From (1), we have

$$\varrho(e)y = [X_\alpha,y] = K(X_\alpha,y)h_\alpha = 0.$$

From 1.8.5, y is a linear combination of eigenvectors of $\varrho(h) = \mathrm{ad}_\mathfrak{g} H_\alpha$ for integral eigenvalues $\geqq 0$. Now

$$[H_\alpha,y] = -\alpha(H_\alpha)y = -2y,$$

which is contradictory. Hence, $\dim \mathfrak{g}^\alpha = 1$. Then 1.9.5 (iii) establishes the final assertion of (iii) and the proofs of (i) and (iii) are complete. Moreover, the assertions of (v) now result from what was stated in (g).

1.10.3. For $\lambda \in \mathfrak{h}^*$, we shall retain throughout the notation h_λ from the preceding proof. The mapping $\lambda \mapsto h_\lambda$ is an isomorphism of the vector space $\mathfrak{h}^*$ onto the vector space $\mathfrak{h}$.

We shall frequently denote the Killing form of $\mathfrak{g}$ by $\langle \cdot,\cdot \rangle$, and, for $\lambda, \mu \in \mathfrak{h}^*$, we shall set

$$\langle \lambda,\mu \rangle = \langle h_\lambda,h_\mu \rangle = \lambda(h_\mu) = \mu(h_\lambda),$$

so that $\langle \cdot,\cdot \rangle$ is a non-degenerate symmetric bilinear form on $\mathfrak{h}^*$.

1.10.4. The notation H_α from 1.10.2 (ii) will also be retained throughout. We have $\langle h_\alpha, h_\alpha\rangle = \alpha(h_\alpha) \neq 0$, and

$$H_\alpha = \frac{2h_\alpha}{\langle h_\alpha, h_\alpha\rangle},$$

whence

$$h_\alpha = \frac{2H_\alpha}{\langle H_\alpha, H_\alpha\rangle}.$$

1.10.5. For all $\alpha \in R$ and for any choice of a non-zero X_α in $\mathfrak{g}^\alpha$, we identify $\mathfrak{g}^\alpha + \mathfrak{g}^{-\alpha} + kH_\alpha$ with $\mathfrak{sl}(2,k)$ under the isomorphism of 1.10.2 (v).

1.10.6. Proposition ($\mathfrak{g}$ semi-simple). (i) *Let x be a generic element of $\mathfrak{g}$. Then x is semi-simple, and the only Cartan subalgebra containing x is the centralizer $\mathfrak{g}^x$ of x in $\mathfrak{g}$.*

(ii) *Let $\mathfrak{h}$ be a Cartan subalgebra of $\mathfrak{g}$. Then $\mathfrak{h}$ is a maximal commutative Lie subalgebra of $\mathfrak{g}$. All its elements are semi-simple in $\mathfrak{g}$. The Lie algebra $\mathfrak{h}$ is reductive in $\mathfrak{g}$.*

(iii) *Let $\mathscr{E}$ be the set of commutative subalgebras of $\mathfrak{g}$ all of whose elements are semi-simple. Then the Cartan subalgebras are the maximal elements of $\mathscr{E}$.*

(iv) *Let x be a semi-simple element of $\mathfrak{g}$. Then x belongs to a Cartan subalgebra of $\mathfrak{g}$, and x is generic if and only if* $\dim \mathfrak{g}^x$ *is equal to the rank of $\mathfrak{g}$.*

To prove (i) and (ii), we may assume k to be algebraically closed. Then (i) follows from 1.9.9 and 1.10.2, and (ii) follows from 1.9.4 and 1.10.2.

Every Cartan subalgebra of $\mathfrak{g}$ is a maximal element of $\mathscr{E}$, from (ii). Let $\mathfrak{h} \in \mathscr{E}$. Let $\mathfrak{c}$ and $\mathfrak{n}$ be its centralizer and its normalizer in $\mathfrak{g}$, so that $\mathfrak{h} \subset \mathfrak{c} \subset \mathfrak{n}$. Then $[\mathfrak{h}, \mathfrak{n}] \subset \mathfrak{h}$, whence $[\mathfrak{h},\mathfrak{n}] = 0$ since $\mathfrak{h}$ is reductive in $\mathfrak{g}$; hence $\mathfrak{n} = \mathfrak{c}$. From 1.7.7, $\mathfrak{c}$ is reductive in $\mathfrak{g}$, and hence $\mathfrak{c} = \mathfrak{h} \times \mathfrak{c}'$, where $\mathfrak{c}'$ is a reductive Lie algebra in $\mathfrak{g}$.

If $\mathfrak{c}' \neq 0$, there exists in $\mathfrak{c}'$ a commutative subalgebra $\mathfrak{h}'$ which is reductive in $\mathfrak{g}$ and non-null (cf. for example 1.7.13), and we have $\mathfrak{h} \times \mathfrak{h}' \in \mathscr{E}$. Hence if $\mathfrak{h}$ is maximal in $\mathscr{E}$, we have $\mathfrak{c}' = 0$, hence $\mathfrak{h} = \mathfrak{n}$, and $\mathfrak{h}$ is a Cartan subalgebra of $\mathfrak{g}$.

Let x be a semi-simple element of $\mathfrak{g}$. From (iii), x belongs to a Cartan subalgebra of $\mathfrak{g}$. The nilspace of ad x is $\mathfrak{g}^x$. Hence x is generic, if and only if $\dim \mathfrak{g}^x$ is equal to the rank of $\mathfrak{g}$.

1.10.7. Proposition. *Let $(\mathfrak{g},\mathfrak{h})$ be a split semi-simple Lie algebra, let $R = R(\mathfrak{g},\mathfrak{h})$, and let $\alpha,\beta \in R$. Then:*

(i) *The scalar $\beta(H_\alpha)$ is a rational integer $a_{\beta\alpha}$.*

(ii) *The set of the* $t \in \mathbf{Z}$ *such that* $\beta + t\alpha \in R \cup \{0\}$ *is an interval* $[-t', t'']$, *where* $t', t'' \geqq 0$. *We have* $a_{\beta\alpha} = t' - t''$.

(iii) $\beta - a_{\beta\alpha}\alpha \in R$.

(iv) *If* $\beta - \alpha \notin R \cup \{0\}$, *then* $a_{\beta\alpha} \leqq 0$, $t' = 0$, $t'' = -a_{\beta\alpha}$.

(v) *If* $\beta + \alpha \in R$, *then* $[\mathfrak{g}^{\alpha}, \mathfrak{g}^{\beta}] = \mathfrak{g}^{\alpha+\beta}$.

(vi) *The only roots proportional to* α *are* α *and* $-\alpha$.

Let ϱ be the representation $x \mapsto \mathrm{ad}_{\mathfrak{g}} x$ of $\mathfrak{g}^{\alpha} + \mathfrak{g}^{-\alpha} + kH_{\alpha}$. Let $\mathfrak{a} = \sum_{t \in \mathbf{Z}} \mathfrak{g}^{\beta + t\alpha}$, which is stable under ϱ; let σ be the subrepresentation of ϱ defined by $\mathfrak{a}$. The eigenvalues of $\sigma(H_{\alpha})$ are the $(\beta + t\alpha)(H_{\alpha}) = \beta(H_{\alpha}) + 2t$ for those $t \in \mathbf{Z}$ such that $\beta + t\alpha \in R \cup \{0\}$. Taking 1.8.5 into account, these eigenvalues are integers (whence (i)), and, since they are congruent modulo 2, they range from a minimum $-r$ to a maximum r by increments of 2. This proves that the set of those $t \in \mathbf{Z}$ such that $\beta + t\alpha \in R \cup \{0\}$ is an interval $[-t', t'']$; since 0 belongs to this interval, we have $t'' \geqq 0$, $t' \geqq 0$. Since $r = a_{\beta\alpha} + 2t''$ and $-r = a_{\beta\alpha} - 2t'$, we have $0 = 2a_{\beta\alpha} + 2t'' - 2t'$, whence (ii).

Let us prove (vi). Assume that there exists $\xi \in k$ with $\beta = \xi\alpha$. Then

$$2\xi = \xi\alpha(H_{\alpha}) = \beta(H_{\alpha}) \in \mathbf{Z}.$$

Exchanging α and β, we see that $2\xi^{-1} \in \mathbf{Z}$. Hence $\xi \in \{\pm\frac{1}{2}, \pm 1, \pm 2\}$. We must exclude $\pm\frac{1}{2}$ and ± 2. Since the negative of a root is a root, it is sufficient to exclude $\frac{1}{2}$ and 2, and even (by exchanging α and β) to exclude 2 alone. Let us assume that $\beta = 2\alpha$. Then

$$\mathfrak{a} = \mathfrak{h} \oplus \mathfrak{g}^{\alpha} \oplus \mathfrak{g}^{-\alpha} \oplus \mathfrak{g}^{2\alpha} \oplus \mathfrak{g}^{-2\alpha}.$$

The non-zero eigenvalues of $\sigma(H_{\alpha})$ are ± 2, ± 4, with multiplicity 1. But $\sigma(\mathfrak{g}^{\alpha}) \cdot \mathfrak{g}^{\alpha} = 0$, while $\mathfrak{g}^{\alpha}$ corresponds to the eigenvalue 2 of $\sigma(H_{\alpha})$; this contradicts 1.8.5, and we have proved (vi).

Since $-t' \leqq t'' = t' \leqq t'$, we have $-a_{\beta\alpha} \in [-t', t'']$, hence

$$\beta - a_{\beta\alpha}\alpha \in R \cup \{0\}.$$

If $\beta - a_{\beta\alpha}\alpha = 0$, we have $\beta = \pm\alpha$, and hence $a_{\beta\alpha} = \pm 2$, which is a contradiction. This proves (iii).

If $\beta - \alpha \notin R \cup \{0\}$, then $t' = 0$, and hence $a_{\beta\alpha} = -t'' \leqq 0$, whence (iv).

If $\beta + \alpha \in R$, then $\beta \neq \alpha$, and hence $\beta + t\alpha \neq 0$ for all $t \in \mathbf{Z}$; all eigenvalues of $\sigma(H_{\alpha})$ have multiplicity 1 and are congruent modulo 2, so that, from 1.8.5, σ is irreducible. Since $t'' \geqq 1$, we have $\sigma(\mathfrak{g}^{\alpha})\mathfrak{g}^{\beta} \neq 0$ from 1.8.4. Since $[\mathfrak{g}^{\alpha}, \mathfrak{g}^{\beta}] \subset \mathfrak{g}^{\alpha+\beta}$ and $\dim \mathfrak{g}^{\alpha+\beta} = 1$, we have proved (v).

1.10.8. The integers $a_{\beta\alpha} = \beta(H_\alpha)$ of 1.10.7 (i) are termed the *Cartan integers* of $(\mathfrak{g},\mathfrak{h})$. For all $\alpha \in R$, we have $a_{\alpha\alpha} = 2$. On the other hand,

$$a_{\beta\alpha} = \langle h_\beta, H_\alpha\rangle = 2\frac{\langle h_\beta,h_\alpha\rangle}{\langle h_\alpha,h_\alpha\rangle} = 2\frac{\langle \beta,\alpha\rangle}{\langle \alpha,\alpha\rangle} = 2\frac{\langle H_\beta,H_\alpha\rangle}{\langle H_\beta,H_\beta\rangle}.$$

1.10.9. For all $\alpha \in R$, we denote by s_α the endomorphism of the vector space $\mathfrak{h}^*$ defined by

$$s_\alpha(\lambda) = \lambda - \lambda(H_\alpha)\alpha = \lambda - 2\frac{\langle \lambda,\alpha\rangle}{\langle \alpha,\alpha\rangle}\alpha .$$

The vector space $\mathfrak{h}^*$ is the direct sum of $k\alpha$ and the orthogonal subspace of H_α; the restriction of s_α to this orthogonal subspace is the identity mapping, and $s_\alpha(\alpha) = -\alpha$; s_α is said to be the *reflexion relative to* α. We have $s_\alpha^2 = 1$, and s_α preserves the form $\langle\cdot,\cdot\rangle$ on $\mathfrak{h}^*$. If $\beta \in R$, then, from 1.10.7 (iii), $s_\alpha(\beta) = \beta - a_{\beta\alpha}\alpha \in R$, hence $s_\alpha(R) = R$.

1.10.10. The group of automorphisms of $\mathfrak{h}^*$ generated by the s_α $(\alpha \in R)$ is termed the *Weyl group* of $(\mathfrak{g},\mathfrak{h})$ and denoted by $W(\mathfrak{g},\mathfrak{h})$. We have $w(R) = R$ for all $w \in W(\mathfrak{g},\mathfrak{h})$, and hence $W(\mathfrak{g},\mathfrak{h})$ is finite.

1.10.11. It follows from 1.10.2, 1.10.7 and 1.10.9 that, in the sense of 11.1.1, R is a reduced root system in $\mathfrak{h}^*$, and that the set of the H_α is the inverse root system in $\mathfrak{h}$. We shall therefore avail ourselves of the properties of root systems; they are recalled in section 11.1.

In particular, if $\mathfrak{h}_{\mathbf{Q}} = \sum_{\alpha\in R}\mathbf{Q}H_\alpha$ and $\mathfrak{h}^*_{\mathbf{Q}} = \sum_{\alpha\in R}\mathbf{Q}\alpha$, then $\mathfrak{h}$ can be identified with $\mathfrak{h}_{\mathbf{Q}} \otimes_{\mathbf{Q}} k$, $\mathfrak{h}^*$ with $\mathfrak{h}^*_{\mathbf{Q}} \otimes_{\mathbf{Q}} k$, and $\mathfrak{h}^*_{\mathbf{Q}}$ with the dual of $\mathfrak{h}_{\mathbf{Q}}$. We set $\mathfrak{h}_{\mathbf{R}} = \mathfrak{h}_{\mathbf{Q}} \otimes_{\mathbf{Q}} \mathbf{R}$ and $\mathfrak{h}^*_{\mathbf{R}} = \mathfrak{h}^*_{\mathbf{Q}} \otimes_{\mathbf{Q}} \mathbf{R}$. The Weyl chambers are defined in $\mathfrak{h}_{\mathbf{R}}$ and $\mathfrak{h}^*_{\mathbf{R}}$ (11.1.5).

1.10.12. The Weyl group operates in $\mathfrak{h}^*$ and $\mathfrak{h}^*_{\mathbf{Q}}$, and hence, by transport of the structure, in $\mathfrak{h}$ and $\mathfrak{h}_{\mathbf{Q}}$, hence in $\mathfrak{h}_{\mathbf{R}}$ and $\mathfrak{h}^*_{\mathbf{R}}$.

1.10.13. From 1.10.2 (iii), 1.10.4 and 1.10.7 (i), $\langle\cdot,\cdot\rangle$ takes rational values on $\mathfrak{h}_{\mathbf{Q}}$ and $\mathfrak{h}^*_{\mathbf{Q}}$. Since $\langle h,h\rangle = \sum_{\alpha\in R}\alpha(h)^2$ for all $h \in \mathfrak{h}$, the forms $\langle\cdot,\cdot\rangle$ on $\mathfrak{h}_{\mathbf{Q}}$ and $\mathfrak{h}^*_{\mathbf{Q}}$ are positive and non-degenerate. The bilinear forms on $\mathfrak{h}_{\mathbf{R}}$ and $\mathfrak{h}^*_{\mathbf{R}}$, again denoted by $\langle\cdot,\cdot\rangle$, which we can canonically deduce from them are also positive and non-degenerate. Furthermore, all these forms are invariant under $W(\mathfrak{g},\mathfrak{h})$.

1.10.14. Let us choose a basis B of R (11.1.6). Let R_+ and R_- be the corresponding sets of positive and negative roots respectively (11.1.7). Let $\mathfrak{n}_+ = \bigoplus_{\alpha\in R_+}\mathfrak{g}^\alpha$ and $\mathfrak{n}_- = \bigoplus_{\alpha\in R_-}\mathfrak{g}^\alpha$. Obviously, $\mathfrak{n}_+$ and $\mathfrak{n}_-$ are Lie subalgebras of $\mathfrak{g}$ such that $\mathfrak{g} = \mathfrak{h} \oplus \mathfrak{n}_+ \oplus \mathfrak{n}_-$; this decomposition is termed the *triangular decomposition* of $\mathfrak{g}$ defined by B. Since $[\mathfrak{g}^\alpha,\mathfrak{g}^\beta] \subset \mathfrak{g}^{\alpha+\beta}$, $\mathrm{ad}_\mathfrak{g} x$ is nilpotent for all $x \in \mathfrak{n}_+$ and for all $x \in \mathfrak{n}_-$. In particular, $\mathfrak{n}_+$ and $\mathfrak{n}_-$ are nilpotent Lie algebras.

We set $\mathfrak{b}_+ = \mathfrak{h} \oplus \mathfrak{n}_+$ and $\mathfrak{b}_- = \mathfrak{h} \oplus \mathfrak{n}_-$. Since $[\mathfrak{h},\mathfrak{n}_+] = \mathfrak{n}_+, \mathfrak{b}_+$ is a solvable Lie subalgebra of $\mathfrak{g}$ such that $[\mathfrak{b}_+,\mathfrak{b}_+] = \mathfrak{n}_+$. Similarly, $\mathfrak{b}_-$ is a solvable Lie subalgebra of $\mathfrak{g}$ such that $[\mathfrak{b}_-,\mathfrak{b}_-] = \mathfrak{n}_-$; $\mathfrak{b}_+$ is said to be the *Borel subalgebra* defined by $\mathfrak{h}$ and B.

1.10.15. PROPOSITION. *We retain the notation of* 1.10.14.

(i) *If α and β are distinct elements of B, then* $[\mathfrak{g}^\alpha,\mathfrak{g}^{-\beta}] = 0$.

(ii) *The Lie algebras $\mathfrak{n}_+$ and $\mathfrak{n}_-$ are generated by the $\mathfrak{g}^\alpha$ for $\alpha \in B$ and $-\alpha \in B$ respectively.*

(iii) *The Lie algebras $\mathfrak{b}_+$ and $\mathfrak{b}_-$ are maximal solvable Lie algebras of $\mathfrak{g}$; each is equal to its normalizer in $\mathfrak{g}$.*

If α and β are distinct elements of B, then $\alpha - \beta \notin R$ from 11.1.7, and hence $[\mathfrak{g}^\alpha,\mathfrak{g}^{-\beta}] = 0$.

Let $\alpha \in R_+$. From 11.1.10, there exist $\alpha_1, \ldots, \alpha_p \in B$ such that:

(1) $\alpha = \alpha_1 + \cdots + \alpha_p$;

(2) $\alpha_1 + \cdots + \alpha_i \in R_+$ for $i = 1, \ldots, p$.

From 1.10.7 (v), $\mathfrak{g}^\alpha = [\mathfrak{g}^{\alpha_1},[\mathfrak{g}^{\alpha_2},[\ldots, \mathfrak{g}^{\alpha_p}]\ldots]]$. Hence $\mathfrak{n}_+$ is generated by the $\mathfrak{g}^\alpha$ for $\alpha \in B$. A similar reasoning applies for $\mathfrak{n}_-$.

Let $\mathfrak{k}$ be a Lie subalgebra of $\mathfrak{g}$ containing $\mathfrak{b}_+$. Then $\mathfrak{k}$ is stable under $\mathrm{ad}_\mathfrak{g}\mathfrak{h}$, and hence $\mathfrak{k} = \sum_{\alpha\in\mathfrak{h}^*} (\mathfrak{k} \cap \mathfrak{g}^\alpha)$. If $\mathfrak{k} \neq \mathfrak{b}_+$, there thus exists $\alpha \in R_+$ such that $\mathfrak{g}^{-\alpha} \subset \mathfrak{k}$; then $\mathfrak{k} \supset \mathfrak{g}^\alpha + \mathfrak{g}^{-\alpha} + kH_\alpha$, and hence $\mathfrak{k}$ is not solvable and $[\mathfrak{k},\mathfrak{b}_+] \supset \mathfrak{g}^{-\alpha}$. Thus $\mathfrak{b}_+$ is a maximal solvable Lie subalgebra of $\mathfrak{g}$ which is equal to its normalizer in $\mathfrak{g}$. A similar reasoning applies for $\mathfrak{b}_-$.

1.10.16. PROPOSITION (k algebraically closed, $\mathfrak{g}$ semi-simple). *Let $\mathfrak{k}$ be a solvable Lie subalgebra of $\mathfrak{g}$. Then $\mathfrak{k}$ is contained in a Borel subalgebra of $\mathfrak{g}$.*

Let $\mathscr{A}$ be the adjoint group of $\mathfrak{g}$ (1.5.10). Let D be the set of flags in $\mathfrak{g}$, that is, the set of increasing sequences $(\mathfrak{d}_0,\mathfrak{d}_1,\ldots,\mathfrak{d}_n)$ of vector subspaces of $\mathfrak{g}$ of dimension $0, 1, 2, \ldots, \dim \mathfrak{g}$. It is a complete algebraic manifold (BO, p. 34) in which $\mathscr{A}$ operates (BO, p. 241). There exists in D a closed $\mathscr{A}$-orbit E (BO, p. 98); this is a complete manifold.

Let $\mathfrak{b}$ be a Borel subalgebra of $\mathfrak{g}$. Let $\mathscr{K}$ and $\mathscr{B}$ be the smallest algebraic subgroups of $\mathscr{A}$ whose Lie algebras contain $\mathrm{ad}\,\mathfrak{k}$ and $\mathrm{ad}\,\mathfrak{b}$ respectively. The Lie algebra of $\mathscr{K}$ possesses a derived ideal which is contained in $\mathrm{ad}\,\mathfrak{k}$

(BO, p. 195), and hence is solvable; hence $\mathscr{K}$ is solvable (CH′, p. 121). Similarly, $\mathscr{B}$ is solvable.

In E there exist fixed flags $(\mathfrak{d}_0,\mathfrak{d}_1, \ldots, \mathfrak{d}_n)$ and $(\mathfrak{d}'_0,\mathfrak{d}'_1, \ldots, \mathfrak{d}'_n)$ for $\mathscr{K}$ and $\mathscr{B}$ respectively (BO, p. 242). By substituting $a(\mathfrak{b})$ for $\mathfrak{b}$ for a suitably chosen in $\mathscr{A}$, we may assume that $\mathfrak{d}'_i = \mathfrak{d}_i$ for all i. Let $\mathscr{L}$ be the set of those $a \in \mathscr{A}$ under which $\mathfrak{d}_0,\mathfrak{d}_1, \ldots, \mathfrak{d}_n$ are stable. Then $\mathscr{L}$ is a solvable algebraic subgroup of $\mathscr{A}$ whose Lie algebra contains ad $\mathfrak{b}$, and so is equal to ad $\mathfrak{b}$ (1.10.15 (iii)). Moreover, $\mathscr{L} \supset \mathscr{K}$, hence $\mathfrak{b} \supset \mathfrak{k}$.

1.10.17. LEMMA ($\mathfrak{g}$ semi-simple). *Let* $\mathfrak{g} = \mathfrak{h} \oplus \mathfrak{n}_+ \oplus \mathfrak{n}_-$ *be a triangular decomposition of* $\mathfrak{g}$, *and let* $h \in \mathfrak{h}$ *and* $n \in \mathfrak{n}_+$. *Then* $\mathrm{ad}_\mathfrak{g} h$ *and* $\mathrm{ad}_\mathfrak{g}(h+n)$ *have the same characteristic polynomial.*

Let R_+ be the set of positive roots of $(\mathfrak{g},\mathfrak{h})$ corresponding to the given triangular decomposition. There exists a total ordering on $\mathfrak{h}^*_\mathbf{Q}$ which is compatible with its vector space structure and such that the elements of R_+ are >0. If $v \in \mathfrak{g}^\mu$, then $(\mathrm{ad}_\mathfrak{g}\mathfrak{h})v \in \mathfrak{g}^\mu$ and $(\mathrm{ad}_\mathfrak{g} n)v \in \mathfrak{g}^{\mu_1} + \cdots + \mathfrak{g}^{\mu_p}$ with $\mu_1, \ldots, \mu_p > \mu$. Hence, with respect to a suitable basis for $\mathfrak{g}$, the matrix of $\mathrm{ad}_\mathfrak{g} h$ is diagonal and that of $\mathrm{ad}_\mathfrak{g} n$ is strictly lower triangular, which proves the lemma.

1.10.18. PROPOSITION (k algebraically closed, $\mathfrak{g}$ semi-simple). *Let* $\mathfrak{b}$ *and* $\mathfrak{b}'$ *be Borel subalgebras of* $\mathfrak{g}$. *There exists a Cartan subalgebra contained in* $\mathfrak{b} \cap \mathfrak{b}'$.

Let $\mathfrak{h}$ be a Cartan subalgebra of $\mathfrak{g}$ contained in $\mathfrak{b}$,$n = [\mathfrak{b},\mathfrak{b}]$, $\mathfrak{n}' = [\mathfrak{b}',\mathfrak{b}']$, $\mathfrak{p} = \mathfrak{b} \cap \mathfrak{b}'$, and $\mathfrak{s}$ a vector subspace of $\mathfrak{g}$ which is complementary to $\mathfrak{b} + \mathfrak{b}'$. We denote the orthogonal subspaces of $\mathfrak{s}$, $\mathfrak{b}$ and $\mathfrak{b}'$ with respect to the Killing form of $\mathfrak{g}$ by $\mathfrak{s}^\perp$, $\mathfrak{b}^\perp$ and $\mathfrak{b}'^\perp$ respectively. Setting $l = \dim \mathfrak{h}$, $n = \dim \mathfrak{n}$ and $p = \dim \mathfrak{p}$, we have

$$\dim \mathfrak{b} = \dim \mathfrak{b}' = l + n, \qquad \dim \mathfrak{s}^\perp = \dim(\mathfrak{b} + \mathfrak{b}') = 2(l+n) - p,$$

and hence

$$(1) \qquad \dim(\mathfrak{s}^\perp \cap \mathfrak{p}) \geqq \dim \mathfrak{s}^\perp + \dim \mathfrak{p} - \dim \mathfrak{g}$$

$$= 2(l+n) - p + p - (l+2n) = l.$$

We have $\mathfrak{n} \subset \mathfrak{b}^\perp$, $\mathfrak{n}' \subset \mathfrak{b}'^\perp$. The elements of $\mathfrak{p} \cap \mathfrak{n}$ are nilpotent in $\mathfrak{g}$, and belong to $\mathfrak{b}'$, and hence to $\mathfrak{n}'$. Consequently, $\mathfrak{p} \cap \mathfrak{n} \subset \mathfrak{n} \cap \mathfrak{n}' \subset \mathfrak{b}^\perp \cap \mathfrak{b}'^\perp$, whence $\mathfrak{s}^\perp \cap \mathfrak{p} \cap \mathfrak{n} = 0$. Taking (1) into account, we see that $\mathfrak{s}^\perp \cap \mathfrak{p}$ is complementary to $\mathfrak{n}$ in $\mathfrak{b}$. Let z be an element of $\mathfrak{h}$ which is generic in $\mathfrak{g}$.

There exists $y \in \mathfrak{n}$ such that $x = y + z \in \mathfrak{s}^{\perp} \in \mathfrak{p}$. From 1.10.17, $\mathrm{ad}_{\mathfrak{g}} x$ and $\mathrm{ad}_{\mathfrak{g}} z$ have the same characteristic polynomial. Hence x is a generic element of $\mathfrak{g}$ which belongs to $\mathfrak{b} \cap \mathfrak{b}'$ and consequently is generic in $\mathfrak{b}$ and $\mathfrak{b}'$ (1.9.13). Since $\mathfrak{g}$, $\mathfrak{b}$ and $\mathfrak{b}'$ have the same rank, the nilspaces of $\mathrm{ad}_{\mathfrak{g}} x$, $\mathrm{ad}_{\mathfrak{b}} x$ and $\mathrm{ad}_{\mathfrak{b}'} x$ are equal to a Cartan subalgebra of $\mathfrak{g}$.

1.10.19. PROPOSITION. *Let $(\mathfrak{g},\mathfrak{h})$ be a split semi-simple Lie algebra. Let w be an element of the Weyl group operating in $\mathfrak{h}$. There exists an elementary automorphism θ of $\mathfrak{g}$ such that $\theta|\mathfrak{h} = w$.*

It is sufficient to consider the case where $\alpha \in R$ exists such that $w = s_\alpha$. Let $X_\alpha \in \mathfrak{g}^\alpha$ and $X_{-} \in \mathfrak{g}^{-\alpha}$ be such that $[X_\alpha, X_{-\alpha}] = -H_\alpha$ and let

$$\theta = (\exp \mathrm{ad}\, X_\alpha)(\exp \mathrm{ad}\, X_{-\alpha})(\exp \mathrm{ad}\, X_\alpha) \in \mathrm{Aut}_e(\mathfrak{g}).$$

Let $h \in \mathfrak{h}$, and let us show that $\theta(h) = s_\alpha(h) = h - \alpha(h)H_\alpha$. This is obvious if $\alpha(h) = 0$, and it is thus sufficient to consider the case where $h = H_\alpha$. Now $[X_\alpha, H_\alpha] = -2X_\alpha$, and hence

$$(\exp \mathrm{ad}\, X_\alpha) \cdot H_\alpha = H_\alpha - 2X_\alpha.$$

Then

$$[X_{-\alpha}, H_\alpha - 2X_\alpha] = 2X_{-\alpha} - 2H_\alpha,$$

$$[X_{-\alpha}, [X_{-\alpha}, H_\alpha - 2X_\alpha]] = -4X_{-\alpha},$$

hence

$$(\exp \mathrm{ad}\, X_{-\alpha})(H_\alpha - 2X_\alpha) = H_\alpha - 2X_\alpha + 2X_{-\alpha} - 2H_\alpha - 2X_{-\alpha} = -2X_\alpha - H_\alpha.$$

Finally, $[X_\alpha, -2X_\alpha - H_\alpha] = 2X_\alpha$, and therefore

$$(\exp \mathrm{ad}\, X_\alpha)(-2X_\alpha - H_\alpha) = -2X_\alpha - H_\alpha + 2X_\alpha = -H_\alpha.$$

Thus

$$\theta(H_\alpha) = -H_\alpha = s_\alpha(H_\alpha).$$

1.10.20. PROPOSITION (k algebraically closed, $\mathfrak{g}$ semi-simple). *Let $\mathfrak{b}$ and $\mathfrak{b}'$ be Borel subalgebras of $\mathfrak{g}$. There exists an elementary automorphism of $\mathfrak{g}$ which transforms $\mathfrak{b}$ into $\mathfrak{b}'$.*

The algebras $\mathfrak{b}$, $\mathfrak{b}'$ are defined by Cartan subalgebras $\mathfrak{h}$, $\mathfrak{h}'$ and bases B and B' of $R(\mathfrak{g},\mathfrak{h})$ and $R(\mathfrak{g},\mathfrak{h}')$ respectively. By virtue of 1.9.11, we can restrict ourselves to the case where $\mathfrak{h}' = \mathfrak{h}$. There exists an element of the Weyl group which transforms B into B' (11.1.6). The proposition then follows from 1.10.19.

1.10.21. PROPOSITION (k algebraically closed, $\mathfrak{g}$ semi-simple). *Let N be the set of nilpotent elements of $\mathfrak{g}$, $\mathscr{A}$ the adjoint group of $\mathfrak{g}$, $\mathfrak{g} = \mathfrak{h} \oplus \mathfrak{n}_+ \oplus \mathfrak{n}_-$ a triangular decomposition of $\mathfrak{g}$, and $\mathscr{N}$ the irreducible algebraic subgroup of $\mathscr{A}$ with Lie algebra* $\mathrm{ad}_{\mathfrak{g}}\mathfrak{n}_-$ (CH, p. 181). *Then $\mathscr{A}\mathfrak{n}_+ = N$, and $\mathscr{N}\mathfrak{n}_+$ is dense in N.*

It is obvious that $\mathscr{A}\mathfrak{n}_+ \subset N$. Let $x \in N$ and let us prove that $x \in \mathscr{A}\mathfrak{n}_+$. From 1.10.16, x belongs to a Borel subalgebra. From 1.10.20, we may assume that $x \in \mathfrak{h} \oplus \mathfrak{n}_+$. From 1.10.17, we then have $x \in \mathfrak{n}_+$.

Since $\mathfrak{b}_+$ is equal to its normalizer in $\mathfrak{g}$ (1.10.15), $\mathrm{ad}_{\mathfrak{g}}\mathfrak{b}_+$ is algebraic (CH, p. 172); let $\mathscr{B}$ be the corresponding irreducible algebraic subgroup of $\mathscr{A}$. Let ψ be the mapping $(n,b) \mapsto nb$ of $\mathscr{N} \times \mathscr{B}$ into $\mathscr{A}$. The mapping which is tangent to ψ at $(1,1)$ is bijective. Hence $\mathscr{N}\mathscr{B}$ is dense in $\mathscr{A}$ and consequently $\mathscr{N}\mathscr{B}\mathfrak{n}_+$ is dense in $\mathscr{A}\mathfrak{n}_+ = N$. Now $[\mathfrak{b}_+,\mathfrak{n}_+] \subset \mathfrak{n}_+$, and hence $\mathscr{B}\mathfrak{n}_+ \subset \mathfrak{n}_+$.

1.10.22. Let $\mathfrak{r}$ be a reductive Lie algebra. We denote the centre of $\mathfrak{r}$ by $\mathfrak{c}$, and the Lie algebra $[\mathfrak{r},\mathfrak{r}]$, which is semi-simple, by $\mathfrak{g}$. Then $\mathfrak{r} = \mathfrak{g} \oplus \mathfrak{c}$. The Cartan subalgebras of $\mathfrak{r}$ are the $\mathfrak{h} \oplus \mathfrak{c}$, where $\mathfrak{h}$ is a Cartan subalgebra of $\mathfrak{g}$. $\mathfrak{h} \oplus \mathfrak{c}$ is splitting if and only if $\mathfrak{h}$ is splitting.

Let us assume $\mathfrak{h}$ to be splitting. Let $R = R(\mathfrak{g},\mathfrak{h})$. Then $\mathfrak{g} = \mathfrak{h} \oplus (\oplus_{\alpha \in R}\mathfrak{g}^{\alpha})$. Relative to $\mathfrak{h} \oplus \mathfrak{c}$, we have $\mathfrak{r}^0 = \mathfrak{h} \oplus \mathfrak{c}$; the roots are the linear forms on $\mathfrak{h} \oplus \mathfrak{c}$ which are zero on $\mathfrak{c}$ and extend the elements of R; if β is such a root, then $\mathfrak{r}^{\beta} = \mathfrak{g}^{\beta|\mathfrak{h}}$. The properties for the semi-simple case can thus be extended without difficulty to the reductive case. Let S be the set [again denoted by $R(\mathfrak{r},\mathfrak{h} \oplus \mathfrak{c})$] of the roots of $\mathfrak{r}$ relative to $\mathfrak{h} \oplus \mathfrak{c}$; this is a system of roots, not relative to $(\mathfrak{h} \oplus \mathfrak{c})^*$, but relative to the orthogonal subspace (which can be canonically identified with $\mathfrak{h}^*$) of $\mathfrak{c}$ in $(\mathfrak{h} \oplus \mathfrak{c})^*$. The sum of $\mathfrak{c}$ and of a Borel subalgebra of $\mathfrak{g}$ is termed a Borel subalgebra of $\mathfrak{r}$.

1.11. Regular linear forms

1.11.1. If $f \in \mathfrak{g}^*$, the alternating bilinear form $(x,y) \mapsto f([x,y])$ on $\mathfrak{g}$ is denoted by B_f. For every subset $\mathfrak{a}$ of $\mathfrak{g}$, the orthogonal subspace of $\mathfrak{a}$ with respect to B_f is denoted by $\mathfrak{a}^f$. This orthogonal subspace contains the centralizer of $\mathfrak{a}$ in $\mathfrak{g}$ and every ideal of $\mathfrak{g}$ contained in $\mathrm{Ker}\, f$.

1.11.2. Let $\mathfrak{a}$ be an ideal of $\mathfrak{g}$. Let us consider the representation $x \mapsto (\mathrm{ad}_{\mathfrak{g}}x)|\mathfrak{a}$ of $\mathfrak{g}$ in $\mathfrak{a}$, and its dual in $\mathfrak{a}^*$. If $x \in \mathfrak{g}$, then

$$x \in \mathfrak{a}^f \quad \Leftrightarrow \quad f([x,y]) = 0 \quad \text{for all } y \in \mathfrak{a}$$
$$\Leftrightarrow \quad x \cdot (f \mid \mathfrak{a}) = 0.$$

Hence $\mathfrak{a}^f$ is a Lie subalgebra of $\mathfrak{g}$.

In particular, $\mathfrak{g}^f$ is a Lie subalgebra of $\mathfrak{g}$. Since, by passage to the quotient, B_f defines a non-degenerate alternating bilinear form on $\mathfrak{g}/\mathfrak{g}^f$, the number $\dim \mathfrak{g} - \dim \mathfrak{g}^f$ is even.

1.11.3. Let us assume that $\mathrm{ad}_{\mathfrak{g}}(\mathfrak{g})$ is an algebraic Lie algebra, and let $\mathscr{G}$ be the adjoint group of $\mathfrak{g}$. Let $\mathfrak{a}$ be an ideal of $\mathfrak{g}$. The group $\mathscr{G}$ operates in $\mathfrak{a}$ and hence in $\mathfrak{a}^*$; the corresponding representations of $\mathfrak{g}$ in $\mathfrak{a}$ and $\mathfrak{a}^*$ are the representations deduced from the adjoint representations. Let $f \in \mathfrak{g}^*$. From 1.11.2 and BO, p. 190, $\mathfrak{a}^f$ is the Lie algebra of the stabilizer of $f|\mathfrak{a}$ in $\mathscr{G}$.

In particular, $\mathfrak{g}^f$ is the Lie algebra of the stabilizer of f in $\mathscr{G}$. From BO, p. 39, the $\mathscr{G}$-orbit of f in $\mathfrak{g}^*$ thus has the dimension $\dim \mathfrak{g} - \dim \mathfrak{g}^f$; this dimension is even.

1.11.4. Lemma. *Let W be a vector subspace of $\mathfrak{g}$. The set of the $f \in \mathfrak{g}^*$ such that $W \cap \mathfrak{g}^f \neq 0$ is a closed subset of $\mathfrak{g}^*$.*

Let $(e_1, \ldots, e_n)$ be a basis of $\mathfrak{g}$ such that $(e_1, \ldots, e_p)$ is a basis of W, and $(e_1^*, \ldots, e_n^*)$ the dual basis of $\mathfrak{g}^*$. We set $[e_i, e_j] = \sum_k \gamma_{ijk} e_k$. Let $f = \lambda_1 e_1^* + \cdots + \lambda_n e_n^* \in \mathfrak{g}^*$. The following conditions are equivalent:

(a) $W \cap \mathfrak{g}^f \neq 0$;

(b) there exist $\mu_1, \ldots, \mu_p \in k$, which are not all zero, such that

$$0 = f([\mu_1 e_1 + \cdots + \mu_p e_p, e_j]) = \sum_{l=1}^{p} \sum_{k=1}^{n} \mu_l \gamma_{ljk} \lambda_k \quad (j = 1, \ldots, n);$$

(c) the matrix $(\sum_{k=1}^n \gamma_{ljk}\lambda_k)_{1 \leq l \leq p, 1 \leq j \leq n}$ is of rank $< p$.

The latter condition can be expressed by some polynomial equations in $\lambda_1, \ldots, \lambda_n$.

1.11.5. Proposition. *Let $r = \inf_{f \in \mathfrak{g}^*} \dim \mathfrak{g}^f$. The set of the $f \in \mathfrak{g}^*$ such that $\dim \mathfrak{g}^f = r$ is open in $\mathfrak{g}^*$.*

If $\dim \mathfrak{g}^f > r$, then $\mathfrak{g}^f \cap W \neq 0$ for every vector subspace W of $\mathfrak{g}$ whose dimension is $(\dim \mathfrak{g}) - r$. It is then sufficient to apply 1.11.4.

1.11.6. With the notation of 1.11.5, r is termed the *index* of $\mathfrak{g}$. An element f of $\mathfrak{g}^*$ is said to be *regular* if $\dim \mathfrak{g}^f = r$. The index is invariant under extensions of the base field.

1.11.7. Proposition. *Let f be a regular element of $\mathfrak{g}^*$. Then the Lie algebra $\mathfrak{g}^f$ is commutative.*

We fix a subspace W of $\mathfrak{g}$ such that $\mathfrak{g} = W \oplus \mathfrak{g}^f$, and an elements g of $\mathfrak{g}^*$. If $\lambda \in k$, then, since f is regular,

$$\mathfrak{g} = W \oplus \mathfrak{g}^{f+\lambda g} \quad \Leftrightarrow \quad W \cap \mathfrak{g}^{f+\lambda g} = 0.$$

From 1.11.4, this condition defines an open subset k' of k. We have $0 \in k'$.

Let $(e_1, \ldots, e_n)$ be a basis for $\mathfrak{g}$ such that $(e_1, \ldots, e_p)$ is a basis for W. Let $x = \xi_1 e_1 + \cdots + \xi_n e_n \in \mathfrak{g}$. For $\lambda \in k'$, let $x = x_1(\lambda) + x_2(\lambda)$, with $x_1(\lambda) \in W$ and $x_2(\lambda) \in \mathfrak{g}^{f+\lambda g}$. If we set $x_1(\lambda) = \nu_1 e_1 + \cdots + \nu_p e_p$, the ν_i are determined by the conditions

$$(f + \lambda g)([(\xi_1 - \nu_1)e_1 + \cdots + (\xi_p - \nu_p)e_p + \xi_{p+1}e_{p+1} + \cdots \xi_n e_n e_j]) = 0$$

$$(j = 1, \ldots, n),$$

i.e.,

$$(1) \qquad \sum_{i=1}^{p} \nu_i (f + \lambda g)([e_i, e_j]) = \sum_{i=1}^{n} \xi_i (f + \lambda g)([e_i, e_j]) \qquad (j = 1, \ldots, n).$$

One of the cofactors of p rows and p columns of the matrix $((f + \lambda g)([e_i, e_j]))_{1 \leqq i \leqq p, 1 \leqq j \leqq n}$ is non-zero for $\lambda = 0$, and hence for λ taking values in an open subset k'' of k'. The function $\lambda \mapsto x_2(\lambda)$ is rational over k''. Let $y \in \mathfrak{g}^f$. Then

$$\langle f + \lambda g, [x_2(\lambda), y] \rangle = 0 \qquad \text{for } \lambda \in k'',$$

hence, taking the derivative at $\lambda = 0$,

$$\langle g, [x_2(0), y] \rangle + \langle f, [x_2'(0), y] \rangle = 0.$$

Let us now assume that $x \in \mathfrak{g}^f$. Then $x_2(0) = x$, and hence the foregoing equation reduces to $\langle g, [x, y] \rangle = 0$. Since g is arbitrary, we conclude that $[x, y] = 0$.

1.11.8. Let V be a vector space of finite dimension n. Let $d \in \{0, 1, \ldots, n\}$. We denote by $\mathrm{Gr}(V, d)$ the Grassmannian of V for the dimension d, i.e., the set of d-dimensional vector subspaces of V. This Grassmannian is equipped with the structure of an algebraic manifold. We recall the following points on this subject. Let $b = (e_1, \ldots, e_n)$ be a basis for V. This basis defines an open subset O_b of $\mathrm{Gr}(V, d)$, namely the set of the $W \in \mathrm{Gr}(V, d)$ such that $W \cap (ke_{d+1} + \cdots + ke_n) = 0$. Let $W \in O_b$. For $i = 1, \ldots, d$, let $\lambda_{i1}, \ldots, \lambda_{i,n-d}$ be the scalars such that

$$w_i = e_i + \lambda_{i1} e_{d+1} + \cdots \lambda_{i.n-d} e_n \in W.$$

Then the mapping

$$W \mapsto (\lambda_{11}, \ldots, \lambda_{1,n-d}; \lambda_{21}, \ldots, \lambda_{2,n-d}; \ldots; \lambda_{d1}, \ldots, \lambda_{d,n-d})$$

is an isomorphism of the algebraic manifold O_b onto the algebraic manifold $k^{d(n-d)}$ (BO, p. 240).

The manifold $\mathrm{Gr}(V,d)$ is irreducible and complete.

1.11.9. We retain the notation of 1.11.8, and assume V to be equipped with a Lie algebra structure. The set A of the elements of $\mathrm{Gr}(V,d)$ which are Lie subalgebras of V is closed in $\mathrm{Gr}(V,d)$. In fact, it is sufficient to prove that $A \cap O_b$ is closed in O_b. Now $W \in A \cap O_b$ if and only if, for $i,j = 1, \ldots, d$, $[w_i,w_j]$ is a linear combination of $w_1, \ldots, w_d$. This can be expressed by polynomial conditions on the λ_{uv}, whence our assertion.

The set of the elements of $\mathrm{Gr}(V,d)$ which are solvable Lie subalgebras of V is closed in $\mathrm{Gr}(V,d)$. Indeed, let $w_1^1, w_2^1, w_3^1, \ldots$ be the elements $[w_i,w_j]$ arranged in a certain order, let $w_1^2, w_2^2, w_3^2, \ldots$ be the elements $[w_i^1, w_j^1]$ arranged in a certain order, etc. Then an element W of $A \cap O_b$ is solvable if and only if $w_1^d, w_2^d, w_3^d, \ldots$ are all zero. This can be expressed by polynomial conditions on the $\lambda_{u,v}$, whence our assertion. In the same way, it can be seen that the set of the elements of $\mathrm{Gr}(V,d)$ which are nilpotent (or commutative) Lie subalgebras of V is closed in $\mathrm{Gr}(V,d)$.

Similarly, the set of the $(f,W) \in V^* \times \mathrm{Gr}(V,d)$ such that $f([V,W]) = 0$, or such that $f([W,W]) = 0$, is closed in $V^* \times \mathrm{Gr}(V,d)$.

1.11.10. PROPOSITION. *Let r be the index of $\mathfrak{g}$, and $f \in \mathfrak{g}^*$. There exists an r-dimensional commutative subalgebra of $\mathfrak{g}^f$.*

Let T be an indeterminate, $k' = k(T)$, $\mathfrak{g}' = \mathfrak{g} \otimes k'$, g a regular element of $\mathfrak{g}'$, and $f' = f + Tg \in \mathfrak{g}'^*$. It is easily seen that f' is regular in $\mathfrak{g}'^*$. Let $\mathfrak{h}' = \mathfrak{g}'^{f'}$. Then $\mathfrak{h}'$ is an r-dimensional commutative Lie subalgebra of $\mathfrak{g}'$ (1.11.7), and $f'([\mathfrak{g}',\mathfrak{h}']) = 0$. Let $\mathfrak{k} = \mathfrak{h}' \cap (\mathfrak{g} \otimes k[T])$. Then $\mathfrak{k} \otimes_{k[T]} k(T) = \mathfrak{h}'$; in the free $k[T]$-module $\mathfrak{g} \otimes k[T]$, $\mathfrak{k}$ is a submodule of rank r, and $(\mathfrak{g} \otimes k[T])/\mathfrak{k}$ is torsion-free, so that $\mathfrak{k}$ is a direct factor submodule in $\mathfrak{g} \otimes k[T]$. Let φ be the homomorphism of $k[T]$ onto k such that $\varphi(T) = 0$; let ψ be the homomorphism $1 \otimes \varphi$ of $\mathfrak{g} \otimes k[T]$ onto $\mathfrak{g}$. Then $\mathfrak{h} = \psi(\mathfrak{k})$ is an r-dimensional commutative Lie subalgebra of $\mathfrak{g}$, and $f([\mathfrak{g},\mathfrak{h}]) = 0$.

1.11.11. Let us assume $\mathfrak{g}$ to be semi-simple. Let K be the Killing form of $\mathfrak{g}$, and φ the Killing isomorphism of $\mathfrak{g}$ onto $\mathfrak{g}^*$. Let $x \in \mathfrak{g}$, and $f = \varphi(x)$. Then

$$\begin{aligned} \mathfrak{g}^f &= \{y \in \mathfrak{g} \mid f([y,z]) = 0 \quad \text{for all } z \in \mathfrak{g}\} \\ &= \{y \in \mathfrak{g} \mid K([x,y],z) = 0 \quad \text{for all } z \in \mathfrak{g}\} \\ &= \{y \in \mathfrak{g} \mid [x,y] = 0\}, \end{aligned}$$

and hence

(1) $$g^f = \mathfrak{g}^x.$$

An element of $\mathfrak{g}$ is said to be *regular* if its image under φ is a regular element of $\mathfrak{g}^*$.

1.11.12. PROPOSITION. *If $\mathfrak{g}$ is semi-simple, its rank is equal to its index.*

Let x be an element of $\mathfrak{g}$ which is both generic and regular (such elements exist). From 1.10.6 (i), dim $\mathfrak{g}^x$ is the rank of $\mathfrak{g}$. From 1.11.11, dim $\mathfrak{g}^x$ is the index of $\mathfrak{g}$.

1.11.13. PROPOSITION. *We assume $\mathfrak{g}$ to be semi-simple and of rank r.*

(i) *Let $x \in \mathfrak{g}$. Then x is regular if and only if* dim $\mathfrak{g}^x = r$.

(ii) *Let $x \in \mathfrak{g}$. Then x is generic if and only if x is semi-simple and regular.*

(iii) *For all $y \in \mathfrak{g}, \mathfrak{g}^y$ contains an r-dimensional commutative Lie subalgebra.*

Assertion (i) follows from 1.11.11 and 1.11.12. Taking (i) into account, (ii) follows from 1.10.6 (i) and (iv). Assertion (iii) follows from 1.11.10 and 1.11.11.

1.12. Polarizations

We begin by grouping together all the lemmas which will be useful to us, now or later, concerning alternating bilinear forms.

1.12.1. Let V be a finite-dimensional vector space, and B an alternating bilinear form on V. For every subset W of V, we denote the orthogonal subspace of W with respect to B by $W^\perp$. In particular, $V^\perp$ is termed the *kernel* of B. For every vector subspace W of V, we have

$$\dim W + \dim W^\perp = \dim V + \dim (W \cap V^\perp).$$

A vector subspace W of V is said to be *totally isotropic* if $W \subset W^\perp$. The largest dimension of totally isotropic vector spaces is $\frac{1}{2}(\dim V + \dim V^\perp)$. For a totally isotropic vector subspace W, the following conditions are equivalent:

(a) W is maximal totally isotropic;

(b) $\dim W = \frac{1}{2}(\dim V + \dim V^\perp)$;

(c) $W \supset W^\perp$;

(d) $W = W^\perp$.

If these conditions are satisfied, then $W \supset V^\perp$.

1.12.2. LEMMA. *Let V and B be as in* 1.12.1, *let V' be a hyperplane of V, B' the restriction of B to V', and N and N' the kernels of B and B'.*

(i) *If $N \subset V'$, then N is a hyperplane in N'. Every vector subspace which is maximal totally isotropic with respect to B' is maximal totally isotropic with respect to B.*

(ii) *If $N \not\subset V$; then $N' = N \cap V'$ is a hyperplane in N. If M is a vector subspace which is maximal totally isotropic with respect to B, then $M \cap V'$ is maximal totally isotropic with respect to B', and* $\dim M = 1 + \dim (M \cap V')$.

Let us assume that $N \subset V'$. Then $N \subset N'$. Let $x \in V$ such that $x \notin V'$. Then

$$N = \{u \in N' \mid B(u,x) = 0\},$$

hence

$$\dim N'/N \leqq 1.$$

But $\dim V - \dim N$ and $\dim V' - \dim N'$ are even, and hence $\dim N'/N = 1$. On the other hand,

$$\tfrac{1}{2}(\dim V + \dim N) = \tfrac{1}{2}(\dim V' + \dim N'),$$

whence the final assertion in (i).

Let us assume $N \not\subset V'$, and let $x \in N$ such that $x \notin V'$. If $u \in N'$, u is orthogonal to V' and to x, and hence $u \in N$. Thus $N' \subset N \cap V'$. It is obvious that $N \cap V' \subset N$: hence $N' = N \cap V'$ is a hyperplane in N. Let M be a vector subspace which is maximal totally isotropic with respect to B. Then $x \in N \subset M$. If $y \in V'$ is orthogonal to $M \cap V'$, then y is orthogonal to

$$M = (M \cap V') + kx,$$

and hence $y \in M \cap V'$, whence the final assertion of (ii).

1.12.3. LEMMA. *Let V and B be as in* 1.12.1, *and let $(V_0, V_1, \ldots, V_n)$ be an increasing sequence of vector subspaces of V such that* $\dim V_i = i$ *and $V_n = V$, B_i the restriction of B to V_i, N_i the kernel of B_i, and*

$$P_i = N_1 + N_2 + \cdots + N_i.$$

(i) *P_n is a vector subspace which is maximal totally isotropic with respect to B.*

(ii) *$P_n \cap V_i = P_i$ for all i.*

We reason by induction on n. Let us assume that $N_n \subset V_{n-1}$. Then $N_n \subset N_{n-1}$ (1.12.2), and hence $P_n = P_{n-1}$. From the induction hypothesis, $P_n \cap V_i = P_{n-1} \cap V_i = P_i$ for $i < n$, and $P_n = P_{n-1}$ is maximal totally isotropic with respect to B_{n-1} and hence with respect to B (1.12.2).

Let us assume that $N_n \not\subset V_{n-1}$. Then $N_{n-1} = N_n \cap V_{n-1}$ is a hyperplane in N_n (1.12.2). Hence $P_n \cap V_{n-1} = P_{n-1} + (N_n \cap V_{n-1}) = P_{n-1}$, and then $P_n \cap V_i = P_i$ for $i < n$ from the induction hypothesis. On the other hand, $P_n = P_{n-1} + N_n$ is totally isotropic, and $\dim P_n = 1 + \dim P_{n-1}$, hence P_n is maximal totally isotropic (1.12.2).

1.12.4. LEMMA. *Let V and B be as in* 1.12.1, *and let W be a vector subspace of V, $X = W^{\perp}$, and P_1 and P_2 maximal totally isotropic vector subspaces of W and X respectively. Then:*

(i) $X \cap X^{\perp} = (W \cap W^{\perp}) + V^{\perp}$.

(ii) $P_1 \cap P_2 = W \cap W^{\perp}$.

(iii) *$P_1 + P_2$ is a maximal totally isotropic vector subspace of V.*

(iv) *$(P_1 + P_2) \cap W = P_1$ and $(P_1 + P_2) \cap X = P_2$.*

Since $W^{\perp} \supset V^{\perp}$, we have

$$(W \cap W^{\perp}) + V^{\perp} = (W + V^{\perp}) \cap W^{\perp} = X^{\perp} \cap X,$$

whence (i). Next,

$$P_1 \cap P_2 \subset W \cap X = W \cap W^{\perp} \subset P_1 \cap (X \cap X^{\perp}) \subset P_1 \cap P_2,$$

whence (ii). Since P_1 and P_2 are totally isotropic and orthogonal, $P_1 + P_2$ is totally isotropic; and we have

$$2 \dim (P_1 + P_2) = 2 \dim P_1 + 2 \dim P_2 - 2 \dim (P_1 \cap P_2)$$

$$= \dim W + \dim (W \cap W^{\perp}) + \dim X + \dim (X \cap X^{\perp}) - 2 \dim (P_1 \cap P_2).$$

From (i) and (ii), this number is equal to

$$\dim W + \dim (W \cap W^{\perp}) + \dim X + \dim (W \cap W^{\perp}) + \dim V^{\perp}$$
$$- \dim (W \cap W^{\perp} \cap V^{\perp}) - 2 \dim (W \cap W^{\perp})$$
$$= \dim W + \dim W^{\perp} + \dim V^{\perp} - \dim (W \cap V^{\perp}).$$

Now, if S is complementary to $W \cap V^{\perp}$ in W, we have

$$\dim S + \dim S^{\perp} = \dim V, \qquad S^{\perp} = W^{\perp}.$$

Hence

$$2 \dim (P_1 + P_2) = \dim V + \dim V^{\perp},$$

Whence (iii).

It is obvious that $(P_1 + P_2) \cap W \supset P_1$, and $(P_1 + P_2) \cap W$ is totally isotropic, therefore $(P_1 + P_2) \cap W = P_1$. Similarly, $(P_1 + P_2) \cap X = P_2$.

1.12.5. Lemma. *Let V and B be as in* 1.12.1, *and let W be a vector subspace of V, $X = W^{\perp}$, and P a maximal totally isotropic vector subspace of V. Then*

$$\tfrac{1}{2}(\dim W + \dim (W \cap W^{\perp})) - \dim (W \cap P)$$
$$= \tfrac{1}{2}(\dim X + \dim (X \cap X^{\perp})) - \dim (X \cap P).$$

Since $(P + W)^{\perp} = P^{\perp} \cap X$ and $P \supset V^{\perp}$, we have

$$\dim V + \dim V^{\perp} = \dim (P + W) + \dim (P \cap X)$$
$$= \dim P + \dim W - \dim (P \cap W) + \dim (P \cap X).$$

Since $\dim P = \frac{1}{2}(\dim V + \dim V^{\perp})$, we deduce that

$$\tfrac{1}{2}(\dim V + \dim V^{\perp}) = \dim W - \dim (P \cap W) + \dim (P \cap X).$$

On the other hand, from 1.12.4,

$$\tfrac{1}{2}(\dim V + \dim V^{\perp}) = \tfrac{1}{2}(\dim W + \dim (W \cap W^{\perp}))$$
$$+ \tfrac{1}{2}(\dim X + \dim (X \cap X^{\perp})) - \dim (W \cap W^{\perp})$$

hence

$$\dim (P \cap X) - \dim (P \cap W) = \tfrac{1}{2}(\dim X + \dim (X \cap X^{\perp}))$$
$$- \tfrac{1}{2}(\dim W + \dim (W \cap W^{\perp})).$$

1.12.6. Lemma. *Let V, B, W, X and P be as in* 1.12.5. *The following conditions are equivalent:*

(i) $P = (P \cap W) + (P \cap X)$;

(ii) *$P \cap W$ is a maximal totally isotropic vector subspace of W;*

(iii) *$P \cap X$ is a maximal totally isotropic vector subspace of X;*

(iv) *P is the sum of a maximal totally isotropic vector subspace of W and a maximal totally isotropic vector subspace of X.*

(ii) $\Leftrightarrow$ (iii). This follows from 1.12.5.

(iv) $\Rightarrow$ (ii). This follows from 1.12.4.

(ii) and (iii) $\Rightarrow$ (i). If conditions (ii) and (iii) are satisfied, $(P \cap W) + (P \cap X)$ is maximal totally isotropic (1.12.4) and contained in P, and hence equal to P.

(i) $\Rightarrow$ (iv). Let us assume that $P = (P \cap W) + (P \cap X)$. Let P_1 and P_2 be maximal totally isotropic vector subspaces of W and X containing $P \cap W$ and $P \cap X$ respectively. Then $P \subset P_1 + P_2$ and $P_1 + P_2$ is totally isotropic, whence $P = P_1 + P_2$.

1.12.7. Let $f \in \mathfrak{g}^*$. A Lie subalgebra $\mathfrak{h}$ of $\mathfrak{g}$ is said to be *subordinate* to f if it is totally isotropic with respect to B_f, i.e., if $f([\mathfrak{h},\mathfrak{h}]) = 0$, or again, if $f|\mathfrak{h}$ is a one-dimensional representation of $\mathfrak{h}$.

1.12.8. Let $f \in \mathfrak{g}^*$. A Lie subalgebra of $\mathfrak{g}$ subordinate to f of dimension $\frac{1}{2}(\dim \mathfrak{g} + \dim \mathfrak{g}^f)$, in other words a Lie subalgebra of $\mathfrak{g}$ which is a maximal totally isotropic vector subspace of $\mathfrak{g}$ (equipped with B_f), is termed *a polarization* of $\mathfrak{g}$ at f. We denote the set of polarizations of $\mathfrak{g}$ at f by $\mathrm{P}(f;\mathfrak{g})$, or by $\mathrm{P}(f)$. We denote the set of solvable polarizations of $\mathfrak{g}$ at f by $\mathrm{PR}(f;\mathfrak{g})$, or $\mathrm{PR}(f)$.

1.12.9. We shall see that if $\mathfrak{g}$ is completely solvable, there exist polarizations of $\mathfrak{g}$ at every point of $\mathfrak{g}^*$ (1.12.10). On the other hand, if $\mathfrak{g}$ is solvable, or semi-simple, this result is far from accurate (1.14.7).

Let us assume k to be algebraically closed (the solvable case then being identical with the completely solvable case). Here again, even if $\mathfrak{g}$ is semi-simple, there may exist $f \in \mathfrak{g}^*$ at which $\mathfrak{g}$ has no polarization (1.14.9). However, these f are exceptions. In fact we shall see that for $\mathfrak{g}$ arbitrary and f regular, $\mathfrak{g}$ has a polarization, and even a solvable polarization, at f (1.12.16).

1.12.10. PROPOSITION. *Let* $(\mathfrak{g}_0, \mathfrak{g}_1, \ldots, \mathfrak{g}_n)$ *be an increasing sequence of ideals of* $\mathfrak{g}$ *such that* $\dim \mathfrak{g}_i = i$, $\mathfrak{g}_n = \mathfrak{g}$. *Let* $f \in \mathfrak{g}^*$, $f_i = f|\mathfrak{g}_i$, *and* $\mathfrak{p}_i = \mathfrak{g}_1^{f_1} + \cdots + \mathfrak{g}_i^{f_i}$ *Then:*

(i) $\mathfrak{p}_n \in \mathrm{P}(f)$.

(ii) $\mathfrak{p}_n \cap \mathfrak{g}_i = \mathfrak{p}_i$.

(iii) *Let* D *be a derivation of* $\mathfrak{g}$ *under which the* $\mathfrak{g}_i$ *are stable and such that* $f(D\mathfrak{g}) = 0$. *Then* $D(\mathfrak{p}_n) \subset \mathfrak{p}_n$.

Let $x \in \mathfrak{g}_i^{f_i}$ and $y \in \mathfrak{g}_j^{f_j}$ with $i \geqq j$. Then $[x,y] \in \mathfrak{g}_j$. If $u \in \mathfrak{g}_j$, then

$$f_j([[x,y],u]) = f([[x, u], y]) + f([x,[y,u]]) \in f(\mathfrak{g}_j,\mathfrak{g}_j^{f_j}) + f(\mathfrak{g}_i^{f_i},\mathfrak{g}_i) = 0.$$

Hence $[x,y] \in \mathfrak{g}_j^{f_j}$. This proves that $\mathfrak{p}_n$ is a Lie subalgebra of $\mathfrak{g}$, and thus $\mathfrak{p}_n \in P(f)$ from 1.12.3. We have $\mathfrak{p}_n \cap \mathfrak{g}_i = \mathfrak{p}_i$ (1.12.3). Finally, let $x \in \mathfrak{g}_i^{f_i}$, $y \in \mathfrak{g}_i$; then

$$f([Dx,y]) = f(D([x,y])) - f([x,Dy]) = -f([x,Dy]) \in f(\mathfrak{g}_i^{f_i},\mathfrak{g}_i) = 0,$$

and hence $Dx \in \mathfrak{g}_i^{f_i}$. This proves (iii).

1.12.11. With the notation of 1.12.10, if s represents the sequence $(\mathfrak{g}_0, \mathfrak{g}_1, \ldots, \mathfrak{g}_n)$, the polarization $\mathfrak{p}_n$ is denoted by $\mathfrak{p}(f,s)$.

1.12.12. LEMMA (k algebraically closed, $\mathfrak{g}$ semi-simple). *Let n be the dimension of* $\mathfrak{g}$, *r its rank, K its Killing form, and* $x \in \mathfrak{g}$. *There exists a solvable Lie subalgebra* $\mathfrak{b}$ *of* $\mathfrak{g}$ *of dimension* $\frac{1}{2}(n+r)$ *such that* $K(x,[\mathfrak{b},\mathfrak{b}]) = 0$.

Let p be the canonical projection of $\mathfrak{g} \times \mathrm{Gr}(\mathfrak{g},\frac{1}{2}(n+r))$ onto $\mathfrak{g}$. Let C be the set of the $(y,\mathfrak{b}) \in \mathfrak{g} \times \mathrm{Gr}(\mathfrak{g},\frac{1}{2}(n+r))$ such that $\mathfrak{b}$ is a solvable Lie subalgebra of $\mathfrak{g}$ and such that $K(y,[\mathfrak{b},\mathfrak{b}]) = 0$. From 1.11.9, C is closed in $\mathfrak{g} \times \mathrm{Gr}(\mathfrak{g},\frac{1}{2}(n+r))$, and hence $p(C)$ is closed in $\mathfrak{g}$. Let A be the set of generic elements of $\mathfrak{g}$. If $y \in A$, let $\mathfrak{h}$ be the Cartan subalgebra $\mathfrak{g}^y$; let B be an basis for $R(\mathfrak{g},\mathfrak{h}), \mathfrak{g} = \mathfrak{h} \oplus \mathfrak{n}_+ \oplus \mathfrak{n}_-$ the corresponding triangular decomposition, and $\mathfrak{b} = \mathfrak{h} \oplus \mathfrak{n}_+$. Then $\mathfrak{b}$ is a solvable Lie subalgebra of $\mathfrak{g}$ of dimension $\frac{1}{2}(n+r)$, $[\mathfrak{b},\mathfrak{b}] = \mathfrak{n}_+$, and $K(\mathfrak{h},\mathfrak{n}_+) = 0$; hence $(y,\mathfrak{b}) \in C$. This proves that $p(C) \supset A$, whence $p(C) = \mathfrak{g}$, which establishes the lemma.

1.12.13. LEMMA (k algebraically closed). *Let* $g \in \mathfrak{g}^*$, $\mathfrak{r}$ *a solvable ideal of* $\mathfrak{g}, \mathfrak{h} = r^g$, $r = g|\mathfrak{r}$, $h = g|\mathfrak{h}$ *and* $\mathfrak{l} \in \mathrm{PR}(\mathfrak{h})$.

(i) *There exists* $\mathfrak{p} \in \mathrm{P}(r)$ *such that* $[\mathfrak{l},\mathfrak{p}] \subset \mathfrak{p}$.

(ii) *Let* $\mathfrak{p} \in \mathrm{P}(r)$ *such that* $[\mathfrak{l},\mathfrak{p}] \subset \mathfrak{p}$. *Then* $\mathfrak{l} + \mathfrak{p} \in \mathrm{PR}(g)$.

The mapping $x \mapsto (\mathrm{ad}_{\mathfrak{g}} x)|\mathfrak{r}$ of $\mathfrak{l}$ into the set of derivations of $\mathfrak{r}$ defines a semi-direct product of $\mathfrak{l}$ and $\mathfrak{r}$. This semi-direct product is solvable; from 1.3.12, there thus exists an increasing sequence of ideals of $\mathfrak{r}$ of dimensions $0, 1, \ldots, \dim \mathfrak{r}$, which are stable under $\mathrm{ad}_{\mathfrak{g}}\mathfrak{l}$. Taking 1.12.10 into account, this proves (i).

Let $\mathfrak{p} \in P(\mathfrak{r})$ such that $[\mathfrak{l},\mathfrak{p}] \subset \mathfrak{p}$. Then $\mathfrak{q} = \mathfrak{l} + \mathfrak{p}$ is a solvable Lie subalgebra of $\mathfrak{g}$. From 1.12.4 (iii), we have $\mathfrak{q} \in \mathrm{PR}(g)$.

1.12.14. THEOREM (k algebraically closed). *Let* $\mathfrak{a}$ *be a solvable ideal of* $\mathfrak{g}$. *Let* $g \in \mathfrak{g}^*$ *be such that* $\dim \mathfrak{g}^g \leq \dim \mathfrak{g}^{g'}$ *for all* $g' \in \mathfrak{g}^*$ *such that* $g'|\mathfrak{a} = g|\mathfrak{a}$. *There exists a solvable polarization of* $\mathfrak{g}$ *at* g.

This is obvious if $\dim \mathfrak{g} \leq 1$. We reason by induction on $\dim \mathfrak{g}$. Let $\mathfrak{r}$ be the radical of $\mathfrak{g}, \mathfrak{h}$ the Lie subalgebra $\mathfrak{r}^g$ of $\mathfrak{g}$, and $\mathfrak{s} = \mathfrak{r} \cap \mathrm{Ker}\, g$.

Let us assume that $\mathfrak{h} = \mathfrak{g}$ and $\mathfrak{s} \neq 0$. Then $\mathfrak{s}$ is an ideal of $\mathfrak{g}$ and it is sufficient to apply the induction hypothesis to $\mathfrak{g}/\mathfrak{s}$, $(\mathfrak{a}+\mathfrak{s})/\mathfrak{s}$, and to the linear form on $\mathfrak{g}/\mathfrak{s}$ deduced from g by passing to the quotient.

Let us assume that $\mathfrak{h} = \mathfrak{g}$ and $\mathfrak{s} = 0$. Then $\mathfrak{r}$ is of dimension ≤ 1 and central in $\mathfrak{g}$, hence $\mathfrak{g}$ is semi-simple or the product of a semi-simple Lie algebra with a one-dimensional Lie algebra. We immediately return to the case where $\mathfrak{g}$ is semi-simple (hence $\mathfrak{a} = 0$). It is then sufficient to apply 1.12.12.

Let us assume that $\mathfrak{h} \neq \mathfrak{g}$. Let $h = g|\mathfrak{h}$ and $r = g|\mathfrak{r}$. Then $\mathfrak{r}^r = \mathfrak{r} \cap \mathfrak{r}^g = \mathfrak{r} \cap \mathfrak{h}$, and

(1) $$\mathfrak{h}^h = \mathfrak{r}^r + \mathfrak{g}^g,$$

(from 1.12.4 (i)), and

(2) $$\mathfrak{r}^r \cap \mathfrak{g}^g = \{x \in \mathfrak{r} \mid r([x,y]) = 0 \quad \text{for all } y \in \mathfrak{g}\}.$$

Let $h' \in \mathfrak{h}^*$ be such that $h'|\mathfrak{r}^r = h|\mathfrak{r}^r$. There exists a $g' \in \mathfrak{g}^*$ which extends r and h'. Then $g'|\mathfrak{a} = g|\mathfrak{a}$, and hence $\dim \mathfrak{g}^g \leqq \dim \mathfrak{g}^{g'}$; now, from (2), $\mathfrak{r}^r \cap \mathfrak{g}^g = \mathfrak{r}^r \cap \mathfrak{g}^{g'}$, and therefore, taking (1) into account, $\dim \mathfrak{h}^h \leqq \dim \mathfrak{h}^{h'}$. From the induction hypothesis, there exists $\mathfrak{l} \in \mathrm{PR}(h)$. From 1.12.13, there exists a solvable polarization of $\mathfrak{g}$ at g.

1.12.15. For k algebraically closed and $\mathfrak{g}$ solvable, taking $\mathfrak{a} = \mathfrak{g}$, we recover the existence of a polarization of $\mathfrak{g}$ at any point of $\mathfrak{g}^*$ (cf. 1.12.10 (i)). On the other hand, for any $\mathfrak{g}$, we obtain, by making $\mathfrak{a} = 0$, the following corollary:

1.12.16. Corollary (k algebraically closed). *If f is a regular form on $\mathfrak{g}$, there exists a solvable polarization of $\mathfrak{g}$ at f.*

1.12.17. Corollary (k algebraically closed). *Let n be the dimension of $\mathfrak{g}$, and r its index. For all $f \in \mathfrak{g}^*$, there exists a Lie subalgebra of $\mathfrak{g}$ which is subordinate to f, solvable and of dimension $\frac{1}{2}(n + r)$.*

This is true for f regular (1.12.16). We pass to the general case, as in 1.12.12, making use of the fact that the Grassmannians are complete.

1.12.18. Proposition (k algebraically closed, $\mathfrak{g}$ semi-simple). *Let $g \in \mathfrak{g}^*$.*

(i) $\mathrm{PR}(g) \neq \emptyset$, *if and only if g is regular.*

(ii) *Let us assume g to be regular, and let x be the element of $\mathfrak{g}$ corresponding to g under the Killing isomorphism. Then a Lie subalgebra $\mathfrak{p}$ of $\mathfrak{g}$ belongs to* $\mathrm{PR}(g)$ *if and only if $\mathfrak{p}$ is a Borel subalgebra containing x.*

Let n be the dimension of $\mathfrak{g}$, and r its rank. If $\mathfrak{p} \in \mathrm{PR}(g)$, then $\dim \mathfrak{p} = \frac{1}{2}(n + r)$ (1.10.16), hence $\dim \mathfrak{g}^g \leqq r$, and consequently $\dim \mathfrak{g}^g = r$; thus g is regular. Let us assume g to be regular, and let x be as in (ii). We have $\dim \mathfrak{g}^g = r$. From 1.12.16, $\mathrm{PR}(g) \neq \emptyset$. We have therefore proved (i). If $\mathfrak{p} \in \mathrm{PR}(g)$, then $\dim \mathfrak{p} = \frac{1}{2}(n + r)$, and hence, from 1.10.16, $\mathfrak{p}$ is a Borel subalgebra of $\mathfrak{g}$. Moreover, $\mathfrak{p}$ contains $\mathfrak{g}^g = \mathfrak{g}^x$, and hence x. Conversely, let $\mathfrak{b}$ be a Borel subalgebra of $\mathfrak{g}$ containing x. Since $\mathfrak{b}$ is orthogonal to $[\mathfrak{b},\mathfrak{b}]$, we have $g([\mathfrak{b},\mathfrak{b}]) = 0$. Now $\dim \mathfrak{b} = \frac{1}{2}(n + r)$, and hence $\mathfrak{b} \in \mathrm{PR}(g)$.

1.13. Symmetric semi-simple Lie algebras

1.13.1. We term a pair $(\mathfrak{g},\theta)$, where $\mathfrak{g}$ is a Lie algebra and θ an automorphism of $\mathfrak{g}$ such that $\theta^2 = 1$, a *symmetric Lie algebra.* Let $\mathfrak{k}$ be the set of the $x \in \mathfrak{g}$ such that $\theta x = x$, and let $\mathfrak{p}$ be the set of the $x \in \mathfrak{g}$ such that $\theta x = -x$. It is obvious that

$$\mathfrak{g} = \mathfrak{k} \oplus \mathfrak{p}, \qquad [\mathfrak{k},\mathfrak{k}] \subset \mathfrak{k}, \qquad [\mathfrak{k},\mathfrak{p}] \subset \mathfrak{p}, \qquad [\mathfrak{p},\mathfrak{p}] \subset \mathfrak{k}.$$

In particular, $\mathfrak{k}$ is a Lie subalgebra of $\mathfrak{g}$. Let K be the Killing form of $\mathfrak{g}$. It is invariant under every automorphism of $\mathfrak{g}$; consequently, if $x \in \mathfrak{k}$ and $y \in \mathfrak{p}$, we have $K(x,y) = K(\theta x,\theta y) = K(x,-y)$, whence $K(x,y) = 0$; hence $\mathfrak{k}$ and $\mathfrak{p}$ are orthogonal with respect to K.

1.13.2. Furthermore, let us assume $\mathfrak{g}$ to be semi-simple. Then K is non-degenerate, and hence $\mathfrak{k}$ and $\mathfrak{p}$ are orthogonal to each other. $\mathfrak{g} = \mathfrak{k} \oplus \mathfrak{p}$ is said to be the *symmetric decomposition* of $\mathfrak{g}$ defined by θ. The restrictions of K to $\mathfrak{k}$ and $\mathfrak{p}$ are non-degenerate. A given $\mathfrak{k}$ determines $\mathfrak{p}$ and hence θ; θ is said to be the automorphism defined by $\mathfrak{k}$.

A Lie subalgebra is said to be *symmetrizing* if it is the set of fixed points of an automorphism θ of $\mathfrak{g}$ such that $\theta^2 = 1$.

1.13.3. PROPOSITION ($\mathfrak{g}$ semi-simple). *Let $\mathfrak{k}$ be a symmetrizing subalgebra of $\mathfrak{g}$. Then $\mathfrak{k}$ is reductive in $\mathfrak{g}$ and equal to its normalizer in $\mathfrak{g}$.*

Let θ be the automorphism of $\mathfrak{g}$ defined by $\mathfrak{k}$. Let $x \in \mathfrak{g}$, and s and n be its semi-simple and nilpotent components. If $x \in \mathfrak{k}$, then $\theta x = x$, hence, by transport of structure, $\theta s = s$ and $\theta n = n$, and hence $s \in \mathfrak{k}$ and $n \in \mathfrak{k}$. Taking 1.13.2 and 1.7.6 into account, it can be seen that $\mathfrak{k}$ is reductive in $\mathfrak{g}$.

Let $\mathfrak{g} = \mathfrak{k} \oplus \mathfrak{p}$ be the symmetric decomposition of $\mathfrak{g}$ defined by θ. The normalizer of $\mathfrak{k}$ in $\mathfrak{g}$ is of the form $\mathfrak{k} \oplus \mathfrak{r}$, where $\mathfrak{r} \subset \mathfrak{p}$. We have $[\mathfrak{k},\mathfrak{r}] \subset \mathfrak{k} \cap \mathfrak{p} = 0$. Since $\mathfrak{k}$ is reductive in $\mathfrak{g}$, there exists a complement s of $\mathfrak{r}$ in $\mathfrak{p}$ such that $[\mathfrak{k},\mathfrak{s}] \subset \mathfrak{s}$. Then

$$[\mathfrak{g},\mathfrak{k} + \mathfrak{s}] = [\mathfrak{k} + \mathfrak{r} + \mathfrak{s}, \mathfrak{k} + \mathfrak{s}] \subset \mathfrak{k} + [\mathfrak{k},r + \mathfrak{s}] \subset \mathfrak{k} + \mathfrak{s},$$

so that $\mathfrak{k} + \mathfrak{s}$ is an ideal of $\mathfrak{g}$. Since $[\mathfrak{r},\mathfrak{r}] \subset \mathfrak{k}$, $\mathfrak{g}/\mathfrak{k} + \mathfrak{s}$ is commutative and hence null, whence $\mathfrak{r} = 0$.

1.13.4. Let $(\mathfrak{g}, \theta)$ be a symmetric semi-simple Lie algebra, and let $\mathfrak{g} = \mathfrak{k} \oplus \mathfrak{p}$ be the corresponding symmetric decomposition. Every element x of $\mathfrak{p}$ such that

$$\dim (\mathfrak{g}^0(x) \cap \mathfrak{p}) \leqq \dim (\mathfrak{g}^0(y) \cap \mathfrak{p})$$

for all $y \in \mathfrak{p}$ is termed a *generic element* of $(\mathfrak{g},\theta)$. This is also the set of the $x \in \mathfrak{p}$ such that $(\operatorname{ad} x)^2|\mathfrak{p}$ possesses a nilspace of minimal dimension. The set of generic elements of $(\mathfrak{g},\theta)$ is thus open and non-empty in $\mathfrak{p}$.

1.13.5. Let $(\mathfrak{g},\theta)$ be a symmetric semi-simple Lie algebra, and let $\mathfrak{g} = \mathfrak{k} \oplus \mathfrak{p}$ be the corresponding symmetric decomposition. A commutative Lie subalgebra $\mathfrak{a}$ of $\mathfrak{g}$ which is reductive in $\mathfrak{g}$, contained in $\mathfrak{p}$, and such that the centralizer of $\mathfrak{a}$ in $\mathfrak{p}$ is equal to $\mathfrak{a}$ is termed a *Cartan subspace* of $(\mathfrak{g},\theta)$, or of $\mathfrak{p}$.

1.13.6. THEOREM. *Let* $(\mathfrak{g},\theta)$ *be a symmetric semi-simple Lie algebra. There exist Cartan subspaces of* $(\mathfrak{g},\theta)$.

Let $\mathfrak{g} = \mathfrak{k} \oplus \mathfrak{p}$ be the symmetric decomposition defined by θ, and x a generic element of $(\mathfrak{g},\theta)$, $\mathfrak{a} = \mathfrak{g}^0(x) \cap \mathfrak{p}$. Then $\theta x = -x$, and hence, by transport of structure,

$$\theta(\mathfrak{g}^0(x)) = \mathfrak{g}^0(x), \qquad \theta(\mathfrak{g}^*(x)) = \mathfrak{g}^*(x).$$

Consequently, $\mathfrak{g}^0(x) = \mathfrak{a} \oplus (\mathfrak{g}^0(x) \cap \mathfrak{k})$. If $y \in \mathfrak{a}$, $\mathfrak{g}^0(x)$ and $\mathfrak{g}^*(x)$ are stable under $\operatorname{ad}_{\mathfrak{g}} y$ (1.9.6), and hence $\mathfrak{a}$ and $\mathfrak{g}^*(x) \cap \mathfrak{p}$ are stable under $(\operatorname{ad}_{\mathfrak{g}} y)^2$. Let S be the set of the $y \in \mathfrak{a}$ such that $(\operatorname{ad}_{\mathfrak{g}} y)^2|\mathfrak{g}^*(x) \cap \mathfrak{p}$ is bijective. Then S is open in $\mathfrak{a}$, and $x \in S$. Let R be the set of the $y \in \mathfrak{a}$ such that $(\operatorname{ad}_{\mathfrak{g}} y)^2|\mathfrak{a}$ is not nilpotent. Then R is open in $\mathfrak{a}$. If $R \neq \emptyset$, there exists $y \in R \cap S$, and then the nilspace of $(\operatorname{ad}_{\mathfrak{g}} y)^2|\mathfrak{p}$ has dimension $< \dim \mathfrak{a}$, which contradicts the assumption that x is generic. Thus, for all $y \in \mathfrak{a}$, there exists an integer n such that $(\operatorname{ad}_{\mathfrak{g}} y)^n|\mathfrak{a} = 0$. Since

$$(\operatorname{ad}_{\mathfrak{g}} y)(\mathfrak{g}^0(x) \cap \mathfrak{k}) \subset \mathfrak{g}^0(x) \cap \mathfrak{p},$$

it can be seen that $\operatorname{ad}_{\mathfrak{g}} y|\mathfrak{g}^0(x)$ is nilpotent. Let L be the Killing form of $\mathfrak{g}^0(x)$. From the above, $L(y,y) = 0$ for $y \in \mathfrak{a}$. Hence $L|\mathfrak{a} \times \mathfrak{a} = 0$ by polarization. On the other hand, from 1.13.1 applied to $\mathfrak{g}^0(x)$, $\mathfrak{a}$ is orthogonal to $\mathfrak{g}^0(x) \cap \mathfrak{k}$ with respect to L. Therefore $\mathfrak{a}$ is orthogonal to $\mathfrak{g}^0(x)$ with respect to L. Now $\mathfrak{g}^0(x)$ is reductive in $\mathfrak{g}$ (1.9.7), and hence $\mathfrak{a}$ is contained in the centre of $\mathfrak{g}^0(x)$. It can be seen at the same time that $\mathfrak{a}$ is a commutative subalgebra of $\mathfrak{g}$ and that $\mathfrak{a}$ is reductive in $\mathfrak{g}$ (1.6.4). If $z \in \mathfrak{p}$ commutes with $\mathfrak{a}$, then $z \in \mathfrak{g}^0(x)$, and hence $z \in \mathfrak{a}$. Thus $\mathfrak{a}$ is a Cartan subspace of $(\mathfrak{g},\theta)$.

1.13.7. PROPOSITION. *Let* $(\mathfrak{g},\theta)$ *be a symmetric semi-simple Lie algebra,* $\mathfrak{g} = \mathfrak{k} \oplus \mathfrak{p}$ *the corresponding symmetric decomposition,* $\mathfrak{a}$ *a Cartan subspace of* $(\mathfrak{g},\theta)$, $\mathfrak{m}$ *the centralizer of* $\mathfrak{a}$ *in* $\mathfrak{k}$, *and* K *the Killing form of* $\mathfrak{g}$.

(i) *The centralizer of* $\mathfrak{a}$ *in* $\mathfrak{g}$ *is* $\mathfrak{a} \oplus \mathfrak{m}$.

(ii) *The restrictions of* K *to* $\mathfrak{a}$ *and* $\mathfrak{m}$ *are non-degenerate.*

(iii) *The Lie algebra* $\mathfrak{m}$ *is reductive in* $\mathfrak{g}$.

(iv) *Let* $\mathfrak{l}$ *be a Cartan subalgebra of* $\mathfrak{m}$, *and* $\mathfrak{h} = \mathfrak{a} \oplus \mathfrak{l}$. *Then* $\mathfrak{h}$ *is a Cartan subalgebra of* $\mathfrak{g}$.

Let $y \in \mathfrak{p}$ and $z \in \mathfrak{k}$. Then $[y,\mathfrak{a}] \subset \mathfrak{k}$, $[z,\mathfrak{a}] \subset \mathfrak{p}$; hence, if $[y + z,\mathfrak{a}] = 0$, then $[y,\mathfrak{a}] = 0$ (whence $y \in \mathfrak{a}$) and $[z,\mathfrak{a}] = 0$ (whence $z \in \mathfrak{m}$). This proves (i). From 1.7.7, the restriction of K to $\mathfrak{a} \oplus \mathfrak{m}$ is non-degenerate. Since $\mathfrak{a}$ and $\mathfrak{m}$ are orthogonal with respect to K, this proves (ii). From 1.7.7, $\mathfrak{a} \oplus \mathfrak{m}$ is reductive in $\mathfrak{g}$; taking into account 1.6.4 applied in both directions, we deduce (iii) from it. The Lie algebra $\mathfrak{l}$ is commutative and reductive in $\mathfrak{g}$, from (iii) and 1.10.6; the same applies to $\mathfrak{a}$; finally, $\mathfrak{a}$ and $\mathfrak{l}$ commute. Thus $\mathfrak{h}$ is commutative and reductive in $\mathfrak{g}$. Let x be an element of $\mathfrak{g}$ which commutes with $\mathfrak{h}$. From (i), we have $x = y + z$ with $y \in \mathfrak{a}$, $z \in \mathfrak{m}$. Moreover, $[z,\mathfrak{l}] = 0$ and hence $z \in \mathfrak{l}$. Thus, $x \in \mathfrak{h}$, so that $\mathfrak{h}$ is a Cartan subalgebra of $\mathfrak{g}$ (1.10.6 (iii)).

1.13.8. Let $\mathfrak{g}$, θ, $\mathfrak{k}$, $\mathfrak{p}$, $\mathfrak{a}$, $\mathfrak{m}$, $\mathfrak{l}$ and $\mathfrak{h}$ be as in 1.13.7. Let $\bar{k}$ be an algebraic closure of k, $\bar{\mathfrak{g}} = \mathfrak{g} \otimes \bar{k}$, $\bar{\mathfrak{h}} = \mathfrak{h} \otimes \bar{k}$, etc., $\bar{\theta}$ be the $\bar{k}$-linear extension of θ to $\bar{\mathfrak{g}}$, $R = R(\bar{\mathfrak{g}},\bar{\mathfrak{h}})$, and R' and R'' be the sets of elements of R which are zero and non-zero on $\mathfrak{a}$, respectively. Then $\bar{\theta}(\bar{\mathfrak{h}}) = \bar{\mathfrak{h}}$; let us denote the transpose of $\bar{\theta}|\bar{\mathfrak{h}}$ by $\bar{\theta}$; then $\bar{\theta}(R) = R$, and R' is the set of elements of R which are invariant under $\bar{\theta}$. From 11.1.16, there exists a basis B of R with the following properties (we denote the set of positive and negative roots with respect to B by R_+ and R_- respectively; we set $R''_+ = R'' \cap R_+$ and $R''_- = R'' \cap R_-$):

(a) $\bar{\theta}(R''_+) = R''_-$;

(b) if $\alpha \in R''_+$, $\gamma \in R$, and $\gamma - \alpha|\mathfrak{a} = 0$, then $\gamma \in R''_+$;

(c) $(R''_+ + R''_+) \cap R \subset R''_+$;

(d) $B \cap R'$ is a basis for R'.

1.13.9. Proposition. *Retaining the notation of* 1.13.8, *we have:*

(i) $R'|\bar{\mathfrak{l}}$ *is the system of roots of* $\bar{\mathfrak{m}}$ *with respect to* $\bar{\mathfrak{l}}$.

(ii) *If* $\alpha \in R'$, *then* $\bar{\mathfrak{m}}^{\alpha|\bar{\mathfrak{l}}} = \bar{\mathfrak{g}}^{\alpha}$.

(iii) $\bar{\mathfrak{m}} = \bar{\mathfrak{l}} \oplus (\oplus_{\alpha \in R'} \bar{\mathfrak{g}}^{\alpha})$.

Let $\alpha \in R'$ and $x \in \bar{\mathfrak{g}}^{\alpha}$. Then x commutes with $\mathfrak{a}$, hence $x = y + z$ with $y \in \bar{\mathfrak{a}}$ and $z \in \bar{\mathfrak{m}}$. For $l \in \bar{\mathfrak{l}}$, we have $\alpha(l)(y + z) = [l,y + z] = [l,z]$. Hence $y = 0$, $x \in \bar{\mathfrak{m}}$, $\alpha|\bar{\mathfrak{l}}$ is a root of $\bar{\mathfrak{m}}$ with respect to $\bar{\mathfrak{l}}$, and $\bar{\mathfrak{g}}^{\alpha} = \bar{\mathfrak{m}}^{\alpha|\bar{\mathfrak{l}}}$. Conversely, let β be a root of $\bar{\mathfrak{m}}$ with respect to $\bar{\mathfrak{l}}$, x a non-zero element of $\bar{\mathfrak{m}}^{\beta}$ and α the linear form on $\bar{\mathfrak{h}}$ which extends β and is zero on $\bar{\mathfrak{a}}$. For $a \in \bar{\mathfrak{a}}$ and $l \in \bar{\mathfrak{l}}$, we have $[a + l,x] = [l,x] = \beta(l)x = \alpha(a + l)x$, hence $\alpha \in R'$. We have thus proved (i) and (ii), and (iii) follows from (i) and (ii).

1.13.10. Let $(\mathfrak{g},\theta)$ be a symmetric semi-simple Lie algebra. A Cartan subspace $\mathfrak{a}$ of $(\mathfrak{g},\theta)$ is said to be *splitting* if, for all $x \in \mathfrak{a}$, $\mathrm{ad}_{\mathfrak{g}} x$ is triangularizable (hence diagonalizable). Then, with the notation of 1.13.8, $\alpha|\mathfrak{a}$ takes its values in k for all $\alpha \in R$.

1.13.11. PROPOSITION. *We retain the notation of* 1.13.8, *and assume that* $\mathfrak{a}$ *is splitting. Let* $\bar{\mathfrak{n}} = \sum_{\alpha \in R''_+} \bar{\mathfrak{g}}^{\alpha}$.

(i) *There exists one and only one vector subspace* $\mathfrak{n}$ *of* $\mathfrak{g}$ *such that* $\bar{\mathfrak{n}} = \mathfrak{n} \otimes \bar{k}$.

(ii) *The subspace* $\mathfrak{n}$ *is a nilpotent Lie subalgebra of* $\mathfrak{g}$.

(iii) $\mathfrak{g} = \mathfrak{k} \oplus \mathfrak{a} \oplus \mathfrak{n}$.

(iv) *The orthogonal subspace of* $\mathfrak{a}$ *in* $\mathfrak{g}$ *is* $\mathfrak{k} \oplus \mathfrak{n}$.

(v) *Let* $\mathfrak{q} = \mathfrak{m} \oplus \mathfrak{a} \oplus \mathfrak{n}$. *Then* $\mathfrak{q}$ *is a Lie subalgebra of* $\mathfrak{g}$ *which has* $\mathfrak{n}$ *as an ideal.*

Let Γ be the Galois group of $\bar{k}$ over k. Then Γ operates on $\bar{\mathfrak{g}}$, on $\bar{\mathfrak{h}}$, and hence on $\bar{\mathfrak{h}}^*$ leaving R stable. Let $g \in \Gamma$ and $\alpha \in R$. Since $\mathfrak{a}$ is splitting, we have $(g\alpha - \alpha)|\mathfrak{a} = 0$. Therefore, from 1.13.8, $g(R''_+) = R''_+$, and consequently $g(\bar{\mathfrak{n}}) = \bar{\mathfrak{n}}$. This proves (i).

We have $(R''_+ + R''_+) \cap R \subset R''_+$ (1.13.8 (c)); and, if n is sufficiently large, the sum of any n elements of R''_+ belongs to the complement of R. Hence $\bar{\mathfrak{n}}$ is a nilpotent Lie subalgebra of $\bar{\mathfrak{g}}$. This proves (ii).

From 1.13.9, we have

$$\bar{\mathfrak{g}} = \bar{\mathfrak{a}} + \bar{\mathfrak{m}} + \sum_{\alpha \in R_+} \bar{\mathfrak{g}}^{\alpha} + \sum_{\alpha \in R''_-} \bar{\mathfrak{g}}^{\alpha} \subset \bar{\mathfrak{k}} + \bar{\mathfrak{a}} + \bar{\mathfrak{n}} + \sum_{\alpha \in R''_-} \bar{\mathfrak{g}}^{\alpha}.$$

On the other hand, if $\alpha \in R''_-$ and $u \in \bar{\mathfrak{g}}^{\alpha}$, it is obvious that

$$u = (u + \bar{\theta}u) - \bar{\theta}u \in \bar{\mathfrak{k}} + \bar{\mathfrak{g}}^{\bar{\theta}\alpha} \subset \bar{\mathfrak{k}} + \bar{\mathfrak{n}}.$$

Hence $\bar{\mathfrak{g}} = \bar{\mathfrak{k}} + \bar{\mathfrak{a}} + \bar{\mathfrak{n}}$. Let $x \in \bar{\mathfrak{k}}$, $y \in \bar{\mathfrak{a}}$, $z \in \bar{\mathfrak{n}}$, with $x + y + z = 0$. Then

$$0 = \bar{\theta}(x + y = z) = x - y + \bar{\theta}z,$$

thus

$$2y + z - \bar{\theta}z = 0.$$

But $\bar{\theta}(\bar{\mathfrak{n}}) = \sum_{\alpha \in R''_-} \bar{\mathfrak{g}}^\alpha$, hence the sum $\bar{\mathfrak{a}} + \bar{\mathfrak{n}} + \bar{\theta}(\bar{\mathfrak{n}})$ is direct, so that $y = z = 0$. Then $x = 0$. Thus $\bar{\mathfrak{g}} = \bar{\mathfrak{k}} \oplus \bar{\mathfrak{a}} \oplus \bar{\mathfrak{n}}$, which proves (iii).

From 1.10.2 (iii), $\mathfrak{a}$ is orthogonal to $\mathfrak{n}$. On the other hand, $\mathfrak{a} \subset \mathfrak{p}$. Hence $\mathfrak{a}$ is orthogonal to $\mathfrak{k} + \mathfrak{n}$. Taking (iii) into account, this proves (iv).

It is obvious that $[\bar{\mathfrak{a}} + \bar{\mathfrak{l}}, \bar{\mathfrak{n}}] \subset \bar{\mathfrak{n}}$. Let $\alpha \in R''_+$ and $\beta \in R'$. If $\alpha + \beta \in R$, then $\alpha + \beta \in R''_+$ from 1.13.8, hence $[\bar{\mathfrak{g}}^\alpha, \bar{\mathfrak{g}}^\beta] \subset \bar{\mathfrak{g}}^{\alpha+\beta} \subset \bar{\mathfrak{n}}$. Taking 1.13.9 into account, this proves that $[\bar{\mathfrak{m}}, \bar{\mathfrak{n}}] \subset \bar{\mathfrak{n}}$, whence $[\bar{\mathfrak{m}} + \bar{\mathfrak{a}}, \bar{\mathfrak{n}}] \subset \bar{\mathfrak{n}}$.

1.13.12. With the notation of 1.13.11, $\mathfrak{g} = \mathfrak{k} \oplus \mathfrak{a} \oplus \mathfrak{n}$ is termed the *Iwasawa decomposition* of $\mathfrak{g}$ defined by $\mathfrak{k}$, $\mathfrak{a}$, $\mathfrak{h}$, B.

1.13.13. From 1.13.3 and CH, p. 172, $\mathrm{ad}_{\mathfrak{g}}\mathfrak{k}$ is the Lie algebra of an irreducible algebraic subgroup K of the adjoint group of $\mathfrak{g}$. Since $[\mathfrak{k}, \mathfrak{p}] \subset \mathfrak{p}$, we have $K(\mathfrak{p}) \subset \mathfrak{p}$. We then have:

PROPOSITION. *Let* $\mathfrak{a}$ *be a Cartan subspace of* $\mathfrak{p}$. *Then* $K(\mathfrak{a})$ *is dense in* $\mathfrak{p}$.

We use the notation of 1.13.8. Let $y \in \mathfrak{a}$ be such that no root of R'' is zero at y. Then $\mathfrak{g}^0(y) = \mathfrak{a} \oplus \mathfrak{m}$, hence $(\mathrm{ad}_{\mathfrak{g}} y)^2(\mathfrak{g})$ is complementary to $\mathfrak{a} \oplus \mathfrak{m}$ in $\mathfrak{g}$, and $(\mathrm{ad}_{\mathfrak{g}} y)^2(\mathfrak{p})$ is complementary to $\mathfrak{a}$ in $\mathfrak{p}$. But $\mathfrak{k} \supset (\mathrm{ad}_{\mathfrak{g}} y)(\mathfrak{p})$, hence

$$\mathfrak{p} = \mathfrak{a} + [\mathfrak{k}, y].$$

Let f be the mapping $(k,a) \mapsto ka$ of $K \times \mathfrak{a}$ into $\mathfrak{p}$. Let T be the linear mapping which is tangent to f at $(1,y)$. Then $f(1,a) = a$, hence $T(0 \times \mathfrak{a}) = \mathfrak{a}$, and $f(k,y) = ky$, hence $T(\mathrm{ad}_{\mathfrak{g}}\mathfrak{k} \times 0) = [\mathfrak{k}, y]$. Equation (1) proves that T is surjective, and the proposition follows.

1.13.14. Let $\mathfrak{v}$ be a semi-simple Lie algebra. Let $\mathfrak{g} = \mathfrak{v} \times \mathfrak{v}$. Let θ be the automorphism $(x,y) \mapsto (y,x)$ of $\mathfrak{v} \times \mathfrak{v}$. $(\mathfrak{g}, \theta)$ is termed the *diagonal symmetric Lie algebra* defined by $\mathfrak{v}$ (cf. 1.14.15). Let $\mathfrak{k}$ be the set of the (x,x) for $x \in \mathfrak{v}$, $\mathfrak{p}$ the set of $(x,-x)$ for $x \in \mathfrak{v}$; then $\mathfrak{g} = \mathfrak{k} \oplus \mathfrak{p}$ is the symmetric decomposition of $\mathfrak{g}$ defined by θ.

Let $\mathfrak{w}$ be a Cartan subalgebra of $\mathfrak{v}$. The set $\mathfrak{a}$ of the $(x,-x)$ for $x \in \mathfrak{w}$ is a Cartan subspace of $\mathfrak{p}$. The centralizer $\mathfrak{m}$ of $\mathfrak{a}$ in $\mathfrak{k}$ is the set of the (x,x) for $x \in \mathfrak{w}$. Let $\mathfrak{h} = \mathfrak{a} \oplus \mathfrak{m} = \mathfrak{w} \times \mathfrak{w}$; it is a Cartan subalgebra of $\mathfrak{g}$. We identity $\mathfrak{h}^*$ with $\mathfrak{w}^* \times \mathfrak{w}^*$.

Let us assume that $\mathfrak{w}$ is splitting. Then $\mathfrak{a}$ and $\mathfrak{h}$ are splitting. Let $S = R(\mathfrak{v}, \mathfrak{w})$ and $R = R(\mathfrak{g}, \mathfrak{h})$. Then $R = (S \times 0) \cup (0 \times S)$. Consequently, with the notation of 1.13.8, R' is empty.

Let C be a basis for S. Then $B = (C\times 0) \cup (0\times(-C))$ is a basis for R which possesses the properties of 1.13.8. We define S_+, S_- and R_+, R_- in the obvious way. We have $R_+ = (S_+\times 0) \cup (0\times(-S_+))$. Let $\mathfrak{v} = \mathfrak{w} \oplus \mathfrak{x} \oplus \mathfrak{x}_-$. be the triangular decomposition of $\mathfrak{v}$ defined by C. Let $\mathfrak{n} = \mathfrak{x}\times\mathfrak{x}_-$ and $\mathfrak{n}_- = \mathfrak{x}_-\times\mathfrak{x}$. Then $\mathfrak{g} = \mathfrak{h} \oplus \mathfrak{n} \oplus \mathfrak{n}_-$ is the triangular decomposition of $\mathfrak{g}$ defined by B, and $\mathfrak{g} = \mathfrak{k} \oplus \mathfrak{a} \oplus \mathfrak{n}$ is the Iwasawa decomposition of $\mathfrak{g}$ defined by $\mathfrak{k}$, $\mathfrak{a}$, $\mathfrak{h}$, B.

1.14. Supplementary remarks

1.14.1. Sections 1.1 to 1.10 restate what can be found in a large number of books and articles; cf. for example [16], [71], [117] and [118]. Sections 1.11 and 1.12 expound recently published notions; cf. [4], [9], [18], [30], [31], [36], [46], [47], [49] and [124]. Their chief interest for us will become apparent during the construction of simple induced representations. Section 1.13 is standard (cf. for example [80]) when $k = \mathbf{R}$ and $\mathfrak{g} = \mathfrak{k} \oplus \mathfrak{p}$ is a Cartan decomposition, that is, a symmetric decomposition such that the restrictions of the Killing form to $\mathfrak{k}$ or $\mathfrak{p}$ are negative or positive, respectively. Above all, we shall make use of the symmetric decompositions for k algebraically closed. If $k = \mathbf{C}$, it follows easily from 1.13.7 that the symmetric decompositions of $\mathfrak{g}$ may be obtained in the following way: we select a real form $\mathfrak{g}_{\mathbf{R}}$ of $\mathfrak{g}$, a Cartan decomposition $\mathfrak{g}_{\mathbf{R}} = \mathfrak{k}_{\mathbf{R}} \oplus \mathfrak{p}_{\mathbf{R}}$ of $\mathfrak{g}_{\mathbf{R}}$, and we take for $\mathfrak{k}, \mathfrak{p}$ the complexifications of $\mathfrak{k}_{\mathbf{R}}$, $\mathfrak{p}_{\mathbf{R}}$.

The elements termed "generic" in 1.9.8 were for a long time termed "regular". In the last ten years, the expression "regular element" has acquired a new sense in the semi-simple case (cf. 1.11.13), so it seemed preferable to alter the terminology. The terms "index", "symmetric decomposition", "symmetrizing Lie subalgebra" and "diagonal" are new.

The following supplementary remarks (except 1.14.14 and 1.14.15) refer solely to sections 1.11 and 1.12.

Proposition 1.11.7 is due to Duflo and Vergne ([43], [49]). The notion of polarization appeared for the first time in Kirillov's thesis [75]; it has been much developed subsequently, notably by Auslander and Kostant [4]. When $k = \mathbf{R}$, there are numerous interesting results which are not discussed in this book; a general account of most of them can be found in [9], ch. IV. Proposition 1.12.10 is due to Vergne [124], and theorem 1.12.14 to Duflo [47]. The principle behind the proof in 1.11.10 is taken from [16], ch. VII.

1.14.2. Let $f \in \mathfrak{g}^*$, and V be the orthogonal subspace of $\mathfrak{g}^f$ in $\mathfrak{g}^*$, so that V can be canonically identified with the dual of $\mathfrak{g}/\mathfrak{g}^f$. The mapping $x \mapsto {}^t(\mathrm{ad}\, x)f$ of $\mathfrak{g}$ into $\mathfrak{g}^*$ is a $\mathfrak{g}^f$-homomorphism, and has $\mathfrak{g}^f$ as its kernel and V as its image, whence we have a $\mathfrak{g}^f$-isomorphism φ of $\mathfrak{g}/\mathfrak{g}^f$ onto V. Let B be the non-degenerate alternating bilinear form on $\mathfrak{g}/\mathfrak{g}^f$ deduced from B_f by passage to the quotient. Then $B(\xi,\eta) = \langle\varphi\xi,\eta\rangle$ for $\xi,\eta \in \mathfrak{g}/\mathfrak{g}^f$. The forms B_f and B are $\mathfrak{g}^f$-invariant.

1.14.3. Let V be a vector subspace of $\mathfrak{g}$, and $f \in V^*$. Let $g \in \mathfrak{g}^*$ such that $g|V = f$ and such that $\dim \mathfrak{g}^g \leq \dim \mathfrak{g}^{g'}$ for all $g' \in \mathfrak{g}^*$ such that $g'|V = f$. Then $[\mathfrak{g}^g,\mathfrak{g}^g] \subset V$. (The proof, attributed to Carmona, occurs in [47]; this proof has been expounded for a special case in 1.11.7.)

1.14.4. We use the notation of 1.12.10 and 1.12.11. Let $f' \in \mathfrak{g}^*$ be such that $f'|\mathfrak{p}(f,s) = 0$. Then $\mathfrak{p}(f+f',s) = \mathfrak{p}(f,s)$ ([9], p. 75).

1.14.5. Adopt the assumptions of 1.12.14, and, in addition, take a decreasing sequence of ideals $\mathfrak{a} = \mathfrak{a}_0 \supset \mathfrak{a}_1 \supset \cdots \supset \mathfrak{a}_p$ of $\mathfrak{g}$. There exists $\mathfrak{h} \in \mathrm{PR}(g)$ such that $\mathfrak{h} \cap \mathfrak{a}_i \in \mathrm{P}(g|\mathfrak{a}_i)$ for all i. (Duflo, unpublished).

1.14.6. Let $\mathfrak{g}$ be the four-dimensional nilpotent Lie algebra with a basis (x,y,z,t) such that $[x,y] = z, [x,z] = t$, $[y,z] = 0$, $[\mathfrak{g},t] = 0$. Let $f \in \mathfrak{g}^*$ be such that $f(x) = f(y) = f(z) = 0$ and $f(t) = 1$. Then $\mathfrak{h} = kx + kt$ is maximal in the set of Lie subalgebras subordinate to f, but $\mathfrak{h} \notin \mathrm{P}(f)$. (In this example, we have $\mathfrak{h} \not\supset \mathfrak{g}^f$. By making the example more complex, we can arrive at $\mathfrak{h} \supset \mathfrak{g}^f$).

1.14.7. (a) Let $\mathfrak{g}$ be the four-dimensional real solvable Lie algebra with a basis (x, y, z, t) such that $[x,y] = z$, $[x,z] = -y$, $[y,z] = t$, $[\mathfrak{g},t] = 0$. Let $f \in \mathfrak{g}^*$ be such that $f(t) \neq 0$. There exists no polarization of $\mathfrak{g}$ at f.

(b) Let $\mathfrak{g}$ be the three-dimensional real semi-simple Lie algebra with a basis (x, y, z) such that $[x,y] = z$, $[y,z] = x$, $[z,x] = y$. Let $f \in \mathfrak{g}^* - \{0\}$. There exists no polarization of $\mathfrak{g}$ at f.

1.14.8. Assume $\mathfrak{g}$ to be real and solvable.

(a) If no Lie subalgebra of $\mathfrak{g}$ has a quotient isomorphic to the algebra of 1.14.7 (a), then, for all $f \in \mathfrak{g}^*$, we have $\mathrm{P}(f) \neq \emptyset$.

(b) $\mathfrak{g}$ is said to be *exponential* if, for all $x \in \mathfrak{g}$, ad x has no non-zero purely imaginary eigenvalue. A completely solvable Lie algebra is exponential, but the converse is false. Let us assume $\mathfrak{g}$ to be exponential; then the condition in (a) is satisfied; hence, for all $f \in \mathfrak{g}^*$, we have $\mathrm{P}(f) \neq \emptyset$ ([9], pp. 83, 87).

1.14.9. Let $(\mathfrak{g},\mathfrak{h})$ be a splitting simple Lie algebra of type B_2 and (α,β) a basis of $R(\mathfrak{g},\mathfrak{h})$ such that the positive roots are α,β, $\alpha+\beta$, $\alpha+2\beta$. Let $x \in \mathfrak{g}^{\alpha+2\beta} - \{0\}$. Let $f \in \mathfrak{g}^*$ be the image of x under the Killing isomorphism. Then $\mathrm{P}(f) = \emptyset$. (Steinberg; cf. [9], p. 63.)

1.14.10. Let $\mathfrak{g}$ be the three-dimensional nilpotent Lie algebra with a basis (x, y, z) such that $[x,y] = z$, $[\mathfrak{g},z] = 0$. Let e, f, h be as in 1.8.1. Let $\mathfrak{h}$ be the semi-direct product of $\mathfrak{sl}(2,k)$ and $\mathfrak{g}$ such that the adjoint action of $\mathfrak{sl}(2,k)$ in $\mathfrak{g}$ annihilates z and induces the identity representation of $\mathfrak{sl}(2,k)$ in $kx + ky = k^2$. Let A (or B) be the set of points of $\mathfrak{h}^*$ at which the elements $x^2 - 2ez$, $xy + hz$, $y^2 + 2fz$ (or x, y, z) of $S(\mathfrak{h})$ are zero. Let $g \in \mathfrak{h}^*$. If $g \notin A \cup B$ (or $g \in B - \{0\}$), there exists a solvable polarization of dimension 4 (or 5) at g. If $g \in A - (A \cap B)$, then $\mathrm{P}(g) = \emptyset$.

1.14.11. Let $\mathfrak{g}$ be the semi-direct product Lie algebra of $\mathfrak{sl}(2,k)$ and k^2 corresponding to the identity representation of $\mathfrak{sl}(2,k)$. We denote by (e,f,h) the basis for $\mathfrak{sl}(2,k)$ used in 1.8.1 and by (u,v) the canonical basis for k^2, so that (e,f,h,u,v) is a basis for $\mathfrak{g}$.

Let $\varphi \in \mathfrak{g}^*$ be such that $\varphi(e) = \varphi(h) = \varphi(v) = 0$ and $\varphi(f) = \varphi(u) = 1$. Then $\mathfrak{g}^\varphi = k(f + 2u)$. The only polarization of $\mathfrak{g}$ at φ is $kf + ku + kv$, which is nilpotent.

Let $\psi \in \mathfrak{g}^*$ be such that $\psi(e) = \psi(h) = \psi(u) = 0$ and $\psi(f) = \psi(u) = 1$. Then $\mathfrak{g}^\psi = ke$. Some polarizations of $\mathfrak{g}$ at ψ are $\mathfrak{h}_1 = ke + k(h-u) + k(f-v)$, which is semi-simple, $\mathfrak{h}_2 = ke + kh + ku$, which is solvable and non-nilpotent, and $\mathfrak{h}_3 = ke + ku + kv$, which is nilpotent. Let $\mathscr{A}$ be the adjoint group of $\mathfrak{g}$ and $\mathfrak{h}_1^\perp, \mathfrak{h}_2^\perp, \mathfrak{h}_3^\perp$ the orthogonal subspaces of $\mathfrak{h}_1, \mathfrak{h}_2, \mathfrak{h}_3$ in $\mathfrak{g}^*$ respectively. Then $\psi + \mathfrak{h}_i^\perp$ is contained in $\mathscr{A} \cdot \psi$ for $i = 3$, but not for $i = 1$ or $i = 2$.

1.14.12. If $\mathrm{tr}(\mathrm{ad}\, x) = 0$ for all $x \in \mathfrak{g}$, then $\mathfrak{g}^f \neq 0$ for all $f \in \mathfrak{g}^*$ (M. Vergne, unpublished).

1.14.13. Let $(x_1, x_2, \ldots, x_n)$ be a basis for $\mathfrak{g}$. Let us consider $([x_i,x_j])_{1 \leq i,j \leq n}$ as a matrix over the ring $S(\mathfrak{g})$, and let p be its rank. Then the index of $\mathfrak{g}$ is $n - p$.

1.14.14. We use the notation of 1.13.8.

(a) If $\alpha \in R$, then $\alpha + \bar{\theta}\alpha \notin R$. (Let $X \in \mathfrak{g}^\alpha - \{0\}$, whence $\bar{\theta}X \in \bar{\mathfrak{g}}^{\bar{\theta}\alpha} - \{0\}$. If $\beta = \alpha + \bar{\theta}\alpha \in R$, then $[X,\bar{\theta}X] \in \mathfrak{g}^\beta - \{0\}$. Now $\beta(\mathfrak{a}) = 0$, so that $\bar{\mathfrak{g}}^\beta \subset \mathfrak{k}$

(1.13.9), hence $[X,\bar{\theta}X] = \bar{\theta}\ ([X,\bar{\theta}X]) = [\bar{\theta}X,X]$, and $[X,\bar{\theta}X] = 0$, which is a contradiction.)

(b) The theory of root systems enables us to deduce from (a) that the set of restrictions to $\bar{\mathfrak{a}}$ of the elements of R'' is a root system S in $\bar{\mathfrak{a}}^*$ (cf. [1]). Let $\lambda \in \bar{\mathfrak{a}}^* - \{0\}$. The following are equivalent:

(i) $\lambda \in S$;

(ii) there exists a non-zero element X of $\bar{\mathfrak{g}}$ such that $[H,X] = \lambda(H)X$ for all $H \in \bar{\mathfrak{a}}$ [88].

The Weyl group of S is termed the Weyl group of $(\mathfrak{g}, \theta, \mathfrak{a})$.

1.14.15. Let $\mathfrak{v}$ be a complex semi-simple Lie algebra, $\mathfrak{v}_0$ the real semi-simple Lie algebra deduced from $\mathfrak{v}$ by restriction of the base field to **R**, $\mathfrak{c}_0$ a compact form of $\mathfrak{v}$, and σ the conjugation of $\mathfrak{v}$ with respect to $\mathfrak{c}_0$. Let $\mathfrak{g} = \mathfrak{v} \times \mathfrak{v}$, let $\mathfrak{k}$ be the set of (x,x) for $x \in \mathfrak{v}$, and let i be the mapping $x \mapsto (\sigma(x),x)$ of $\mathfrak{v}$ into $\mathfrak{g}$. Then $\mathfrak{g}$ is the complexification of $i(\mathfrak{v}_0)$ and $\mathfrak{k}$ is the complexification of $i(\mathfrak{c}_0)$.

1.14.16. There exists a complex Lie algebra $\mathfrak{g}$ with the following property: for every $f \in \mathfrak{g}^*$, and every ball V centered at f, there exists $f' \in V$ such that $\mathfrak{g}^f$ and $\mathfrak{g}^{f'}$ are not conjugate by Aut $\mathfrak{g}$ [163].

1.14.17. Let $\mathfrak{g}$ be a complex semi-simple Lie algebra, $f \in \mathfrak{g}^*$ and $\mathfrak{p}$ a polarization of $\mathfrak{g}$ at f. Then $\mathfrak{p}$ contains a Borel subalgebra of $\mathfrak{g}$ [175]. This remains true for k algebraically closed (P. Tauvel; see [200]).

CHAPTER 2

ENVELOPING ALGEBRAS

2.1. The Poincaré—Birkhoff—Witt theorem

2.1.1. Let T be the tensor algebra of the vector space $\mathfrak{g}$. We recall that

$$T = T^0 \oplus T^1 \oplus \cdots \oplus T^n \oplus \cdots,$$

where $T^n = \mathfrak{g} \oplus \mathfrak{g} \otimes \cdots \otimes \mathfrak{g}$ (n times); in particular, $T^0 = k \cdot 1$ and $T^1 = \mathfrak{g}$; the product in T is simply tensor multiplication.

Let J be the two-sided ideal of T generated by the tensors

$$x \otimes y - y \otimes x - [x,y],$$

where $x,y \in \mathfrak{g}$. The associative algebra T/J is termed the *enveloping algebra of* $\mathfrak{g}$ (or occasionally the *universal* enveloping algebra of $\mathfrak{g}$) and is denoted by $U(\mathfrak{g})$. The composite mapping σ of the canonical mappings $\mathfrak{g} \to T \to U(\mathfrak{g})$ is termed the canonical mapping of $\mathfrak{g}$ into $U(\mathfrak{g})$; for all values of $x,y \in \mathfrak{g}$, we have

$$\sigma(x)\sigma(y) - \sigma(y)\sigma(x) = \sigma([x,y]).$$

The centre of $U(\mathfrak{g})$ is denoted by $Z(\mathfrak{g})$.

If $\mathfrak{g}$ is commutative, then $U(\mathfrak{g}) = S(\mathfrak{g})$ (the symmetric algebra of the vector space $\mathfrak{g}$).

2.1.2. Let $T_+ = T^1 \oplus T^2 \oplus \cdots$, which is a two-sided ideal of T. We denote the canonical image of T_+ in $U(\mathfrak{g})$ by $U_+(\mathfrak{g})$. Then

$$T = T^0 \oplus T_+ = k \cdot 1 \oplus T_+, \qquad J \subset T_+;$$

hence, if U^0 represents the canonical image of T^0 in $U(\mathfrak{g})$, we have $U(\mathfrak{g}) = U^0 \oplus U_+(\mathfrak{g})$, and $U^0 = k \cdot 1$ is one-dimensional; we often identify U^0 with k. If $u \in U(\mathfrak{g})$, its component in $U^C = k$ is called the *constant term* of u. The associative algebra $U(\mathfrak{g})$ is generated by 1 and the canonical image of $\mathfrak{g}$ in $U(\mathfrak{g})$.

2.1.3. LEMMA. *Let* σ *be the canonical mapping of* $\mathfrak{g}$ *into* $U(\mathfrak{g})$, *let* A *be an algebra with unity, and let* τ *be a linear mapping of* $\mathfrak{g}$ *into* A *such that*

$$\tau(x)\tau(y) - \tau(y)\tau(x) = \tau([x,y])$$

for all $x,y \in \mathfrak{g}$. *There exists one and only one homomorphism* τ' *of* $U(\mathfrak{g})$ *into* A *such that* $\tau'(1) = 1$ *and* $\tau' \circ \sigma = \tau$.

Since the algebra $U(\mathfrak{g})$ is generated by 1 and $\sigma(\mathfrak{g})$, τ' is unique. On the other hand, let φ be the unique homomorphism of T into A which extends τ and such that $\varphi(1) = 1$. For $x,y \in \mathfrak{g}$, we have

$$\varphi(x \otimes y - y \otimes x - [x,y]) = \tau(x)\tau(y) - \tau(y)\tau(x) - \tau([x,y]) = 0;$$

hence $\varphi(J) = 0$ and, by passage to the quotient, φ defines a homomorphism τ' of $U(\mathfrak{g})$ into A such that $\tau'(1) = 1$ and $\tau' \circ \sigma = \tau$.

2.1.4. Up to Section 2.1.8, we fix a basis $(x_1, \ldots, x_n)$ for $\mathfrak{g}$. We denote the canonical image of x_i in $U(\mathfrak{g})$ by y_i. For every finite sequence $I = (i_1, \ldots, i_p)$ of integers between 1 and n, we set $y_I = y_{i_1} y_{i_2} \cdots y_{i_p} \in U(\mathfrak{g})$. If i is an integer, we write $i \leqq I$ when $i \leqq i_1, \ldots, i \leqq i_p$. We denote the canonical image in $U(\mathfrak{g})$ of $T^0 + T^1 + \cdots + T^q$ by $U_q(\mathfrak{g})$.

2.1.5. LEMMA. *Let* $a_1, \ldots, a_p \in \mathfrak{g}$, σ *the canonical mapping of* $\mathfrak{g}$ *into* $U(\mathfrak{g})$, *and* π *be a permutation of* $\{1, \ldots, p\}$. *Then*

$$\sigma(a_1) \cdots \sigma(a_p) - \sigma(a_{\pi(1)}) \cdots \sigma(a_{\pi(p)}) \in U_{p-1}(\mathfrak{g}).$$

It is sufficient to prove this when π is the transposition of j and $j+1$. In this case, the lemma follows from the equality

$$\sigma(a_j)\sigma(a_{j+1}) - \sigma(a_{j+1})\sigma(a_j) = \sigma([a_j,a_{j+1}]).$$

2.1.6. LEMMA. *The* y_I, *for all increasing sequences* I *of length* $\leqq p$, *generate the vector space* $U_p(\mathfrak{g})$.

It is obvious that the vector space $U_p(\mathfrak{g})$ is generated by the y_I, for all sequences I of length $\leqq p$. It is then sufficient to apply 2.1.5.

2.1.7. Let P be the algebra $k[z_1, \ldots, z_n]$ of polynomials in n indeterminates $z_1, \ldots, z_n$. For every integer $i \geqq 0$, let P_i be the set of elements of P of degree $\leqq i$. If $I = (i_1, \ldots, i_p)$ is a sequence of integers between 1 and n, we set $z_I = z_{i_1} z_{i_2} \cdots z_{i_p}$. The following lemma will become natural when we have proved 2.1.11 and defined (in 2.2.21) the left regular representation.

LEMMA. *For every integer* $p \geqq 0$, *there exists a unique linear mapping* f_p *of the vector space* $\mathfrak{g} \otimes P_p$ *into* P *which satisfies the following conditions:*

(A_p) $f_p(x_i \otimes z_I) = z_i z_I$ *for* $i \leqq I$, $z_I \in P_p$;

(B_p) $f_p(x_i \otimes z_I) - z_i z_I \in P_q$ *for* $z_I \in P_q$, $q \leqq p$;

(C_p) $f_p(x_i \otimes f_p(x_j \otimes z_J)) = f_p(x_j \otimes f_p(x_i \otimes z_J)) + f_p([x_i,x_j] \otimes z_J)$ *for* $z_J \in P_{p-1}$. [*The terms in* (C_p) *are meaningful by virtue of* (B_p)].

Moreover, the restriction of f_p *to* $\mathfrak{g} \otimes P_{p-1}$ *is* f_{p-1}.

For $p = 0$, the condition (A_0) imposes $f_0(x_i \otimes 1) = z_i \otimes 1$, and the conditions ($B_0$) and ($C_0$) are then satisfied. Let us assume the existence and uniqueness of f_{p-1}. If f_p exists, then $f_p|\mathfrak{g} \otimes P_{p-1}$ satisfies (A_{p-1}), (B_{p-1}), (C_{p-1}) and is hence equal to f_{p-1}. Thus everything depends on proving that f_{p-1} has one and only one linear extension f_p to $\mathfrak{g} \otimes P_p$ which satisfies (A_p), (B_p), (C_p).

We must define $f_p(x_i \otimes z_I)$ for an increasing sequence I of p elements. If $i \leqq I$, the choice is dictated by (A_p). Otherwise, I can be written as (j,J) with $j < i$, $j \leqq J$. Then

$$f_p(x_i \otimes z_I) = f_p(x_i \otimes f_{p-1}(x_j \otimes z_J)) \qquad \text{from } (A_{p-1})$$

$$= f_p(x_j \otimes f_{p-1}(x_i \otimes z_J)) + f_{p-1}([x_i,x_j] \otimes z_J) \qquad \text{from } (C_p).$$

Now $f_{p-1}(x_i \otimes z_J) = z_i z_J + w$, with $w \in P_{p-1}$ from (B_{p-1}). Hence

$$f_p(x_j \otimes f_{p-1}(x_i \otimes z_J)) = z_j z_i z_J + f_{p-1}(x_j \otimes w) \qquad \text{from } (A_p)$$

$$= z_i z_I + f_{p-1}(x_j \otimes w).$$

The above defines in a unique way a linear extension f_p of f_{p-1} to $\mathfrak{g} \otimes P_p$, and this extension satisfies (A_p) and (B_p). It remains for us to prove that f_p, when defined in this way, satisfies (C_p).

The condition (C_p) is satisfied by construction if $j \leqq i$ and $j \leqq J$. Since $[x_j,x_i] = -[x_i,x_j]$, it is also satisfied if $i < j$ and $i \leqq J$. Since (C_p) is trivially satisfied if $j = i$, it can be seen that (C_p) is satisfied if $i \leqq J$ or $j \leqq J$. Otherwise, $J = (k,K)$, where $k \leqq K$, $k < i$, $k < j$. Henceforth writing $f_p(x \otimes z) = xz$ for $x \in \mathfrak{g}$ and $z \in P_p$ for brevity, we have, from the induction hypothesis,

$$x_j z_J = x_j(x_k z_K) = x_k(x_j z_K) + [x_j,x_k] z_K. \tag{1}$$

Now $x_j z_K$ is of the form $z_j z_K + w$, where $w \in P_{p-2}$. We can apply (C_p) to $x_i(x_k(z_j z_K))$ because $k \leqq K$ and $k < j$, and to $x_i(x_k w)$ from the induction hypothesis, hence to $x_i(x_k(x_j z_K))$; taking (1) into account, this yields

$$x_i(x_j z_J) = x_k(x_i(x_j z_K)) + [x_i,x_k](x_j z_K) + [x_j,x_k](x_i z_K) + [x_i,[x_j,x_k]] z_K.$$

Interchanging i and j, and cancelling term by term,

$$\begin{aligned} x_i(x_j z_J) - x_j(x_i z_J) &= x_k(x_i(x_j z_K) - x_j(x_i z_K)) + [x_i,[x_j,x_k]]z_K - [x_j,[x_i,x_k]]z_K \\ &= x_k([x_i,x_j]z_K) + (x_i,[x_j,x_k]]z_K + [x_j,[x_k,x_i]]z_K \\ &= [x_i,x_j]x_k z_K + [x_k,[x_i,x_j]]z_K + [x_i,[x_j,x_k]]z_K + [x_j,[x_k,x_i]]z_K \\ &= [x_i,x_j]x_k z_K = [x_i,x_j]z_J. \end{aligned}$$

2.1.8. Lemma. *The y_I, for every increasing sequence I, form a basis for the vector space $U(\mathfrak{g})$.*

From 2.1.7, whose notation we shall use, there exists a bilinear mapping f of $\mathfrak{g} \times P$ into P such that $f(x_i,z_I) = z_i z_I$ for $i \leqq I$ and

$$f(x_i,f(x_j,z_J)) = f(x_j,f(x_i,z_J)) + f([x_i,x_j],z_J),$$

for all i, j, J. In other words, there exists a representation ϱ of $\mathfrak{g}$ in P such that $\varrho(x_i)z_I = z_i z_I$ for $i \leqq I$. From 2.1.3, there exists a homomorphism φ of $U(\mathfrak{g})$ into End(P) such that $\varphi(y_i)z_I = z_i z_I$ for $i \leqq I$. From this we deduce step by step that, if $i_1 \leqq i_2 \leqq \cdots \leqq i_p$, we have

$$\varphi(y_{i_1}y_{i_2}\cdots y_{i_p})\cdot 1 = z_{i_1}z_{i_2}\cdots z_{i_p}.$$

Hence, the y_I, for I increasing, are linearly independent. From 2.1.6, they generate $U(\mathfrak{g})$.

2.1.9. Proposition. *The canonical mapping of $\mathfrak{g}$ into $U(\mathfrak{g})$ is injective.*

This follows immediately from 2.1.8.

2.1.10. Henceforth we shall identify every element of $\mathfrak{g}$ with its canonical image in $U(\mathfrak{g})$. Hence $\mathfrak{g}$ is embedded in $U(\mathfrak{g})$ and *we need not concern ourselves further with the canonical mapping of $\mathfrak{g}$ into $U(\mathfrak{g})$.*

2.1.11. Theorem. *Let $(x_1, \ldots, x_n)$ be a basis for the vector space $\mathfrak{g}$. Then the $x_1^{\nu_1}x_2^{\nu_2}\cdots x_n^{\nu_n}$, where $\nu_1, \ldots, \nu_n \in \mathbf{N}$, form a basis for $U(\mathfrak{g})$.*

This follows immediately from 2.1.8.

2.1.12. If, contrary to our general conventions, k is a commutative ring, and $\mathfrak{g}$ is a Lie algebra over k in an obvious sense, with as a basis $(x_1, \ldots, x_n)$ a k-module, then what was stated in 2.1, and in particular 2.1.11, remains valid.

References: [12], [16], [71], [132].

2.2. The functor U

2.2.1. Taking 2.1.10 into account, Lemma 2.1.3 can be restated as follows:

PROPOSITION. *Let A be an algebra with unity, τ a linear mapping of $\mathfrak{g}$ into A such that $\tau(x)\tau(y) - \tau(y)\tau(x) = \tau([x,y])$ for all $x,y \in \mathfrak{g}$. Then τ can be uniquely extended to a homomorphism of $U(\mathfrak{g})$ into A which transforms* 1 *into* 1.

2.2.2. COROLLARY. *Let V be a vector space, and $\mathscr{R}$ and $\mathscr{R}'$ the sets of representations of $\mathfrak{g}$ and $U(\mathfrak{g})$ in V respectively. For all $\varrho \in \mathscr{R}$, there exists one and only one $\varrho' \in \mathscr{R}'$ which extends ϱ, and the mapping $\varrho \mapsto \varrho'$ is a bijection of $\mathscr{R}$ onto $\mathscr{R}'$.*

2.2.3. With the notation of 2.2.2, the vector subspaces of V which are stable under ϱ are the vector subspaces of V which are stable under ϱ'. If ϱ_1 and ϱ_2 are representations of $\mathfrak{g}$, then ϱ_1 is equivalent to ϱ_2 if and only if ϱ_1' is equivalent to ϱ_2'. In short, the study of ϱ amounts to that of ϱ', and, subsequently, we shall often use the same symbol to denote a representation of $\mathfrak{g}$ and the corresponding representation of $U(\mathfrak{g})$. This will become ambiguous when we are discussing the kernel of the representation, but in that case, unless otherwise indicated, we shall be referring to *the kernel in* $U(\mathfrak{g})$.

2.2.4. PROPOSITION. *Let $\mathfrak{g}$ and $\mathfrak{g}'$ be Lie algebras and φ a homomorphism of $\mathfrak{g}$ into $\mathfrak{g}'$. There exists one and only one homomorphism ψ of $U(\mathfrak{g})$ into $U(\mathfrak{g}')$ which is an extension of φ and such that $\psi(1) = 1$.*

This follows from 2.2.1.

2.2.5. With the notation of 2.2.4, we set $\psi = U(\varphi)$. If $\varphi' : \mathfrak{g}' \to \mathfrak{g}''$ is a Lie algebra homomorphism, then

$$U(\varphi' \circ \varphi) = U(\varphi') \circ U(\varphi).$$

2.2.6. Let $\mathfrak{g}'$ be a Lie subalgebra of $\mathfrak{g}$, and i the canonical injection of $\mathfrak{g}'$ into $\mathfrak{g}$. Then the homomorphism $U(i)$ of $U(\mathfrak{g}')$ into $U(\mathfrak{g})$ *is injective*. Indeed, let $(x_1, \ldots, x_n)$ be a basis for $\mathfrak{g}$ such that $(x_1, \ldots, x_p)$ is a basis for $\mathfrak{g}'$; the $x_1^{\nu_1} \cdots x_p^{\nu_p}$ calculated in $U(\mathfrak{g}')$ (where $\nu_1, \ldots, \nu_p \in \mathbf{N}$) form a basis for $U(\mathfrak{g}')$ (2.1.11) and their images under $U(i)$ are the $x_1^{\nu_1} \cdots x_p^{\nu_p}$ calculated in $U(\mathfrak{g})$ and are thus linearly independent (2.1.11).

We identify $U(\mathfrak{g}')$ with $i(U(\mathfrak{g}'))$ by virtue of i. Thus, $U(\mathfrak{g}')$ is the subalgebra of $U(\mathfrak{g})$ generated by 1 and $\mathfrak{g}'$.

2.2.7. PROPOSITION. *Let $\mathfrak{g}'$ be a Lie subalgebra of $\mathfrak{g}$, and $(y_1, \ldots, y_q)$ a basis for a complement of $\mathfrak{g}'$ in $\mathfrak{g}$. Then the $y_1^{\nu_1} \cdots y_q^{\nu_q}$, where $\nu_1, \ldots, \nu_q \in \mathbf{N}$, form a basis for $U(\mathfrak{g})$ considered as a left or right module over $U(\mathfrak{g}')$.*

Let $(y_{q+1}, \ldots, y_n)$ be a basis for $\mathfrak{g}'$. The $y_1^{\nu_1} \cdots y_n^{\nu_n}$, where $\nu_1, \ldots, \nu_n \in \mathbf{N}$, form a basis for the vector space $U(\mathfrak{g})$. Hence every element of $U(\mathfrak{g})$ can be uniquely written in the form

$$\sum_{\nu_1, \ldots, \nu_q \in \mathbf{N}} y_1^{\nu_1} \cdots y_q^{\nu_q} x_{,\nu_1, \ldots, \nu_q},$$

where the $x_{\nu_1, \ldots, \nu_q}$ belong to $U(\mathfrak{g}')$. Consequently, the $y_1^{\nu_1} \cdots y_q^{\nu_q}$, where $\nu_1, \ldots, \nu_q \in \mathbf{N}$, form a basis for the right $U(\mathfrak{g}')$-module $U(\mathfrak{g})$. The same reasoning applies for the left $U(\mathfrak{g}')$-module $U(\mathfrak{g})$.

2.2.8. Let us assume, contrary to our general conventions, that k is a commutative ring, $\mathfrak{g}$ a Lie algebra over k, and $\mathfrak{g}'$ a Lie subalgebra of $\mathfrak{g}$, that $\mathfrak{g}'$, as a k-module, has a finite basis and that $(y_1, \ldots, y_q)$ is a basis for a complement of $\mathfrak{g}'$ in $\mathfrak{g}$. Then, 2.2.7 remains valid, with the same proof.

2.2.9. PROPOSITION. *Let $\mathfrak{h}$ snd $\mathfrak{k}$ be Lie subalgebras of $\mathfrak{g}$ such that $\mathfrak{g} = \mathfrak{h} + \mathfrak{k}$. Let $\mathfrak{l} = \mathfrak{h} \cap \mathfrak{k}$. Let us consider $U(\mathfrak{h})$ as a right $U(\mathfrak{l})$-module and $U(\mathfrak{k})$ as a left* $U(\mathfrak{l})$-module. *There exists one and only one linear mapping f of $U(\mathfrak{h}) \otimes_{U(\mathfrak{l})} U(\mathfrak{k})$ into $U(\mathfrak{g})$ such that $f(v \otimes w) = vw$ for $v \in U(\mathfrak{h})$, $w \in U(\mathfrak{k})$. This mapping is bijective.*

Let g be the bilinear mapping $(v,w) \mapsto vw$ of $U(\mathfrak{h}) \times U(\mathfrak{k})$ into $U(\mathfrak{g})$. If $v \in U(\mathfrak{h})$, $w \in U(\mathfrak{k})$, $z \in U(\mathfrak{l})$, then $g(vz,w = \mathfrak{g}(v,zw)$. This proves the existence and uniqueness of f. Let $(a_1, \ldots, a_m)$ be a basis for $\mathfrak{h}$, and $(b_1, \ldots, b_n)$ a basis for a supplement of $\mathfrak{l}$ in $\mathfrak{k}$. From 2.2.7, the

$$a_1^{\mu_1} \cdots a_m^{\mu_m} \otimes b_1^{\nu_1} \cdots b_n^{\nu_n},$$

where $\mu_1, \ldots, \nu_n \in \mathbf{N}$, form a basis for the vector space $U(\mathfrak{h}) \otimes_{U(\mathfrak{l})} U(\mathfrak{k})$. The image of this basis under f is a basis for the vector space $U(\mathfrak{g})$ (2.1.11). Hence f is bijective.

2.2.10. PROPOSITION. *Let $\mathfrak{g}_1, \ldots, \mathfrak{g}_n$ be Lie subalgebras of $\mathfrak{g}$ such that $g = \mathfrak{g}_1 \oplus \cdots \oplus g_n$. There exists one and only one linear mapping f of $U(\mathfrak{g}_1) \otimes \cdots \otimes U(\mathfrak{g}_n)$ into $U(\mathfrak{g})$ such that $f(u_1 \otimes \cdots \otimes u_n) = u_1 \cdots u_n$ for $u_1 \in U(\mathfrak{g}_1), \ldots, u_n \in U(\mathfrak{g}_n)$. This mapping is bijective.*

The proof is similar to that of 2.2.9.

2.2.11. The isomorphism f of 2.2.10 is said to be *canonical* and by virtue of it we sometimes identify the vector spaces $U(\mathfrak{g}_1) \otimes \cdots \otimes U(\mathfrak{g}_n)$ and $U(\mathfrak{g})$ with each other. If, in addition, $\mathfrak{g}_1, \ldots, \mathfrak{g}_n$ pairwise commute in $U(\mathfrak{g})$, f is an algebra isomorphism. From this we deduce:

2.2.12. COROLLARY. *Let* $\mathfrak{g}_1, \ldots, \mathfrak{g}_n$ *be Lie algebras, and* $\mathfrak{g}$ *be their product. The multilinear mapping* $(u_1, \ldots, u_n) \mapsto u_1 \cdots u_n$ *of* $U(\mathfrak{g}_1) \times \cdots \times U(\mathfrak{g}_n)$ *into* $U(\mathfrak{g})$ *defines a linear mapping* f *of* $U(\mathfrak{g}_1) \otimes \cdots U(\mathfrak{g}_n)$ *into* $U(\mathfrak{g})$. *Then* f *is an algebra isomorphism.*

2.2.13. With the hypotheses of 2.2.12, the algebras $U(\mathfrak{g}_1) \otimes \cdots \otimes U(\mathfrak{g}_n)$ and $U(\mathfrak{g}_1 \times \cdots \times \mathfrak{g}_n)$ can be identified with each other by virtue of f.

2.2.14. PROPOSITION. *Let* $\mathfrak{h}$ *be an ideal of* $\mathfrak{g}$.

(i) *The left ideal* R *of* $U(\mathfrak{g})$ *generated by* $\mathfrak{h}$ *coincides with the right ideal of* $U(\mathfrak{g})$ *generated by* $\mathfrak{h}$.

(ii) *Let* j *be the canonical homomorphism of* $\mathfrak{g}$ *onto* $\mathfrak{g}/\mathfrak{h}$. *Then the homomorphism* $U(j)$ *of* $U(\mathfrak{g})$ *into* $U(\mathfrak{g}/\mathfrak{h})$ *is surjective with kernel* R.

Let $(x_1, \ldots, x_m)$ be a basis for a complement of $\mathfrak{h}$ in $\mathfrak{g}$. For $i = 1, \ldots, m$, let $y_i = j(x_i)$. From 2.2.7 we have

$$U(\mathfrak{g}) = \bigoplus_{\nu_1,\ldots,\nu_m \in \mathbf{N}} x_1^{\nu_1} \cdots x_m^{\nu_m} U(\mathfrak{h}).$$

For all $v \in U(\mathfrak{h})$, let $\varepsilon(v)$ be its constant term. If

$$u = \sum_{\nu_1,\ldots,\nu_m \in \mathbf{N}} x_1^{\nu_1} \cdots x_m^{\nu_m} v_{\nu_1 \cdots \nu_m},$$

where the $v_{\nu_1 \cdots \nu_m}$ are in $U(\mathfrak{h})$, then

$$U(j)u = \sum_{\nu_1,\ldots,\nu_m \in \mathbf{N}} y_1^{\nu_1} \cdots y_m^{\nu_m} \varepsilon(v_{\nu_1 \cdots \nu_m}).$$

Hence

$$\operatorname{Ker} U(j) = \bigoplus_{\nu_1,\ldots,\nu_m \in \mathbf{N}} x_1^{\nu_1} \cdots x_m^{\nu_m} U_+(\mathfrak{h}).$$

On the other hand,

$$R = \sum_{\nu_1 \ldots \nu_m \in \mathbf{N}} x_1^{\nu_1} \cdots x_m^{\nu_m} U(\mathfrak{h})\mathfrak{h} = \sum_{\nu_1,\ldots,\nu_m \in \mathbf{N}} x_1^{\nu_1} \cdots x_m^{\nu_m} U_+(\mathfrak{h})$$

and hence $R = \operatorname{Ker} U(j)$. If R' is the right ideal of $U(\mathfrak{g})$ generated by $\mathfrak{h}$, it can likewise be seen that $R' = \operatorname{Ker} U(j)$. Finally, $U(j)(U(\mathfrak{g})) \supset j(\mathfrak{g}) = \mathfrak{g}/\mathfrak{h}$. and hence $U(j)$ is surjective.

2.2.15. Proposition 2.2.14 defines an isomorphism, termed *canonical*, of the algebra $U(\mathfrak{g})/R$ onto the algebra $U(\mathfrak{g}/\mathfrak{h})$. We identify $U(\mathfrak{g})/R$ with $U(\mathfrak{g}/\mathfrak{h})$ by means of this isomorphism.

2.2.16. Let $\mathfrak{g}'$ be the opposite Lie algebra of $\mathfrak{g}$. Let A be the opposite algebra of $U(\mathfrak{g}')$. Let τ be the canonical injection of $\mathfrak{g}$ into A and let $x,y \in \mathfrak{g}$. Then $\tau(x)\tau(y) - \tau(y)\tau(x)$ is equal to $yx - xy$ calculated in $U(\mathfrak{g}')$, hence to $[y,x]$ calculated in $\mathfrak{g}'$, and hence to $[x,y]$ calculated in $\mathfrak{g}$. As a consequence of (2.2.1), there exists a homomorphism φ of the algebra $U(\mathfrak{g})$ into the algebra A which extends the identity mapping of $\mathfrak{g}$. From 2.1.11, φ transforms a basis for $U(\mathfrak{g})$ into a basis for A, and is therefore an isomorphism. We identify $U(\mathfrak{g})$ with A by means of this isomorphism.

2.2.17. PROPOSITION. *There exists one and only one anti-automorphism ψ of the algebra $U(\mathfrak{g})$ such that $\psi(x) = -x$ for all $x \in \mathfrak{g}$.*

With the notation of 2.2.16, the mapping $x \mapsto -x$ of $\mathfrak{g}'$ into $\mathfrak{g}$ is a Lie algebra isomorphism, and can hence be extended to an isomorphism ψ of $U(\mathfrak{g}')$ onto $U(\mathfrak{g})$. Then ψ is an anti-isomorphism of $A = U(\mathfrak{g})$ onto $U(\mathfrak{g})$.

2.2.18. The anti-automorphism ψ of 2.2.17 is termed the *principal anti-automorphism* of $U(\mathfrak{g})$, and is denoted by $u \mapsto u^{\mathsf{T}}$. If $x_1, \ldots, x_n \in \mathfrak{g}$, we have

$$(x_1 x_2 \cdots x_n)^{\mathsf{T}} = (-1)^n \, x_n x_{n-1} \cdots x_1 .$$

2.2.19. Let ϱ be a representation of $\mathfrak{g}$ and ϱ^* the dual representation. The mapping $u \mapsto {}^t\varrho(u^{\mathsf{T}})$, where u runs through $U(\mathfrak{g})$, is a representation of $U(\mathfrak{g})$, and it extends ϱ^*. Hence, for all $u \in U(\mathfrak{g})$ we have

$$\varrho^*(u) = {}^t\varrho(u^{\mathsf{T}}).$$

2.2.20. Let k' be an extension of k. Let us apply 2.2.1 to the canonical injection of $\mathfrak{g}$ into $U(\mathfrak{g} \otimes k')$. We obtain a homomorphism φ of the k-algebra $U(\mathfrak{g})$ into the k-algebra $U(\mathfrak{g} \otimes k')$ such that $\varphi(x) = x$ for all $x \in \mathfrak{g}$. Next φ can be extended to a homomorphism $\varphi' : U(\mathfrak{g}) \otimes k' \to U(\mathfrak{g} \otimes k')$ of k'-algebras. From 2.1.11, φ' transforms a k'-basis for $U(\mathfrak{g}) \otimes k'$ into a k'-basis for $U(\mathfrak{g} \otimes k')$. Hence φ' is an isomorphism by means of which we can identify the k'-algebras $U(\mathfrak{g}) \otimes k'$ and $U(\mathfrak{g} \otimes k')$ with each other.

Consequently, $Z(\mathfrak{g}) \otimes k'$ can be identified with $Z(\mathfrak{g} \otimes k')$.

2.2.21. For all $u \in U(\mathfrak{g})$, let $L(u)$ and $R(u)$ be the mappings $v \mapsto uv$ and $v \mapsto vu$ of $U(\mathfrak{g})$ into itself. We know that the mapping $u \mapsto L(u)$ is a representation of $U(\mathfrak{g})$ in $U(\mathfrak{g})$, termed the left regular representation of $U(\mathfrak{g})$. The corresponding representation of $\mathfrak{g}$, i.e. the mapping $x \mapsto L(x)$ $(x \in \mathfrak{g})$, is termed the *left regular representation of* $\mathfrak{g}$ *in* $U(\mathfrak{g})$. The mapping $u \mapsto R(u^{\mathsf{T}})$ is a representation of $U(\mathfrak{g})$ in $U(\mathfrak{g})$. The corresponding representation of $\mathfrak{g}$, i.e. the mapping $x \mapsto -R(x)$ ($x \in \mathfrak{g}$) is termed the *right regular representation of* $\mathfrak{g}$ *in* $U(\mathfrak{g})$.

Since $R(u)L(u) = L(u)R(u)$ for all $u \in U(\mathfrak{g})$, the mapping $x \mapsto \varrho(x) = L(x) - R(x)$ ($x \in \mathfrak{g}$) is again a representation of $\mathfrak{g}$ in $U(\mathfrak{g})$, termed the *adjoint representation of* $\mathfrak{g}$ *in* $U(\mathfrak{g})$. If $x \in \mathfrak{g}$ and $u \in U(\mathfrak{g})$, we have $\varrho(x)u = [x,u]$. (The corresponding representation of $U(\mathfrak{g})$ cannot conveniently be made explicit in a nice way.) Let I and J be sub-$\mathfrak{g}$-modules of $U(\mathfrak{g})$ for the adjoint representation (for example, two-sided ideals of $U(\mathfrak{g})$). Let us assume that $I \supset J$. The representation deduced from ϱ in I/J is termed the *adjoint representation of* $\mathfrak{g}$ *in* I/J. Subsequently, when we are considering I/J (and, for example, $U(\mathfrak{g})$ itself) as a $\mathfrak{g}$-module, *we shall always be concerned with the adjoint representation, unless otherwise indicated.*

2.2.22. For example, if $\mathfrak{k}$ is an ideal of $\mathfrak{g}$, then $U(\mathfrak{k})$ is a sub-$\mathfrak{g}$-module of $U(\mathfrak{g})$, and we can consider the adjoint representation ε of $\mathfrak{g}$ in $U(\mathfrak{k})$.

LEMMA. *Let* α *be the principal anti-automorphism of* $U(\mathfrak{g})$, $\delta = \varepsilon \circ \alpha$, $y_1, y_2, \ldots, y_p \in \mathfrak{g}$, $\nu_1, \ldots, \nu_p \in \mathbf{N}$, and $z \in U(\mathfrak{k})$. *Then*

$$zy_1^{\nu_1} \cdots y_p^{\nu_p} = \sum_{0 \leq \mu_i \leq \nu_i} \binom{\nu_1}{\mu_1} \cdots \binom{\nu_p}{\mu_p} y_1^{\mu_1} \cdots y_p^{\mu_p} (\delta(y_1^{\nu_1-\mu_1} \cdots y_p^{\nu_p-\mu_p})z).$$

This is clear for $p = 0$; let us assume the lemma to be proved for $p - 1$. Then the formula is true for $\nu_1 = 0$; let us assume it to be proved for $\nu_1 - 1$. Since $zy_1 = y_1 z + \delta(y_1)z$, we have

$$zy_1^{\nu_1} \cdots y_p^{\nu_p} =$$

$$= \sum_{\substack{0 \leq \mu_1 \leq \nu_1 - 1,\\ 0 \leq \mu_2 \leq \nu_2, \ldots}} \binom{\nu_1 - 1}{\mu_1} \binom{\nu_2}{\mu_2} \cdots y_1^{\mu_1+1} y_2^{\mu_2} \cdots (\delta(y_1^{\nu_1-1-\mu_1} y_2^{\nu_2-\mu_2} \cdots)z)$$

$$+ \sum_{\substack{0 \leq \mu_1 \leq \nu_1 - 1,\\ 0 \leq \mu_2 \leq \nu_2, \ldots}} \binom{\nu_1 - 1}{\mu_1} \binom{\nu_2}{\mu_2} \cdots y_1^{\mu_1} y_2^{\mu_2} \cdots (\delta(y_1^{\nu_1-\mu_1} y_2^{\nu_2-\mu_2} \cdots)z).$$

$$= \sum_{0 \leq \mu_i \leq \nu_i} \beta_{\mu_1, \ldots, \mu_p} y_1^{\mu_1} \cdots y_p^{\mu_p} (\delta(y_1^{\nu_1-\mu_1} \cdots y_p^{\nu_p-\mu_p})z,$$

where

$$\beta_{\mu_1, \ldots, \mu_p} = \left(\binom{\nu_1 - 1}{\mu_1 - 1} + \binom{\nu_1 - 1}{\mu_1}\right) \binom{\nu_2}{\mu_2} \cdots \binom{\nu_p}{\mu_p} = \binom{\nu_1}{\mu_1} \binom{\nu_2}{\mu_2} \cdots \binom{\nu_p}{\mu_p}.$$

2.2.23. Let $f \in \mathfrak{g}^*$ be a one-dimensional representation of $\mathfrak{g}$. The mapping $x \mapsto x + f(x)$ of $\mathfrak{g}$ into $U(\mathfrak{g})$ is a Lie algebra homomorphism. Hence there exists one and only one homomorphism α_f of the algebra $U(\mathfrak{g})$ into itself such that $\alpha_f(x) = x + f(x)$ for all $x \in \mathfrak{g}$. We have

$$\alpha_f \circ \alpha_{-f} = \alpha_{-f} \circ \alpha_f = \mathrm{id}_{U(\mathfrak{g})},$$

and hence α_f is an automorphism of $U(\mathfrak{g})$.

Let ϱ be a representation of $\mathfrak{g}$ in W. Let ϱ' be the representation $\varrho \otimes f$, which also operates in W. For $w \in W$ and $x \in \mathfrak{g}$, we have

$$\varrho'(x)w = \varrho(x)w + f(x)w = \varrho(\alpha_f(x))w,$$

hence

$$\varrho' = \varrho \circ \alpha_f.$$

References: [13], [16], [71].

2.3. The filtration of the enveloping algebra

2.3.1. Let n be an integer $\geqq 0$. The vector subspace of $U(\mathfrak{g})$ generated by the products $x_1 x_2 \cdots x_p$, where $x_1, \ldots, x_p \in \mathfrak{g}$ and $p \leqq n$ is denoted by $U_n(\mathfrak{g})$ (cf. 2.1.4). The sequence $(U_n(\mathfrak{g}))_{n \geqq 0}$ is increasing and its union is $U(\mathfrak{g})$; we have

$$U_0(\mathfrak{g}) = k \cdot 1, \qquad U_1(\mathfrak{g}) = k \cdot 1 \oplus \mathfrak{g}, \qquad U_n(\mathfrak{g})\, U_p(\mathfrak{g}) \subset U_{n+p}(\mathfrak{g}).$$

This sequence $(U_n(\mathfrak{g}))_{n \geqq 0}$ is termed the *canonical filtration* of $U(\mathfrak{g})$. If u is a non-zero element of $U(\mathfrak{g})$, the smallest integer n such that $u \in U_n(\mathfrak{g})$ is termed the *filtration of u*; the elements of $U_p(\mathfrak{g})$ are those of filtration $\leqq p$.

2.3.2. Let $(e_1, \ldots, e_r)$ be a basis for $\mathfrak{g}$. The $e_1^{\nu_1} \cdots e_r^{\nu_r} \in U(\mathfrak{g})$ such that $\nu_1 + \cdots + \nu_r \leqq n$ form a basis for $U_n(\mathfrak{g})$; indeed, they belong to $U_n(\mathfrak{g})$ and are linearly independent; and every product $x_1 \cdots x_p$ (where $x_1, \ldots, x_p \in \mathfrak{g}$ and $p \leqq n$) is a linear combination of these elements.

From this it follows that, if $\mathfrak{g}'$ is a Lie subalgebra of $\mathfrak{g}$, then

$$U_n(\mathfrak{g}') = U_n(\mathfrak{g}) \cap U(\mathfrak{g}').$$

2.3.3. For all $n \geqq 0$, $U_n(\mathfrak{g})$ is a finite-dimensional sub-$\mathfrak{g}$-module of $U(\mathfrak{g})$. If $\mathfrak{g}$ is semi-simple, $U(\mathfrak{g})$ is thus the sum of finite-dimensional simple sub-$\mathfrak{g}$-modules (1.6.3). If $\mathfrak{g}$ is completely solvable, the adjoint representation of $\mathfrak{g}$ in $\mathfrak{g}$ is triangularizable; it is readily deduced from 2.1.11 that each $U_n(\mathfrak{g})$ is a triangularizable $\mathfrak{g}$-module. Similarly, if $\mathfrak{g}$ is nilpotent, each $U_n(\mathfrak{g})$ is a strictly triangularizable $\mathfrak{g}$-module.

2.3.4. Let us recall how to construct the *graded algebra G associated with the filtered algebra* $U(\mathfrak{g})$. Let G^n be the vector space $U_n(\mathfrak{g})/U_{n-1}(\mathfrak{g})$ and G the vector space $G^0 \oplus G^1 \oplus \cdots$ (we define $U_{-1}(\mathfrak{g}) = 0$). The multiplication in $U(\mathfrak{g})$ defines by passage to the quotient a bilinear mapping of $G^m \times G^n$ into G^{m+n}; whence, by linearity, a multiplication in G which makes G an (associative) algebra with unity. We have $G^0 = k \cdot 1 = k$, and G^1 can be canonically identified with $\mathfrak{g}$. Clearly, the products of n elements of $G^1 = \mathfrak{g}$ generate the vector space G^n. Then, from 2.1.5, the algebra G is *commutative*.

2.3.5. Since G is commutative, the canonical injection of $\mathfrak{g}$ into G can be uniquely extended to a homomorphism φ of the symmetric algebra $S(\mathfrak{g})$ of $\mathfrak{g}$ into G, such that $\varphi(1) = 1$. We say that φ is the *canonical homomorphism* of $S(\mathfrak{g})$ into G. If $S^n(\mathfrak{g})$ denotes the set of elements of $S(\mathfrak{g})$ which are homogeneous of degree n, then $\varphi(S^n(\mathfrak{g})) \subset G^n$.

2.3.6. PROPOSITION. *The canonical homomorphism of $S(\mathfrak{g})$ into G is an isomorphism.*

Let $(x_1, \ldots, x_n)$ be a basis for $\mathfrak{g}$. For $\nu = (\nu_1, \ldots, \nu_n) \in \mathbf{N}^n$, let X^ν be the product $x_1^{\nu_1} \cdots x_n^{\nu_n}$ calculated in $S(\mathfrak{g})$, x^ν the product $x_1^{\nu_1} \cdots x_n^{\nu_n}$ calculated in $U(\mathfrak{g})$, and x'^ν the canonical image of x^ν in $G^{|\nu|}$ (in conformity with our usual convention, we set $|\nu| = \nu_1 + \cdots + \nu_n$). From 2.3.2, the x'^ν, for $|\nu| = p$, form a basis for G^p. Hence $(x'^\nu)_{\nu \in N^n}$ is a basis for G. Since $\varphi(X^\nu) = x'^\nu$, it can be seen that φ is bijective.

2.3.7. By virtue of the isomorphism in 2.3.6, we can identify G with $S(\mathfrak{g})$. Thus we can state that the symmetric algebra $S(\mathfrak{g})$ is the graded algebra associated with the filtered algebra $U(\mathfrak{g})$.

2.3.8. A ring is said to be *Noetherian* if it satisfies the maximal condition for left ideals and for right ideals.

COROLLARY. *The algebra $U(\mathfrak{g})$ is Noetherian.*

Indeed, from 2.3.7, the graded algebra associated with $U(\mathfrak{g})$ is Noetherian (cf. AC III, pp. 42, 44).

2.3.9. COROLLARY. (i) *Let u and v be non-zero elements of $U(\mathfrak{g})$, and n and p their filtrations. The filtration of uv is $n + p$.*

(ii) *The algebra $U(\mathfrak{g})$ is integral* (*cf.* 3.1.2).

Let u' and v' be the canonical images of u and v in G^n and G^p respectively. Then $u' \neq 0$ and $v' \neq 0$, and hence $u'v' \neq 0$ from 2.3.7. Consequently, $uv \in U_{n+p}(\mathfrak{g})$ and $uv \notin U_{n+p-1}(\mathfrak{g})$; this proves both (i) and (ii).

2.3.10. PROPOSITION. *Let I be a two-sided ideal of $U(\mathfrak{g})$, $A = U(\mathfrak{g})/I$, and A_n be the canonical image of $U_n(\mathfrak{g})$ in A, so that A is filtered by the A_n. Let $G = \oplus_{n \geqq 0} U_n(\mathfrak{g})/U_{n-1}(\mathfrak{g})$ be the graded algebra associated with $U(\mathfrak{g})$, and $I^{\mathrm{gr}} = \oplus_{n \geqq 0}(I \cap U_n(\mathfrak{g}))/(I \cap U_{n-1}(\mathfrak{g}))$ the graded ideal associated with I. Then the graded algebra associated with A is canonically isomorphic to G/I^{gr}.*

Let us set $I_n = I \cap U_n(\mathfrak{g})$. Then

$$\begin{aligned} A_n/A_{n-1} &= (U_n(\mathfrak{g}) + I)/(U_{n-1}(\mathfrak{g}) + I) \\ &= U_n(\mathfrak{g})/I_n + U_{n-1}(\mathfrak{g}) = (U_n(\mathfrak{g})/U_{n-1}(\mathfrak{g}))/(I_n/I_{n-1}), \end{aligned}$$

hence $\oplus_{n \geqq 0}(A_n/A_{n-1}) = G/I^{\mathrm{gr}}$, and this identification is compatible with the multiplicative structures.

References: [16], [71].

2.4. The canonical mapping of the symmetric algebra into the enveloping algebra

2.4.1. Let n be an integer $\geqq 0$, $T^n(\mathfrak{g}) = \mathfrak{g} \otimes \mathfrak{g} \otimes \cdots \otimes \mathfrak{g}$ (n factors), $S^n(\mathfrak{g})$ be the set of homogeneous elements of degree n in the symmetric algebra $S(\mathfrak{g})$, and $G^n(\mathfrak{g}) = U_n(\mathfrak{g})/U_{n-1}(\mathfrak{g})$. Let us consider the diagram

$$(1) \qquad \begin{array}{ccc} T^n(\mathfrak{g}) & \xrightarrow{\psi_n} & U_n(\mathfrak{g}) \\ {\scriptstyle \tau_n}\downarrow & & \downarrow{\scriptstyle \theta_n} \\ S^n(\mathfrak{g}) & \xrightarrow{\varphi_n} & G^n(\mathfrak{g}) \end{array}$$

(We recall that $U(\mathfrak{g})$ and $S(\mathfrak{g})$ are quotients of the tensor algebra of $\mathfrak{g}$, which gives a meaning to ψ_n and τ_n; θ_n is the canonical mapping of $U_n(\mathfrak{g})$ onto $G^n(\mathfrak{g}) = U_n(\mathfrak{g})/U_{n-1}(\mathfrak{g})$; lastly, φ_n has been defined in 2.3.5.)

2.4.2. LEMMA. *Diagram* (1) *is commutative.*

Let $x_1, \ldots, x_n \in \mathfrak{g}$. Then $\psi_n(x_1 \otimes \cdots \otimes x_n)$ is the product $x_1 \cdots x_n$ calculated in $U(\mathfrak{g})$, and hence $\theta_n(\psi_n(x_1 \otimes \cdots \otimes x_n))$ is the product $x_1 \cdots x_n$ calculated in $G = G^0(\mathfrak{g}) \oplus G^1(\mathfrak{g}) \oplus \cdots$. Similarly, $\tau_n(x_1 \otimes \cdots \otimes x_n)$ is the product $x_1 \cdots x_n$ calculated in $S(\mathfrak{g})$, hence $\varphi_n(\tau_n(x_1 \otimes \cdots \otimes x_n))$ is the product $x_1 \cdots x_n$ calculated in G.

2.4.3. An element of $U(\mathfrak{g})$ is said to be *symmetric homogeneous of degree n* if it is the canonical image in $U(\mathfrak{g})$ of a tensor which is homogeneous symmetric of degree n over $\mathfrak{g}$. The set of elements of $U(\mathfrak{g})$ which are symmetric homogeneous of degree n is denoted by $U^n(\mathfrak{g})$.

2.4.4. PROPOSITION. *We have* $U_n(\mathfrak{g}) = U_{n-1}(\mathfrak{g}) \oplus U^n(\mathfrak{g})$.

Let us use the notation of 2.4.1. Let $T'^n(\mathfrak{g})$ be the set of symmetric elements of $T^n(\mathfrak{g})$. Then $\tau_n|T'^n(\mathfrak{g})$ is a bijection of $T'^n(\mathfrak{g})$ onto $S^n(\mathfrak{g})$. Moreover, φ_n is bijective (2.3.6). From 2.4.2, $\theta_n \circ \psi_n|T'^n(\mathfrak{g})$ is a bijection of $T'^n(\mathfrak{g})$ onto $G^n(\mathfrak{g})$, and hence $\psi_n|T'^n(\mathfrak{g})$ is a bijection of $T'^n(\mathfrak{g})$ onto a complement of $U_{n-1}(\mathfrak{g})$ in $U_n(\mathfrak{g})$.

2.4.5. Diagram (1) hence defines a *commutative diagram of bijections*

(2)
$$\begin{array}{ccc} T'^n(\mathfrak{g}) & \longrightarrow & U^n(\mathfrak{g}) \\ \downarrow & & \downarrow \\ S^n(\mathfrak{g}) & \longrightarrow & G^n(\mathfrak{g}) \end{array}$$

In particular, we obtain a bijection ω_n, termed *canonical*, of $S^n(\mathfrak{g})$ onto $U^n(\mathfrak{g})$. If $x_1, \ldots, x_n \in \mathfrak{g}$, then

(3)
$$\omega_n(x_1 x_2 \cdots x_n) = \frac{1}{n!} \sum_{\pi \in \mathfrak{S}_n} x_{\pi(1)} x_{\pi(2)} \cdots x_{\pi(n)}.$$

(The products are calculated in $S(\mathfrak{g})$ for the left-hand side and in $U(\mathfrak{g})$ for the right-hand side.) In particular, $\omega_n(x^n) = x^n$ for all $x \in \mathfrak{g}$, and this is sufficient to characterize the linear mapping ω_n since $S^n(\mathfrak{g})$ is generated, as a vector space, by the nth powers of the elements of $\mathfrak{g}$.

2.4.6. From 2.4.4, we have

$$U(\mathfrak{g}) = U^0(\mathfrak{g}) \oplus U^1(\mathfrak{g}) \oplus U^2(\mathfrak{g}) \oplus \cdots,$$

with, moreover $U^0(\mathfrak{g}) = U_0(\mathfrak{g}) = k$ and $U^1(\mathfrak{g}) = \mathfrak{g}$. The direct sum of the canonical bijections $\omega_n: S^n(\mathfrak{g}) \to U^n(\mathfrak{g})$ is a bijection, again termed *canonical*, of $S(\mathfrak{g})$ onto $U(\mathfrak{g})$. This is sometimes called the *symmetrization*.

2.4.7. Let $\mathfrak{g}$ and $\mathfrak{g}'$ be Lie algebras, $\omega: S(\mathfrak{g}) \to U(\mathfrak{g})$ and $\omega': S(\mathfrak{g}') \to U(\mathfrak{g}')$ the canonical bijections, $\eta: \mathfrak{g} \to \mathfrak{g}'$ a homomorphism, and $S(\eta)$ the canonical extension of η to $S(\mathfrak{g})$. From formula (3) of 2.4.5, we have

$$U(\eta)U^n(\mathfrak{g}) \subset U^n(\mathfrak{g}') \qquad U(\eta) \circ \omega = \omega' \circ S(\eta).$$

In particular, if $\mathfrak{g}'$ is a Lie subalgebra of $\mathfrak{g}$, then $\omega|S(\mathfrak{g}')$ is the canonical bijection of $S(\mathfrak{g}')$ onto $U(\mathfrak{g}')$.

2.4.8. PROPOSITION. *If* $u \in U^n(\mathfrak{g})$, *then* $u^{\mathsf{T}} = (-1)^n u$.

If $x_1, \ldots, x_n \in \mathfrak{g}$, then

$$\Big(\sum_{\pi \in \mathfrak{S}_n} x_{\pi(1)} x_{\pi(2)} \cdots x_{\pi(n)}\Big)^{\mathsf{T}} = (-1)^n \sum_{\pi \in \mathfrak{S}_n} x_{\pi(n)} x_{\pi(n-1)} \cdots x_{\pi(1)}$$
$$= (-1)^n \sum_{\pi \in \mathfrak{S}_n} x_{\pi(1)} x_{\pi(2)} \cdots x_{\pi(n)}.$$

2.4.9. PROPOSITION. *Let D be a derivation of $\mathfrak{g}$.*

(i) *There exists one and only one derivation D' of $U(\mathfrak{g})$ which extends D.*

(ii) *For every integer $n \geq 0$, we have $D'(U_n(\mathfrak{g})) \subset U_n(\mathfrak{g})$ and $D'(U^n(\mathfrak{g})) \subset U^n(\mathfrak{g})$.*

(iii) *If there exists $x \in \mathfrak{g}$ such that $D = \mathrm{ad}_{\mathfrak{g}} x$, then $D'(u) = xu - ux$ for all $u \in U(\mathfrak{g})$.*

(iv) *Let D'' be the unique derivation of the algebra $S(\mathfrak{g})$ which extends D. Let φ be the canonical mapping of $S(\mathfrak{g})$ into $U(\mathfrak{g})$. Then $D'' \circ \varphi = \varphi \circ D'$.*

Let T be the tensor algebra of $\mathfrak{g}$, and Δ the unique derivation of T which extends D. If $x,y \in \mathfrak{g}$, then

$$\Delta(x \otimes y - y \otimes x - [x,y]) =$$
$$= Dx \otimes y + x \otimes Dy - Dy \otimes x - y \otimes Dx - D([x,y])$$
$$= (Dx \otimes y - y \otimes Dx - [Dx,y]) + (x \otimes Dy - Dy \otimes x - [x,Dy]).$$

With the notation of 2.1.1, we thus have $\Delta(J) \subset J$, whence the existence of a derivation D' of $U(\mathfrak{g}) = T/J$ which extends D; its uniqueness is obvious since $U(\mathfrak{g})$ is generated by 1 and $\mathfrak{g}$. The set of homogeneous tensors of degree n and the set of symmetric homogeneous tensors of degree n are stable under Δ, whence (ii). If $x \in \mathfrak{g}$, the mapping $u \mapsto xu - ux$ of $U(\mathfrak{g})$ into $U(\mathfrak{g})$ is a derivation of $U(\mathfrak{g})$ which extends $\mathrm{ad}_{\mathfrak{g}} x$, whence (iii). With the notation of 2.4.1, there exists a derivation D_1' of the algebra $G = G^0(\mathfrak{g}) \oplus G^1(\mathfrak{g}) \oplus \cdots$ such that $D_1'(G^n(\mathfrak{g})) \subset G^n(\mathfrak{g})$ and such that $D_1' | G^n(\mathfrak{g})$ can be deduced from $D' | U_n(\mathfrak{g})$ by passage to the quotient. Thus there exists a derivation D_2' of the algebra $S(\mathfrak{g})$ which can be deduced from D' by virtue of the canonical bijection of $S(\mathfrak{g})$ onto $U(\mathfrak{g})$. In particular, $D_2' | \mathfrak{g} = D$, so that D_2' is the derivation D'' of the proposition; this proves (iv).

2.4.10. Let us recall (1.2.14 and 2.2.21) that $S(\mathfrak{g})$ and $U(\mathfrak{g})$ are $\mathfrak{g}$-modules in a natural way. Having stated this, we have:

PROPOSITION. *The canonical bijection of $S(\mathfrak{g})$ onto $U(\mathfrak{g})$ (or of $S^n(\mathfrak{g})$ onto $U^n(\mathfrak{g})$) is a $\mathfrak{g}$-module isomorphism.*

This follows from 2.4.9 (iii) and (iv).

2.4.11. The set of invariants of the $\mathfrak{g}$-module $S(\mathfrak{g})$, i.e. the set of the elements of $S(\mathfrak{g})$ which are annihilated under the representation deduced in $S(\mathfrak{g})$ from the adjoint representation, is denoted by $Y(\mathfrak{g})$. Having stated this, we have:

COROLLARY. *Let φ be the canonical bijection of $S(\mathfrak{g})$ onto $U(\mathfrak{g})$. Then $\varphi(Y(\mathfrak{g})) = Z(\mathfrak{g})$.*

2.4.12. We retain the notation of 2.4.11. Since $U(\mathfrak{g})$ is not in general commutative, φ is not in general an algebra isomorphism. No more is it generally true that $\varphi|Y(\mathfrak{g})$ is an algebra isomorphism of $Y(\mathfrak{g})$ onto $Z(\mathfrak{g})$ (4.9.6 (b)), but we shall encounter results to this effect (4.8.12, 10.4.5, 6.6.9).

2.4.13. COROLLARY. *Let $\mathfrak{k}$ be a Lie subalgebra of $\mathfrak{g}$ which is reductive in $\mathfrak{g}$. Let C be the commutant of $\mathfrak{k}$ in $U(\mathfrak{g})$. Then C is a Noetherian algebra of finite type.*

Let D be the canonical image of C in $S(\mathfrak{g})$. From 2.4.10, we have $D = S(\mathfrak{g})^{\mathfrak{k}}$ when we consider $\mathfrak{g}$ (and hence $S(\mathfrak{g})$) as a $\mathfrak{k}$-module. The algebra D is of finite type (1.7.10). Let $C_n = C \cap U_n(\mathfrak{g})$; the C_n constitute a filtration of C. If we identify $S(\mathfrak{g})$ with the graded algebra associated with $U(\mathfrak{g})$, then D can be identified with the graded algebra C' associated with C. Hence C' is an algebra of finite type, so that C is a Noetherian algebra of finite type.

2.4.14. PROPOSITION. *Let $\mathfrak{m}_1, \ldots, \mathfrak{m}_r$ be vector subspaces of $\mathfrak{g}$ such that $\mathfrak{g} = \mathfrak{m}_1 \oplus \cdots \oplus \mathfrak{m}_r$. Let φ be the canonical bijection of $S(\mathfrak{g})$ onto $U(\mathfrak{g})$, and φ_i the restriction of φ to $S(\mathfrak{m}_i)$. The multilinear mapping $(p_1, \ldots, p_r) \mapsto \varphi_1(p_1) \cdots \varphi_r(p_r)$ of $S(\mathfrak{m}_1) \times \cdots \times S(\mathfrak{m}_r)$ into $U(\mathfrak{g})$ defines an isomorphism of the vector space $S(\mathfrak{m}_1) \otimes \cdots \otimes S(\mathfrak{m}_r)$ onto the vector space $U(\mathfrak{g})$.*

Let $(x_1, \ldots, x_m)$ be a basis for $\mathfrak{m}_1$, $(y_1, \ldots, y_n)$ a basis for $\mathfrak{m}_2, \ldots,$ and $(z_1, \ldots, z_p)$ a basis for $\mathfrak{m}_r$. For $\alpha = (\alpha_1, \ldots, \alpha_m) \in \mathbf{N}^m$, let us denote the product $x_1^{\alpha_1} \cdots x_m^{\alpha_m}$ calculated in $S(\mathfrak{m}_1)$ by x^α. Let us define the $y^\beta, \ldots,$ and the z^γ in a similar way. The x^α form a basis for $S(\mathfrak{m}_1), \ldots,$ and the z^γ form a basis for $S(\mathfrak{m}_r)$. If $|\alpha| + |\beta| + \cdots + |\gamma| = s$, then $\varphi_1(x^\alpha)\varphi_2(y^\beta) \cdots \varphi_r(z^\gamma) \in U_s(\mathfrak{g})$, and the image in $U_s(\mathfrak{g})/U_{s-1}(\mathfrak{g})$ of this element is the same as that of the product

$$x_1^{\alpha_1} \cdots x_m^{\alpha_m} y_1^{\beta_1} \cdots y_n^{\beta_n} \cdots z_1^{\gamma_1} \cdots z_p^{\gamma_p}$$

calculated in $U(\mathfrak{g})$. The images in $U_s(\mathfrak{g})/U_{s-1}(\mathfrak{g})$ of

$$\varphi_1(x^\alpha)\varphi_2(y^\beta)\cdots\varphi_r(z^\gamma),$$

where $|\alpha| + \cdots + |\gamma| = s$, thus form a basis for $U_s(\mathfrak{g})/U_{s-1}(\mathfrak{g})$. It follows from this that the $\varphi_1(x^\alpha)\varphi_2(y^\beta)\cdots\varphi_r(z^\gamma)$, for any $\alpha, \beta, \ldots, \gamma$, form a basis for $U(\mathfrak{g})$, whence the proposition follows.

2.4.15. PROPOSITION. *Let* $\mathfrak{k}$ *be a Lie subalgebra of* $\mathfrak{g}$*; we assume that there exists a complement* $\mathfrak{s}$ *of* $\mathfrak{k}$ *in* $\mathfrak{g}$ *such that* $[\mathfrak{k},\mathfrak{s}] \subset \mathfrak{s}$. *Let* φ *be the canonical bijection of* $S(\mathfrak{g})$ *onto* $U(\mathfrak{g})$, σ *the adjoint representation of* $\mathfrak{k}$ *in* $U(\mathfrak{k})$, ϱ *the representation of* $\mathfrak{k}$ *in* $S(\mathfrak{s})$ *deduced from the representation* $x \mapsto \mathrm{ad}_{\mathfrak{g}}x|\mathfrak{s}$ *of* $\mathfrak{k}$ *in* $\mathfrak{s}$, *and* τ *the restriction to* $\mathfrak{k}$ *of the adjoint representation of* $\mathfrak{g}$ *in* $U(\mathfrak{g})$. *The bilinear mapping* $(p,q) \mapsto \varphi(p)q$ *of* $S(\mathfrak{s}) \times U(\mathfrak{k})$ *into* $U(\mathfrak{g})$ *defines a linear mapping* ζ *of* $S(\mathfrak{s}) \otimes U(\mathfrak{k})$ *into* $U(\mathfrak{g})$.

(i) *The mapping* ζ *is a* $\mathfrak{k}$*-module isomorphism (for the representations* $\varrho \otimes \sigma$ *and* τ*).*

(ii) *If* I *is a two-sided ideal of* $U(\mathfrak{k})$, *then* $\zeta(S(\mathfrak{s}) \otimes I) = U(\mathfrak{g})I$.

The fact that ζ is bijective follows from 2.4.7 and 2.4.14. If $k \in \mathfrak{k}$, $p \in S(\mathfrak{s})$ and $q \in U(\mathfrak{k})$, then

$$\zeta(k \cdot (p \otimes q)) = \zeta(k \cdot p \otimes q + p \otimes [k, q]) = \varphi(k \cdot p)q + \varphi(p)[k,q]$$

$$[k,\varphi(p)]q + \varphi(p)[k,q] \quad \text{from 2.4.10}$$

$$= k\varphi(p)q - \varphi(p)qk = [k,\zeta(p \otimes q)],$$

whence (i). Finally,

$$\zeta(S(\mathfrak{s}) \otimes I) = \varphi(S(\mathfrak{s}))I = \varphi(S(\mathfrak{s}))U(\mathfrak{k})I = U(\mathfrak{g})I.$$

2.4.16. Let $\mathscr{H}$ be an algebraic group, $\mathfrak{h}$ its Lie algebra, π a rational homomorphism of $\mathscr{H}$ in $\mathrm{Aut}(\mathfrak{g})$ and π' the corresponding homomorphism of $\mathfrak{h}$ into the Lie algebra of derivations of $\mathfrak{g}$. For all $h \in \mathscr{H}$, let $\varrho(h)$ be the automorphism of the algebra $S(\mathfrak{g})$ which extends $\pi(h)$; for all $x \in \mathfrak{h}$, let $\varrho'(x)$ be the derivation of the algebra $S(\mathfrak{g})$ which extends $\pi'(x)$. For all $n \in \mathbf{N}$, $h \mapsto \varrho(h)|S^n(\mathfrak{g})$ is a rational representation of $\mathscr{H}$, and the corresponding representation of $\mathfrak{h}$ is $x \mapsto \varrho'(x)|S^n(\mathfrak{g})$ (BO, p. 137).

Let φ be the canonical bijection of $S(\mathfrak{g})$ onto $U(\mathfrak{g})$. For all $h \in \mathscr{H}$, $\sigma(h) = \varphi\varrho(h)\varphi^{-1}$ is the automorphism of the algebra $U(\mathfrak{g})$ which extends $\pi(h)$ (2.4.7). For all $x \in \mathfrak{h}$, $\sigma'(x) = \varphi\varrho'(x)\varphi^{-1}$ is the derivation of the algebra $U(\mathfrak{g})$ which extends $\pi'(x)$ (2.4.9). From the previous paragraph it follows that the mapping $h \mapsto \sigma(h)|U_n(\mathfrak{g})$ is a rational representation of $\mathscr{H}$, and that the corresponding representation of $\mathfrak{h}$ is $x \mapsto \sigma'(x)|U_n(\mathfrak{g})$.

2.4.17. PROPOSITION. *Let $\mathscr{A}$ be the adjoint algebraic group of $\mathfrak{g}$. For all $a \in \mathrm{Aut}(\mathfrak{g})$, let a_U be the automorphism of $U(\mathfrak{g})$ which extends a. Let I be a two-sided ideal of $U(\mathfrak{g})$. Then $a_U(I) = I$ for all $a \in \mathscr{A}$.*

Let $\mathfrak{d}$ be the Lie algebra of derivations of $\mathfrak{g}$. For all $d \in \mathfrak{d}$, let d_U be the derivation of $U(\mathfrak{g})$ which extends d; we have

$$d_U(I) \subset I \Leftrightarrow d_U(I \cap U_n(\mathfrak{g})) \subset I \cap U_n(\mathfrak{g}) \quad \text{for all } n.$$

Let $\mathfrak{e}$ be the set of those $d \in \mathfrak{d}$ which satisfy this condition. Similarly, if $a \in \mathrm{Aut}(\mathfrak{g})$, we have

$$a_U(I) = I \Leftrightarrow a_U(I \cap U_n(\mathfrak{g})) = I \cap U_n(\mathfrak{g}) \quad \text{for all } n.$$

Let $\mathscr{B}$ be the set of those $a \in \mathrm{Aut}(\mathfrak{g})$ which satisfy this condition. From 2.4.16, $\mathscr{B}$ is an algebraic subgroup of $\mathrm{Aut}(\mathfrak{g})$ with Lie algebra $\mathfrak{e}$. Now, $\mathrm{ad}_{\mathfrak{g}}\mathfrak{g} \subset \mathfrak{e}$, hence $\mathscr{A} \subset \mathscr{B}$.

References: [16], [71].

2.5. The existence of finite-dimensional representations

2.5.1. LEMMA. *Let $I_1, \ldots, I_m$ be right (or left) ideals of finite codimension of $U(\mathfrak{g})$. Then the product ideal $I_1 I_2 \cdots I_m$ has finite codimension.*

By induction on m, it is sufficient to consider the case of two (e.g. right) ideals. The right $U(\mathfrak{g})$-module I_1 is generated by a finite number of elements $u_1, \ldots, u_p$ (2.3.8). Let $v_1, \ldots, v_q$ be elements of $U(\mathfrak{g})$ which generate a complement of I_2 in $U(\mathfrak{g})$. Then each element of I_1 is congruent modulo $I_1 I_2$ $I_1 I_2$ to a linear combination of the $u_i v_j$. Consequently,

$$\dim (U(\mathfrak{g})/I_1 I_2) = \dim (U(\mathfrak{g})/I_1) + \dim (I_1/I_1 I_2) < +\infty.$$

2.5.2. LEMMA. *Let $\mathfrak{a}$ be an ideal of $\mathfrak{g}$, $\mathfrak{b}$ a vector subspace of $\mathfrak{g}$ such that $\mathfrak{g} = \mathfrak{a} + \mathfrak{b}$, and σ a finite-dimensional representation of $\mathfrak{g}$. Assume that $\sigma(x)$ is nilpotent for all $x \in \mathfrak{a} \cup \mathfrak{b}$. Then $\sigma(x)$ is nilpotent for all $x \in \mathfrak{g}$.*

By considering a Jordan—Hölder series of σ, we return to the case where σ is simple. Then $\sigma(\mathfrak{a}) = 0$ (1.4.5), hence $\sigma(\mathfrak{g}) = \sigma(\mathfrak{b})$.

2.5.3. LEMMA. *Let $\mathfrak{a}$ be an ideal of $\mathfrak{g}$, $\mathfrak{b}$ a Lie subalgebra of $\mathfrak{g}$ such that $\mathfrak{g} = \mathfrak{a} \oplus \mathfrak{b}$, π the left regular representation of $\mathfrak{a}$ in $U(\mathfrak{a})$ and φ the adjoint repre-*

sentation of $\mathfrak{g}$ *in* $U(\mathfrak{a})$. *The linear mapping* ψ *of* $\mathfrak{g}$ *into* $\mathrm{End}(U(\mathfrak{a}))$ *such that* $\psi|\mathfrak{a} = \pi$ *and* $\psi|\mathfrak{b} = \varphi|\mathfrak{b}$ *is a representation of* $\mathfrak{g}$.

Let $x \in \mathfrak{a}$, $y \in \mathfrak{b}$. With the notation of 2.2.21, we have (on writing $U(\mathfrak{a}) = U$)

$$\begin{aligned}[\psi(x),\psi(y)] &= [L(x) \mid U,(L(y) - R(y)) \mid U] = [L(x),L(y)] \mid U \\ &= L([x,y]) \mid U = \psi[x,y]) \qquad \text{since } [x,y] \in \mathfrak{a}.\end{aligned}$$

2.5.4. Lemma. *Let* $\mathfrak{a}$ *be an ideal of* $\mathfrak{g}$, $\mathfrak{b}$ *a Lie subalgebra of* $\mathfrak{g}$ *such that* $\mathfrak{g} = \mathfrak{a} \oplus \mathfrak{b}$, V *a finite-dimensional vector space, and* ϱ *a representation of* $\mathfrak{a}$ *in* V *whose largest nilpotency ideal contains* $[\mathfrak{b},\mathfrak{a}]$.

(i) *There exists a finite-dimensional representation* σ *of* $\mathfrak{g}$, *whose largest nilpotency ideal contains* $\mathfrak{n}$, *such that* ϱ *is a quotient representation of* $\sigma|\mathfrak{a}$.

(ii) *If, for all* $y \in \mathfrak{b}$, $\mathrm{ad}_{\mathfrak{g}}y|\mathfrak{a}$ *is nilpotent, we can choose* σ *so that, in addition, the largest nilpotency ideal of* σ *contains* $\mathfrak{b}$.

We write $U(\mathfrak{a}) = U$. Let $V^1, \ldots, V^r$ be sub-$\mathfrak{a}$-modules of V, with sum V, such that, for all i, the U-module V^i is generated by a single element. Since the $\mathfrak{a}$-module V is a quotient of the $\mathfrak{a}$-module $V^1 \oplus \cdots \oplus V^r$, it suffices to prove the proposition with V replaced by V^i. Henceforth, we shall therefore assume that the U-module V is monogeneous. Let I be the kernel of ϱ (in U); it has finite codimension in U. By virtue of the left regular representation, we consider U as an $\mathfrak{a}$-module. Then U/I is an $\mathfrak{a}$-module which has V as quotient $\mathfrak{a}$-module.

Let $(V_0, V_1, \ldots, V_n)$ be a Jordan—Hölder series of the $\mathfrak{a}$-module V. Let ϱ_i be the representation of $\mathfrak{a}$ in V_i/V_{i+1}. Let I' be the intersection of the kernels (in U) of the ϱ_i. Then $I'^n \subset I \subset I'$ and $I' \cap \mathfrak{a} = \mathfrak{n}$. We introduce the notation of 2.5.3 and hence consider U as a $\mathfrak{g}$-module. If $x \in \mathfrak{b}$, then $\varphi(x)$ is a derivation of U which maps $\mathfrak{a}$ into $[\mathfrak{b},\mathfrak{a}] \subset I'$, and under which I' and I'^n are hence stable. Consequently, I'^n is a sub-$\mathfrak{g}$-module of U. The $\mathfrak{g}$-module U/I'^n defines a representation σ of $\mathfrak{g}$, which has finite dimension (2.5.1). For $x \in I' \cap \mathfrak{a}$ we have $x^n U \in I'^n$, hence $\sigma(x)^n = 0$; thus the largest nilpotency ideal of σ contains $\mathfrak{n}$. We then have the surjective $\mathfrak{a}$-module homomorphisms $U/I'^n \to U/I \to V$, whence (i).

Let us assume that $\mathrm{ad}_{\mathfrak{g}}y|\mathfrak{a}$ is nilpotent for all $y \in \mathfrak{b}$. Then for all $y \in \mathfrak{b}$, $\varphi(y)$ is locally nilpotent, hence $\sigma(y)$ is nilpotent. Since $[\mathfrak{b},\mathfrak{n}] \subset [\mathfrak{b},\mathfrak{a}] \subset \mathfrak{n}$, $\sigma(z)$ is nilpotent for all $z \in \mathfrak{b} + \mathfrak{n}$ (2.5.2). Now

$$[\mathfrak{g},\mathfrak{b} + \mathfrak{n}] \subset [\mathfrak{b},\mathfrak{b}] + [\mathfrak{b},\mathfrak{a}] + [\mathfrak{a},\mathfrak{n}] \subset \mathfrak{b} + \mathfrak{n},$$

hence the largest nilpotency ideal of σ contains $\mathfrak{b} + \mathfrak{n}$.

2.5.5. THEOREM. *Let* $\mathfrak{n}$ *be the largest nilpotent ideal of* $\mathfrak{g}$. *There is an finite-dimensional injective representation* ϱ *of* $\mathfrak{g}$ *whose largest nilpotency ideal contains* $\mathfrak{n}$.

(a) The one-dimensional Lie algebra k has the injective representation $\lambda \mapsto \begin{pmatrix} 0 & 0 \\ \lambda & 0 \end{pmatrix}$. Hence every commutative Lie algebra has a finite-dimensional injective representation whose image consists of nilpotent endomorphisms.

(b) Let $\mathfrak{c}$ be the centre of $\mathfrak{g}$. Then $\mathfrak{c} \subset \mathfrak{n}$. There exist ideals $\mathfrak{n}_0, \mathfrak{n}_1, \ldots, \mathfrak{n}_p$ of $\mathfrak{n}$ such that $\mathfrak{c} = \mathfrak{n}_0 \subset \mathfrak{n}_1 \subset \cdots \subset \mathfrak{n}_p = \mathfrak{n}$ and such that $\dim \mathfrak{n}_i/\mathfrak{n}_{i-1} = 1$ for $1 \leqq i \leqq p$. Then $\mathfrak{n}_i$ is the sum of $\mathfrak{n}_{i-1}$ and a one-dimensional Lie subalgebra. From (a), $\mathfrak{c}$ has a finite-dimensional injective representation φ whose image consists of nilpotent endomorphisms. Applying 2.5.4 in stages, we obtain a finite-dimensional representation ψ of $\mathfrak{n}$ such that every element of $\psi(\mathfrak{n})$ is nilpotent, and such that φ is a quotient of $\psi|\mathfrak{c}$.

(c) Let $\mathfrak{r}$ be the radical of $\mathfrak{g}$. Then $[\mathfrak{r},\mathfrak{r}] \subset \mathfrak{n} \subset \mathfrak{r}$ (1.7.1). Let $(\mathfrak{r}_0, \mathfrak{r}_1, \ldots, \mathfrak{r}_q)$ be a sequence of vector subspaces of $\mathfrak{r}$ such that $\mathfrak{n} = \mathfrak{r}_0 \subset \mathfrak{r}_1 \subset \cdots \subset \mathfrak{r}_q = \mathfrak{r}$, and such that $\dim \mathfrak{r}_i/\mathfrak{r}_{i-1} = 1$ for $1 \leqq i \leqq q$. The $\mathfrak{r}_i$ are ideals of $\mathfrak{r}$. The algebra $\mathfrak{r}_i$ is the sum of $\mathfrak{r}_{i-1}$ and a one-dimensional Lie subalgebra. Since $[\mathfrak{r},\mathfrak{r}] \subset \mathfrak{n}$, by applying 2.5.4 in stages we obtain a finite-dimensional representation τ of $\mathfrak{r}$ such that every element of $\tau(\mathfrak{n})$ is nilpotent, and such that ψ is a quotient of $\tau|\mathfrak{n}$.

(d) There exists a Lie subalgebra $\mathfrak{s}$ of $\mathfrak{g}$ such that $\mathfrak{g} = \mathfrak{s} \oplus \mathfrak{r}$ (1.6.9). We have $[\mathfrak{s},\mathfrak{r}] \subset \mathfrak{n}$ (1.7.1). From 2.5.4, there exists a finite-dimensional representation σ of $\mathfrak{g}$, such that every element of $\sigma(\mathfrak{n})$ is nilpotent, and such that τ is a quotient of $\sigma|\mathfrak{r}$. Then φ is a quotient of $\sigma|\mathfrak{c}$, hence $\sigma|\mathfrak{c}$ is injective.

(e) Let ϱ be the direct sum of σ and the adjoint representation. Then $\operatorname{Ker} \varrho = (\operatorname{Ker} \sigma) \cap \mathfrak{c} = 0$, hence ϱ is injective and finite-dimensional. Every element of $\varrho(\mathfrak{n})$ is nilpotent.

2.5.6. Let us return to the representation ϱ of theorem 2.5.5. The mapping $x \mapsto -\operatorname{tr} \varrho(x)$ of $\mathfrak{g}$ into k is a one-dimensional representation ϱ' of $\mathfrak{g}$, and $(\varrho \oplus \varrho')(x)$ has trace zero for all $x \in \mathfrak{g}$. Thus we see that every Lie algebra is isomorphic to a Lie subalgebra of $\mathfrak{sl}(V)$ for a vector space V of suitable finite dimension.

2.5.7. THEOREM. *Let* u *be a non-zero element of* $U(\mathfrak{g})$. *There exists a finite-dimensional representation* π *of* $\mathfrak{g}$ *such that* $\pi(u) \neq 0$.

(a) From 2.2.6 and 2.5.6, we may assume that $\mathfrak{g} = \mathfrak{sl}(V)$, where V is a finite-dimensional vector space. Let $(x_0, x_1, \ldots, x_p)$ be a basis for $\mathfrak{gl}(V)$ such that $x_0 = 1$ and $(x_1, \ldots, x_p)$ is a basis for $\mathfrak{sl}(V)$.

(b) Let n be the filtration of u. Then $u = u_1 + u_2$ with $u_1 \in U^n(\mathfrak{g})$, $u_2 \in U_{n-1}(\mathfrak{g})$, $u_1 \neq 0$, and

$$u_1 = \sum_{1 \leqq i_1 \leqq i_2 \leqq \cdots \leqq i_n \leqq p} \alpha_{i_1 \cdots i_n} \left(\sum_{\tau \in \mathfrak{S}_n} x_{i_{\tau(1)}} x_{i_{\tau(2)}} \cdots x_{i_{\tau(n)}} \right)$$

(c) Let $V_n = V \otimes V \otimes \cdots \otimes V$ (n factors), so that $\mathrm{End}(V_n)$ can be identified with $(\mathrm{End}\, V) \otimes \cdots \otimes (\mathrm{End}\, V)$ (n factors). The $x_{i_1} \otimes \cdots \otimes x_{i_n}$, where $0 \leqq i_1, \ldots, i_n \leqq p$, constitute a basis for the vector space $\mathrm{End}(V_n)$. Let F be the vector subspace of $\mathrm{End}(V_n)$ generated by the $u_1 \otimes \cdots \otimes u_n$, where $u_1, \ldots, u_n \in \mathrm{End}(V)$ and $u_j = 1$ for at least one j. The $x_{i_1} \otimes \cdots \otimes x_{i_n}$ such that one of the indices $i_1, \ldots, i_n$ is equal to 0 constitute a basis for F.

(d) Let σ be the identical representation of $\mathfrak{g}$ in V, and $\sigma_n = \sigma \otimes \cdots \otimes \sigma$ (n factors). If $x \in \mathfrak{g}$, then

$$\sigma_n(x) = x \otimes 1 \otimes \cdots \otimes 1 + 1 \otimes x \otimes 1 \otimes \cdots \otimes 1 + \cdots + 1 \otimes \cdots \otimes 1 \otimes x.$$

From this we deduce that $\sigma_n(u_2) \in F$. On the other hand,

$$\sigma_n(x_{i_1} \cdots x_{i_n}) \in \sum_{\tau \in \mathfrak{S}_n} x_{i_{\tau(1)}} \otimes x_{i_{\tau(2)}} \otimes \cdots \otimes x_{i_{\tau(n)}} + F$$

hence

$$\sigma_n \left(\sum_{\tau \in \mathfrak{S}_n} x_{i_{\tau(1)}} \cdots x_{i_{\tau(n)}} \right) \in n! \sum_{\tau \in \mathfrak{S}_n} x_{i_{\tau(1)}} \otimes \cdots \otimes x_{i_{\tau(n)}} + F.$$

The $\sum_{\tau \in \mathfrak{S}_n} x_{i_{\tau(1)}} \otimes \cdots \otimes x_{i_{\tau(n)}}$, for all sequences $(i_1, \ldots, i_n)$ such that $1 \leqq i_1 \leqq \cdots \leqq i_n \leqq p$, are elements of $\mathrm{End}(V_n)$ which, from what we said in (c), are linearly independent modulo F. It can thus be seen that $\sigma_n(u_1) \notin F$, whence $\sigma_n(u) \neq 0$,

Reference: [16], [62], [71].

2.6. The commutant of a simple module

2.6.1. Let n be an integer >0. For $\nu = (\nu_1, \ldots, \nu_n) \in \mathbf{N}^n$, let us set, as usual, $|\nu| = \nu_1 + \cdots + \nu_n$. For $r = 0, 1, 2, \ldots$, let J_r be the set of the $\nu \in \mathbf{N}^n$ such that $|\nu| = r$. There exists a unique ordering on $\mathbf{N}^n$ such that:

(1) $J_0 < J_1 < J_2 < \cdots$;

(2) on each J_r, the induced ordering is the lexicographic ordering.

For this ordering, which we denote by $\leqq$, we have the following properties:

(a) the ordered set $\mathbf{N}^n$ is isomorphic to $\mathbf{N}$;

(b) if $\nu, \nu' \in \mathbf{N}^n$, then $|\nu| < |\nu'| \Rightarrow \nu < \nu'$;
(c) if $\nu, \nu', \nu'' \in \mathbf{N}^n$, then $\nu \leqq \nu' \Leftrightarrow \nu + \nu'' \leqq \nu' + \nu''$.

2.6.2. We shall also use on $\mathbf{N}^n$ the product ordering of the natural ordering on the factors $\mathbf{N}$. For this ordering we have the following property:

LEMMA. *Let us order $\mathbf{N}^n$ by the product ordering. Let $S \subset \mathbf{N}^n$, and let S_0 be the set of minimal elements of S. Then S_0 is finite. Every element of S is larger than some element of S_0.*

Let us assume that the elements of S_0 can be arranged in an infinite sequence $(s_1, s_2, \ldots)$ of pairwise distinct elements. Every infinite sequence of distinct positive integers has an increasing infinite subsequence. By taking successive subsequences we could thus find an increasing infinite subsequence of $(s_1, s_2, \ldots)$, which is a contradiction. Hence S_0 is finite. The final assertion of the lemma is obvious.

2.6.3. LEMMA. *Let A be a commutative integral ring, B a commutative A-algebra with unity of finite type, and M a B-module of finite type. There exists $f \in A - \{0\}$ such that $M \otimes_A A_f$ is a free A_f-module.*

(The subring of the field of fractions of A generated by A and f^{-1} is denoted by A_f.)

By induction on the number of generators of M, we return to the case where M is monogeneous and hence is the quotient of B by an ideal; since such a quotient is a commutative A-algebra of finite type, it is sufficient to prove the existence of $f \in A - \{0\}$ such that $B \otimes_A A_f$ is a free A_f-module.

Let $(x_1, \ldots, x_n)$ be a generating system of the A-algebra B. For $\nu = (\nu_1, \ldots, \nu_n) \in \mathbf{N}^n$, let us set $x^\nu = x_1^{\nu_1} \cdots x_n^{\nu_n}$. Let us order $\mathbf{N}^n$ as in 2.6.1, and let us set

$$B_\nu = \sum_{\nu' \leqq \nu} x^{\nu'} A, \qquad B_\nu^- = \sum_{\nu' < \nu} x^{\nu'} A.$$

Let x_*^ν be the image of x^ν in B_ν / B_ν^-, and I_ν the annihilator of x_*^ν in A. For $i = 1, \ldots, n$, let ε_i be the element of $\mathbf{N}^n$ all of whose coordinates are zero except the ith which is equal to 1. If $a \in I_\nu$, we have $x^\nu a \in B_\nu^-$, hence

$$x^{\nu+\varepsilon_i} a = x_i x^\nu a \in x_i B_\nu^- \subset B_{\nu+\varepsilon_i}^-,$$

and consequently $I_\nu \subset I_{\nu+\varepsilon_i}$. Let Λ be the set of the $\nu \in \mathbf{N}^n$ such that $I_\nu \neq 0$. There exist $\nu_1, \ldots, \nu_r \in \Lambda$ such that every non-null I_ν contains one of the I_{ν_i} (2.6.2).

There exists a non-zero element f in $I_{\nu_1} \cap \cdots \cap I_{\nu_r}$. If $\nu \in \Lambda$, we have $f \in I_\nu$, and hence $(B_\nu/B_\nu^-) \otimes_A A_f = 0$. If $\nu \notin \Lambda$, the module B_ν/B_ν^- is isomorphic to A, hence $(B_\nu/B_\nu^-) \otimes_A A_f$ is isomorphic to A_f. Thus, B_f is a successive extension of free A_f-modules, and hence is a free A_f-module.

2.6.4. LEMMA. *Let C be an algebra provided with an increasing filtration $(C_0, C_1, \ldots)$. We assume that the associated graded algebra* $\mathrm{gr}(C)$ *is of finite type and commutative. Let M be a simple C-module. Let D be the set of C-endomorphisms of M (the set D is a field from Schur's lemma). Let $x \in D$. Then x is algebraic over k.*

Let A be the subalgebra $k[x]$ of D generated by 1 and x. Let us assume that x is transcendental over k. Then A can be identified with the algebra of polynomials in one indeterminate x over k.

Let V be the algebra $A \otimes X$. There exists on M one and only one V-module structure such that $(p \otimes c)m = pcm = cpm$ for $m \in M$, $p \in A$, $c \in C$. We choose a non-zero element m_0 of M; then $M = V \cdot m_0$. For $r = 0, 1, \ldots$, we set $V_r = A \otimes C_r$ and $M_r = V_r m_0$. Thus M becomes a filtered module over the filtered ring V. The graded $\mathrm{gr}(V)$-module $\mathrm{gr}(M)$ is monogeneous, and $\mathrm{gr}(V)$ is a commutative algebra of finite type over A. There exists $f \in A - \{0\}$ such that $\mathrm{gr}(M) \otimes_A A_f$ is free over A_f (2.6.3). Since A_f is principal, each $(M_r/M_{r-1}) \otimes_A A_f$ is free over A_f (AL VII, p. 97). Then $M \otimes_A A_f$ is a successive extension of free modules over A_f, and hence is free over A_f.

Let g be an element of $A - \{0\}$ which does not divide a power of f. The multiplication by g in A_f is not surjective. Hence the homothety η with ratio g in the A_f-module $M \otimes_A A_f$ is not surjective. Now, $\eta(m \otimes a) = m \otimes ga = gm \otimes a$ for all $m \in M$ and all $a \in A_f$. Since D is a field, the mapping $m \mapsto gm$ of M into M is bijective, whence we have a contradiction.

2.6.5. PROPOSITION. *Let W be a $\mathfrak{g}$-module. The following conditions are equivalent:*

(i) *W is absolutely simple;*

(ii) *there exists an algebraically closed extension k' of k such that $W \otimes k'$ is a simple $(\mathfrak{g} \otimes k')$-module;*

(iii) *W is simple and every $\mathfrak{g}$-endomorphism of W is scalar;*

(iv) *$W \neq 0$, and, for any $x_1, \ldots, x_n, y_1, \ldots, y_n \in W$ with $x_1, \ldots, x_n$ linearly independent, there exists $u \in U(\mathfrak{g})$ such that*

$$ux_1 = y_1, \ldots, ux_n = y_n.$$

Let D be the set of $\mathfrak{g}$-endomorphisms of W, and D' the commutant of D in $\mathrm{End}(W)$.

(i) $\Rightarrow$ (ii). Obvious.

(ii) $\Rightarrow$ (iii). We assume that (ii) is satisfied. It is clear that W is simple. Let φ be a $\mathfrak{g}$-endomorphism of W. From 2.6.4, there exists $\lambda \in k'$ such that the $(\mathfrak{g} \otimes k')$-endomorphism $\varphi \otimes 1$ of $W \otimes k'$ is the homothety with ratio λ. Since W is stable under $\varphi \otimes 1$, we have $\lambda \in k$.

(iii) $\Rightarrow$ (iv). We assume that W is simple and that $D = k$. Then $D' = \mathrm{End}(W)$. Let $x_1, \ldots, x_n, y_1, \ldots, y_n$ be as in (iv). There exists $f \in \mathrm{End}(W)$ such that $f(x_1) = y_1, \ldots, f(x_n) = y_n$. From the density theorem (AL VIII, p. 39), and since $f \in D'$, there exists $u \in U(\mathfrak{g})$ such that $ux_1 = y_1, \ldots, ux_n = y_n$.

(iv) $\Rightarrow$ (i). We assume that (iv) is satisfied. Let k_1 be an extension of k, and $w \in (W \otimes k_1) - \{0\}$. Then $w = x_1 \otimes \lambda_1 + \cdots + x_n \otimes \lambda_n$, where $\lambda_1, \ldots, \lambda_n$ are non-zero elements of k_1, and $x_1, \ldots, x_n$ are linearly independent elements of W. Let $y \in W$. There exists $u \in U(\mathfrak{g})$ such that $ux_1 = y$, $ux_2 = \cdots = ux_n = 0$. Then $uw = y \otimes \lambda_1$. Hence $U(\mathfrak{g} \otimes k_1) \cdot w = W \otimes k_1$, so that the $(\mathfrak{g} \otimes k_1)$-module $W \otimes k_1$ is simple.

2.6.6. Corollary. *If k is algebraically closed, every simple $\mathfrak{g}$-module is absolutely simple.*

2.6.7. Let V be a $\mathfrak{g}$-module. If, for all $z \in Z(\mathfrak{g})$, z_V is a homothety with ratio $\chi(z) \in k$, we say that the $\mathfrak{g}$-module V *admits a central character*. The mapping χ is then a homomorphism of the algebra $Z(\mathfrak{g})$ into k, termed the *central character* of the $\mathfrak{g}$-module V, or of the corresponding representation.

2.6.8. Proposition (k algebraically closed). *Every simple $\mathfrak{g}$-module admits a central character.*

This follows from 2.6.5.

2.6.9. Proposition. *Let W be a simple $\mathfrak{g}$-module, D the set of $\mathfrak{g}$-endomorphisms of W (D is a field from Schur's lemma), and k' an extension of k. Then:*

(i) $\dim_k D < +\infty$.

(ii) *The $(\mathfrak{g} \otimes k')$-module $W \otimes k'$ is a finite sum of simple $(\mathfrak{g} \otimes k')$-modules.*

(a) Let N' be a sub-$(\mathfrak{g} \otimes k')$-module of $W \otimes k'$. Then there exists a sub-extension of finite type $k'' \subset k'$ of k and a sub-$(\mathfrak{g} \otimes k'')$-module N'' of $W \otimes k''$ such that $N' = N'' \otimes_{k''} k'$. Indeed, $W \otimes k'$ is a monogeneous $U(\mathfrak{g} \otimes k')$-module, and hence Noetherian. Consequently, N' is generated by a finite number of elements $x_1, \dots, x_p$. There exists a subextension of finite type $k'' \subset k'$ of k such that $x_1, \dots, x_p \in W \otimes k''$. Let N'' be the sub-$(\mathfrak{g} \otimes k'')$-module of $W \otimes k''$ generated by $x_1, \dots, x_p$. Then $N'' \otimes_{k''} k' = N'$.

(b) We assume that k' is algebraic over k. Let N' be a sub-$(\mathfrak{g} \otimes k')$-module of $W \otimes k'$. We introduce k'' and N'' with the properties of (a). Then k'' is of finite degree over k. From 1.2.19 (d), there exists a sub-$(\mathfrak{g} \otimes k'')$-module N_1'' of $W \otimes k''$ which is complementary to N''. Then $N_1'' \otimes_{k''} k'$ is a sub-$(\mathfrak{g} \otimes k')$-module of $W \otimes k'$ which is complementary to N'. This proves that $W \otimes k'$ is a semi-simple $(\mathfrak{g} \otimes k')$-module.

(c) We assume that k' is the algebraic closure of k. From (b), there exists in $W \otimes k'$ a simple sub-$(\mathfrak{g} \otimes k')$-module N'. We introduce k'' and N'' with the properties of (a). We may assume k'' to be Galois over k. Let Γ be the Galois group of k'' over k. Then $\sum_{\gamma \in \Gamma} (1 \otimes \gamma)(N'')$ is a non-null sub-$(\mathfrak{g} \otimes k'')$-module of $W \otimes k''$ which is invariant under $1 \otimes \Gamma$, and hence equal to $W \otimes k''$ since W is simple. For all $\gamma \in \Gamma$, let $\bar{\gamma}$ be a k-automorphism of k' extending γ. Then

$$W \otimes k' = \sum_{\gamma \in \Gamma} (1 \otimes \bar{\gamma})(N'),$$

which proves that $W \otimes k'$ is a finite sum of simple $(\mathfrak{g} \otimes k')$-modules. From 2.6.5, every endomorphism of a simple $(\mathfrak{g} \otimes k')$-module is scalar. Hence $D \otimes k' = \mathrm{End}_{\mathfrak{g} \otimes k'}(W \otimes k')$ is finite-dimensional over k', which proves (i).

(d) We have

$$W \otimes k' = (W \otimes_D D) \otimes_k k' = W \otimes_D (D \otimes_k k').$$

The sub-$(\mathfrak{g} \otimes k')$-modules of $W \otimes k'$ are of the form $W \otimes_D I$, where I is a left ideal of $D \otimes_k k'$ (AL VIII, p. 43). Since $\dim_k D < +\infty$, the algebra $D \otimes_k k'$ is semi-simple (AL VIII, p. 85). This proves (ii).

References: [30], [35], [102].

2.7. The dual of the enveloping algebra

2.7.1. We identify $U(\mathfrak{g} \times \mathfrak{g})$ with $U(\mathfrak{g}) \otimes U(\mathfrak{g})$ (2.2.12). Let d be the diagonal mapping $x \mapsto (x,x) = x \otimes 1 + 1 \otimes x$ of $\mathfrak{g}$ into $\mathfrak{g} \times \mathfrak{g}$; it is a Lie algebra homomorphism. We can thus consider $c = U(d)$, which is a homomorphism of the algebra $U(\mathfrak{g})$ into the algebra $U(\mathfrak{g}) \otimes U(\mathfrak{g})$; this homomorphism c is termed the *coproduct of* $U(\mathfrak{g})$. We have

$$c(x) = x \otimes 1 + 1 \otimes x \qquad \text{for all } x \in \mathfrak{g},$$

$$c(u^{\mathsf{T}}) = (c(u))^{\mathsf{T}} \qquad \text{for all } u \in U(\mathfrak{g}).$$

If $\varphi : \mathfrak{g} \to \mathfrak{g}'$ is a Lie algebra homomorphism, and we denote the coproduct of $U(\mathfrak{g}')$ by c', then

$$c' \circ U(\varphi) = (U(\varphi) \otimes U(\varphi)) \circ c.$$

2.7.2. Proposition. *Let* $(e_1, \ldots, e_n)$ *be a basis for* $\mathfrak{g}$. *For* $\nu = (\nu_1, \ldots, \nu_n) \in \mathbf{N}^n$, *we set* $e_\nu = e_1^{\nu_1} \cdots e_n^{\nu_n} / \nu_1! \cdots \nu_n!$. *Then, for all* $\nu \in \mathbf{N}^n$, *we have*

$$c(e_\nu) = \sum_{\lambda+\mu=\nu} e_\lambda \otimes e_\mu .$$

Indeed,

$$\frac{1}{\nu_1!} c(e_1^{\nu_1}) = \frac{1}{\nu_1!} c(e_1)^{\nu_1} = \frac{1}{\nu_1!} (e_1 \otimes 1 + 1 \otimes e_1)^{\nu_1}$$
$$= \sum_{\lambda_1+\mu_1=\nu_1} \frac{1}{\lambda_1!\mu_1!} e_1^{\lambda_1} \otimes e_1^{\mu_1},$$

hence

$$c(e_\nu) = \prod_{i=1}^{n} \sum_{\lambda_i+\mu_i=\nu_i} \frac{1}{\lambda_i!\mu_i!} e_i^{\lambda_i} \otimes e_i^{\mu_i}$$
$$= \sum_{\substack{\lambda_1+\mu_1=\nu_1,\ldots,\\ \lambda_n+\mu_n=\nu_n}} \frac{1}{\lambda_1!\mu_1! \ldots \lambda_n!\mu_n!} e_1^{\lambda_1} \cdots e_n^{\lambda_n} \otimes e_1^{\mu_1} \cdots e_n^{\mu_n}$$
$$= \sum_{\substack{\lambda,\mu\in\mathbf{N}^n,\\ \lambda+\mu=\nu}} e_\lambda \otimes e_\mu .$$

2.7.3. Let ϱ_1 and ϱ_2 be representations of $\mathfrak{g}$ in V_1 and V_2, ϱ their tensor product, and σ the representation of $\mathfrak{g}\times\mathfrak{g}$ in $V_1 \otimes V_2$ such that

$$\sigma(u_1 \otimes u_2) = \varrho_1(u_1) \otimes \varrho_2(u_2) \quad \text{for } u_1, u_2 \in U(\mathfrak{g}).$$

Then

$$\varrho = \sigma \circ c.$$

Indeed, for all $x \in \mathfrak{g}$, we have

$$(\sigma \circ c)(x) = \sigma(x \otimes 1 + 1 \otimes x) = \varrho_1(x) \otimes 1 + 1 \otimes \varrho_2(x) = \varrho(x).$$

2.7.4. We denote the dual vector space of $U(\mathfrak{g})$ by $U(\mathfrak{g})^*$. The transpose ${}^t c$ ot c defines by restriction a linear mapping of $U(\mathfrak{g})^* \otimes U(\mathfrak{g})^*$ into $U(\mathfrak{g})^*$. The vector space $U(\mathfrak{g})^*$ is thus equipped with the structure of an *algebra* (which, as we shall see, is associative). If $\varphi : \mathfrak{g} \to \mathfrak{g}'$ is a Lie algebra homomorphism, then ${}^t U(\varphi) : U(\mathfrak{g}')s \to U(\mathfrak{g})^*$ is an algebra homomorphism from 2.7.1.

2.7.5. Proposition. *Let* $(e_1, \ldots, e_n)$ *be a basis for* $\mathfrak{g}$, *and* $k[[X_1, \ldots, X_n]]$ *the algebra of formal series over* k *in* n *indeterminates* $X_1, \ldots, X_n$. *For* $\nu = (\nu_1, \ldots, \nu_n) \in \mathbf{N}^n$, *let us define* e_ν *as in* 2.7.2, *and let* $X^\nu = X_1^{\nu_1} \cdots X_n^{\nu_n}$.

If $f \in U(\mathfrak{g})^*$, *we denote the formal series* $\sum_{\nu \in \mathbf{N}^n} f(e_\nu) X^\nu$ *by* s_f. *Then* $f \mapsto s_f$ *is an isomorphism of the algebra* $U(\mathfrak{g})^*$ *onto the algebra* $k[[X_1, \ldots, X_n]]$.

Since $(e_\nu)_{\nu \in \mathbf{N}^n}$ is a basis for $U(\mathfrak{g})$ (2.1.11), the mapping $f \mapsto s_f$ of $U(\mathfrak{g})^*$ into $k[[X_1, \ldots, X_n]]$ is bijective. Moreover, if $f,g \in U(\mathfrak{g})^*$, then

$$\begin{aligned} s_{fg} &= \sum_{\nu \in \mathbf{N}^n} \langle fg, e_\nu \rangle X^\nu = \sum_{\nu \in \mathbf{N}^n} \langle f \otimes g, c(e_\nu) \rangle X^\nu \\ &= \sum_{\nu \in \mathbf{N}^n} \langle f \otimes g, \sum_{\substack{\lambda,\mu \in \mathbf{N}^n, \\ \lambda + \mu = \nu}} e_\lambda \otimes e_\mu \rangle X^\nu \qquad \text{from 2.7.2} \\ &= \sum_{\lambda,\mu \in \mathbf{N}^n} \langle f, e_\lambda \rangle \langle g, e_\mu \rangle X^{\lambda+\mu} = s_f s_g . \end{aligned}$$

2.7.6. In particular, the algebra $U(\mathfrak{g})^*$ is associative and commutative, and its unity is the linear form ε on $U(\mathfrak{g})$ such that $\operatorname{Ker} \varepsilon = U_+(\mathfrak{g})$ and $\varepsilon(1) = 1$. From 2.7.1, the transpose of the principal anti-automorphism of $U(\mathfrak{g})$ is an automorphism of the algebra $U(\mathfrak{g})^*$, denoted by $f \mapsto f^{\mathsf{T}}$ and termed the *principal automorphism of* $U(\mathfrak{g})^*$.

2.7.7. For $u \in U(\mathfrak{g})$, we have defined in 2.2.21 the endomorphisms $L(u)$ and $R(u)$ of the vector space $U(\mathfrak{g})$. Let us consider their transposes ${}^tL(u)$ and ${}^tR(u)$ in $U(\mathfrak{g})^*$. From 2.2.21, the mappings $u \mapsto {}^tL(u^{\mathsf{T}})$, $u \mapsto {}^tR(u)$ are representations of $U(\mathfrak{g})$. Their restrictions to $\mathfrak{g}$ are termed *left and right coregular representations* of $\mathfrak{g}$; they are interchanged by the principal automorphism of $U(\mathfrak{g})^*$.

If $x \in \mathfrak{g}$, ${}^tL(x)$ and ${}^tR(x)$ are derivations of the algebra $U(\mathfrak{g})^*$; indeed, let c be the coproduct of $U(\mathfrak{g})$; for $f,g \in U(\mathfrak{g})^*$ and $u \in U(\mathfrak{g})$, we have

$$\begin{aligned} \langle {}^tL(x)(fg), u \rangle &= \langle fg, xu \rangle = \langle f \otimes g, (x \otimes 1 + 1 \otimes x) c(u) \rangle \\ &= \langle {}^tL(x) f \otimes g, c(u) \rangle + \langle f \otimes {}^tL(x) g, c(u) \rangle \\ &= \langle {}^tL(x) f \cdot g + f \cdot {}^tL(x) g, u \rangle , \end{aligned}$$

whence results our assertion concerning ${}^tL(x)$. A similar reasoning applies for ${}^tR(x)$.

2.7.8. Let V be a vector space, and ϱ a representation of $\mathfrak{g}$, or of $U(\mathfrak{g})$, in V. For $v \in V$ and $v' \in V^*$, we denote the linear form $u \mapsto \langle \varrho(u)v, v' \rangle$ on $U(\mathfrak{g})$ by $\theta^\rho(v,v')$, or simply by $\theta(v,v')$. Hence we have $\theta(v,v') \in U(\mathfrak{g})^*$.

The $\theta(v,v')$, for $v \in V$ and $v' \in V^*$, are termed the *coefficients* of ϱ. The vector subspace of $U(\mathfrak{g})^*$ generated by the coefficients of ϱ is denoted by $C(\varrho)$. Clearly, $\operatorname{Ker} \varrho$ is the orthogonal subspace of $C(\varrho)$ in $U(\mathfrak{g})$. If ϱ is finite-dimensional, $C(\varrho)$ is finite-dimensional, and hence, from the classical theory of duality, $C(\varrho)$ is the orthogonal subspace of Ker ϱ in $U(\mathfrak{g})^*$.

2.7.9. LEMMA. *Let $\mathfrak{g}_1$ be a Lie algebra, V_1 a vector space, ϱ_1 a representation of $\mathfrak{g}_1$ in V_1, $v_1 \in V_1$ and $v_1' \in V_1^*$. Let $\mathfrak{g}_2$, V_2, ϱ_2, v_2 and v_2' be defined similarly. Let σ be the representation of*

$$U(\mathfrak{g}_1 \times \mathfrak{g}_2) = U(\mathfrak{g}_1) \otimes U(\mathfrak{g}_2) \quad in \quad V_1 \otimes V_2$$

such that

$$\sigma(u_1 \otimes u_2) = \varrho_1(u_1) \otimes \varrho_2(u_2) \quad for\ u_1 \in U(\mathfrak{g}_1), \quad u_2 \in U(\mathfrak{g}_2).$$

Then

$$\theta^\sigma(v_1 \otimes v_2, v_1' \otimes v_2') = \theta^{\varrho_1}(v_1, v_1') \otimes \theta^{\varrho_2}(v_2, v_2').$$

Indeed, if $u_1 \in U(\mathfrak{g}_1)$ and $u_2 \in U(\mathfrak{g}_2)$, we have

$$\begin{aligned}\langle \theta(v_1 \otimes v_2, v_1' \otimes v_2'), u_1 \otimes u_2 \rangle &= \langle \sigma(u_1 \otimes u_2)(v_1 \otimes v_2), v_1' \otimes v_2' \rangle \\ &= \langle \varrho_1(u_1)v_1, v_1' \rangle \langle \varrho_2(u_2)v_2, v_2' \rangle \\ &= \langle \theta(v_1, v_1'), u_1 \rangle \langle \theta(v_2, v_2'), u_2 \rangle \\ &= \langle \theta(v_1, v_1') \otimes \theta(v_2, v_2'), u_1 \otimes u_2 \rangle\end{aligned}$$

2.7.10. PROPOSITION. *Let V_1 and V_2 be vector spaces, let ϱ_1 and ϱ_2 be representations of $\mathfrak{g}$ in V_1 and V_2 let, $v_1 \in V_1$, $v_1' \in V_1^*$, $v_2 \in V_2$, $v_2' \in V_2^*$ and let ϱ be the tensor product of ϱ_1 and ϱ_2. Then:*

(i) $\theta^\varrho(v_1 \otimes v_2, v_1' \otimes v_2') = \theta^{\varrho_1}(v_1, v_1')\, \theta^{\varrho_2}(v_2, v_2')$;

(ii) $C(\varrho) = C(\varrho_1)C(\varrho_2)$;

(iii) *If ϱ is finite-dimensional, then* $C(\varrho)^{\mathsf{T}} = C(\varrho^*)$.

Let σ be as in 2.7.9, and let c be the coproduct of $U(\mathfrak{g})$. For $u \in U(\mathfrak{g})$, we have

$$\begin{aligned}&\langle \theta^{\varrho_1}(v_1, v_1')\, \theta^{\varrho_2}(v_2, v_2'), u \rangle = \\ &\quad = \langle \theta^{\varrho_1}(v_1, v_1') \otimes \theta^{\varrho_2}(v_2, v_2'), c(u) \rangle \\ &\quad = \langle \theta^\sigma(v_1 \otimes v_2, v_1' \otimes v_2'), c(u) \rangle && \text{from 2.7.9} \\ &\quad = \langle \sigma(c(u))(v_1 \otimes v_2), v_1' \otimes v_2' \rangle \\ &\quad = \langle \varrho(u)(v_1 \otimes v_2), v_1' \otimes v_2' \rangle && \text{from 2.7.3} \\ &\quad = \langle \theta^\varrho(v_1 \otimes v_2, v_1' \otimes v_2'), u \rangle.\end{aligned}$$

This proves (i), and (ii) follows from (i). Assertion (iii) is obvious.

2.7.11. PROPOSITION. *Let V be a vector space, and ϱ a representation of $\mathfrak{g}$ in V.*

(i) *If $v' \in V^*$, the mapping $v \mapsto \theta(v,v')$ of V into $U(\mathfrak{g})^*$ equipped with the right coregular representation is a $\mathfrak{g}$-module homomorphism $h_{v'}$. The mapping $v' \mapsto h_{v'}$ of V^* into $\mathrm{Hom}_{\mathfrak{g}}(V,U(\mathfrak{g})^*)$ is bijective.*

(ii) *If $v \in V$, the mapping $v' \mapsto \theta(v,v')$ of V^* into $U(\mathfrak{g})^*$ equipped with the left coregular representation is a $\mathfrak{g}$-module homomorphism h_v. The mapping $v \mapsto h_v$ of V into $\mathrm{Hom}(V^*,U(\mathfrak{g})^*)$ is injective; if V is finite-dimensional then it is surjective.*

(iii) *$C(\varrho)$ is a sub-$\mathfrak{g}$-module of $U(\mathfrak{g})^*$ for the left and right coregular representations.*

Let $v \in V$, $v' \in V^*$, and $u,u' \in U(\mathfrak{g})$. Then

$$\begin{aligned}\langle \theta(\varrho(u)v,v'),u'\rangle &= \langle \varrho(u')\varrho(u)v,v'\rangle = \langle \varrho(u'u)v,v'\rangle \\ &= \langle \theta(v,v'),u'u\rangle = \langle {}^{t}R(u)\theta(v,v'),u'\rangle,\end{aligned}$$

hence $v \mapsto \theta(v,v')$ is a $\mathfrak{g}$-homomorphism $h_{v'}$ of V into $U(\mathfrak{g})^*$ equipped with the right coregular representation. If $h_{v'} = 0$, then

$$\langle \varrho(1)v,v'\rangle = \langle h_{v'}(v),1\rangle = 0 \quad \text{for all } v \in V,$$

and hence $v' = 0$. Let $\varphi : V \to U(\mathfrak{g})^*$ be a $\mathfrak{g}$-homomorphism for the right coregular representation. Let v_0' be the element $v \mapsto \langle \varphi(v),1\rangle$ of V^*. For all $v \in V$ and all $u \in U(\mathfrak{g})$, we have

$$\begin{aligned}\langle \theta(v,v_0'),u\rangle &= \langle \varrho(u)v,v_0'\rangle = \langle \varphi(uv),1\rangle \\ &= \langle {}^{t}R(u)\varphi(v),1\rangle = \langle \varphi(v),u\rangle,\end{aligned}$$

and hence $\varphi(v) = \theta(v,v_0')$ and $\varphi = h_{v_0'}$. This proves (i).

Again let $v \in V$, $v' \in V^*$ and $u,u' \in U(\mathfrak{g})$. Then

$$\begin{aligned}\langle \theta(v,\varrho^*(u)v'),u'\rangle &= \langle \varrho(u')v,\varrho^*(u)v'\rangle = \langle \varrho(u^{\mathsf{T}}u')v,v'\rangle \\ &= \langle \theta(v,v'),u^{\mathsf{T}}u'\rangle = \langle {}^{t}L(u^{\mathsf{T}})\theta(v,v'),u'\rangle,\end{aligned}$$

and hence $v' \mapsto \theta(v,v')$ is a $\mathfrak{g}$-homomorphism h_v of V^* into $U(\mathfrak{g})^*$ equipped with the left coregular representation. If $h_v = 0$, then

$$\langle \varrho(1)v,v'\rangle = \langle h_v(v'),1\rangle = 0 \quad \text{for all } v' \in V^*,$$

and hence $v = 0$. Let us assume that V is finite-dimensional, and let $\varphi : V^* \to U(\mathfrak{g})^*$ be a $\mathfrak{g}$-homomorphism for the left coregular representation. The mapping $v' \mapsto \langle \psi(v'),1\rangle$ of V^* into k can be identified with an element

v_0 of V. For all $v' \in V^*$ and all $u \in U(\mathfrak{g})$, we have

$$\langle \theta(v_0,v'),u\rangle = \langle \varrho(u)v_0,v'\rangle = \langle v_0,\varrho^*(u^{\mathsf{T}})v'\rangle = \langle \psi(u^{\mathsf{T}}v'),1\rangle$$
$$= \langle {}^{\mathrm{t}}L(u)\psi(v'),1\rangle = \langle \psi(v'),u\rangle ,$$

hence $\psi(v') = \theta(v_0,v')$ and $\psi = h_{v_0}$. This proves (ii).

Assertion (iii) follows from (i) and (ii).

2.7.12. PROPOSITION. *Let Φ and Ψ be the sums of the $C(\varrho)$ when ϱ runs through the sets of finite-dimensional representations and finite-dimensional simple representations of $\mathfrak{g}$ respectively.*

(i) *Φ is the sum of the finite-dimensional sub-$\mathfrak{g}$-modules of $U(\mathfrak{g})^*$ equipped with the left coregular representation, or with the right coregular representation.*

(ii) *Ψ is the sum of the finite-dimensional simple sub-$\mathfrak{g}$-modules of $U(\mathfrak{g})^*$ equipped with the left coregular representation, or with the right coregular representation.*

(iii) *Φ and Ψ are subalgebras of $U(\mathfrak{g})^*$. Using the notation of 2.7.6 and denoting the one-dimensional null representation of $\mathfrak{g}$ by ϱ_0, we have $C(\varrho_0)=k\cdot\varepsilon$.*

Assertions (i) and (ii) follow from 2.7.11. From 2.7.10 (ii) (or 2.7.10 (ii) and 1.7.8), Φ (or Ψ) is a subalgebra of $U(\mathfrak{g})^*$. Clearly, $C(\varrho_0) = k \cdot \varepsilon$.

2.7.13. With the notation of 2.7.12, we have

$$\Psi = \bigoplus_{\varrho\in\mathfrak{g}^\wedge} C(\varrho) = k \cdot \varepsilon \oplus \bigoplus_{\substack{\varrho\in\mathfrak{g}^\wedge,\\ \varrho\neq\varrho_0}} C(\varrho).$$

Let $\varphi_0 : \Psi \to k$ be the linear form which is zero on $\oplus_{p\in\mathfrak{g}^\wedge,p\neq p_0}(\varrho)$ and takes the value 1 at ε. Let $B_0 : \Psi\times\Psi \to k$ be the bilinear form $(f,g) \mapsto \varphi_0(fg)$. We say that φ_0 is the *fundamental linear form* on Ψ and β_0 is the *fundamental bilinear form* on $\Psi\times\Psi$.

The form φ_0 is obviously invariant under both coregular representations. Taking 2.7.7 into account, we deduce that β_0 is invariant under both coregular representations.

2.7.14. LEMMA. *Let V be a finite-dimensional, $\mathfrak{g}$-module $(e_1, \ldots, e_n)$ a basis for V, $(e_1^*, \ldots, e_n^*)$ the dual basis for V^*, $v \in V$, and $v^* \in V^*$. With the notation of 2.7.6, we have*

$$\sum_i \theta(e_i,v^*)\, \theta(e_i^*,v) = \langle v,v^*\rangle \cdot \varepsilon .$$

Let $x \in \mathfrak{g}$, $u \in U(\mathfrak{g})$, and let $\sum u_j \otimes u_j'$ be the image of u under the coproduct. From 2.7.7 and 2.7.11, we have, for the right coregular representation,

$$x \cdot \Big(\sum_i \theta(e_i,v^*)\theta(e_i^*,v)\Big) = \sum_i \theta(xe_i,v^*)\,\theta(e_i^*,v) + \theta(e_i,v^*)\,\theta(xe_i^*,v),$$

hence

$$\langle x \cdot \Big(\sum_i \theta(e_i,v)\,\theta(e_i^*,v)\Big),u\rangle =$$

$$\begin{aligned} &= \sum_{i,j} \langle\theta(xe_i,v^*),u_j\rangle\langle\theta(e_i^*,v),u_j'\rangle + \langle\theta(e_i,v^*),u_j\rangle\langle\theta(xe_i^*,v),u_j'\rangle \\ &= \sum_j \Big(\sum_i \langle u_jxe_i,v^*\rangle\langle u_j'e_i^*,v\rangle + \langle u_je_i,v^*\rangle\langle u_j'xe_i^*,v\rangle\Big) \\ &= \sum_j \Big(\sum_i \langle e_i, -xu_j^{\mathsf{T}}v^*\rangle\langle e_i^*,u_j'^{\mathsf{T}}v\rangle + \langle e_i,u_j^{\mathsf{T}}v^*\rangle\langle e_i^*, -xu_j'^{\mathsf{T}}v\rangle\Big) \\ &= \sum_j (\langle u_j'^{\mathsf{T}}v, -xu_j^{\mathsf{T}}v^*\rangle + \langle -xu_j'^{\mathsf{T}}v,u_j^{\mathsf{T}}v^*\rangle) = 0. \end{aligned}$$

Consequently, there exists a $\lambda \in k$ such that

$$\sum_i \theta(e_i,v^*)\,\theta(e_i^*,v) = \lambda\varepsilon.$$

Then

$$\begin{aligned} \lambda = \langle\lambda\varepsilon,1\rangle &= \sum_i \langle\theta(e_i,v^*)\,\theta(e_i^*,v),1\rangle \\ &= \sum_i \langle e_i,v^*\rangle\langle e_i^*,v\rangle = \langle v,v^*\rangle. \end{aligned}$$

2.7.15. Proposition (notation as in 2.7.13).

(i) *Let* $\varrho,\varrho' \in \mathfrak{g}^\wedge$ *such that* $\varrho' \neq \varrho^*$. *Then* $C(\varrho)$ *and* $C(\varrho')$ *are orthogonal with respect to* β_0.

(ii) *The restriction of* β_0 *to* $C(\varrho)\times C(\varrho^*)$ *is non-degenerate.*

Let $\varrho,\varrho' \in \mathfrak{g}^\wedge$, and let us assume that the restriction of β_0 to $C(\varrho)\times C(\varrho')$ is non-zero. Let V and V' be the spaces of ϱ and ϱ'. From 2.7.11, there exists a non-zero $\mathfrak{g}$-invariant bilinear form on $V\times V'$. Hence there exists a non-zero $\mathfrak{g}$-homomorphism of V into V'^*. Since ϱ and ϱ' are simple, we deduce that $\varrho' = \varrho^*$. This proves (i).

Let us assume that the restriction of β_0 to $C(\varrho)\times C(\varrho^*)$ is degenerate. Let W be the orthogonal subspace of $C(\varrho^*)$ in $C(\varrho)$. Then $W \neq 0$, and W is a sub-$\mathfrak{g}$-module of $C(\varrho)$ for both coregular representations (since β_0 is $\mathfrak{g}$-invariant). Let V be the space of ϱ. From 2.7.11, there exists $v_0' \in V^*$

such that the mapping $v \mapsto \theta(v,v'_0)$ of V into Ψ is an isomorphism of V onto a sub-$\mathfrak{g}$-module of W (for the right coregular representation). Let $(e_1, \ldots, e_n)$ be a basis for V, $(e_1^*, \ldots, e_n^*)$ the dual basis for V^*, and $v_0 \in V$ such that $\langle v'_0, v_0 \rangle \neq 0$. From 2.7.14, we have

$$\sum_i \beta_0(\theta(e_i, v'_0), \theta(e_i^*, v_0)) \neq 0.$$

Now this is impossible because $\theta(e_i, v'_0) \in W$ and $\theta(e_i^*, v_0) \in C(\varrho^*)$ for all i.

2.7.16. Let $\mathfrak{h}$ be a Lie subalgebra of $\mathfrak{g}$. Let T be the set of the $f \in U(\mathfrak{g})^*$ such that $f(\mathfrak{h}U(\mathfrak{g})) = 0$. It follows directly that T is a sub-$\mathfrak{g}$-module of $U(\mathfrak{g})^*$ for the right coregular representation, and that T is a subalgebra of $U(\mathfrak{g})^*$. We require the following generalisation of 2.7.15 (ii).

2.7.17. Proposition (notation as in 2.7.13 and 2.7.16). *Assume $\mathfrak{h}$ to be reductive in $\mathfrak{g}$. Let $\varrho \in \mathfrak{g}^\wedge$. Then the restriction of β_0 to $(T \cap C(\varrho)) \times (T \cap C(\varrho^*))$ is non-degenerate.*

If $v' \in V^{*\mathfrak{h}}$, the mapping $h_{v'}$: $v \mapsto \theta(v,v')$ of V into $U(\mathfrak{g})^*$ is a $\mathfrak{g}$-homomorphism of V into T. Resuming the proof of 2.7.11, we see that $v' \mapsto h_{v'}$ is a bijection of $V^{*\mathfrak{h}}$ onto $\mathrm{Hom}_{\mathfrak{g}}(V,T)$. On the other hand, $\varrho|\mathfrak{h}$ is semi-simple since $\mathfrak{h}$ is reductive in $\mathfrak{g}$, and hence $V^{*\mathfrak{h}}$ can be identified with the dual of $V^{\mathfrak{h}}$. Let W be the orthogonal subspace of $T \cap C(\varrho^*)$ in $T \cap C(\varrho)$, and let us assume that $W \neq 0$. We may resume the proof of 2.7.15, with $v'_0 \in V^{*\mathfrak{h}}$; this enables us to assume that $v_0 \in V^{\mathfrak{h}}$, and then, for all i, $\theta(e_i^*, v_0) \in T \cap C(\varrho^*)$, whence we have a contradiction.

References: [13], [16], [68], [69], [70].

2.8. Supplementary remarks

2.8.1. Theorem 2.5.5 is Ado's famous theorem. Theorem 2.5.7 is due to Harish-Chandra [62]. Lemma 2.6.4 is due to Quillen [102]. Proposition 2.6.9 is due to Gabriel (unpublished).

2.8.2. The vector space $U^n(\mathfrak{g})$ is generated by the $x^n (x \in \mathfrak{g})$.

2.8.3. Let $u,v \in U(\mathfrak{g})$. Then $uv - vu \in U'_+(\mathfrak{g})$. In particular, we cannot have $uv - vu = 1$. On the other hand, $uv - vu$ can be a non-zero central element; cf., however, 7.8.4 (a).

2.8.4. We adopt the notation of 2.4.1. Let W be a vector subspace of $T^n(\mathfrak{g})$. If $\tau_n|W$ is an isomorphism of W onto $S^n(\mathfrak{g})$, then $\varphi_n|W$ is an isomorphism of W onto a complement of $U_{n-1}(\mathfrak{g})$ in $U_n(\mathfrak{g})$ ([16], ch. I, p. 33).

2.8.5. Let $(x_1, \ldots, x_n)$ be a basis for $\mathfrak{g}$. For all $\nu = (\nu_1, \ldots, \nu_n) \in \mathbf{N}^n$, let y_ν be a product $a_1 a_2 \cdots a_{|\nu|}$, where $\nu_1, \ldots, \nu_n$ of the a_i are equal to $x_1, \ldots, x_n$, respectively. Then $(y_\nu)_{\nu \in \mathbf{N}^n}$ is a basis for $U(\mathfrak{g})$ [17].

2.8.6. Let ϱ be the left regular representation of $\mathfrak{g}$ in $U(\mathfrak{g})$. We assume that $\mathfrak{g} \neq 0$. Then $U_+(\mathfrak{g})$ is stable under ϱ but does not have a complement in $U(\mathfrak{g})$ which is stable under ϱ; in particular, ϱ is not semi-simple.

2.8.7. Let π_m be the canonical mapping of $U_m(\mathfrak{g})$ onto $S^m(\mathfrak{g})$. Let $p \in S^m(\mathfrak{g})$ and $q \in S^n(\mathfrak{g})$. Let $p^\sim \in U_m(\mathfrak{g})$ and $q^\sim \in U_n(\mathfrak{g})$ be such that $\pi_m(p^\sim) = p$ and $\pi_n(q^\sim) = q$. Then

$$[p^\sim, q^\sim] \in U_{m+n-1}(\mathfrak{g})$$

and $\pi_{m+n-1}([p^\sim, q^\sim])$ only depends on p and q; let us denote this element by $[p,q]$. By linearity, we obtain a Lie algebra structure on $S(\mathfrak{g})$ which extends that of $\mathfrak{g}$. Let $R(\mathfrak{g}) = \operatorname{Fract} S(\mathfrak{g})$. There exists one and only one Lie algebra structure on $R(\mathfrak{g})$ which extends that of $S(\mathfrak{g})$ and such that $[r_1 r_2, r_3] = [r_1, r_3] r_2 + r_1 [r_2, r_3]$ for all $r_1, r_2, r_3 \in R(\mathfrak{g})$. The bracket in $R(\mathfrak{g})$ is called the *Poisson bracket* ([5], [125]).

2.8.8. Assume $\mathfrak{g}$ to be semi-simple. Let I be a two-sided ideal of finite codimension in $U(\mathfrak{g})$. Then $I^2 = I$. (For, $U(\mathfrak{g})/I^2$ is finite-dimensional from 2.5.1, and every finite-dimensional representation of $U(\mathfrak{g})/I^2$, which can be identified with a representation of $\mathfrak{g}$, is semi-simple. Hence the algebra $U(\mathfrak{g})/I^2$ is semi-simple, so that its nilpotent ideal I/I^2 is null.)

2.8.9. Let A be an algebra of denumerable dimension, and M a simple A-module. If k is non-denumerable, every A-endomorphism of M is algebraic over k ([30]; cf. problem 23).

2.8.10. Let $\mathfrak{g}_0$ be a real Lie algebra and $\mathfrak{g}$ its complexification, so that $U(\mathfrak{g})$ is the complexification of $U(\mathfrak{g}_0)$. Let $u \mapsto \bar{u}$ be the conjugation of $U(\mathfrak{g})$ with respect to $U(\mathfrak{g}_0)$. For all $u \in U(\mathfrak{g})$, we set $u^* = \bar{u}^{\mathsf{T}}$. Then $u \mapsto u^*$ is an involution of $U(\mathfrak{g})$ (termed *principal*), i.e., for $u, v \in U(\mathfrak{g})$ and $\lambda \in \mathbf{C}$, we have

$$(u+v)^* = u^* + v^*, \qquad (\lambda u)^* = \bar{\lambda} u^*, \qquad (uv)^* = v^* u^*, \qquad u^{**} = u.$$

2.8.11. Let ϱ be a representation of $\mathfrak{g}$. Generally, the orthogonal space of $\operatorname{Ker} \varrho$ in $U(\mathfrak{g})^*$ is distinct from $C(\varrho)$. (Take $\mathfrak{g}$ semi-simple and take for ϱ the sum of the simple representations of $\mathfrak{g}$.)

2.8.12. (a) If $x_1, \ldots, x_n \in \mathfrak{g}$, we define $[x_1, \ldots, x_n]$ by recurrence on n, setting $[x_1, \ldots, x_n] = [[x_1, \ldots, x_{n-1}], x_n]$ for $n > 1$, and $[x_1] = x_1$.

If σ is a permutation of $\{1, \ldots, n\}$, we denote the set of intervals $\{p,p+1, \ldots, q\}$ of $\{1, \ldots, n\}$ such that $\sigma p < \sigma(p+1) < \cdots < \sigma q$ by C_σ; we denote the set of maximal elements of C_σ by D_σ; the interval $\{1, \ldots, n\}$ can be uniquely decomposed into a union of pairwise disjoint elements of D_σ. Let $G_{n,m}$ be the set of permutations σ of $\{1, \ldots, n\}$ such that Card $D_\sigma = m$.

Let T be the tensor algebra of $\mathfrak{g}$. There exists one and only one linear mapping θ of T into $\mathfrak{g}$ such that

$$\theta(x_1 \otimes \cdots \otimes x_n) = \sum_{m=1}^{n} (-1)^{m-1} \frac{1}{nm} \binom{n}{m}^{-1} \sum_{\sigma \in G_{n,m}} [x_{\sigma 1}, \ldots, x_{\sigma n}]$$

for any $x_1, \ldots, x_n \in \mathfrak{g}$. Then

$$\theta(x_1) = x_1,$$

$$\theta(x_1 \otimes x_2) = \tfrac{1}{2}[x_1,x_2],$$

$$\theta(x_1 \otimes x_2 \otimes x_3) = \tfrac{1}{6}([[x_1,x_2],x_3] + [[x_3,x_2],x_1]).$$

(b) Let J be the two-sided ideal of T defined in 2.1.1, and let $S \subset T$ be the set of symmetric tensors. Then $T = S \oplus J$, whence we have a projection φ of T onto S. If $x_1, \ldots, x_n \in \mathfrak{g}$, and P is a subset $\{i_1, \ldots, i_r\}$ of $\{1, \ldots, n\}$ $(i_1 < \cdots < i_r)$, for brevity we write Px instead of $x_{i_1} \otimes \cdots \otimes x_{i_r}$. Let $\mathfrak{P}((n,s)$ be the set of sequences $(P_1, \ldots, P_s)$ of disjoint non-empty subsets of $\{1, \ldots, n\}$ with union $\{1, \ldots, n\}$. Then

$$\varphi(x_1 \otimes \cdots \otimes x_n) = \sum_{1 \leq s \leq n} \frac{1}{s!} \sum_{(P_1, \ldots, P_s) \in \mathfrak{P}(n,s)} \theta(P_1 x) \otimes \cdots \otimes \theta(P_s x).$$

For example,

$$\begin{aligned}\varphi(x_1 \otimes x_2 \otimes x_3) = {} & \tfrac{1}{6} \sum_{\sigma \in \mathfrak{S}_3} x_{\sigma 1} \otimes x_{\sigma 2} \otimes x_{\sigma 3} \\ & + \tfrac{1}{2}(\theta(x_1 \otimes x_2) \otimes x_3 + x_3 \otimes \theta(x_1 \otimes x_2) \\ & + \theta(x_1 \otimes x_3) \otimes x_2 + x_2 \otimes \theta(x_1 \otimes x_3) \\ & + \theta(x_2 \otimes x_3) \otimes x_1 + x_1 \otimes \theta(x_2 \otimes x_3)) \\ & + (x_1 \otimes x_2 \otimes x_3).\end{aligned}$$

Let us identify $\theta \mid \otimes^n \mathfrak{g}$ with an n-linear mapping of $\mathfrak{g}^n$ into $\mathfrak{g}$. If $x_1, \ldots, x_n \in \mathfrak{g}$, we denote the element $(1/n!) \sum_{\sigma \in \mathfrak{S}_n} x_{\sigma 1} x_{\sigma 2} \cdots x_{\sigma n}$ of $U^n(\mathfrak{g})$ by $(x_1, \ldots, x_n)$. The above then yields the components of $x_1 x_2 \cdots x_n$ in $U^n(\mathfrak{g})$,

$U^{n-1}(\mathfrak{g})$, $U^{n-2}(\mathfrak{g})$, ...; for example,

$$x_1x_2x_3 = (x_1,x_2,x_3)$$
$$+ (\theta(x_1,x_2),x_3) + (\theta(x_1,x_3),x_2) + (\theta(x_2,x_3),x_1) + \theta(x_1,x_2,x_3).$$

(cf. [119].)

(c) If $x,x_1,\ldots,x_n \in \mathfrak{g}$, then, in $U(\mathfrak{g})$,

$$x(x_1,\ldots,x_n) = (x,x_1,\ldots,x_n) - \frac{b_1}{1!}\sum_h ([x,x_h],x_1,\ldots,\hat{x}_h,\ldots,x_n)$$
$$+ \frac{b_2}{2!}\sum_{h\neq l} ([[x,x_h],x_l],x_1,\ldots,\hat{x}_h,\ldots,\hat{x}_l,\ldots,x_n) - \cdots,$$

where the b_i are the Bernoulli numbers ($b_1 = -\frac{1}{2}$, $b_2 = \frac{1}{6}$, $b_3 = 0$, $b_4 = -\frac{1}{30}$, ...), and where the sign ^ above a letter indicates that it should be omitted. (Writing $(\ldots,\hat{x}_h,\ldots,\hat{x}_l,\ldots)$ does not mean that we have imposed $h < l$.) (Cf. [5], [60].)

2.8.13. We denote the coproducts in $U(\mathfrak{g})$ and $S(\mathfrak{g})$ by c and c', and the symmetrization by $\omega : S(\mathfrak{g}) \to U(\mathfrak{g})$. Then:

(i) $(\omega \otimes \omega)(c'(x^n)) = c(\omega(x^n))$ for all $x \in \mathfrak{g}$. Consequently,

$$(\omega \otimes \omega) \circ c' = c \circ \omega.$$

(ii) Let $\omega' : U(\mathfrak{g})^* \to S(\mathfrak{g})^*$ be the transpose of ω. We deduce from (i) that ω' is an algebra isomorphism.

2.8.14. For every integer $n \geqq 0$, the orthogonal subspace of $U_n(\mathfrak{g})$ in $U(\mathfrak{g})^*$ is an ideal of the algebra $U(\mathfrak{g})^*$. (Use 2.7.5.)

2.8.15. For all $u \in U(\mathfrak{g})$, let $f(u)$ be the constant term of u. Equipped with its coproduct and the co-unity f, the algebra $U(\mathfrak{g})$ becomes a cocommutative bigebra (AL III, p. 149).

2.8.16. (a) We adopt the notation of 2.7.12. The mapping of $U(\mathfrak{g}) \otimes U(\mathfrak{g})$ into $U(\mathfrak{g})$ deduced from the multiplication has transpose $\gamma : U(\mathfrak{g})^* \to (U(\mathfrak{g}) \otimes (U(\mathfrak{g}))^*$. For every finite-dimensional representation ϱ of $\mathfrak{g}$, we have $\gamma(C(\varrho)) \subset C(\varrho) \otimes C(\varrho)$. Hence $\gamma(\Phi) \subset \Phi \otimes \Phi$. The mapping γ of Φ into $\Phi \otimes \Phi$ is an algebra homomorphism. Equipped with γ and the co-unity $f \mapsto f(1)$, the algebra Φ becomes a commutative bigebra, which is termed the *Hopf algebra* of $\mathfrak{g}$, and is denoted by $H(\mathfrak{g})$.

(b) The invertible elements of $H(\mathfrak{g})$ are the non-zero scalar multiples of the homomorphisms of $U(\mathfrak{g})$ into k.

(c) We assume that the radical $\mathfrak{r}$ of $\mathfrak{g}$ is nilpotent and that k is algebraically closed. Let Φ' and Φ'' be the sums of the $C(\varrho)$ when ϱ runs through the set of finite-dimensional representations of $\mathfrak{g}$ whose largest nilpotency ideal is $\mathfrak{r}$, and the set of one-dimensional representations of $\mathfrak{g}$, respectively. Then Φ' and Φ'' are sub-bigebras of $H(\mathfrak{g})$ which are stable under the mapping $f \mapsto f^{\mathsf{T}}$. The algebra Φ' is of finite type, and the algebra Φ is the tensor product of the algebras Φ' and Φ''.

(d) We assume k to be algebraically closed. The following conditions are equivalent:

(i) $\mathfrak{g} = [\mathfrak{g},\mathfrak{g}]$;

(ii) the algebra $H(\mathfrak{g})$ is of finite type;

(iii) every invertible element of $H(\mathfrak{g})$ is a scalar multiple of the unity of $H(\mathfrak{g})$ ([68], [69], [70]).

2.8.17. (a) We make $U(\mathfrak{g})$ into a topological algebra by taking the family (I_α) of two-sided ideals of finite codimension of $U(\mathfrak{g})$ as a fundamental system of neighbourhoods of zero. The continuous linear forms on $U(\mathfrak{g})$ (k being equipped with the discrete topology) are then the elements of $H(\mathfrak{g})$ (2.8.16). The completion $U(\mathfrak{g})^\sim$ of $U(\mathfrak{g})$, i.e., the projective limit of the $U(\mathfrak{g})/I_\alpha$, can be canonically identified with the dual vector space of $H(\mathfrak{g})$. The coproduct of $U(\mathfrak{g})$ can be extended to a continuous homomorphism c of $U(\mathfrak{g})^\sim$ into the completed algebra of $U(\mathfrak{g}) \otimes U(\mathfrak{g})$.

(b) If $\mathfrak{g} = [\mathfrak{g},\mathfrak{g}]$, the set of the $u \in U(\mathfrak{g})^\sim$ such that $c^\sim(u) = u \otimes 1 + 1 \otimes u$ is $\mathfrak{g}$.

(c) Let G be the set of the $u \in 1 + \mathfrak{g}U(\mathfrak{g})^\sim$ such that $c^\sim(u) = u \otimes u$. It is a subgroup of the group of invertible elements of $U(\mathfrak{g})^\sim$ and it can be identified with the set of homomorphisms of the algebra $H(\mathfrak{g})$ into k. The elements of $H(\mathfrak{g})$ define on G an algebra F of functions with values in k. Then, if $\mathfrak{g} = [\mathfrak{g},\mathfrak{g}]$ and k is algebraically closed, G, equipped with the algebra F of functions, is a simply connected affine algebraic group with Lie algebra $\mathfrak{g}$ ([68], [69], [70]).

2.8.18. Let ϱ be a representation of $\mathfrak{g}$ in V. The mapping $(v,v') \mapsto \theta(v,v')$ of $V \times V^*$ into $U(\mathfrak{g})^*$ defines a linear mapping ζ of $V \otimes V^*$ into $U(\mathfrak{g})^*$. Then ζ is a $\mathfrak{g}$-homomorphism if $U(\mathfrak{g})^*$ is provided with the dual representation of the adjoint representation. If ϱ is absolutely simple, ζ is an isomorphism of $V \otimes V^*$ onto $C(\varrho)$. If ϱ is finite-dimensional and absolutely simple, then there exists a canonical isomorphism of $\mathrm{End}(V)$ onto $C(\varrho)$ (whence $C(\varrho)$ has an algebra structure).

CHAPTER 3

TWO SIDED IDEALS IN ENVELOPING ALGEBRAS

As we explained in the introduction, our principal aim is the study of the primitive ideals of $U(\mathfrak{g})$. For several reasons, we are led to the study of the prime ideals of $U(\mathfrak{g})$ at the same time; this is scarcely surprising if we consider the commutative case.

Related to these two central notions are some less important but technically useful notions, viz. semi-prime ideals, as in the commutative case, maximal ideals (we will be gratified when the primitive ideals are maximal), and completely prime ideals (we will be gratified when the prime ideals are completely prime).

Throughout this chapter, A denotes a ring (with unity).

3.1. Primitive ideals and prime ideals

3.1.1. Let I be a two-sided ideal of A. We say that I is *prime* if it satisfies the following equivalent conditions:

(i) $I \neq A$, and, in A/I, the product of two non-null two-sided ideals is non-null;

(ii) $I \neq A$, if a and b are elements of A which do not belong to I, then $aAb \not\subset I$.

(Condition (ii) may be written $(AaA)(AbA) \not\subset I$, whence the equivalence of (i) and (ii) follows directly.)

3.1.2. A is said to be *integral* if $A \neq 0$ and if, in A, the product of two non-zero elements is non-zero. I is said to be *completely prime* if A/I is integral.

3.1.3. I is said to be *semi-prime* if $I \neq A$ and if, in A/I, every nilpotent two-sided ideal is null. The intersection of a non-empty family of semi-prime two-sided ideals is a semi-prime two-sided ideal.

3.1.4. I is said to be *primitive* if it is the the annihilator of a simple left A-module. The set of primitive ideals of A is denoted by Prim (A).

3.1.5. I is said to be *maximal* if it is maximal in the set of two-sided ideals of A distinct from A. Every two-sided ideal of A distinct from A is contained in a maximal two-sided ideal.

3.1.6. We have the following table of implications:

$$\begin{array}{ccccc} I \text{ maximal} & & I \text{ completely prime} & & \\ \Downarrow 1 & & \Downarrow 3 & & \\ I \text{ primitive} & \overset{2}{\Rightarrow} & I \text{ prime} & \overset{4}{\Rightarrow} & I \text{ semi-prime}. \end{array}$$

Implications 3 and 4 are obvious. If I is maximal, let J be a maximal left ideal containing I. Then A/J is a simple A-module under the regular representation ϱ of A in A/J, and Ker $\varrho \supset I$ whence Ker $\varrho = I$. This proves implication 1. Finally, let us assume that I is the annihilator of a simple left A-module M. Let I_1 and I_2 be two-sided ideals of A containing I and distinct from I. Then $I_2 M \neq 0$ and $AI_2 M \subset I_2 M$, hence $I_2 M = M$, and consequently $(I_1 I_2) M = I_1 M \neq 0$. It can thus be seen that $I_1 I_2 \not\subset I$ which proves implication 2.

3.1.7. The converses of the implications of 3.1.6 are all false, even in enveloping algebras (3.7.2, 6.6.3). Let us assume, however, that A is commutative. It is then clear that a prime ideal is completely prime. On the other hand, let I be a primitive ideal; it is the annihilator of a module A/J, where J is a maximal ideal of A. Then $I = J$ because A is commutative. Thus a primitive ideal of A is maximal.

3.1.8. PROPOSITION. *We assume that A is Noetherian* (*cf.* 2.3.8). *Let I be a two-sided ideal of A, and $\mathscr{E}$ the set of two-sided ideals of A, a power of which is contained in I. Then:*

(i) *$\mathscr{E}$ has a largest element J.*

(ii) *If $I \neq A$, J is the smallest semi-prime two-sided ideal of A containing I.*

Let J be the sum of the elements of $\mathscr{E}$. Since A is Noetherian, there exist $I_1, \ldots, I_r \in \mathscr{E}$ such that $J = I_1 + \cdots + I_r$. A sufficiently large power of $I_1 + \cdots I_r$ is contained in I, and hence $J \in \mathscr{E}$. Then J is the largest element of $\mathscr{E}$. If $I \neq A$, then J is clearly semi-prime. If J' is a semi-prime two-sided ideal containing I, the image of J in A/J' is null, and hence $J \subset J'$.

3.1.9. The semi-prime two-sided ideal J associated with I by 3.1.8 is called the *root* of I.

3.1.10. PROPOSITION. *Assume that A is Noetherian. Let I be a two-sided ideal of A, and $\mathscr{P}$ the set of prime two-sided ideals of A containing I.*

(i) *$\mathscr{P}$ possesses only a finite number of minimal elements; let these elements be $I_1, \ldots, I_r$, pairwise distinct.*

(ii) *The root of I is $I_1 \cap \cdots \cap I_r$.*

(iii) *Every element of $\mathscr{P}$ contains one of the I_j.*

(iv) *None of the I_j contains the intersection of the others.*

We may assume that I is semi-prime.

Let $\mathscr{S}$ be the set of semi-prime ideals of A which are not the intersection of a finite number of prime ideals of A. Let us assume that $\mathscr{S} \neq \emptyset$. Then $\mathscr{S}$ has a maximal element M. Since M is not prime, there exist two-sided ideals M_1 and M_2 of A containing M such that $M_1 \neq M$, $M_2 \neq M$ and $M_1 M_2 \subset M$. Let N_1 and N_2 be the roots of M_1 and M_2. There exists an integer n such that

$$(N_1 \cap N_2)^n \subset M_1 \cap M_2,$$

whence

$$(N_1 \cap N_2)^{2n} \subset (M_1 \cap M_2)^2 \subset M_1 M_2 \subset M,$$

hence $N_1 \cap N_2 = M$ since M is semi-prime. From the maximality of M, N_1 and N_2 are finite intersections of prime ideals, and hence also M, which is contradictory.

Hence $\mathscr{S} = \emptyset$, and we may write $I = I_1 \cap I_2 \cap \cdots \cap I_r$, where $I_1, \ldots, I_r$ are prime ideals. We may assume that none of the I_j contains the intersection of the others. Every element of $\mathscr{P}$ contains $I_1 I_2 \cdots I_r$, and hence contains one of the I_j. This proves that $I_1, I_2, \ldots, I_r$ are the minimal elements of $\mathscr{P}$.

3.1.11. The intersection of the primitive ideals of A is termed the *radical* of A. Since the annihilator of a module is the intersection of the annihilators of its elements, the radical of A is also the intersection of the maximal left ideals of A.

3.1.12. PROPOSITION. *Let R be the radical of A, $x \in A$, and I the two-sided ideal generated by x. The following conditions are equivalent:*

(i) *$x \in R$;*

(ii) *for all $y \in I$, $1 + y$ is invertible.*

Let us assume that $x \in R$. If a maximal left ideal contains $A(1 + x)$, it contains x and $1 + x$, hence 1, which is impossible. Consequently, $A(1 + x) = A$, and there exists $x_1 \in A$ such that $x_1(1 + x) = 1$. Then $x_1 = 1 - x_1 x$ and $-x_1 x \in R$, hence x_1 is left invertible from the foregoing. Thus, x_1 is invertible and consequently $1 + x$ is invertible. This proves that (i) $\Rightarrow$ (ii).

Let us assume that $x \notin R$. There exists a maximal left ideal M of A such that $x \notin M$. Then $Ax + M = A$, hence there exist $a \in A$ and $m \in M$ such that $ax + m = 1$. Thus, $1 - ax$ is non-invertible. This proves that not (i) $\Rightarrow$ not (ii).

3.1.13. For all $a \in A$, we denote the set of the $x \in A$ such that $xa = 0$ (or $ax = 0$) by $\mathfrak{l}(a)$ (or $\mathfrak{r}(a)$). It is a left (or right) ideal of A.

3.1.14. Lemma. *Assume that A is Noetherian. Let L be a left ideal of A all of whose elements are nilpotent. Then L is nilpotent. If 0 is a semi-prime ideal of A, then $L = 0$.*

By virtue of 3.1.8 applied to the ideal 0, we return to the case where 0 is a semi-prime ideal of A. Let us assume that $L \neq 0$. We choose from among the non-zero elements of L an element x in such a way that $\mathfrak{r}(x)$ is maximal. Let $y \in A$; we prove that $xyx = 0$. This is obvious if $yx = 0$. Otherwise, let k be the integer >1 such that $(yx)^{k-1} \neq 0$, $(yx)^k = 0$. Then $(yx)^{k-1} \in L$ and $\mathfrak{r}((yx)^{k-1}) \supset \mathfrak{r}(x)$, whence $\mathfrak{r}((yx)^{k-1}) = \mathfrak{r}(x)$. Since $yx \in \mathfrak{r}(yx)^{k-1})$, we have $yx \in \mathfrak{r}(x)$, hence $xyx = 0$. Thus $xAx = 0$, which contradicts the fact that the ideal 0 is semi-prime.

3.1.15. Proposition. *Let I be a two-sided ideal of $U(\mathfrak{g})$ distinct from $U(\mathfrak{g})$. The following conditions are equivalent:*

(i) *I is semi-prime;*
(ii) *I is an intersection of primitive ideals.*

(ii) $\Rightarrow$ (i). This follows from 3.1.3 and 3.1.6.

(i) $\Rightarrow$ (ii). We assume that I is semi-prime. Let us set $B = U(\mathfrak{g})/I$, and let R be the radical of B. Let X be an indeterminate and C the algebra $B \otimes k[X]$. Let $a \in R$, and assume that $C(1 - aX) \neq C$. There exists a simple C-module M and a non-zero element m_0 of M such that $(1 - aX) \cdot m_0 = 0$. Let x be the mapping $m \mapsto X_M m$ of M into M; since X is in the centre of C, x is an endomorphism of the C-module M, which is non-zero since $a_M x(m_0) = m_0$. Now, from 2.2.12, C is a quotient of the enveloping algebra of $\mathfrak{g} \times k$. Then x is invertible and algebraic over k (2.6.4). If we set $y = x^{-1}$, there then exists $p \in k[X]$ such that $x = p(y)$. Then $a_M(m_0) = y(m_0)$, whence

$$(1 - ap(a))m_0 = (1 - yp(y))(m_0) = 0.$$

Since $ap(a) \in R$, this contradicts 3.1.12.

Hence $C(1 - aX) = C$. There exist $a_0, a_1, \ldots, a_n \in B$ such that

$$(a_0 + a_1X + \cdots + a_nX^n)(1 - aX) = 1,$$

whence $a_0 = 1$, $a_1 = a, \ldots, a_n = a^n$, $a^{n+1} = 0$. Thus, every element of R is nilpotent, so that $R = 0$ (3.1.14). The ideal 0 of B is hence an intersection of primitive ideals.

3.1.16. In particular, the ideal 0 of $U(\mathfrak{g})$ is an intersection of primitive ideals. Moreover, Proposition 3.1.15 can be immediately extended to the ideals of the *quotient* algebras of $U(\mathfrak{g})$.

Reference: [48].

3.2. The space of primitive ideals

3.2.1. For every subset T of Prim (A), let $I(T)$ be the intersection of the elements of T; this set is a two-sided ideal of A. Let T^- be the set of primitive ideals of A containing $I(T)$.

LEMMA. (i) $\emptyset^- = \emptyset$.

(ii) *If* $T \subset$ Prim A, *then* $T \subset T^-$.

(iii) *If* $T \subset$ Prim A, *then* $T^{--} = T^-$.

(iv) *If* $T_1, T_2 \subset$ Prim A, *then* $(T_1 \cup T_2)^- = T_1^- \cup T_2^-$.

Assertions (i) and (ii) are obvious. Clearly, $I(T^-) = I(T)$, whence $T^{--} = T^-$. Let $I_1 = I(T_1)$ and $I_2 = I(T_2)$. Then $I(T_1 \cup T_2) = I_1 \cap I_2$. Hence $(T_1 \cup T_2)^-$ is the set of primitive ideals containing $I_1 \cap I_2$, i.e. (3.1.6) containing I_1 or I_2. This proves (iv).

3.2.2. There thus exists one and only one topology on Prim (A) such that, for all $T \subset$ Prim (A), T^- is the closure of T for this topology. This is known as the *Jacobson topology* over Prim (A). (If A is commutative, it is also known as the Zariski topology). When we consider Prim (A) as a topological space, we will always be concerned with the Jacobson topology.

3.2.3. PROPOSITION. *Let* $T \subset$ Prim (A). *The following conditions are equivalent:*

(i) *T is closed;*

(ii) *there exists a subset B of A such that T is the set of primitive ideals of A containing B.*

(i) $\Rightarrow$ (ii) is obvious.

(ii) $\Rightarrow$ (i). If T is the set of primitive ideals of A containing B, we have $I(T) \supset B$, hence $T^- \subset T$ and $T^- = T$.

3.2.4. Let $\mathscr{F}$ be the set of closed subsets of Prim (A). Let $\mathscr{I}$ be the set of two-sided ideals of A which are intersections of primitive ideals. If $F \in \mathscr{F}$, let $\varphi(F)$ be the intersection of the elements of F. If $I \in \mathscr{I}$, let $\psi(I)$ be the set of elements of Prim (A) which contain I. Then φ and ψ are reciprocal bijections of $\mathscr{F}$ onto $\mathscr{I}$ and of $\mathscr{I}$ onto $\mathscr{F}$; these bijections are decreasing.

If A is a quotient algebra of $U(\mathfrak{g})$, we thus obtain, from 3.1.15, a bijection between the set of non-empty closed subsets of Prim (A) and the set of semi-prime ideals of A.

3.2.5. We recall that a topological space X is termed *irreducible* if every finite intersection of non-empty open subsets of X is non-empty, or, equivalently, if every non-empty open subset of X is dense in X.

PROPOSITION. *Let F be a closed subset of* Prim (A), *and I the intersection of the elements of F. The following conditions are equivalent:*

(i) *F is irreducible;*

(ii) *I is prime.*

Let us assume that F is irreducible. Let $a,b \in A$ such that $aAb \subset I$. Let G and H be the sets of the $J \in F$ such that $a \in J$ and $b \in J$ respectively. Then G and H are closed subsets of F. If $J \in F$, then $I \subset J$, hence $aAb \subset J$, hence $a \in J$ or $b \in J$, so that $J \in G$ or $J \in H$. Thus $F = G \cup H$. Since F is irreducible, we have for example $F = G$. Hence $a \in J$ for all $J \in F$, and consequently $a \in I$. This proves that I is prime.

Let us assume that F is not irreducible. Then $F = G \cup H$, where G and H are closed subsets of Prim (A) distinct from F. Let J and K be the two-sided ideals corresponding to G and H. Then $I = J \cap K$, $I \neq J$, $I \neq K$, and hence I is not prime.

3.2.6. Let I be a two-sided ideal of A, and F the set of elements of Prim (A) containing I. For every $J \in F$, let $\psi(J)$ be the primitive ideal J/I of A/I. It follows directly that ψ is a homeomorphism of F onto Prim (A/I), by means of which we can identify these two spaces.

3.3. The passage to an ideal of $\mathfrak{g}$

3.3.1. Let $\mathfrak{k}$ be a Lie subalgebra of $\mathfrak{g}$ and I a two-sided ideal of $U(\mathfrak{g})$. If I is completely prime, then $I \cap U(\mathfrak{k})$ is a completely prime two-sided ideal of $U(\mathfrak{k})$; but, apart from this observation, the properties of $I \cap U(\mathfrak{k})$ do not appear to be closely linked to the properties of I (cf. 3.8.6). Nevertheless, we shall see in 3.3.4 that the situation is improved if $\mathfrak{k}$ is an ideal of $\mathfrak{g}$.

3.3.2. LEMMA. *We assume that A is an algebra (recall that k is of characteristic 0). Let I be a two-sided ideal of A, $\mathscr{D}$ a set of derivations of A, and J the set of the $x \in A$ such that $D_1 D_2 \cdots D_n x \in I$ for all $D_1, \ldots, D_n \in \mathscr{D}$, $n \geqq 0$.*

(i) *J is the largest two-sided ideal of A contained in I and stable under $\mathscr{D}$.*

(ii) *If I is prime, then J is prime.*

Assertion (i) is obvious. Let us assume that I is prime. Let $a,b \in A$ be such that $aAb \subset J$.

Let $D_1, \ldots, D_p \in \mathscr{D}$ and $m_1, \ldots, m_p \in \mathbf{N}$ such that $D_1^{m_1} \cdots D_p^{m_p}\, b \notin I$, and let us show that $D_1^{n_1} \cdots D_p^{n_p} a \in I$ for all $n_1, \ldots, n_p \in \mathbf{N}$. Let us provide $\mathbf{N}^p$ with the ordering defined in 2.6.1. Let $(s_1, \ldots, s_p)$ be the smallest element of $\mathbf{N}^p$ such that $D_1^{s_1} \cdots D_p^{s_p} b \notin I$. For all $x \in A$, we have

$$D_1^{n_1+s_1} \cdots D_p^{n_p+s_p}(axb) =$$

$$= \sum_{i_1+j_1+l_1=n_1+s_1,\ldots} \alpha(i_1,j_1,l_1,\ldots)(D_1^{i_1} \cdots D_p^{i_p} a)(D_1^{j_1} \cdots D_p^{j_p} x)(D_1^{l_1} \cdots D_p^{l_p} b),$$

where the $\alpha(i_1,j_1,l_1,\ldots)$ are integers <0. Hence

$$D_1^{n_1+s_1} \cdots D_p^{n_p+s_p}(axb) =$$

$$= \alpha(n_1,0,s_1,\ldots)(D_1^{n_1} \cdots D_p^{n_p})x(D_1^{s_1} \cdots D_p^{s_p} b) + r,$$

where r is a sum of terms of the form

$$\alpha(i_1,j_1,l_1,\ldots)(D_1^{i_1} \cdots a)(D_1^{j_1} \cdots x)\, D_1^{l_1} \cdots b)$$

such that $(i_1, \ldots, i_p) < (n_1, \ldots, n_p)$ or $(l_1, \ldots, l_p) < s_1, \ldots, s_p)$. If we reason by induction on $(n_1, \ldots, n_p)$, r belongs to I by the definition of $(s_1, \ldots, s_p)$. The first member belongs to I. Hence

$$(D_1^{n_1} \cdots D_p^{n_p} a)\, A(D_1^{s_1} \cdots D_p^{s_p} b) \subset I,$$

which proves our assertion.

Having established this, let us assume that $b \notin J$, and let us show that $a \in J$. Let $D_1, \ldots, D_u \in \mathscr{D}$. There exist $D_{u+1}, \ldots, D_v \in \mathscr{D}$ such that

$D_{u+1} \cdots D_v b \notin I$, which can be written as $D_1^0 \cdots D_u^0 D_{u+1}^1 \cdots D_v^1 b \notin I$. From the above, we have

$$D_1^1 \cdots D_u^1 D_{u+1}^0 \cdots D_v^0 a \in I.$$

Hence $a \in J$. In the same way, it can be seen that $a \notin J \Rightarrow b \in J$. This proves (ii).

3.3.3. LEMMA. *We assume that A is a Noetherian algebra. Let I be a two-sided ideal of A, J its root, and $P_1, \ldots, P_s$ the minimal prime ideals of A containing I. Let D be a derivation of A under which I is stable. Then $J, P_1, \ldots, P_s$ are stable under D.*

Let Q_i be the largest two-sided ideal of A which is contained in P_i and is stable under D. Then $I \subset Q_i$, and Q_i is prime (3.3.2), hence $Q_i = P_i$ and $D(P_i) \subset P_i$. Since $J = P_1 \cap \cdots \cap P_s$ (3.1.10), we have $D(J) \subset J$.

3.3.4. PROPOSITION. *Let $\mathfrak{k}$ be an ideal of $\mathfrak{g}$, I a two-sided ideal of $U(\mathfrak{g})$, and K the two-sided ideal $I \cap U(\mathfrak{k})$ of $U(\mathfrak{k})$.*

(i) *If I is semi-prime, then K is semi-prime.*

(ii) *If I is prime, then K is prime.*

Let us assume that K is not semi-prime. Let K' be the root of K. There exists a power K'' of K' such that $K'' \not\subset K$ and $K''^2 \subset K$. We have $[\mathfrak{g}, K] \subset K$, hence $[\mathfrak{g}, K'] \subset K'$ (3.3.3) and $[\mathfrak{g}, K''] \subset K''$, whence $U(\mathfrak{g})K'' = K''U(\mathfrak{g})$. Let J be the two-sided ideal $U(\mathfrak{g})K''U(\mathfrak{g})$ of $U(\mathfrak{g})$. Then $K'' \subset J \cap U(\mathfrak{k})$, hence $J \cap U(\mathfrak{k}) \not\subset K$ and consequently $J \not\subset I$. But

$$J^2 = U(\mathfrak{g})K''U(\mathfrak{g})K''U(\mathfrak{g}) = U(\mathfrak{g})K''^2U(\mathfrak{g}) \subset U(\mathfrak{g})KU(\mathfrak{g}) \subset I,$$

so that I is not semi-prime. This proves (i).

Let us assume that I is prime. Then, from (i), K is semi-prime. Let $K_1, \ldots, K_s$ be the minimal elements, pairwise distinct, in the set of prime ideals of $U(\mathfrak{k})$ containing K. We assume that K is not prime. Then $s > 1$; we set $K_0 = K_2 \cap \cdots \cap K_s$. We have $K_0 K_1 \subset K$, $K_0 \neq K$ and $K_1 \neq K$. From 3.3.3, $[\mathfrak{g}, K_i] \subset K_i$ for all i, whence

$$U(\mathfrak{g})K_i = K_iU(\mathfrak{g}) = U(\mathfrak{g})K_iU(\mathfrak{g}).$$

Then

$$U(\mathfrak{g})K_0U(\mathfrak{g}) \not\subset I, \qquad U(\mathfrak{g})K_1U(\mathfrak{g}) \not\subset I,$$

$$(U(\mathfrak{g})K_0)(K\, U(\mathfrak{g})) \subset I,$$

which is a contradiction. This proves (ii).

3.3.5. Let us briefly study an operation inverse to that of 3.3.4.

PROPOSITION. *Let $\mathfrak{k}$ be an ideal of $\mathfrak{g}$, and $(x_1, \ldots, x_n)$ a basis for a complement of $\mathfrak{k}$ in $\mathfrak{g}$. For $\nu = (\nu_1, \ldots, \nu_n) \in \mathbf{N}^n$, let us set $x^\nu = x_1^{\nu_1} \cdots x_n^{\nu_n}$, so that $U(\mathfrak{g}) \oplus_{\nu \in \mathbf{N}^n} x^\nu U(\mathfrak{k})$. Let K be a two-sided ideal of $U(\mathfrak{k})$ such that $[\mathfrak{g},K] \subset K$. Then $I = U(\mathfrak{g})K = \oplus_{\nu \in \mathbf{N}^n} x^\nu K$ is the two-sided ideal of $U(\mathfrak{g})$ generated by K (cf. 3.8.9).*

In fact, $I\mathfrak{g} = U(\mathfrak{g})K(\mathfrak{g}) \subset U(\mathfrak{g})\mathfrak{g}K + U(\mathfrak{g})K = I$, hence I is indeed a two-sided ideal.

3.3.6. With the notation of 3.3.5, it is clear that $I \cap U(\mathfrak{k}) = K$. On the other hand, if we start with a two-sided ideal J of $U(\mathfrak{g})$ and form $L = J \cap U(\mathfrak{k})$, the two-sided ideal of $U(\mathfrak{g})$ generated by L is in general strictly contained in J. However, we have the useful result given in 3.3.8 below.

3.3.7. LEMMA. *If elements x,y of an algebra with unity satisfy $[x,y] = y$, then $yx^n = (x-1)^n y$ for $n \geqq 0$.*

This is obvious for $n = 0$, and, if it is true for n, then:

$$yx^{n+1} = (x-1)^n yx = (x-1)^n(xy - y) = (x-1)^{n+1}y.$$

3.3.8. PROPOSITION. *Let I be a prime ideal of $U(\mathfrak{g})$, ε the adjoint representation of $\mathfrak{g}$ in $U(\mathfrak{g})/I$, $\lambda \in \mathfrak{g}^*$, and u a non-zero element of $U(\mathfrak{g})/I$ such that $\varepsilon(x)u = \lambda(x)u$ for all $x \in \mathfrak{g}$. Let $\mathfrak{g}' = \operatorname{Ker} \lambda$, which is an ideal of $\mathfrak{g}$, and $I' = I \cap U(\mathfrak{g}')$. Then I is the two-sided ideal of $U(\mathfrak{g})$ generated by I'.*

We may assume that $\lambda \neq 0$. Let $x \in \mathfrak{g}$ such that $\lambda(x) = 1$. Each $a \in U(\mathfrak{g}) - \{0\}$ can be uniquely written as $x^n a_n + x^{n-1}a_{n-1} + \cdots + a_0$ with $a_n, \ldots, a_0 \in U(\mathfrak{g}')$, $a_n \neq 0$. We assume that $a \in I$, and show that $a_0, \ldots, a_n \in I'$. This is obvious if $n = 0$. Assume that $n > 0$ and that the assertion has been proved for all integers $<n$. It is then sufficient to prove that $a_n \in I'$. Let z_0 be a representative of u in $U(\mathfrak{g})$. Then

(1) $$[x,z_0] \in z_0 + I,$$

(2) $$[y,z_0] \in I \quad \text{for } y \in \mathfrak{g}', \text{ hence for } y \in U(\mathfrak{g}').$$

It is clear that

(3) $$[\ldots [[a,z_0], z_0] \ldots, z_0] \in I.$$

We show that the first term of (3), with p factors z_0, is congruent modulo I to

$$n(n-1) \cdots (n-p+1)\, x^{n-p} z_0^p a_n +$$
$$+ x^{n-p-1}b_{n-p-1} + x^{n-p-2}b_{n-p-2} + \cdots + b_0,$$

where $b_{n-p-1}, \ldots, b_0$ are elements of $U(\mathfrak{g})$ which are permutable to z_0 modulo I. This is clear, taking (2) into account, if $p = 0$. Let us assume it for p. From (1), (2) and 3.3.7, we have, modulo I,

$$[n(n-1)\cdots(n-p+1)\,x^{n-p}z_0^p a_n + x^{n-p-1}b_{n-p-1} + \cdots + b_0, z_0] \equiv$$

$$\equiv n(n-1)\cdots(n-p+1)\left(\binom{n-1}{p}x^{n-p-1}z_0 - \binom{n-p}{2}x^{n-p-2}z_0 + \cdots\right.$$

$$\left. + (-1)^{n-p-1}z_0\right) z_0^p a_n$$

$$+ \binom{n-p-1}{1}x^{n-p-2}z_0 - \binom{n-p-1}{2}x^{n-p-3}z_0 + \cdots$$

$$+ (-1)^{n-p}\, z_0)\, b_{n-p-1} + \cdots$$

$$+ z_0 b_1$$

$$= n(n-1)\cdots(n-p)\,x^{n-p-1}z_0^{p+1}a_n + x^{n-p-2}b'_{n-p-2} + \cdots + b'_0,$$

where $b'_{n-p-1}, \ldots, b'_0$ are permutable with z_0 modulo I. The assertion has thus been established for $p+1$, and hence for all integers. In particular, for $p = n$, relation (3) becomes

$$n!z_0^n a_n \in I.$$

We conclude that $a_n \in I$ (hence that $a_n \in I'$) by virtue of the following lemma:

3.3.9. LEMMA. *Let I, ε, λ, u be as in* 3.3.8. *Then u is not a divisor of zero in $U(\mathfrak{g})/I$.*

Let $v \in U(\mathfrak{g})/I$. Let us assume that, for example, $uv = 0$ and prove that $v = 0$. Let u', v' be representatives of u, v in $U(\mathfrak{g})$. Then $u'v' \in I$. If $u'\mathfrak{g}^n v' \subset I$, we deduce that

$$u'\mathfrak{g}^{n+1}v' \subset \mathfrak{g}u'\mathfrak{g}^n v' + [\mathfrak{g}, u']\mathfrak{g}^n v'$$

$$\subset \mathfrak{g}(u'\mathfrak{g}^n v') + (ku' + I)\mathfrak{g}^n v' \subset I.$$

Hence $u'U(\mathfrak{g})v' \subset I$, and consequently $u' \in I$ or $v' \in I$ since I is prime. As $u \neq 0$, we have indeed $v = 0$.

3.3.10. PROPOSITION. *Let $\mathfrak{k}$ be an ideal of $\mathfrak{g}$, K and L two-sided ideals of $U(\mathfrak{k})$, F_K the closed subset of* Prim $U(\mathfrak{k})$ *corresponding to K, and $\mathscr{A}$ the adjoint algebraic group of $\mathfrak{g}$. The following conditions are equivalent:*

(i) *K is the set of the $u \in U(\mathfrak{k})$ such that* $(\mathrm{ad}\, x_1) \cdots (\mathrm{ad}\, x_n) u \in L$ *for all* $n \geqq 0$ *and* $x_1, \ldots, x_n \in \mathfrak{g}$;

(ii) *K is the largest two-sided ideal of $U(\mathfrak{k})$ contained in L such that* $[\mathfrak{g}, K] \subset K$;

(iii) $K = \bigcap_{a \in \mathscr{A}} a(L)$.

If L is primitive, these conditions are moreover equivalent to the following:

(iv) *The orbit $\mathscr{A}(L)$ is dense in F_K, and K is prime.*

The equivalence (i) $\Leftrightarrow$ (ii) is clear. From 2.4.16, (ii) means that K is the largest two-sided ideal of $U(\mathfrak{k})$ contained in L and $\mathscr{A}$-stable, and hence that $K = \bigcap_{a \in \mathscr{A}} a(L)$. Let us assume that L is primitive. If conditions (ii) and (iii) are satisfied, K is prime (3.3.2) and $\mathscr{A}(L)$ is dense in F_K. Conversely, if $\mathscr{A}(L)$ is dense in F_K, and if we set $K' = \bigcap_{a \in \mathscr{A}} a(L)$, the primitive ideals of $U(\mathfrak{k})$ containing K' are the same as those containing K; if moreover K is prime, K is the intersection of the primitive ideals containing it (3.1.15), hence $K = K' = \bigcap_{a \in \mathscr{A}} a(L)$.

3.3.11. If conditions (i), (ii), (iii) of 3.3.10 are satisfied, we say that L is *generic for K* (relative to $\mathfrak{g}$).

References: [16], [31], [35], [51].

3.4. Extension of the scalar field

3.4.1. PROPOSITION. *Let k' be an extension of k, $\mathfrak{g}' = \mathfrak{g} \otimes k'$, I' a two-sided ideal of $U(\mathfrak{g}')$, and $I = I' \cap U(\mathfrak{g})$.*

(i) *If I' is semi-prime, then I is semi-prime.*

(ii) *If I' is prime, then I is prime.*

If $U(\mathfrak{g})/I$ possesses a non-null nilpotent two-sided ideal N, then $k'N$ is a non-null nilpotent two-sided ideal of $U(\mathfrak{g}')/I'$. This proves (i). If in $U(\mathfrak{g})/I$ there exist non-null two-sided ideals A and B such that $AB = 0$, then the non-null two-sided ideals $k'A$ and $k'B$ of $U(\mathfrak{g}')/I'$ have product null. This proves (ii).

3.4.2. PROPOSITION. *Let k' be an extension of k, $\mathfrak{g}' = \mathfrak{g} \otimes k'$, I a two-sided ideal of $U(\mathfrak{g})$, I' the two-sided ideal $I \otimes k'$ of $U(\mathfrak{g}')$, and $P'_1, \ldots, P'_s$ the minimal prime ideals containing I' in $U(\mathfrak{g}')$.*

(i) *If I is is semi-prime, then I' is semi-prime.*

(ii) *If I is prime, then there exists j such that $I = U(\mathfrak{g}) \cap P'_j$. If, addition-*

ally, k' is Galois over k with Galois group Γ, then Γ operates transitively on $\{P'_1, \ldots, P'_s\}$.

(iii) *If I is primitive, then $P'_1, \ldots, P'_s$ are primitive.*

Let us assume that I is semi-prime. Let R' be the root of I'. If k' is Galois over k with Galois group Γ, then R' is invariant under Γ, and hence R' is of the form $R \otimes k'$, where R is a two-sided ideal of $U(\mathfrak{g})$ containing I. A power of R is contained in I, hence $R = I$ and $R' = I'$. In the general case, let k'' be an extension of k' which is Galois over k. From the above, $I \otimes k''$ is semi-prime in $U(\mathfrak{g} \otimes k'')$. Hence I' is semi-prime in $U(\mathfrak{g}')$ (3.4.1).

Let us assume that I is prime. From (i), we have $I' = P'_1 \cap \cdots \cap P'_s$. Then

$$I = (P'_1 \cap U(\mathfrak{g})) \cap \cdots \cap (P'_s \cap U(\mathfrak{g})),$$

hence $I = P'_j \cap U(\mathfrak{g})$ for some j. Let us assume that k' is Galois over k with Galois group Γ. For all $\gamma \in \Gamma$, $\gamma(P'_j)$ is minimal in the set of prime ideals of $U(\mathfrak{g}')$ containing I', and hence is one of the P'_i. Let $K' = \bigcap_{\gamma \in \Gamma} \gamma(P'_j)$. Then K' is of the form $K \otimes k'$, where K is a two-sided ideal of $U(\mathfrak{g})$, and

$$K = K' \cap U(\mathfrak{g}) = P'_j \cap U(\mathfrak{g}) = I,$$

whence $K' = I'$. We may assume that $P'_1, \ldots, P'_s$ are pairwise distinct, and thus none of them contains the intersection of the others. It can thus be seen that Γ operates transitively on $\{P'_1, \ldots, P'_s\}$.

Let us assume that I is primitive. Let V be a simple $\mathfrak{g}$-module with annihilator I. Then $V \otimes k'$ is the direct sum of simple sub-$\mathfrak{g}'$-modules $W'_1, \ldots, W'_n$ (2.6.9). Let J'_i be the annihilator of W'_i. Then $I' = J'_1 \cap \cdots \cap J'_n$. Every P'_i contains $J'_1 \cap \cdots \cap J'_n$, hence contains a J'_i, and hence is equal to it and thus primitive.

References: [51].

3.5. The Krull dimension

3.5.1. Let E be an ordered set. If $a,b \in E$, we denote the set of the $x \in E$ such that $a \leqq x \leqq b$ by $[a,b]_E$ or simply by $[a,b]$. E is said to be *discrete* if the relations $a \in E$, $b \in E$, $a \leqq b$ imply that $a = b$. E is said to be *Artinian* if every decreasing sequence of elements of E is stationary. For every ordered set E, we define by induction the *deviation* of E, denoted by dev E, which is an element of $\mathbf{N} \cup \{-\infty, +\infty\}$, in the following way:

(i) dev $E = -\infty$ if E is discrete;

(ii) dev $E = 0$ if E is Artinian and not discrete;

(iii) let $n \in \mathbf{N}$; then dev $E \leqq n$ if and only if the following condition is satisfied:

for every strictly decreasing infinite sequence $(a_1, a_2, \ldots)$ of elements of E, we have dev $[a_{i+1}, a_i] < n$ for i sufficiently large.

(We can obviously replace 'strictly decreasing' by 'decreasing' in this condition.)

(iv) If there exists no $n \in \mathbf{N}$ such that dev $E \leqq n$, we set dev $E = +\infty$.

3.5.2. LEMMA. *Let E and F be ordered sets, and f a strictly increasing mapping of E into F. Then* dev $E \leqq$ dev F.

We may assume that dev F is finite and that E is not discrete so that F is not discrete. Let $n \in \mathbf{N}$. Let us assume that the lemma has been established for dev $F < n$, and consider the case where dev $F = n$. Let $(a_1, a_2, \ldots)$ be a strictly decreasing sequence of elements of E. Then dev $[f(a_{i+1}), f(a_i)] < n$ for i sufficiently large. From the induction hypothesis, dev $[a_{i+1}, a_i] < n$ for i sufficiently large, hence dev $E \leqq n$.

3.5.3. LEMMA. *Let E and F be non-empty ordered sets, and let $E \times F$ be provided with the product ordering. Then* dev $(E \times F) =$ sup (dev E, dev F).

We have dev $E \leqq$ dev $(E \times F)$ and dev $F \leqq$ dev $(E \times F)$ from 3.5.2. Let us prove that dev $(E \times F) \leqq$ sup (dev E, dev F). This is obvious for E and F Artinian. Let us assume, therefore, that the property has been established for sup (dev E, dev F) $< n$, and consider the case where sup (dev E, dev F) $= n$. Let $((a_i, b_i))_{i=1,2,\ldots}$ be a strictly decreasing sequence of elements of $E \times F$. Then

$$\text{dev } [a_{i+1}, a_i] < n, \qquad \text{dev } [b_{i+1}, b_i] < n \quad \text{for } i \geqq i_0$$

hence

$$\text{dev } [(a_{i+1}, b_{i+1}), (a_i, b_i)] = \text{dev } ([a_{i+1}, a_i] \times [b_{i+1}, b_i]) < n \quad \text{for } i > i_0$$

from the induction hypothesis.

3.5.4. LEMMA. *Let E be an ordered set, and S the set of stationary infinite sequences of elements of E; we order S by agreeing that $(e_1, e_2, \ldots) \leqq (f_1 f_2, \ldots)$ if $e_1 \leqq f_1$, $e_2 \leqq f_2, \ldots$ Let C be the set of increasing sequences belonging to S. Then* dev $C =$ dev $S = 1 +$ dev E.

(i) dev $C \leqq$ dev S. This is obvious.

(ii) dev $S \leqq 1 +$ dev E. This is obvious if E is discrete. Let $n \in \mathbf{N}$. We assume that the assertion has been proved when dev $E < n$ and consider the case where dev $E = n$. Let $(s^1, s^2, \ldots)$ be a decreasing sequence of elements of S. Set $s^i = (s_1^i, s_2^i, \ldots)$; let s_∞^i be the element to which all the s_p^i are equal for p sufficiently large. For $i \geqq i_0$, we have dev $[s_\infty^{i+1}, s_\infty^i] < n$. Let S_i be the set of stationary infinite sequences of elements of $[s_\infty^{i+1}, s_\infty^i]$. From the induction hypothesis, we have dev $S_i < 1 + n$ for $i \geqq i_0$. Now

$$[s^{i+1}, s^i]_S = [s_1^{i+1}, s_1^i] \times [s_2^{i+1}, s_2^i] \times \cdots \times [s_{p_i}^{i+1}, s_{p_i}^i] \times S_i$$

for a certain integer p_i. From 3.5.3, we have dev $[s^{i+1}, s^i]_S < 1 + n$ for $i \geqq i_0$. Hence dev $S \leqq 1 + n$.

(iii) $1 +$ dev $E \leqq$ dev C. This is obvious if E is Artinian. Let $n \geqq 1$. Let us assume that our assertion has been established when dev $E < n$, and consider the case where dev $E = n$. There exists a strictly decreasing infinite sequence $(a_1, a_2, \ldots)$ of elements of E such that dev $[a_{i+1}, a_i] \geqq n - 1$ for an infinite number of values of i. Let C_i be the set of stationary increasing infinite sequences of elements of $[a_{i+1}, a_i]$. From the induction hypothesis and the obvious inequality dev $E \leqq$ dev C, we have dev $C_i \geqq n$ for an infinite number of values of i. Now, if s^i denotes the infinite sequence all of whose elements are equal to a_i, then $[s^{i+1}, s^i]_C = C_i$. Hence dev $C \geqq 1 + n$.

3.5.5. We return to the ring A, and let E be the set of left ideals of A, ordered by inclusion. We term the deviation of E the *Krull dimension of A* and denote it by Kdim A.

3.5.6. LEMMA. *Let us assume that A is a Noetherian algebra. Let X be an indeterminate and $B = A \otimes k[X]$. Then* Kdim $B = 1 +$ Kdim A.

Let E and F be the sets of left ideals of A and B, respectively. In F, we have the strictly decrasing sequence $(B, BX, BX^2, \ldots)$. Let $n \in \mathbf{N}$. The mapping $I \mapsto IX^n + BX^{n+1}$ of E into F is strictly increasing, and hence dev $[BX^{n+1}, BX^n] \geqq$ dev E. Consequently, Kdim $B \geqq 1 +$ Kdim A.

Every element b of B can be written in a unique way in the form

$$\alpha_0(b) + \alpha_1(b)X + \alpha_2(b)X^2 + \cdots,$$

where the $\alpha_i(b)$ belong to A. Let B_p be the set of the $b \in B$ such that

$$0 = \alpha_{p+1}(b) = \alpha_{p+2}(b) = \cdots$$

For all $J \in F$, let $f_p(J)$ be the set of the $\alpha_p(b)$ for $b \in J \cap B_p$. We have $f_p(J) \in E$. Let us set $(f_0(J), f_1(J), \ldots) = f(J)$. Let C be the set of increasing stationary infinite sequences of elements of E. Since A is Noetherian, we have $f(J) \in C$. Let $J, J' \in F$ such that $J \subset J'$ and $f(J) = f(J')$. Then $J \cap B_0 = J' \cap B_0$. Let us assume that $J \cap B_p = J' \cap B_p$ has been proved. Let $b' \in J' \cap B_{p+1}$.

Since $f_{p+1}(J) = f_{p+1}(J')$, there exists $b \in J \cap B_{p+1} \subset J' \cap B_{p+1}$ such that

$$\alpha_{p+1}(b) = \alpha_{p+1}(b');$$

then $b - b' \in J' \cap B_p = J \cap B_p$, whence $b' \in J \cap B_{p+1}$ and

$$J' \cap B_{p+1} = J \cap B_{p+1}.$$

Hence $J = J'$. This proves that f is a strictly increasing mapping of F into C. From 3.5.2 and 3.5.4, we have $\operatorname{dev} F \leqq 1 + \operatorname{dev} E$, i.e., Kdim Kdim $B \leqq 1 + \operatorname{Kdim} A$.

3.5.7. Proposition. *Let* $n = \dim \mathfrak{g}$. *Then* Kdim $U(\mathfrak{g}) \leqq n$.

Let G be the graded algebra associated with the algebra $U(\mathfrak{g})$ filtered by the $U_n(\mathfrak{g})$. From 2.3.6 and 3.5.6, we have Kdim $G = n$. Let E and F be the sets of left ideals of $U(\mathfrak{g})$ and G respectively. For all $I \in E$, let $f(I)$ be the element

$$\bigoplus_{n \geqq 0} (I \cap U_n(\mathfrak{g}))/(I \cap U_{n-1}(\mathfrak{g}))$$

of F. From 3.5.2, it is sufficient to prove that f is strictly increasing. Hence let $I, I' \in E$ such that $I \subset I'$ and $f(I) = f(I')$. Then $I \cap U_0(\mathfrak{g}) = I' \cap U_0(\mathfrak{g})$. Let us assume that

$$I \cap U_p(\mathfrak{g}) = I' \cap U_p(\mathfrak{g})$$

has been proved. Let $u' \in I' \cap U_{p+1}(\mathfrak{g})$. Since $f(I) = f(I')$, there exists $u \in I \cap U_{p+1}(\mathfrak{g}) \subset I' \cap U_{p+1}(\mathfrak{g})$ such that $u' - u \in U_p(\mathfrak{g})$. Then

$$u' - u \in I' \cap U_p(\mathfrak{g}) = I \cap U_p(\mathfrak{g}),$$

whence $u' \in I \cap U_{p+1}(\mathfrak{g})$ and

$$I' \cap U_{p+1}(\mathfrak{g}) = I \cap U_{p+1}(\mathfrak{g}).$$

Hence $I = I'$.

3.5.8. Lemma. *Assume that A is Noetherian. Let $a \in A$. There exists an integer $n \geqq 0$ such that $\mathfrak{l}(a^n) = \mathfrak{l}(a^{n'})$ for $n' \geqq n$. For such an integer n, we have $Aa^n \cap \mathfrak{l}(a^n) = 0$.*

We have $\mathfrak{l}(1) \subset \mathfrak{l}(a) \subset \mathfrak{l}(a^2) \subset \cdots$, whence the first assertion. If $x \in A$ is such that $xa^n \in \mathfrak{l}(a^n)$, we have $xa^{2n} = 0$, hence $xa^n = 0$, whence the second assertion.

3.5.9. A left ideal L of A is termed *essential* if for every non-null left deal L' of A we have $L' \cap L \neq 0$. It amounts to the same to say that for all $x \in A - \{0\}$ we have $Ax \cap L \neq 0$. It follows from this that the intersection of two essential left ideals is an essential left ideal. We define essential right ideals in a similar fashion.

3.5.10. LEMMA. *Assume that A is Noetherian and that the ideal* 0 *is semi-prime. Let L be an essential left ideal of A. There exists in L an element which is not a right divisor of zero* (*cf.* 3.6.11).

(a) From 3.1.14 and 3.5.8, there exists in any non-null left ideal of A a non-zero element a such that $Aa \cap \mathfrak{l}(a) = 0$.

(b) Let us assume that non-zero elements $a_1, \ldots, a_n$ of L have been constructed such that the sum

$$Aa_1 + \cdots + Aa_n + (\mathfrak{l}(a_1) \cap \cdots \cap \mathfrak{l}(a_n) \cap L)$$

is direct. If $L' = \mathfrak{l}(a_1) \cap \cdots \cap \mathfrak{l}(a_n) \cap L \neq 0$, there exists $a_{n+1} \in L' - \{0\}$ such that

$$Aa_{n+1} \cap \mathfrak{l}(a_{n+1}) = 0.$$

Then the sum $Aa_1 + \cdots + Aa_{n+1} + (\mathfrak{l}(a_1) \cap \cdots \cap \mathfrak{l}(a_{n+1}) \cap L)$ is direct. Now the construction under consideration cannot be pursued indefinitely because A is Noetherian. There thus exist non-zero elements $a_1, \ldots, a_n \in L$ such that the sum $Aa_1 + \cdots + Aa_n$ is direct and $\mathfrak{l}(a_1) \cap \cdots \cap \mathfrak{l}(a_n) \cap L = 0$. Since L is essential, $\mathfrak{l}(a_1) \cap \cdots \cap \mathfrak{l}(a_n) = 0$. If $x \in A$ is such that $x(a_1 + \cdots + a_n) = 0$, we have $xa_1 = \cdots = xa_n = 0$, and hence $x = 0$.

3.5.11. LEMMA. *Assume that A is Noetherian. Let $(P_0, P_1, \ldots, P_n)$ be a strictly increasing sequence of prime ideals of A. Then $n <$* Kdim A.

We may assume that Kdim $A < +\infty$. By passing to the quotient by P_0, we are led to prove the following: let us assume that 0 is a prime ideal of A, and let I be a non-null two-sided ideal of A; then Kdim $A >$ Kdim(A/I).

If J is a left ideal of A such that $J \cap I = 0$, then $IAJ \subset J \cap I = 0$, hence $J = 0$ since 0 is prime and $I \neq 0$. From 3.5.10, there exists in I an element s which is not a right divisor of zero. The sequence of left ideals $(A, As, As^2, \ldots)$ is decreasing. If $As^n = As^{n+1}$, there exists $a \in A$ such that $s^n = as^{n+1}$, whence $1 = as$ and $I = A$ in which case our assertion is obvious. We may thus assume that the sequence $(A, As, As^2, \ldots)$ is strictly decreasing. Now the mapping $x \mapsto xs^n$ of A into As^n defines a surjective homomorphism φ of the left A-module A/As into the left A-module As^n/As^{n+1}; if $xs^n \in As^{n+1}$, then $x \in As$, and hence φ is bijective. Since $As \subset I$, there exists a homomorphism of the left A-module As^n/As^{n+1} onto the left A-module A/I. Then we indeed have Kdim $A >$ Kdim (A/I).

3.5.12. THEOREM. *Let* $n = \dim \mathfrak{g}$. *Let* $(P_0, P_1, \ldots, P_r)$ *be a strictly increasing sequence of prime ideals of* $U(\mathfrak{g})$. *Then* $r \leqq n$.

This follows from 3.5.7 and 3.5.11.

References: [50], [59], [110].

3.6. Rings of fractions

3.6.1. Let S be a subset of A. S is said *to allow of an arithmetic of fractions* if the following conditions are satisfied:

(i) $1 \in S$;
(ii) the product of two elements of S belongs to S;
(iii) the elements of S are not divisors of zero in A;
(iv) for $s \in S$ and $a \in A$, there exist $t \in S$ and $b \in A$ such that $at = sb$;
(v) for $s \in S$ and $a \in A$, there exist $t' \in S$ and $b' \in A$ such that $t'a = b's$.

3.6.2. Let S be a subset of A allowing of an arithmetic of fractions. Let $(a,s) \in A \times S$ and $(b,t) \in A \times S$. We write $(a,s) \sim (b,t)$ if there exist $c,d \in A$ such that $ac = bd$, $sc = td \in S$. We note that, if this is so, then $ac' = bd'$ for any $c',d' \in A$ such that $sc' = td' \in S$. (For, there exist $e \in A$, $e' \in S$ such that $(sc)e = (sc')e'$, whence $(td)e = (td')e'$, $ce = c'e'$, $de = d'e'$, $ac'e' = ace = bde = bd'e'$, $ac' = bd'$.) Given this, the relation $\sim$, which is obviously reflexive and symmetric, is transitive; for let

$$(a,s) \sim (b,t) \sim (c,u);$$

there exist $d,e,f \in A$ such that $sd = te = uf \in S$; from the preceding remark, we have $ad = be = cf$, whence $(a,s) \sim (c,u)$.

Let B be the quotient set of $A \times S$ under $\sim$. We shall provisionally denote the equivalence class of (a,s) by a/s.

Let $a/s, b/t \in B$. There exist $c,d \in A$ such that $sc = td = e \in S$. We set

$$(a/s) + (b/t) = (ac + bd)/e.$$

Let a'; s', b', t', c', d', e' with

$$a/s = a'/s', \quad b/t = b'/t', \quad s'c' = t'd' = e' \in S.$$

There exist $x,y \in A$ with $ex = e'y \in S$. Whence

$$scx = s'c'y \in S, \quad tdx = t'd'y \in S,$$

$$acx = a'c'y, \quad bdx = b'd'y, \quad (ac + bd)x = (a'c' + b'd')y,$$

$$(ac + bd)/e = (a'c' + b'd')/e'.$$

This justifies the definition of addition. Since two fractions can be reduced to the same denominator, we can at once verify that the addition is associative and commutative, that $0/1$ is the zero element, and that the additive inverse of a/s is $(-a)/s$.

Let $a/s, b/t \in B$. There exist $c \in A$, $u \in S$ such that $bu = sc$. We set

$$(a/s)(b/t) = ac/tu.$$

Note that, if $c' \in A$, $u' \in S$ are such that $bu' = sc'$, there exist x, x' such that $ux = u'x' \in S$, whence

$$tux = tu'x' \in S, \qquad bux = bu'x', \qquad scx = sc'x',$$

$$cx = c'x', \qquad acx = ac'x', \qquad ac/tu = ac'/tu'.$$

Given this, let $a/s = a'/s'$ and $b/t = b'/t'$. There exist u, u' such that $su = s'u' \in S$, whence $au = a'u'$; next, there exist $v, v' \in S$ and c, c' such that $bv = suc$ and $b'v' = s'u'c'$. Then

$$(a/s)(b/t) = auc/tv, \qquad (a'/s')(b'/t') = a'u'c'/t'v'.$$

There exist x, x' such that $tvx = t'v'x' \in S$. Then

$$bvx = b'v'x', \qquad sucx = s'u'c'x', \qquad cx = c'x',$$

$$aucx = a'u'c'x', \qquad auc/tv = a'u'c'/t'v'.$$

This justifies the definition of multiplication.

Let $a/s, b/t, c/u \in B$. Let $d \in A$, $v \in S$ be such that $bv = sd$. Then let $e \in A$, $w \in S$ be such that $cw = tve$. Then

$$((a/s)(b/t))(c/u) = (ad/tv)(c/u) = ade/uw,$$

$$(a/s)((b/t)(c/u)) = (a/s)(bve/uw) = ade/uw,$$

whence the multiplication is associative. We verify at once that $1/1$ is the unity, and that, for all $s \in S$, $s/1$ and $1/s$ are inverse to each other. On the other hand, $a/s = (a/1)(1/s) = (a/1)(s/1)^{-1}$. To verify the distributive law, it is thus sufficient to verify the equations

$$(a/1)(b/t + b'/t) = (a/1)(b/t) + (a/1)(b'/t),$$

$$(b/t + b'/t)(a/1) = (b/t)(a/1) + (b'/t)(a/1),$$

which is easy.

Thus, B is a ring (with unity).The mapping $a \mapsto a/1$ of A into B is an injective ring homomorphism, under which we can identify A with its image in B. Then every element of S is invertible in B, and $a/s = as^{-1}$. Henceforth, we shall drop the notation a/s.

Let $a \in A$ and $s \in S$. There exist $t \in S$ and $b \in A$ such that $ta = bs$, whence $as^{-1} = t^{-1}b$. Thus, every element of B can also be put into the form $t^{-1}b$, where $b \in A$, $t \in S$.

3.6.3. The ring B is termed the *ring of fractions of A defined by S*, and is denoted by A_S. (It follows from the proof of 3.6.2 that if $b_1, \ldots, b_n \in B$, there exist $a_1, \ldots, a_n, a'_1, \ldots, a'_n \in A$ and $s, s' \in S$ such that $b_i = a_i s^{-1} = s'^{-1} a'_i$ for $i = 1, \ldots, n$.)

For example, if z is a central element of A which is not a divisor of zero in A, then the set $S = \{1, z, z^2, \ldots\}$ allows of an arithmetic of fractions in A, and A_S is then denoted by A_z.

Let S_0 be the set of elements which are not divisors of zero in A. When S_0 allows of an arithmetic of fractions in A (i.e. satisfies conditions (iv) and (v) of 3.6.1), A_{S_0} is simply termed the *ring of fractions* of A and is denoted by Fract(A). (For example, this is the case if A is commutative. See also 3.6.12.)

If A is integral and $S_0 = A - \{0\}$ allows of an arithmetic of fractions, then Fract(A) is obviously a (skew) field, termed the *field of fractions* of A.

3.6.4. Let S be a subset of A allowing of an arithmetic of fractions. If A is an algebra, there exists on A_S one and only one algebra structure such that $\lambda(as^{-1}) = (\lambda a)s^{-1}$ for $a \in A$ and $s \in S$.

3.6.5. PROPOSITION. *Let A and S be as in* 3.6.2, *A' be a ring, and φ be a homomorphism of A into A' such that every element of $\varphi(S)$ is invertible. Then there exists one and only one homomorphism of A_S into A' which extends φ.*

(Of course this general property characterises A_S and the injection $A \to A_S$ up to isomorphism.)

We verify that by setting $\psi(as^{-1}) = \varphi(a)\varphi(s)^{-1}$ for $a \in A$ and $s \in S$, we obtain the unique homomorphism ψ of A_S into A' extending φ.

3.6.6. PROPOSITION. *Let A and S be as in* 3.6.2, *and C be a ring containing A as a subring. We assume every element of S to be invertible in C. Let i be the canonical injection of A into C. Then i can be uniquely extended to an isomorphism ψ of A_S onto a subring of C.*

From 3.6.5, i can be uniquely extended to a homomorphism ψ of A_S into C. Let $a \in A$ and $s \in S$ be such that the element $b = as^{-1}$, calculated in A_S, belongs to the kernel of ψ. Since $\psi(b)$ is equal to as^{-1} calculated in C, we have $a = 0$ and hence $b = 0$.

3.6.7. Under the hypotheses of 3.6.6, we identify A_S with the subring $\psi(A_S)$ of C.

Let C be a ring, A a subring and S a subset of A consisting of invertible elements of C, satisfying conditions (i) and (ii) of 3.6.1. We assume that every element of C can be put into the forms as^{-1} ($a \in A$, $s \in S$) and $s'^{-1}a'$ ($a' \in A$, $s' \in S$). Then S satisfies condition (iii) of 3.6.1. On the other hand, for $s \in S$ and $a \in A$, there exists $s' \in S$ and $a' \in A$ such that $s^{-1}a = a's'^{-1}$, whence $as' = sa'$; thus S satisfies condition (iv) of 3.6.1, and, similarly, condition (v). Hence we can form A_S, which, from 3.6.6, can be identified with C.

3.6.8. PROPOSITION. *Let A and S be as in 3.6.2, and let $a \in A$ and $s \in S$. Then as^{-1} belongs to the centre of A_S, if and only if $axs = sxa$ for all $x \in A$.*

If as^{-1} is central, s commutes with as^{-1} and hence with a; for all $x \in A$, we have

$$0 = xas^{-1} - as^{-1}x = s^{-1}(sxa - axs)s^{-1},$$

whence $sxa = axs$. Conversely, if the condition of the proposition is satisfied, then a and s commute, and

$$xas^{-1} - as^{-1}x = s^{-1}(sxa - axs)s^{-1} = 0$$

for all $x \in A$, so that as^{-1} is central.

3.6.9. LEMMA. *Assume that A i Noetherian. Let x be an element of A which is not a right divisor of zero. Then Ax is an essential left ideal of A.*

Let I be a left ideal of A such that $I \cap Ax = 0$. Let $a_0, \ldots, a_n \in I$ be such that $a_0 + a_1x + \cdots + a_nx^n = 0$. Then $a_0 = 0$, and hence

$$a_1 + a_2x + \cdots + a_nx^{n-1} = 0.$$

Proceeding in a stepwise fashion, all the a_i are zero. Hence the sum $I + Ix + Ix^2 + \cdots$ is direct, which requires that $I = 0$ since A is Noetherian.

3.6.10. Let Z be the set of the $a \in A$ such that $\mathfrak{l}(a)$ is essential. Stating that $a \in Z$ means that for all non-zero x in A there exists $y \in A$ such that $yx \neq 0$, $yxa = 0$.

LEMMA. (i) *Z is a two-sided ideal of A.*
(ii) *If A is Noetherian and the ideal 0 is semi-prime, then $Z = 0$.*

It follows from 3.5.9 that Z is an additive subgroup of A, and clearly Z is a right ideal. Let $a \in Z$, $b \in A$ and $x \in A - \{0\}$. If $xb = 0$, then $1 \cdot x(ba) = 0$. If $xb \neq 0$, there exists $y \in A$ such that $y(xb) \neq 0$ and $y(xb)a = 0$. Hence $ba \in Z$.

Let us assume that A is Noetherian and that the ideal 0 is semi-prime. Let $z \in Z$. There exists $n > 0$ such that $Az^n \cap \mathfrak{l}(z^n) = 0$ (3.5.8). Now $\mathfrak{l}(z^n)$ is essential, and hence $z^n = 0$. Then $Z = 0$ (3.1.14).

3.6.11. Lemma. *Assume that A is Noetherian and that the ideal* 0 *is semi-prime.*

(i) *An element of A is not a right divisor of zero if and only if it is not a left divisor of zero.*

(ii) *Every essential left ideal of A contains an element which is not a divisor of zero.*

Let x be an element of A which is not a right divisor of zero. Then Ax is essential (3.6.9). Hence $\mathfrak{r}(x)$ is contained in the set Z of 3.6.10, and consequently is zero. Thus x is not a left divisor of zero. We deduce (i) from this, and (ii) then follows from 3.5.10.

3.6.12. Theorem. *Assume that A is Noetherian and the ideal* 0 *semi-prime. Let S be the set of elements of A which are not divisors of zero. Then:*

(i) *S allows of an arithmetic of fractions.*

(ii) *A_S is semi-simple and Artinian.*

(iii) *If the ideal* 0 *of A is prime, then A_S is simple and Artinian.*

(iv) *If If A is integral, then A_S is a (skew) field.*

(In this book, a ring is said to be *simple* if its two-sided ideals are trivial. Such a ring is not in general Artinian, contrary to the terminology of *AL*. An Artinian semi-simple ring is a finite product of Artinian simple rings.)

Let $a \in A$ and $s \in S$. Let L be the left ideal of A consisting of those $t \in A$ such that $ta \in As$. Let $x \in A - \{0\}$. There exists $y \in A$ such that $yx \in L$ and $yx \neq 0$; this is obvious if $xa = 0$, and if $xa \neq 0$ then there exists $y \in A$ such that $yxa \neq 0$ and $yxa \in As$ since As is essential (3.6.9). Hence L is essential. From 3.6.11, there exists $t \in S \cap L$, so that condition (v) of 3.6.1 is satisfied. Passing to the opposite ring, condition (iv) of 3.6.1 is also satisfied. This proves (i).

Let L be an essential left ideal of A_S. Then $L \cap A$ is essential in A, and hence contains an element of S (3.6.11), which is invertible in A_S; consequently, $L = A_S$. Given this, let M be a left ideal of A_S. The set of left ideals of A_S whose intersection with M is null possesses, from Zorn's theorem, a maximal element M'. If M'' is a non-null left ideal of A_S, then $M'' \cap (M + M') \neq 0$; hence $M + M'$ is essential, and finally, from the above, $A_S = M \oplus M'$. Thus the left regular representation of A_S is semi-simple, which proves (ii). (We recall that, in a decomposition $A_S \oplus L_i$ of A_S into minimal left ideals, the set of indices is finite as can be proved by the decomposition of 1 according to the L_i (cf. AL VIII, p. 46).)

If J is a non-null two-sided ideal of A_S, then $J \cap A \neq 0$. If there exist in A_S two non-null two-sided ideals with intersection null, it can be seen that 0 is not prime in A. Consequently, (iii) follows from (ii), and (iv) was noted in 3.6.3.

3.6.13. Since $U(\mathfrak{g})$ is Noetherian and integral, it possesses a field of fractions, termed the *enveloping field* of $\mathfrak{g}$. We denote it by $K(\mathfrak{g})$. The mapping $x \mapsto \mathrm{ad}_{K(\mathfrak{g})}x$ $(x \in \mathfrak{g})$ is termed the *adjoint representation of* $\mathfrak{g}$ *in* $K(\mathfrak{g})$.

More generally, let I be a semi-prime ideal of $U(\mathfrak{g})$, let

$$B = \mathrm{Fract}(U(\mathfrak{g})/I),$$

and let φ be the canonical mapping of $U(\mathfrak{g})$ onto $U(\mathfrak{g})/I$. The mapping $x \mapsto \mathrm{ad}_B\varphi(x)$ is termed the *adjoint representation of* $\mathfrak{g}$ *in* B.

3.6.14. LEMMA. *Let A and S be as in* 3.6.2, *and I be a two-sided ideal of A such that the conditions $a \in A$, $s \in S$, $sa \in I$ imply that $a \in I$. Let I' and I'' be the sets of elements of A_S of the form is^{-1} and $s^{-1}i$ respectively, where $\in I$, $s \in S$. Then:*

(i) $I'' \subset I'$.

(ii) *I' is a two-sided ideal of A_S.*

(i) Let $i \in I$, $s \in S$. There exist $a \in A$ and $t \in S$ such that $s^{-1}i = at^{-1}$. Then $sa = it \in I$, hence $a \in I$ and $s^{-1}i \in I'$. This proves (i).

(ii) Let $i_1, i_2 \in I$, and $s_1, s_2 \in S$. There exist $t_1 \in S$ and $t_2 \in A$ such that $s_1t_1 = s_2t_2$. We set $s = s_1t_1 \in S$. Then

$$s_1^{-1} = t_1s^{-1}, \qquad s_2^{-1} = t_2s^{-1}$$

hence

$$i_1s_1^{-1} + i_2s_2^{-1} = (i_1t_1 + i_2t_2)s^{-1} \in I'.$$

Thus, $I' + I' \subset I'$. On the other hand,

$$I'A_S = IS^{-1}AS^{-1} = IAS^{-1}S^{-1} \subset IS^{-1} = I',$$

and

$$A_SI' = AS^{-1}IS^{-1} = AI''S^{-1} \subset AI'S^{-1} = AIS^{-1}S^{-1} \subset IS^{-1} = I',$$

whence (ii).

3.6.15. Proposition. *Let A and S be as in 3.6.2, $\mathscr{I}'$ be the set of two-sided ideals of A_S, and $\mathscr{I}$ be the set of two-sided ideals of A satisfying the following conditions:*

(a) *if $a \in A$, $s \in S$ and $as \in I$, then $a \in I$;*

(b) *if $a \in A$, $s \in S$ and $sa \in I$, then $a \in I$.*

Then:

(i) *If $I \in \mathscr{I}$, the set I_S of the is^{-1} where $i \in I$ and $s \in S$ is equal to the set of the $s^{-1}i$ where $i \in I$ and $s \in S$. We have $I_S \in \mathscr{I}'$.*

(ii) *The mappings $I \mapsto I_S$ and $I' \mapsto I' \cap A$ are reciprocal bijections of $\mathscr{I}$ onto $\mathscr{I}'$ and of $\mathscr{I}'$ onto $\mathscr{I}$.*

(iii) *Let $I \in \mathscr{I}$, and let T be the canonical image of S in A/I. Then T allows of an arithmetic of fractions in A/I, and the canonical injection of A/I into A_S/I_S can be uniquely extended to a homomorphism of $(A/I)_T$ into A_S/I_S; this homomorphism is bijective.*

(iv) *If $I \in \mathscr{I}$ and I is prime, then I_S is prime.*

Assertion (i) follows directly from 3.6.14.

Let $I \in \mathscr{I}$. Obviously, we have $I_S \cap A \supset I$. If $a \in I_S \cap A$, there exist $i \in I$ and $s \in S$ such that $a = is^{-1}$, whence $as \in I$ and $a \in I$. Hence $I_S \cap A = I$.

Let $I' \in \mathscr{I}'$ and $I = I' \cap A$. Let $a \in A$ and $s \in S$ be such that $as \in I$. Then $a \in Is^{-1} \subset I'$ and hence $a \in I$. Similarly, the condition $sa \in I$ implies that $a \in I$. Hence $I \in \mathscr{I}$. Clearly, $I_S \subset I'$. Let as^{-1} ($a \in A$, $s \in S$) be an element of I'. Then $a = (as^{-1})\,s \in I' \cap A = I$, hence $as^{-1} \in I_S$ and $I' = I_S$. The above proves (ii).

The assertions of (iii) follow directly from 3.6.7.

Let $I \in \mathscr{I}$ be a prime ideal. Let $a,a' \in A$ and $s,s' \in S$, and let us assume that $(s^{-1}a)A_S(a's'^{-1}) \subset I_S$. Then $aAa' \subset sI_Ss' \subset I_S$, hence $aAa' \subset I$. Then $a \in I$ or $a' \in I$, hence $s^{-1}a \in I_S$ or $a's'^{-1} \in I_S$. This proves (iv).

3.6.16. The notation I_S of 3.6.15 (for the ideals $I \in \mathscr{I}$) will be retained subsequently. If S is the set of powers of a central element z of A, we write I_z instead of I_S.

3.6.17. Proposition. *Let A and S be as in 3.6.2, $\mathscr{P}$ be the set of prime ideals of A, and $\mathscr{P}'$ be the set of prime ideals of A_S. We define $\mathscr{I}$ as in 3.6.15, and we assume that A is Noetherian.*

(i) *$\mathscr{P} \cap \mathscr{I}$ is the set of prime ideals of A which do not intersect S.*

(ii) *The mappings $P \mapsto P_S$ and $P' \mapsto P' \cap A$ are reciprocal bijections of $\mathscr{P} \cap \mathscr{I}$ onto $\mathscr{P}'$ and of $\mathscr{P}'$ onto $\mathscr{P} \cap \mathscr{I}$.*

(i) Let $P \in \mathscr{P} \cap \mathscr{I}$. If $s \in S \cap P$, then $s \cdot 1 \in P$ and hence $1 \in P$, which is a contradiction; hence $S \cap P = \emptyset$. Now let $P \in \mathscr{P}$ be such that $S \cap P = \emptyset$, and let us prove that $P \in \mathscr{I}$. Let T be the canonical image of S in $B = A/P$. We must prove that T does not contain a divisor of zero in B. Let

$$R = \{b \in B \mid tb = 0 \quad \text{for some } t \in T\}.$$

Then $B + B \subset B$ and $RB \subset R$. Let $b \in B$ and $t \in T$ such that $tb = 0$, and $b' \in B$. There exist $u \in T$ and $c \in B$ such that $ub' = ct$. Then $ub'b = ctb = 0$, hence $b'b \in R$. Thus R is a two-sided ideal of B. Let us assume that $R \neq 0$. Since 0 is a prime ideal of B, R is an essential left ideal of B, and hence does not contain a divisor of zero (3.6.11). Then $0 \in T$, which contradicts the assumption $S \cap P = \emptyset$. Hence $R = 0$, which proves that T does not contain a divisor of zero in B. We have thus proved (i).

(ii) If $P \in \mathscr{P} \cap \mathscr{I}$, we have $P_S \in \mathscr{P}'$ (3.6.15 (iv)). From 3.6.15 (ii), it remains to prove that $P' \in \mathscr{P}'$ implies $P' \cap A \in \mathscr{P}$. From 3.6.15 (iii), it is sufficient to consider the case where $P' = 0$. Let I and J be two-sided ideals of A such that $IJ = 0$, $I \neq 0$, and let us prove that $J = 0$. We may assume that $I = \{a \in A | aJ = 0\}$. Then the conditions $a \in A$, $s \in S$ and $sa \in I$ imply that $a \in I$. Hence IS^{-1} is a non-null two-sided ideal of A_S (3.6.14). Since 0 is a prime ideal of A_S, IS^{-1} is an essential left ideal of A_S, hence (3.6.11) contains an element is^{-1} ($i \in I$, $s \in S$) which is not a divisor of zero in A_S. Then i is not a divisor of zero in A, and the relation $iJ \subset IJ = 0$ implies that $J = 0$.

3.6.18. Proposition. *Let A and S be as in* 3.6.2, *and D be a derivation of A. There exists one and only one derivation D' of A_S which extends D.*

(a) Let D' be a derivation of A_S extending D. If $s \in S$, then

$$0 = D(ss^{-1}) = (Ds)s^{-1} + s(D's^{-1}),$$

whence

$$D's^{-1} = -s^{-1}(Ds)s^{-1}.$$

This proves the uniqueness of D'.

(b) Let $x,y,u,v \in A$ such that $x,u \in S$ and $x^{-1}y = vu^{-1}$. Then

$$y(Du) + (Dy)u = D(yu) = D(xv) = x(Dv) + (Dx)v,$$

hence, by multiplying on the left by $-x^{-1}$ and on the right by u^{-1},

$$-vu^{-1}(Du)u^{-1} + (Dv)u^{-1} = -x^{-1}(Dx)x^{-1}y + x^{-1}(Dy).$$

We thus define a mapping $\bar{D}$ of A_S into A_S by setting

$$\bar{D}(x^{-1}y) = -x^{-1}(Dx)x^{-1}y + x^{-1}(Dy) = (Dv)u^{-1} - vu^{-1}(Du)u^{-1}.$$

We immediately verify that $\bar{D}$ is additive and extends D. Let $\beta,\gamma \in A_S$. Let $b,c,d \in A$ and $w,z \in S$ such that $\beta = w^{-1}b$ and $\gamma = w^{-1}c = dz^{-1}$. Let us write $bdz^{-1} = y^{-1}e$, where $e \in A$, $y \in S$. Then

$$y^{-1}(De) - y^{-1}(Dy)y^{-1}e = \bar{D}(y^{-1}e) = \bar{D}(bdz^{-1})$$
$$= (Db)dz^{-1} + b(Dd)z^{-1} - bdz^{-1}(Dz)z^{-1}$$

and consequently

$$\begin{aligned} D(\beta\gamma) &= \bar{D}(w^{-1}bdz^{-1}) = \bar{D}((yw)^{-1}e) \\ &= (yw)^{-1}(De) - w^{-1}y^{-1}((Dy)w + y(Dw))w^{-1}y^{-1}e \\ &= w^{-1}(y^{-1}(De) - y^{-1}(Dy)y^{-1}e) - w^{-1}(Dw)w^{-1}y^{-1}e \\ &= w^{-1}((Db)dz^{-1} + b(Dd)z^{-1} - bdz^{-1}(Dz)z^{-1}) \\ &\quad - w^{-1}(Dw)w^{-1}y^{-1}e \\ &= (w^{-1}(Db) - w^{-1}(Dw)w^{-1}b)dz^{-1} \\ &\quad + w^{-1}b((Dd)z^{-1} - dz^{-1}(Dz)z^{-1}) \\ &= (\bar{D}\beta)\gamma + \beta(\bar{D}\gamma). \end{aligned}$$

References: [8], [50], [51], [59].

3.7. Prime ideals in the solvable case

3.7.1. LEMMA. *Let M and N be finite-dimensional triangularizable. $\mathfrak{g}$-modules, and P a sub-$\mathfrak{g}$-module of $M \otimes N$. We assume that there exist $a \in M$, $b \in N$ such that $a \neq 0$, $b \neq 0$, $a \otimes b \in P$. Then there exist $a' \in M$, $b' \in N$ such that $a' \neq 0$, $b' \neq 0$, $a' \otimes b' \in P$, and such that ka' and kb' are sub-$\mathfrak{g}$-modules of M and N respectively.*

(a) In part (a) of the proof, we assume that $\mathfrak{g}$ is commutative. There exist linear forms $\lambda_0, \ldots, \lambda_m$ on $\mathfrak{g}$, pairwise distinct, such that $M = M^{\lambda_0} \oplus \cdots \oplus M^{\lambda_m}$ (1.3.19). Similarly, $N = N^{\mu_0} \oplus \cdots \oplus N^{\mu_n}$. Let $a = a_0 + \cdots + a_m$ and $b = b_0 + \cdots + b_n$ be the corresponding decompositions of a and b. We may assume that $a_0, \ldots, a_m, b_0, \ldots, b_n$ are all non-zero. There exists $x \in \mathfrak{g}$ such that $\lambda_0(x), \ldots, \lambda_m(x)$ are pairwise distinct, and such that $\mu_0(x), \ldots, \mu_n(x)$ are also pairwise distinct. Then, by changing the numbering where necessary,

$$\lambda_0(x) + \mu_0(x) \neq \lambda_i(x) + \mu_j(x) \quad \text{if } (i,j) \neq (0,0).$$

[It is sufficient to choose a basis for k over $\mathbf{Q}$, to order k lexicographically and to ensure that $\lambda_0(x) < \lambda_1(x), \ldots, \lambda_m(x)$ and $\mu_0(x) < \mu_1(x), \ldots, \mu_n(x)$.] We have $M \otimes N = \oplus (M^{\lambda_i} \otimes N^{\mu_j})$. Let us set $\varrho_0 = \lambda_0(x) + \mu_0(x)$, and let $\varrho_1, \ldots, \varrho_r$ be the $\lambda_i(x) + \mu_j(x)$ for $(i,j) \neq (0,0)$. Let s be an integer >0 such that $\prod_{i=0}^{r}(x - \varrho_i)^s$ annihilates $M \otimes N$. There exist $f \in k[X]$ and $g \in k[X]$ such that

$$1 = f(X) \prod_{i=1}^{r} (X - \varrho_i)^S + g(X)(X - \varrho_0)^S.$$

We set

$$h(X) = f(X) \prod_{i=1}^{r} (X - \varrho_i)^s.$$

Then $h(x)(a \otimes b) = a_0 \otimes b_0$. Replacing a by a_0 and b by b_0, we return to the case where $m = n = 0$. We then set $\lambda_0 = \lambda$, $\mu_0 = \mu$.

Let $y \in \mathfrak{g}$. There exist integers u and v such that

$$(y - \lambda(y))^u a \neq 0, \qquad (y - \lambda(y))^{u+1} a = 0,$$

$$(y - \mu(y))^v b \neq 0, \qquad (y - \mu(y))^{v+1} b = 0.$$

Setting $a_1 = (y - \lambda(y))^u a$ and $b_1 = (y - \mu(y))^v b$, we have

$$\begin{aligned}(y - \lambda(y) - \mu(y))^{u+v}(a \otimes b) &= ((y - \lambda(y)) \otimes 1 + 1 \otimes (y - \mu(y)))^{u+v}(a \otimes b)\\ &= \frac{(u+v)!}{u!\,v!}\, a_1 \otimes b_1,\end{aligned}$$

and hence $a_1 \otimes b_1 \in P$. On the other hand, a_1 and b_1 are eigenvectors of y_M and y_N respectively; and, if a and b were eigenvectors of some z_M and z_N respectively where $z \in \mathfrak{g}$, then a_1 and b_1 preserve this property. By choosing a basis for $\mathfrak{g}$ and applying the preceding construction step by step, we obtain a' and b' with the required properties.

(b) Let us turn to the general case. We can assume that $M = U(\mathfrak{g})a$ and $N = U(\mathfrak{g})b$.

There exists a decreasing sequence $(M_0, \ldots, M_m)$ of sub-$\mathfrak{g}$-modules of M such that $M_0 = M$, $M_m = 0$ and $M_i/M_{i+1} = 1$ for all i. Let $N_0, N_1, \ldots, N_n$ in N be defined similarly. Let $\mathfrak{g}' = [\mathfrak{g},\mathfrak{g}]$. Then $\mathfrak{g}'M_i \subset M_{i+1}$ and $\mathfrak{g}'N_i \subset N_{i+1}$ for all i.

We shall argue by induction on $\dim M + \dim N$. If $\mathfrak{g}'M = \mathfrak{g}'N = 0$, we can replace $\mathfrak{g}$ by $\mathfrak{g}/\mathfrak{g}'$, and the lemma has been proved in (a). Hence we shall assume that, for example, $\mathfrak{g}'M \neq 0$. Since $\mathfrak{g}'M = \mathfrak{g}'U(\mathfrak{g})a = U(\mathfrak{g})\mathfrak{g}'a$, there exists $x \in \mathfrak{g}'$ such that $xa \neq 0$. There then exists an integer $r \geqq 1$

such that $x^r a \neq 0$, $x^{r+1}a = 0$. On the other hand, there exists an integer $s \geqq 0$ such that $x^s b \neq 0$ and $x^{s+1}b = 0$. Let us set $a_1 = x^r a$ and $b_1 = x^s b$. Then

$$x^{r+s}(a \otimes b) = \frac{(r+s)!}{r!s!} a_1 \otimes b_1,$$

hence $a_1 \otimes b_1 \in P$. Now $a_1 \in M$, so that

$$\dim U(\mathfrak{g})a_1 + \dim U(\mathfrak{g})b_1 < \dim M + \dim N.$$

It is then sufficient to apply the induction hypothesis to a_1 and b_1.

3.7.2. THEOREM. *The following conditions are equivalent:*
(i) $\mathfrak{g}$ *is solvable;*
(ii) *every primitive ideal of* $U(\mathfrak{g})$ *is completely prime;*
(iii) *every prime ideal of* $U(\mathfrak{g})$ *is completely prime.*

(iii) $\Rightarrow$ (ii). This follows from 3.1.6.

(ii) $\Rightarrow$ (i). Let us assume that $\mathfrak{g}$ is not solvable and prove that (ii) is false. Since $\mathfrak{g}$ has a simple quotient, we return to the case where $\mathfrak{g}$ is simple. Let π be the adjoint representation of $\mathfrak{g}$. Its kernel I (in $U(\mathfrak{g})$) is a primitive ideal. Let x be a generator of $\mathfrak{g}$. Then $\pi(x)$ is semi-simple and $\operatorname{Ker} \pi(x)$ is distinct from 0 and $\mathfrak{g}$. Hence there exists $p \in k[X]$ such that $p(\pi(x))$ is a projection which is distinct from 0 and 1. Then

$$\pi(p(x)) \neq 0, \quad \pi(1 - p(x)) \neq 0 \quad \text{and} \quad \pi(p(x)(1 - p(x))) = 0,$$

so that I is not completely prime.

(i) $\Rightarrow$ (iii). Let I be a prime ideal of $U(\mathfrak{g})$.

(a) Let us assume that $\mathfrak{g}$ is completely solvable. The $\mathfrak{g}$-module $U(\mathfrak{g})/I$ is the union of an increasing sequence of finite-dimensional triangularizable $\mathfrak{g}$-modules (2.3.3). Let a and b be non-zero elements of $U(\mathfrak{g})/I$ such that $ab = 0$. They generate triangularizable sub-$\mathfrak{g}$-modules M and N of $U(\mathfrak{g})/I$. The mapping $(x,y) \mapsto xy$ of $M \times N$ into $U(\mathfrak{g})/I$ defines a $\mathfrak{g}$-homomorphism of $M \otimes N$ into $U(\mathfrak{g})/I$ whose kernel P contains $a \otimes b$. By virtue of 3.7.1, we may assume that ka and kb are sub-$\mathfrak{g}$-modules of $U(\mathfrak{g})/I$. Then a and b are not divisors of zero in $U(\mathfrak{g})/I$ (3.3.9), which is a contradiction. Hence I is completely prime.

(b) Let us assume that $\mathfrak{g}$ is solvable. Let k' be an algebraic closure of k. There exists a prime ideal I' of $U(\mathfrak{g} \otimes k')$ such that $I = I' \cap U(\mathfrak{g})$ (3.4.2). From (a), I' is completely prime. Hence I is completely prime.

3.7.3. From 3.7.2 and 3.6.12, if $\mathfrak{g}$ is solvable and I is a prime ideal of $U(\mathfrak{g})$, then $U(\mathfrak{g})/I$ *admits of a field of fractions.*

References: [15], [31], [51].

3.8. Supplementary remarks

3.8.1. This book, which is devoted to enveloping algebras, is not a treatise on non-commutative algebra. We have therefore not given the best possible results concerning rings in general. In particular, the Noetherian hypotheses of the present chapter could often be given in weaker form.

Proposition 3.1.15 can be found in [31] for k non-denumerable, and in [48] for the general case. Definition 3.5.1 is due to Gabriel and Rentschler [110]. Theorem 3.6.12 is a special case of a celebrated theorem of Goldie. Proposition 3.6.17 is due to Goldie and Michler. Theorem 3.7.2 can be found in [31] for k algebraically closed, and in [51] for the general case; for other proofs, see [15] and [93].

3.8.2. Let $(I_\lambda)_{\lambda \in L}$ be a non-empty decreasing filtered family of prime ideals of A, and $I = \bigcap_{\lambda \in L} I_\lambda$. Then I is prime. By applying Zorn's theorem we deduce that every prime ideal of A contains a minimal prime ideal.

3.8.3. In a finite-dimensional algebra, every primitive two-sided ideal is maximal.

3.8.4. Let $I \in \text{Prim}(A)$. The closure of $\{I\}$ in Prim (A) is the set of primitive ideals of A containing I. Given two distinct points of Prim (A), one of the two points possesses a neighbourhood which does not contain the other.

3.8.5. Let Spec $U(\mathfrak{g})$ be the set of prime ideals of $U(\mathfrak{g})$. We define a topology on Spec $U(\mathfrak{g})$ as in 3.2.1. Then Prim $U(\mathfrak{g})$ is dense in Spec $U(\mathfrak{g})$. The mapping $F \mapsto F \cap \text{Prim}\, U(\mathfrak{g})$ is a bijection of the set of closed subsets of Spec $U(\mathfrak{g})$ onto the set of closed subsets of Prim $U(\mathfrak{g})$.

3.8.6. Let $\mathfrak{g} = \mathfrak{sl}(2,\mathbf{C})$, $\mathfrak{b}$ be the set of upper triangular elements of $\mathfrak{g}$. Let ϱ be the identity representation of $\mathfrak{g}$, I its kernel (in $U(\mathfrak{g})$). Then I is a maximal two-sided ideal of $U(\mathfrak{g})$, but $I \cap U(\mathfrak{b})$ is not a semi-prime ideal of $U(\mathfrak{b})$.

3.8.7. Let $\mathfrak{g},x,y,z$ be as in 1.14.10, $\mathfrak{k} = ky + kz$, I be the two-sided ideal $U(\mathfrak{g})(z-1)$ of $U(\mathfrak{g})$. Then I is a maximal two-sided ideal of $U(\mathfrak{g})$, but $I \cap U(\mathfrak{k})$ is not a primitive ideal of $U(\mathfrak{k})$.

3.8.8. Let $A,I,\mathscr{D},J$ be as in 3.2.2. If I is completely prime, then J is completely prime.

3.8.9. With the notation of 3.3.5, if K is prime (completely prime), then I is prime (completely prime).

3.8.10. (a) Let I be a two-sided ideal of $U(\mathfrak{g})$. I is said to be *absolutely primitive* if, for every extension k' of k, $I \otimes k'$ is primitive in $U(\mathfrak{g} \otimes k')$. Let $\mathscr{E}_1$ be the set of absolutely primitive ideals of $U(\mathfrak{g})$, $\mathscr{E}_2$ the set of kernels of absolutely simple representations of $U(\mathfrak{g})$. We have $\mathscr{E}_2 \subset \mathscr{E}_1$. Cf. 4.9.14.

(b) Let $\mathfrak{g}, x, y, z$ be as in 1.14.7 (b). There exists a homomorphism φ of $U(\mathfrak{g})$ onto $\mathbf{H}$ (the quaternion field over $\mathbf{R}$). Let $I = \operatorname{Ker} \varphi$. For every extension k' of $k = \mathbf{R}$, $\mathbf{H} \otimes k'$ is a simple central algebra, hence $I \otimes k'$ is maximal in $U(\mathfrak{g}) \otimes k'$, and I is absolutely primitive. But $U(\mathfrak{g})/I$ does not have an injective absolutely simple representation (such a representation would have to be two-dimensional, and $\mathbf{M}_2(\mathbf{R})$ does not contain the quaternion field).

3.8.11. If $\mathfrak{g}$ is solvable and n-dimensional, the maximal length of chains of prime ideals of $U(\mathfrak{g})$ is n. [Let k' be an algebraically closed extension of k, $(\mathfrak{g}_0, \mathfrak{g}_1, \ldots, \mathfrak{g}_n)$ an increasing sequence of ideals of $\mathfrak{g} \otimes k'$ such that $\dim \mathfrak{g}_i = i$; then consider the ideals $(\mathfrak{g}_i U(\mathfrak{g} \otimes k')) \cap U(\mathfrak{g})$.] We deduce that $\operatorname{Kdim} U(\mathfrak{g}) = n$. On the other hand, if $\mathfrak{g} = \mathfrak{sl}(2, \mathbf{C})$, the maximal length of chains of prime ideals of $U(\mathfrak{g})$ is 2 [99].

3.8.12. The maximal chains of prime ideals of $U(\mathfrak{g})$ are not all of the same length (this may be easily seen from the example of $\mathfrak{g}$ in 3.8.7).

3.8.13. Assume that A is Noetherian and the ideal 0 semi-prime. Let $P_1, \ldots, P_n$ be the minimal prime ideals of A. Let S be the set of elements of A which are not divisors of zero. Then $P_{1S}, \ldots, P_{nS}$ are the minimal ideals of the semi-simple Artinian ring A_S.

3.8.14. We use the notation of 3.6.15. Let $I \in \mathscr{I}$. Then I is completely prime if and only if I_S is completely prime.

3.8.15. Let $x \in U_n(\mathfrak{g}) - \{0\}$ and $y \in U_n(\mathfrak{g}) - \{0\}$. By comparing the dimensions of $U_p(\mathfrak{g})x$, $U_p(\mathfrak{g})y$ and $U_{n+p}(\mathfrak{g})$, we see that $U_p(\mathfrak{g})x \cap U_p(\mathfrak{g})y \neq 0$ for p sufficiently large. We again deduce that $U(\mathfrak{g})$ admits of a field of fractions [121].

3.8.16. Let I be a two-sided ideal of $U(\mathfrak{g})$ distinct from $U(\mathfrak{g})$, and $I^\omega = \bigcap_{n \geq 1} I^n$. If $\mathfrak{g}$ is nilpotent, then $I^\omega = 0$ (which gives us a new proof of 2.5.5 for this case). If $\mathfrak{k} = \bigcap_{n \geq 1} \mathscr{C}^n \mathfrak{g}$ and $I = U_+(\mathfrak{g})$, then I^ω is the two-sided ideal of $U(\mathfrak{g})$ generated by $\mathfrak{k}$, and hence is $\neq 0$ if $\mathfrak{g}$ is not nilpotent ([91], [99]).

3.8.17. Assume that $\mathfrak{g}$ is nilpotent. Let I be a two-sided ideal of $U(\mathfrak{g})$, and L a left ideal of $U(\mathfrak{g})$. There exists an integer n such that $I^n \cap L \subset IL$ [92].

CHAPTER 4

CENTRES

4.1. Notation

4.1.1. We recall that the centre of $U(\mathfrak{g})$ is denoted by $Z(\mathfrak{g})$. The centre of $K(\mathfrak{g})$ (3.6.13) is denoted by $C(\mathfrak{g})$; it is a commutative field. We have $Z(\mathfrak{g}) = C(\mathfrak{g}) \cap U(\mathfrak{g})$.

4.1.2. Since $Z(\mathfrak{g})$ is integral and commutative, Fract $Z(\mathfrak{g})$ exists and can be identified with a subfield of $C(\mathfrak{g})$, which is generally distinct from $C(\mathfrak{g})$ (4.9.8). Hence we have the following diagram of inclusions:

$$\begin{array}{ccc} U(\mathfrak{g}) & \longrightarrow & K(\mathfrak{g}) \\ \uparrow & & \uparrow \\ Z(\mathfrak{g}) \rightarrow & \text{Fract } Z(\mathfrak{g}) & \rightarrow C(\mathfrak{g}) \end{array}$$

4.1.3. Let B be the subalgebra of $K(\mathfrak{g})$ generated by $U(\mathfrak{g})$ and Fract $Z(\mathfrak{g})$. This is the set of the uz^{-1}, where $u \in U(\mathfrak{g})$ and $z \in Z(\mathfrak{g}) - \{0\}$. Clearly, $S = Z(\mathfrak{g}) - \{0\}$ allows of an arithmetic of fractions in $U(\mathfrak{g})$; from 3.6.7, B can be identified with $U(\mathfrak{g})_S$.

If φ is a homomorphism of $U(\mathfrak{g})$ into a ring R such that every element of $\varphi(S)$ is invertible, then there exists a unique homomorphism ψ of $U(\mathfrak{g}) \otimes_{Z(\mathfrak{g})}$ Fract $Z(\mathfrak{g})$ into R such that $\varphi = \psi \circ i$, where

$$i : U(\mathfrak{g}) \rightarrow U(\mathfrak{g}) \otimes_{Z(\mathfrak{g})} \text{Fract } Z(\mathfrak{g})$$

is the canonical mapping. This general property proves that the canonical homomorphism of $U(\mathfrak{g}) \otimes_{Z(\mathfrak{g})}$ Fract $Z(\mathfrak{g})$ into $B = U(\mathfrak{g})_S$ is an isomorphism. We can thus identify $U(\mathfrak{g})_S$ with $U(\mathfrak{g}) \otimes_{Z(\mathfrak{g})}$ Fract $Z(\mathfrak{g})$.

4.1.4. Let $T = B - \{0\}$. Let $uz^{-1} \in B$ and $u'z^{-1} \in T$ (where $u,u' \in U(\mathfrak{g})$, $z \in Z(\mathfrak{g})$, $u' \neq 0$, $z \neq 0$). There exist $v,v' \in U(\mathfrak{g})$ such that $vu = v'u', v \neq 0$, whence $vuz^{-1} = v'u'z^{-1}$; thus T satisfies in B condition (v) of 3.6.1, and similarly condition (iv). Hence T allows of an arithmetic of fractions on B. From 3.6.7, $K(\mathfrak{g})$ can be identified with the field of fractions of B.

4.1.5. More generally, let I be a two-sided ideal of $U(\mathfrak{g})$. Then the centre of $U(\mathfrak{g})/I$ is denoted by $Z(\mathfrak{g}; I)$. If I is semi-prime, the centre of Fract $(U(\mathfrak{g})/I)$ is termed the *core* of I and is denoted by $C(\mathfrak{g}; I)$; from 3.6.12, it is a finite product of commutative fields, and it is even a commutative field if I is prime.

Let us assume that I is prime. Let $S = Z(\mathfrak{g}; I) - \{0\}$. As in 4.1.3, we see that $(U(\mathfrak{g})/I)_S$ can be identified with $(U(\mathfrak{g})/I) \otimes_{Z(\mathfrak{g};I)} \text{Fract}\, Z(\mathfrak{g}; I)$, and with the subalgebra of Fract $(U(\mathfrak{g})/I)$ consisting of the uz^{-1} with $u \in U(\mathfrak{g})/I$ and $z \in S$.

4.1.6. Lemma. *Let A be a Noetherian ring with unity, M a simple A-module with annihilator* 0, *K the ring of fractions of A, C the centre of K, which is a field* (*cf.* 3.6.12), *and E the ring of A-endomorphisms of M. There exists one and only one injective homomorphism φ of C into the centre of E which possesses the following property: if $z \in C$ and $a \in A$ are such that $az \in A$, then $(az)m = a\varphi(z)m$ for all $m \in M$.*

Let $z \in C$. We write $z = rs^{-1}$ with $r, s \in A$, s not a divisor of zero in A. Let $m_0 \in M$ be such that $sm_0 \neq 0$. There exists $\lambda \in E$ such that $rm_0 = \lambda sm_0$; otherwise, from the density theorem (AL VIII, p. 39), there would exist $x \in A$ such that $x(sm_0) = m_0$ and $x(rm_0) = 0$; then $x(rm_0) = rxsm_0 = sxrm_0$ (3.6.8), whence $rm_0 = 0 = 0 \cdot (sm_0)$, which is a contradiction. For all $m \in M$, there exists $y \in A$ such that $m = y(sm_0)$, whence $rm = rysm_0 = syrm_0 = sy\lambda sm_0$, i.e.,

$$rm = \lambda sm. \tag{1}$$

Let $\mu \in E$. From (1), we have

$$\alpha\lambda sm_0 = \mu rm_0 = r\mu m_0 = \lambda s\mu m_0 = \lambda\mu sm_0,$$

whence $\mu\lambda = \lambda\mu$. Thus λ belongs to the centre of E. Let us show that λ only depends on z and not on the choice of s and r. Let u be an element of A which is not a divisor of zero, so that $z = (ru)(su)^{-1}$; for all $m \in M$, we have

$$(ru)m = r(um) = \lambda s(um) = \lambda(su)m,$$

which proves our assertion. It is easily seen that the mapping $z \mapsto \lambda$ is a homomorphism φ of C into the centre of E, which transforms 1 into 1, and hence is injective since C is a field. Finally, let us retain the above notation, and let $a \in A$ be such that $az \in A$. For all $m \in M$, there exists $y \in A$ such that $m = ysm_0$, whence

$$(az)m = azysm_0 = ayrm_0 = ay\lambda sm_0 = a\lambda ysm_0 = a\varphi(z)m.$$

Lastly, if φ' is another homomorphism possessing the properties of the lemma, we have $\varphi'(z)sm = (sz)m = \varphi(z)sm$ for all $m \in M$, hence $\varphi(z)$ and $\varphi'(z)$ coincide at a non-zero point of M and consequently on the whole of M.

4.1.7. Proposition. *If I is a primitive ideal of $U(\mathfrak{g})$, the field $C(\mathfrak{g};I)$ is an extension of finite degree of k.*

This follows from 4.1.6 and 2.6.9 (i).

References: [15], [31], [104].

4.2. Centre and core in the semi-simple case

4.2.1. Lemma (k algebraically closed). *Let V be a finite-dimensional simple $\mathfrak{g}$-module, ϱ the corresponding representation, and I the kernel of ϱ. Then the mapping $u \mapsto \varrho(u)$ of $U(\mathfrak{g})$ into* End(V) *defines by passage to the quotient an isomorphism of the $\mathfrak{g}$-module $U(\mathfrak{g})/I$ onto the $\mathfrak{g}$-module* End(V).

This mapping is surjective (e.g., from 2.6.5) and obviously it is a $\mathfrak{g}$-homomorphism.

4.2.2. Proposition ($\mathfrak{g}$ semi-simple). *Let J be a non-null two-sided ideal of $U(\mathfrak{g})$. Then $J \cap Z(\mathfrak{g}) \neq 0$.*

We may assume that k is algebraically closed. From 1.6.3 and 2.5.7, there exists a finite-dimensional simple representation ϱ of $\mathfrak{g}$ in V such that, if the kernel of ϱ in $U(\mathfrak{g})$ is denoted by I, we have $J \not\subset I$. Since $U(\mathfrak{g})/I$ is a simple algebra, we have $U(\mathfrak{g}) = I + J$. For the adjoint representation of $\mathfrak{g}$ in $U(\mathfrak{g})$, $U(\mathfrak{g})$ is the sum of finite-dimensional simple sub-$\mathfrak{g}$-modules (2.3.3), and I,J are sub-$\mathfrak{g}$-modules. Hence we can write $J = (I \cap J) \oplus W$, where W is a sub-$\mathfrak{g}$-module of J. Then $U(\mathfrak{g}) = I \oplus W$. Since the $\mathfrak{g}$-module End V possesses a non-zero invariant element (i.e, the endomorphism 1 of V), the same applies to the $\mathfrak{g}$-module W (4.2.1). In other words, $W \cap Z(\mathfrak{g}) \neq 0$.

4.2.3. Corollary ($\mathfrak{g}$ semi-simple). *We have* $C(\mathfrak{g}) =$ Fract $Z(\mathfrak{g})$.

Let $c \in C(\mathfrak{g})$. The set of the $u \in U(\mathfrak{g})$ such that $uc \in U(\mathfrak{g})$ is a non-null two-sided ideal of $U(\mathfrak{g})$. From 4.2.2, there exists $z \in Z(\mathfrak{g})$ such that $z \neq 0$ and $zc \in U(\mathfrak{g})$. Then $zc \in Z(\mathfrak{g})$ and $c = z^{-1}(zc) \in$ Fract $Z(\mathfrak{g})$.

4.2.4. Later on (7.3.8 (ii)) we shall see that, if $\mathfrak{g}$ is semi-simple of rank l, then $Z(\mathfrak{g})$ is isomorphic with the algebra of polynomials in l indeterminates over k. Consequently, $C(\mathfrak{g})$ *is isomorphic with the algebra of rational fractions in* l *indeterminates over* k.

4.2.5. PROPOSITION ($\mathfrak{g}$ semi-simple). *Let* I *be a two-sided ideal of* $U(\mathfrak{g})$, *and* φ *the canonical mapping of* $U(\mathfrak{g})$ *onto* $U(\mathfrak{g})/I$. *Then* $\varphi(Z(\mathfrak{g}))$ *is the centre of* $U(\mathfrak{g})/I$.

This follows from 1.2.11 applied to the adjoint representation of $\mathfrak{g}$ in $I, U(\mathfrak{g})$ and $U(\mathfrak{g})/I$.

References: [100], [120].

4.3. The semi-centre

4.3.1. Let I be a two-sided ideal of $U(\mathfrak{g})$, $A = U(\mathfrak{g})/I$, and let ε be the adjoint representation of $\mathfrak{g}$ in A. Let $\lambda \in \mathfrak{g}^*$. We recall (1.2.13) that A_λ is the set of those $a \in A$ such that $\varepsilon(x)a = \lambda(x)a$ for all $x \in \mathfrak{g}$. We have $A_\lambda A_\mu \subset A_{\lambda+\mu}$, and A_0 is the centre of A. λ is said to be a *distinguished* linear form (relative to I) on $\mathfrak{g}$ if $A_\lambda \neq 0$.

If λ is distinguished, λ is a one-dimensional representation of $\mathfrak{g}$, hence $\lambda([\mathfrak{g},\mathfrak{g}]) = 0$; moreover, λ is zero on the largest nilpotency ideal $\mathfrak{n}$ of $\mathfrak{g}$, since, if $x \in \mathfrak{n}$, then $\mathrm{ad}_{\mathfrak{g}} x$ is nilpotent, and so $\mathrm{ad}_{U(\mathfrak{g})} x$ is locally nilpotent.

4.3.2. Retaining the notation of 4.3.1, the sum of the A_λ is direct and is a subalgebra S of A which contains the centre of A. The algebra S is graded by the A_λ. S is said to be the *semi-centre* of A. If $\mathfrak{g} = [\mathfrak{g},\mathfrak{g}]$, or if $\mathfrak{g}$ is nilpotent, S is equal to the centre of A (4.3.1).

4.3.3. Let $a, b \in A_\lambda$, with $b \neq 0$. If I is prime, b is invertible in Fract(A) (3.3.9), and ab^{-1} belongs to the centre of Fract(A), that is, to the core of I; for, let ε be the adjoint representation of $\mathfrak{g}$ in Fract(A); for all $y \in \mathfrak{g}$, we have

$$\varepsilon(y)(ab^{-1}) = \lambda(y)ab^{-1} - ab^{-1}(\lambda(y)b)b^{-1} = 0.$$

4.3.4. LEMMA. *Let* $\mathfrak{g}'$ *be an ideal of codimension* 1 *of* $\mathfrak{g}$, I *a prime ideal of* $U(\mathfrak{g})$, $I' = I \cap U(\mathfrak{g}')$, $A = U(\mathfrak{g})/I$, $A' = U(\mathfrak{g}')/I'$ *and* $B =$ Fract(A). *We assume that* I *is the two-sided ideal of* $U(\mathfrak{g})$ *generated by* I'. *Let* $\mu \in \mathfrak{g}^*$ *and* $a \in A_\mu$. *There exist* b *in the centre of* B *and* $c \in A_\mu \cap A'$ *such that* $a = bc$.

Let $x \in \mathfrak{g}$ be such that $x \notin \mathfrak{g}'$. We may assume that $a \neq 0$. Let u be a representative of a in $U(\mathfrak{g})$. Then u may be uniquely written in the form

$$u = x^n u_n + x^{n-1} u_{n-1} + \cdots + u_0,$$

with $u_n, \dots, u_0 \in U(\mathfrak{g}')$. The u_i do not all belong to I'. By changing the chosen representative u of a, we may thus assume that $u_n \notin I'$. For $y \in \mathfrak{g}$ and $p \geqq 1$, we have

$$[y,x^p] \in x^{p-1}U(\mathfrak{g}') + x^{p-2}U(\mathfrak{g}') + \cdots + U(\mathfrak{g}'),$$

which can be seen immediately by induction on p (taking into account that $[\mathfrak{g},\mathfrak{g}] \subset \mathfrak{g}'$). We then deduce that, modulo I,

$$\begin{aligned}\mu(y)x^n u_n + \mu(y)x^{n-1}u_{n-1} + \cdots + \mu(y)u_0 = \mu(y)u \equiv [y,u] \\ = x^n[y,u_n] + x^{n-1}v_{n-1} + \cdots + v_0,\end{aligned}$$

with $v_{n-1}, \dots, v_0 \in U(\mathfrak{g}')$. From 3.3.5, $[y,u_n] \equiv \mu(y)u_n \pmod{I'}$. Let c be the canonical image of u_n in A'. Since $u_n \notin I'$, we have $c \neq 0$. Then ac^{-1} exists in B and belongs to the centre of B (4.3.3).

4.3.5. Proposition. *Let I be a prime ideal of $U(\mathfrak{g})$. The semi-centre of $U(\mathfrak{g})/I$ is commutative.*

Let $m = \dim \mathfrak{g}$ and assume that the proposition has been proved for $\dim \mathfrak{g} < m$. If every linear form on $\mathfrak{g}$ which is distinguished relative to I is zero, the semi-centre of $A = U(\mathfrak{g})/I$ is equal to its centre. Therefore, let us henceforth assume that there exist $z \in A$ and $\lambda \in \mathfrak{g}^*$ such that $z \neq 0$, $\lambda \neq 0$, $z \in A_\lambda$. Let $\mathfrak{g}' = \operatorname{Ker} \lambda$, and $I' = I \cap U(\mathfrak{g}')$. From 3.3.8, I is the two-sided ideal of $U(\mathfrak{g})$ generated by I'. Let $A' = U(\mathfrak{g}')/I'$ and $B = \operatorname{Fract}(A)$. Let $a_1 \in A_{\mu_1}$ and $a_2 \in A_{\mu_2}$. From 4.3.4, we have $a_1 = b_1c_1$, $a_2 = b_2c_2$, with b_1,b_2 in the centre of B, $c_1 \in A_{\mu_1} \cap A'$, $c_2 \in A_{\mu_2} \cap A'$. Froom the induction hypothesis, c_1 and c_2 commute. Hence b_1,c_1,b_2,c_2 pairwise commute so that a_1 and a_2 commute.

4.3.6. If $\mathfrak{g}$ is semi-simple or nilpotent, we have $C(\mathfrak{g}) = \operatorname{Fract} Z(\mathfrak{g})$ (4.2.3 and 4.7.1). For arbitrary $\mathfrak{g}$, however, this result may not be true (4.9.8). We shall see that the semi-centre allows us to remedy this situation to a certain extent (4.4.2, 4.9.7).

References: [31], [111].

4.4. Centre and core in the solvable case

4.4.1. Proposition ($\mathfrak{g}$ completely solvable). *Let I be a two-sided ideal of $U(\mathfrak{g})$. Let J be a non-null two-sided ideal of $U(\mathfrak{g})/I$. There exists $\lambda \in \mathfrak{g}^*$ such that $J \cap (U(\mathfrak{g})/I)_\lambda \neq 0$.*

The $\mathfrak{g}$-module $U(\mathfrak{g})/I$ is the union of an increasing sequence of finite-dimensional triangularizable sub-$\mathfrak{g}$-modules (2.3.3), hence the same applies for J. Consequently there exist $\lambda \in \mathfrak{g}^*$ and $x \in J - \{0\}$ such that $x \in (U(\mathfrak{g})/I)_\lambda$.

4.4.2. Corollary ($\mathfrak{g}$ completely solvable). *Let I be a prime ideal of $U(\mathfrak{g})$. Let $c \in C(\mathfrak{g};I)$. There exists $\lambda \in \mathfrak{g}^*$ such that c can be written in the form ab^{-1}, where $a,b \in (U(\mathfrak{g})/I)_\lambda$.*

The set of the $u \in U(\mathfrak{g})/I$ such that $uc \in U(\mathfrak{g})/I$ is a non-null two-sided ideal J of $U(\mathfrak{g})/I$. There exist $\lambda \in \mathfrak{g}^*$ and $b \in (U(\mathfrak{g})/I)_\lambda$ such that $b \in J$ and $b \neq 0$ (4.4.1). Then $bc \in (U(\mathfrak{g})/I)_\lambda$, and b is not a divisor of zero in $U(\mathfrak{g})/I$ (3.3.9), hence $c = (bc)b^{-1}$.

4.4.3. There exist solvable (and even nilpotent) Lie algebras $\mathfrak{g}$ such that $Z(\mathfrak{g})$ is not an algebra of finite type (4.9.20). In spite of this, we shall see that, for $\mathfrak{g}$ solvable, the structure of $C(\mathfrak{g};I)$ is simple (4.4.8 and 4.4.11).

4.4.4. For every ring K with unity and every derivation D of K, we denote by $K_D[X]$ (or simply $K[X]$ if $D = 0$) the ring generated by K and an indeterminate X subject only to the relations $Xa = aX + D(a)$ for $a \in K$. Let f be a non-zero element of $K_D[X]$. It can be written in one and only one way in the form

$$f = a_0 + Xa_1 + \cdots + X^n a_n,$$

where $a_0, \ldots, a_n \in K$ and $a_n \neq 0$. If it can also be written in one and only one way in the form

$$f = b_0 + b_1 X + \cdots + b_p X^p,$$

where $b_0, \ldots, b_p \in K$ and $b_p \in 0$, then we have $n = p$ and $b_p = a_n$. f is said to be of *degree n*, and we set $n = \deg(f)$; the element a_n is termed the *leading coefficient* of f; we stipulate that $\deg(0) = -\infty$.

Let us assume that K is a subring of a ring K', and that $x \in K'$ is such that $[x,a] = D(a)$ for all $a \in K$. There exists one and only one homomorphism φ of $K_D[X]$ into K' such that $\varphi|K = \mathrm{id}_K$ and $\varphi(X) = x$. If this homomorphism is injective, x is said to be *transcendental* over K, and we identify $K_D[X]$ with $\varphi(K_D[X])$ under φ; this defines, for all $y \in \varphi(K_D[X])$, the degree (in x) of y over K (denoted by $\deg_K y$ or $\deg y$) and its leading coefficient.

If K is a field, $K_D[X]$ is integral and Noetherian, and we can thus consider the field of fractions Fract $K_D[X]$. If $f,g \in K_D[X]$, we then have $\deg(fg) = \deg f + \deg g$, and the leading coefficient of fg is the product of the leading coefficients of f and g.

4.4.5. LEMMA. *Let A be an algebra with unity, B a sub-algebra containing* 1, *and x an element of A, such that:*

(1) $[x,B] \subset B$;

(2) *x ix transcendental over B;*

(3) *the algebra A is generated by x and B;*

(4) *A and B have the fields of fractions K and L (so that L is the subfield of K generated by B).*

Then:

(i) *$[x,L] \subset L$, and x is transcendental over L. Let L' be the subalgebra of K generated by x and L. Then L' allows of a field of fractions which is K.*

(ii) *If $u \in A$, then* $\deg_B u = \deg_L u$.

(iii) *Let $P,Q \in L'$ with $Q \neq 0$. There exist unique $R,S \in L'$ such that*

$$P = RQ + S, \qquad \deg_L S < \deg_L Q.$$

There exist unique $R',S' \in L'$ such that

$$P = QR' + S', \qquad \deg_L S' < \deg_L Q.$$

Clearly, $[x,L] \subset L$. Let us assume that

$$\alpha_n x^n + \alpha_{n-1} x^{n-1} + \cdots + \alpha_0 = 0,$$

where $\alpha_0, \ldots, \alpha_n \in L$. There exist $a_0, \ldots, a_n, b \in B$ such that $\alpha_0 = b^{-1} a_0, \ldots, \alpha_n = b^{-1} a_n$. Then

$$a_n x^n + a_{n-1} x^{n-1} + \cdots + a_0 = 0,$$

whence $a_i = 0 = \alpha_i$ for all i; hence x is transcendental over L. Every element of K can be written as ab^{-1} and $c^{-1}d$ with $a,b,c,d \in A \subset L'$; from 3.6.7, L' has a field of fractions, and this field of fractions can be identified with K.

Let $u \in A$. Clearly, $\deg_B u \geqq \deg_L u$. On the other hand, if

$$u = \alpha_n x^n + \alpha_{n-1} x^{n-1} + \cdots + \alpha_0 \quad \text{with } \alpha_i \in L,$$

then necessarily $\alpha_i \in B$ for all i since x is transcendental over L. Hence $\deg_L u \geqq \deg_B u$.

Let us prove the existence of R,S in (iii). This is trivial if $\deg_L Q < \deg_L P$. Let $m = \deg_L P$, $n = \deg_L Q$, and let us assume that $m \geqq n$. Let α and β be the leading coefficients of P and Q, respectively. Then $\deg_L(P - \alpha\beta^{-1}x^{m-n}Q) < m$, whence we have the existence of R,S by induction on m. If (R^*,S^*) satisfies the same properties as (R,S), then $(R^* - R)Q = S - S^*$, which requires that $R^* - R = 0$ for reasons of degree. The same reasoning applies for R', S'.

4.4.6. In the situation of 4.4.5, every element α of K can be written as ab^{-1} $(a,b \in A)$ and also as $c^{-1}d$ $(c,d \in A)$; we have $ca = db$, whence

$$\deg_L a - \deg_L b = \deg_L d - \deg_L c,$$

so that we can define $\deg_L \alpha$ (or $\deg \alpha$) by means of the formula

$$\deg_L \alpha = \deg_L a - \deg_L b.$$

If $\beta \in K$, then

$$\deg_L(\alpha\beta) = \deg_L \alpha + \deg_L \beta.$$

4.4.7. LEMMA. *We retain the hypotheses and notation of* 4.4.5. *Let $\mathscr{D}$ be a set of derivations of A, and $\overline{\mathscr{D}}$ the set of their extensions to derivations of K* (3.6.18). *We assume that $\mathscr{D}(B) \subset B$, whence $\overline{\mathscr{D}}(L) \subset L$, $\overline{\mathscr{D}}(L') \subset L'$. We assume that* $\deg(Dx) \leqq 1$ *for all $D \in \mathscr{D}$. Let K_0, L_0 and L'_0 be the sets of elements of K, L and L' respectively, annihilated by $\overline{\mathscr{D}}$.*

(i) *Let $P,Q,R,S \in L'$ with $P = QR + S$ and* $\deg S < \deg Q$. *If $P,Q \in L'_0$, then $R,S \in L'_0$. Similarly, if $P = RQ + S$, then* $\deg S < \deg Q$.

(ii) *Every element of K_0 can be written as PQ^{-1} with $P,Q \in L'_0$.*

(iii) *Let us assume that $K_0 \neq L_0$, whence $L'_0 \neq L_0$; let P be an element of L'_0 of minimal degree* > 0. *Then every element of L'_0 can be written in a unique way as $a_0 + a_1P + a_2P^2 + \cdots$, with $a_0a_1, \ldots, \in L_0$.*

(i) If $D \in \mathscr{D}$, we have $\deg(Dx) \leqq 1$, hence $\deg(DP) \leqq \deg P$ for all $P \in L'$ and consequently for all $P \in K$.

Let $P,Q \in L'_0$ and $R,S \in L'$, with $P = QR + S$, $\deg S < \deg Q$. If $D \in \mathscr{D}$, then $0 = DP = Q(DR) + DS$. If $DR \neq 0$, we deduce that $\deg(DS) \geqq \deg Q$, which is a contradiction; hence $DR = 0$ and $DS = 0$.

(ii) Let $T \in K_0$. We set $d(T) = \inf_{P,Q \in L', PQ^{-1} = T}(\deg P + \deg Q)$. We shall prove (ii) by induction on $d(T)$. The assertion is obvious if $d(T) = 0$. Let us assume that $d(T) = n$ and that the assertion has been proved for $d(T) < n$. Then $T = PQ^{-1}$ with $P,Q \in L'$, $\deg P + \deg Q = n$. Let us assume that $\deg P \geqq \deg Q$. Let $P = RQ + S$ with $R,S \in L'$, $\deg S < \deg Q$. For all $D \in \mathscr{D}$, we have $0 = DT = DR + D(SQ^{-1})$. If $DR \neq 0$, the inequalities $\deg DR \geqq 0$ and $\deg D(SQ^{-1}) < 0$ lead to a contradiction. Hence $R \in L'_0$ and $SQ^{-1} \in K_0$. Since $\deg S + \deg Q < n$, the induction hypothesis yields $SQ^{-1} = S'Q'^{-1}$ with $S',Q' \in L'_0$. Then $T = (RQ' + S')Q'^{-1}$ and $RQ' + S', Q' \in L'_0$. If $\deg P < \deg Q$, it is sufficient to apply the previous result to T^{-1}.

(iii) Let $Q \in L'_0$; we prove (iii). The uniqueness arises from the fact that P is transcendental over L and hence over L'_0. Let us prove the existence by induction on $\deg Q$. This is obvious for $\deg Q = 0$. Hence let $\deg Q > 0$. We have $Q = RP + S$ with $\deg S < \deg P$. From (i), we have $S \in L'_0$ hence $S \in L_0$. Then $\deg Q = \deg R + \deg P$ and $\deg R < \deg Q$. Since $R \in L'_0$ from (i), it is sufficient to apply the induction hypothesis to R.

4.4.8. PROPOSITION ($\mathfrak{g}$ completely solvable). *The field* $C(\mathfrak{g})$ *is a purely transcendental extension of* k *of degree* $\leqq \dim \mathfrak{g}$.

Let $\mathfrak{g} = \mathfrak{g}_n \supset \mathfrak{g}_{n-1} \supset \cdots \supset \mathfrak{g}_0 = 0$ be a sequence of ideals of $\mathfrak{g}$ such that $\dim \mathfrak{g}_i = i$. Let $U_i = U(\mathfrak{g}_i)$, let K_i be the subfield of $K(\mathfrak{g})$ generated by U_i [this subfield can be identified with $K(\mathfrak{g}_i)$], and let $C_i = C(\mathfrak{g}) \cap K_i$. We shall see that either $C_{i+1} = C_i$ or else C_{i+1} is a purely transcendental extension of C_i of degree 1. This will lead to the proposition.

Let x be an element of $\mathfrak{g}_{i+1}$ which does not belong to $\mathfrak{g}_i$. Then $[x, U_i] \subset U_i$, U_{i+1} is the algebra generated by x and U_i, and x is transcendental over U_i from 2.1.11. Let ε be the adjoint representation of $\mathfrak{g}$ in U_{i+1}, and $\mathscr{D} = \varepsilon(\mathfrak{g})$. Then $\mathscr{D}(U_i) \subset U_i$, and $\deg_{U_i}(Dx) \leqq 1$ for all $D \in \mathscr{D}$. We may then apply 4.4.7. The sets of elements of K_{i+1} and K_i annihilated by $\overline{\mathscr{D}}$ are C_{i+1} and C_i respectively. Let us assume that $C_{i+1} \neq C_i$, and note that C_{i+1} is commutative. From 4.4.7 (ii) and (iii), there exist a subring Z of C_{i+1} containing C_i and a $P \in Z$ such that:

(1) every element of Z may be uniquely written as a polynomial in P with coefficients in C_i;

(2) every element of C_{i+1} is the quotient of two elements of Z.

Hence C_{i+1} is a purely trancendental extension of C_i of degree 1.

4.4.9. LEMMA. *Let* K *be an algebra (over* k*) which is a field. Let* Z *be the centre of* K, *and* D *a non-inner derivation of* K.

(i) *The algebra* $K_D[X]$ *is simple.*

(ii) *The centre of* Fract $K_D[X]$ *is the set of those* $z \in Z$ *such that* $Dz = 0$.

(i) Let I be a two-sided ideal of $K_D[X]$ such that $I \neq 0$ and $I \neq K_D[X]$. Let $n = \inf_{P \in I, P \neq 0} \deg P$. Since $I \neq K_D[X]$, we have $I \cap K = 0$, hence $n > 0$. Since K is a field, there exists an element P of I of the form $X^n + X^{n-1}p_1 + \cdots + Xp_{n-1} + p_n$ ($p_1, \ldots, p_n \in K$). If $l \in K$, then

$$[P, l] = nX^{n-1}\, Dl + X^{n-1}[p_1, l] + r$$

with $\deg r < n - 1$. Hence $\deg[P, l] < n$. Since $[P, l] \in I$, we have $[P, l] = 0$, and consequently $nDl + [p_1, l] = 0$. Thus $D = -(1/n)\,\mathrm{ad}_{p_1}$, which is a contradiction, whence (i).

(ii) Let $P = X^m p_0 + X^{m-1} p_1 + \cdots$ (where $p_0 \neq 0$) be a central element of $K_D[X]$, and let us prove that P is an element z of Z such that $Dz = 0$. Then, for all $a \in K$,

$$(1) \qquad 0 = [a,P] = X^m[a,p_0] - mX^{m-1}(Da)p_0 + X^{m-1}[a,p_1] + X^{m-2}q_2 + \cdots \qquad (q_2, \ldots \in K),$$

$$(2) \qquad 0 = [X,P] = X^m(Dp_0) + X^{m-1}q_1' + \cdots \qquad (q_1', \ldots \in K),$$

whence $p_0 \in Z$ and $Dp_0 = 0$. We may assume that $p_0 = 1$, replacing P by Pp_0^{-1} if necessary. Then (1) implies that $mDa = [a,p_1]$ for all $a \in K$. Since D is not an inner derivation, we have $m = 0$, whence our assertion. Given this, (ii) is obtained by applying 4.4.7 (ii) by taking for $\overline{\mathscr{D}}$ the set of inner derivations of Fract $K_D[X]$ defined by X and by the elements of K.

4.4.10. Lemma. *Let K be an algebra which is a field, $t_0, \ldots, t_n$ elements of K, and K_i the subfield of K generated by $k, t_0, \ldots, t_i$ $(0 \leqq i \leqq n)$. We assume that $K = K_n$ and that $[t_i, K_{i-1}] \subset K_{i-1}$ for $i = 1, \ldots, n$. Then the centre of K is an extension of finite type of k.*

This is obvious for $n = 0$; let us assume that the lemma has been established for all integers $< n$.

First case: There exists $a \in K_{n-1}$ such that $[t_n, x] = [a, x]$ for all $x \in K_{n-1}$. Then $t_n - a$ commutes with K_{n-1} and with $t_n - a$, and so is central in K. Let us set $k' = k(t_n - a)$, so that K is a k'-algebra. Let K_i' be the subfield of K generated by $k', t_0, \ldots, t_i$. Then

$$K_{n-1}' = K, \qquad [t_i, K_{i-1}'] \subset K_{i-1}' \quad \text{for } i = 1, \ldots, n-1.$$

Hence it is sufficient to apply the induction hypothesis.

Second case: The derivation $x \mapsto Dx = [t_n, x]$ of K_{n-1} is not inner. From 4.4.9 (i), the subalgebra of K generated by K_{n-1} and t_n can be identified with $(K_{n-1})_D[X]$, hence K can be identified with Fract $(K_{n-1})_D[X]$. From 4.4.9 (ii), the centre of K is a subfield of the centre Z of K_{n-1}. Now Z is an extension of k of finite type from the induction hypothesis, and a subextension of an extension of finite type is of finite type.

4.4.11. Proposition ($\mathfrak{g}$ solvable). *Let I be a prime ideal of $U(\mathfrak{g})$. The field $C(\mathfrak{g}; I)$ is an extension of finite type of k.*

Let $(\mathfrak{g}_n, \mathfrak{g}_{n-1}, \ldots, \mathfrak{g}_0)$ be a decreasing sequence of Lie subalgebras of $\mathfrak{g}$ with dimensions $n = \dim \mathfrak{g}$, $n-1, \ldots, 0$, such that, for $i = 0, 1, \ldots, n-1$, $\mathfrak{g}_i$ is an ideal of $\mathfrak{g}_{i+1}$. Let x_i be an element of $\mathfrak{g}_i$ not belonging to $\mathfrak{g}_{i-1}$. Let

t_i be the canonical image of x_i in $K = \text{Fract}\, U(\mathfrak{g})/I$. Let K_i be the subfield of K generated by $k, t_1, \ldots, t_i$. Then $K = K_n$, and $[t_i, S_{i-1}] \subset K_{i-1}$. It is then sufficient to apply 4.4.10.

References: [8], [31], [51].

4.5. The characterization of primitive ideals in the solvable case

4.5.1. LEMMA. *Let A be a simple central algebra with unity, B an algebra with unity, $\mathscr{I}$ the set of two-sided ideals of B, and $\mathscr{I}$ the set of two-sided ideals of $A \otimes B$.*

(i) *The mapping $I \mapsto A \otimes I$, where I runs through $\mathscr{I}$, is a bijection of $\mathscr{I}$ onto $\mathscr{I}$.*

(ii) *Let $I \in \mathscr{I}$. Then I is a maximal (or prime) two-sided ideal of B if and only if $A \otimes I$ is a maximal (or prime) two-sided ideal of $A \otimes B$.*

Assertion (i) follows from AL VIII (p. 43) and implies the assertion concerning the maximal ideals. If $I_1, I_2 \in \mathscr{I}$, then

$$A \otimes (I_1 I_2) = (A \otimes I_1)(A \otimes I_2).$$

Hence, if $I \in \mathscr{I}$, the conditions

$$I_1 I_2 \subset I \subset I_1 \cap I_2,$$

$$(A \otimes I_1)(A \otimes I_2) \subset A \otimes I \subset (A \otimes I_1) \cap (A \otimes I_2)$$

are equivalent. Furthermore, $I \neq I_1$ is equivalent to $A \otimes I \neq A \otimes I_1$, and $I \neq I_2$ is equivalent to $A \otimes I \neq A \otimes I_2$. This implies the assertion for the prime ideals.

4.5.2. LEMMA. *Let K be a field, and D an inner derivation of K. Every non-null prime ideal of $K_D[X]$ is maximal.*

We are immediately led to the case where $D = 0$. Let Z be the centre of K. In $Z[X]$, the non-null prime ideals are maximal. Now K is a simple central Z-algebra and $K[X] = K \otimes_Z Z[X]$. Hence it is sufficient to apply 4.5.1.

4.5.3. LEMMA. *Let K be an algebra which is a field. Let $\mathfrak{d}$ be a Lie algebra of derivations of $K[X]$ which satisfies the following conditions:*

(i) *is stable under $\mathfrak{d}$;*

(ii) *the elements of* Fract($K[X]$) *annihilated by the elements of $\mathfrak{d}$ (extended to derivations of* Fract($K[X]$)) *are algebraic over k;*

(iii) *there exist mappings* $\lambda : \mathfrak{d} \to k$ *and* $\nu : \mathfrak{d} \to K$ *such that* $\gamma(X) = \lambda(\gamma)X + \nu(\gamma)$ *for all* $\gamma \in \mathfrak{d}$.
Then there exists at most one $\mathfrak{d}$*-stable prime ideal* I *of* $K[X]$ *such that* $I \neq 0$, $I \neq K[X]$.

Let us assume the existence of such an ideal I. Let $n = \inf_{P \in I, P \neq 0} \deg P$. Then $n > 0$. Since K is a field, there exists in I an element of the form

$$P = X^n + X^{n-1}p_1 + \cdots + p_n \qquad (p_1, \ldots, p_n \in K).$$

If $\gamma \in \mathfrak{d}$, then

$$\gamma(P) = nX^{n-1}(\lambda(\gamma)X + \nu(\gamma)) + (n-1)X^{n-1}\lambda(\gamma)p_1 + X^{n-1}\gamma(p_1) + Q$$

with $\deg Q < n-1$. Hence

$$\gamma(P) - n\lambda(\gamma)P = X^{n-1}(n\nu(\gamma) - \lambda(\gamma)p_1 + \gamma(p_1)) + R$$

with $\deg R < n-1$. Now $\gamma(P) - n\lambda(\gamma)P \in I$, hence $\gamma(P) = n\lambda(\gamma)P$ from the choice of n. We then see that $\gamma(p_1) = \lambda(\gamma)p_1 - n\nu(\gamma)$, whence

$$\gamma(X + (1/n)p_1) = \lambda(\gamma)X + \nu(\gamma) + (1/n)\lambda(\gamma)p_1 - \nu(\gamma) = \lambda(\gamma)(X + (1/n)p_1).$$

For all $l \in K$, we have $[P,l] = X^{n-1}[p_1,l] + S$ with $\deg S < n-1$ and $[P,l] \in I$; hence $[p_1,K] = 0$. Consequently there exists an automorphism of $K[X]$ which transforms $X + (1/n)p_1$ into X and which induces the identity on K. Making use of this automorphism, we are led to the case where $\nu = 0$. We can consider X^{-n} in $\mathrm{Fract}(K[X])$, and we have

$$\gamma(PX^{-n}) = n\lambda(\gamma)PX^{-n} + P(-n)\lambda(\gamma)X^{-n} = 0.$$

From hypothesis (ii) of the lemma, we deduce that $P = X^n$. The elements of $K[X]$ without constant term form a two-sided ideal I' of $K[X]$. We have $I'^n \subset K[X]PK[X] \subset I$. Hence $I' \subset I$ since I is prime, whence $I' = I$ since $I \cap K = 0$.

4.5.4. Lemma. *Let* $\mathfrak{h}$ *be a solvable ideal of* $\mathfrak{g}$, $\mathfrak{h}'$ *an ideal of codimension* 1 *in* $\mathfrak{h}$, I *a prime ideal of* $U(\mathfrak{h})$ *such that* $[\mathfrak{g},I] \subset I$, $I' = I \cap U(\mathfrak{h}')$, $\mathscr{P}$ *the set of prime ideals* P *of* $U(\mathfrak{h})$ *such that* $P \supset I$, $P \neq I$, $[\mathfrak{g},P] \subset P$, *and* $\mathscr{P}'$ *the set of prime ideals* P' *of* $U(\mathfrak{h}')$ *such that* $P' \supset I'$, $P' \neq I'$, $[\mathfrak{g},P'] \subset P'$. *Let* $\hat{I}$ *be the intersection of the elements of* $\mathscr{P}$, $\hat{I}'$ *the intersection the elements of* $\mathscr{P}'$. *We assume that* $\hat{I}' \neq I'$.

Let ε *be the adjoint representation of* $\mathfrak{g}$ *in* $\mathrm{Fract}(U(\mathfrak{h})/I)$. *We assume that the elements of* $\mathrm{Fract}(U(\mathfrak{h})/I)$ *annihilated by* $\varepsilon(\mathfrak{g})$ *are algebraic over* k. *Then* $\hat{I} \neq I$.

Let $M = U(\mathfrak{h})/I$ and $N = U(\mathfrak{h}')/I'$. We identify N with a subalgebra of M. The algebra M is integral (3.7.2). Let K be the field of fractions of N.

Let $P \in \mathscr{P}$. Then $P \cap U(\mathfrak{h}') \supset I'$, and $[\mathfrak{g}, P \cap U(\mathfrak{h}')] \subset P \cap U(\mathfrak{h}')$. Hence, if $P \cap U(\mathfrak{h}') \neq I'$, it can be seen that $P \cap U(\mathfrak{h}') \in \mathscr{P}'$ (3.3.4), and consequently $P \cap U(\mathfrak{h}') \supset \hat{I}'$, whence $P \supset I + U(\mathfrak{h})\hat{I}'U(\mathfrak{h})$. The lemma is thus proved if $P \cap U(\mathfrak{h}') \neq I'$ for all $P \in \mathscr{P}$. Henceforth we shall assume that there exists a $P_0 \in \mathscr{P}$ such that $P_0 \cap U(\mathfrak{h}') = I'$.

Let x be an element of $\mathfrak{h}$ which does not belong to $\mathfrak{h}'$, and $\bar{x}$ be its canonical image in M. Let D be the restriction of $\mathrm{ad}_M \bar{x}$ to N. There exists one and only one homomorphism p of $N_D[X]$ into M which extends the identity mapping of N and which is such that $p(X) = \bar{x}$. This homomorphism is surjective. Let $Q_0 = p^{-1}(P_0/I)$ and $Q_1 = \mathrm{Ker}\, p$. Then Q_0 and Q_1 are completely prime ideals of $N_D[X]$, and

$$Q_0 \supset Q_1, \qquad Q_0 \neq Q_1, \qquad Q_0 \cap N = Q_1 \cap N = 0.$$

The derivation D of N can be extended to a derivation $\bar{D}$ of K. Let us consider $N_D[X]$ as a subalgebra of $K_{\bar{D}}[X]$. Clearly, $K_{\bar{D}}[X]$ is the ring of fractions of $N_D[X]$ relative to $N - \{0\}$. Since Q_0 is completely prime and $Q_0 \cap N = 0$, the set Q_0' of the qn^{-1}, where $q \in Q_0$ and $n \in N - \{0\}$, is a prime ideal of $K_{\bar{D}}[X]$ such that

$$Q_0' \cap N_D[X] = Q_0$$

(3.6.15). Similarly, the set Q_1' of the qn^{-1}, where $q \in Q_1$ and $n \in N - \{0\}$, is a prime ideal of $K_{\bar{D}}[X]$ such that $Q_1' \cap N_D[X] = Q_1$. From 4.4.9 (i), the derivation $\bar{D}$ of K is inner. From 4.5.2, $Q_1' = 0$ whence $Q_1 = 0$. Hence p is an isomorphism by means of which M and $N_D[X]$ and hence $\bar{x}$ and X, can henceforth be identified. Since $\bar{D}$ is an inner derivation, there exists an isomorphism of $K[X]$ onto $K_{\bar{D}}[X]$ which transforms X into an element of the form $X + l$, where $l \in K$.

If $y \in \mathfrak{g}$, then y defines derivations of $\mathfrak{h}$, $\mathfrak{h}'$, $U(\mathfrak{h})$, $U(\mathfrak{h}')$, M, N, K and $\mathrm{Fract}(M)$. By assumption, the elements of $\mathrm{Fract}(M)$ which are annihilated by the derivations of this type are algebraic over k. If γ is one of these derivations, then $\gamma(X + l)$ is of the form $\lambda(\gamma)X + \nu(\gamma)$, where $\lambda(\gamma) \in k$ and $\nu(\gamma) \in K$. From 4.5.3, Q_0' is the only $\mathfrak{g}$-stable non-null prime ideal of $K_{\bar{D}}[X]$.

Let $P \in \mathscr{P}$ such that $P \cap U(\mathfrak{h}') = I'$. Then P/I is a non-null completely prime ideal of $M = N_D[X]$, P/I is $\mathfrak{g}$-stable, and $(P/I) \cap N = 0$. Hence the set $(P/I)'$ of the pn^{-1}, where $p \in P/I$ and $n \in N - \{0\}$, is a $\mathfrak{g}$-stable non-null prime ideal of $K_{\bar{D}}[X]$, and $(P/I)' \cap N_D[X] = P/I$. Hence $(P'I)' = Q_0'$, whence $P/I = Q_0$ and $P = P_0$.

Returning to the beginning of the proof, we can see that, if $P \in \mathscr{P}$, then

$$P \supset P_0 \cap (I + U(\mathfrak{h})\hat{I}'U(\mathfrak{h})).$$

Now $P_0 \neq I$ and $I + U(\mathfrak{h})\hat{I}'U(\mathfrak{h}) \neq I$, hence $P_0 \cap (I + U(\mathfrak{h})\hat{I}'U(\mathfrak{h})) \neq I$ since I is prime.

4.5.5. LEMMA ($\mathfrak{g}$ completely solvable). *Let $\mathfrak{h}$ be an ideal of $\mathfrak{g}$, I a prime ideal of $U(\mathfrak{h})$ such that $[\mathfrak{g},I] \subset I$, $\mathscr{P}$ the set of prime ideals P of $U(\mathfrak{h})$ such that $P \supset I$, $P \neq I$, $[\mathfrak{g},P] \subset P$, and $\hat{I}$ the intersection of the elements of $\mathscr{P}$. Let ε be the adjoint representation of $\mathfrak{g}$ in* Fract$(U(\mathfrak{h})/I)$. *Assume that the elements of* Fract$(U(\mathfrak{h})/I)$ *annihilated by $\varepsilon(\mathfrak{g})$ are algebraic over k. Then $\hat{I} \neq I$.*

This is obvious if $\mathfrak{h} = 0$. Let us assume that $\dim \mathfrak{h} = n > 0$ and that the lemma has been proved for $\dim \mathfrak{h} < n$. There exists an ideal $\mathfrak{h}'$ of $\mathfrak{g}$ of codimension 1 in $\mathfrak{h}$. Let us introduce I', $\mathscr{P}'$, $\hat{I}'$ as in 4.5.4. Let ε' be the adjoint representation of $\mathfrak{g}$ in Fract$(U(\mathfrak{h}')/I')$. The elements of Fract$(U(\mathfrak{h}')/I')$ annihilated by $\varepsilon'(\mathfrak{h})$ are annihilated by $\varepsilon(\mathfrak{g})$ and hence algebraic over k. From the induction hypothesis, we have $\hat{I}' \neq I'$. Hence $\hat{I} \neq I$ from 4.5.4.

4.5.6. LEMMA ($\mathfrak{g}$ solvable). *Let k' be an extension of finite degree of $k, \mathfrak{g}' = \mathfrak{g} \otimes k'$, I a prime ideal of $U(\mathfrak{g})$, and I' a prime ideal of $U(\mathfrak{g}')$ such that $I' \cap U(\mathfrak{g}) = I$. If $C(\mathfrak{g};I)$ is an extension of finite degree of k, then $C(\mathfrak{g}';I')$ is an extension of finite degree of k'.*

We identify $U(\mathfrak{g})/I$ with a sub-k-algebra of $U(\mathfrak{g}')/I'$. Let K' be the field of fractions of $U(\mathfrak{g}')/I'$. The field of fractions K of $U(\mathfrak{g})/I$ can be identified with the subfield of K' generated by $U(\mathfrak{g})/I$. Let $A = k'K \subset K'$. Being a left vector space over K, A is finite-dimensional; hence, if $a \in A$ and $a \neq 0$, then a relation of the form $\lambda_n a^n + \lambda_{n-1} a^{n-1} + \cdots + \lambda_0 = 0$, where $\lambda_n, \ldots, \lambda_0 \in K, \lambda_n \neq 0, \lambda_0 \neq 0$ obtains; it follows from this that a is invertible in A. Hence A is a subfield of K' containing $U(\mathfrak{g}')/I'$, whence $K' = A = k'K$.

Consequently, K' can be identified with a quotient algebra of $K \otimes k'$. Let us set $C(\mathfrak{g};I) = C$ and $C \otimes k' = D$. Then

$$K \otimes k' = K \otimes_C (C \otimes k') = K \otimes_C D,$$

and every two-sided ideal of $K \otimes_C D$ is of the form $K \otimes_C D'$ where D' is an ideal of D (4.5.1). Hence K' can be identified with an algebra of the form $K \otimes_C (D/D')$, so that $C(\mathfrak{g}';I')$ can be identified with $D/D' = (C \otimes k')/D'$, whence the lemma follows.

4.5.7. THEOREM ($\mathfrak{g}$ solvable). *Let I be a prime ideal of $U(\mathfrak{g})$. The following conditions are equivalent:*

(i) *I is primitive;*

(ii) *$C(\mathfrak{g};I)$ is an algebraic extension of k;*

(iii) *$C(\mathfrak{g};I)$ is an algebraic extension of finite degree of k;*

(iv) *the intersection of the prime ideals of $U(\mathfrak{g})$ strictly containing I is distinct from I;*

(v) *the intersection of the primitive ideals of $U(\mathfrak{g})$ strictly containing I is distinct from I.*

(iv) $\Rightarrow$ (v). This is obvious.

(v) $\Rightarrow$ (i). From 3.1.15, I is the intersection of a family (I_λ) of primitive ideals. If condition (v) is satisfied, one of the I_λ is equal to I, and hence I is primitive.

(i) $\Rightarrow$ (ii). This follows from 4.1.7.

(ii) $\Rightarrow$ (iii). This follows from 4.4.11.

(iii) $\Rightarrow$ (iv). Let us assume that condition (iii) is satisfied. Let k' be a Galois extension of finite degree of k such that $\mathfrak{g}' = \mathfrak{g} \otimes k'$ is completely solvable. Let Γ be the Galois group of k' over k. From 3.4.2. there exists a prime ideal I' of $U(\mathfrak{g}')$ such that

$$I \otimes k' = \bigcap_{\gamma \in \Gamma} \gamma(I').$$

For all $\gamma \in \Gamma$, let $\mathscr{P}_\gamma$ be the set of prime ideals P of $U(\mathfrak{g}')$ such that $P \supset \gamma(I')$ and $P \neq \gamma(I')$, and let $\hat{I}'_\gamma$ be the intersection of the elements of $\mathscr{P}_\gamma$. From 4.5.6, $C(\mathfrak{g}';\gamma(I'))$ is an extension of finite degree of k'. From 4.5.5 (where we replace $\mathfrak{g}$ and $\mathfrak{h}$ by $\mathfrak{g}'$), we have $\hat{I}_\gamma \neq \gamma(I')$.

Let us assume that $\bigcap_{\gamma \in \Gamma} \hat{I}'_\gamma = \bigcap_{\gamma \in \Gamma} \gamma(I')$. Let $\gamma_1(I')$ be a minimal element in the set of the $\gamma(I')$. Then $\gamma_1(I') \supset \bigcap_{\gamma \in \Gamma} \hat{I}'_\gamma$, hence there exists $\gamma_2 \in \Gamma$ such that $\gamma_1(I') \supset \hat{I}'_{\gamma_2}$; since $\hat{I}'_{\gamma_2}$ strictly contains $\gamma_2(I')$, this contradicts the minimality of $\gamma_1(I')$. Hence $\bigcap_{\gamma \in \Gamma} \hat{I}'_\gamma$ strictly contains $I \otimes k'$, and consequently $(\bigcap_{\gamma \in \Gamma} \hat{I}'_\gamma) \cap U(\mathfrak{g})$ strictly contains I.

Let Q be a prime ideal of $U(\mathfrak{g})$ strictly containing I. From 3.4.2, there exists a prime ideal Q' of $U(\mathfrak{g}')$ such that $Q \otimes k' = \bigcap_{\gamma \in \Gamma} \gamma(Q')$. We have

$$Q' \supset Q \otimes k' \supset I \otimes k' = \bigcap_{\gamma \in \Gamma} \gamma(I'),$$

hence $Q' \supset \gamma_0(I')$ for some $\gamma_0 \in \Gamma$. If $Q' = \gamma_0(I')$, then

$$Q \otimes k' = \bigcap_{\gamma \in \Gamma} \gamma(I') = I \otimes k',$$

whence $Q = I$, which is a contradiction. Hence $Q' \supset \hat{I}'_{\gamma_0}$ and consequently

$$Q \supset \Big(\bigcap_{\gamma \in \Gamma} \hat{I}'_\gamma\Big) \cap U(\mathfrak{g}).$$

Taking the preceding paragraph into account, we see that condition (iv) is satisfied.

4.5.8. Let us assume that $\mathfrak{g}$ is solvable. We term a prime ideal I of $U(\mathfrak{g})$. such that $C(\mathfrak{g};I) = k$ a *rational ideal* of $U(\mathfrak{g})$. Every rational ideal of $U(\mathfrak{g})$ is primitive (4.5.7). If k is algebraically closed, every primitive ideal of $U(\mathfrak{g})$ is rational (4.5.7).

4.5.9. Proposition ($\mathfrak{g}$ completely solvable). *Let $\mathfrak{k}$ be an ideal of $\mathfrak{g}$, I a primitive ideal of $U(\mathfrak{g})$, and $K = I \cap U(\mathfrak{k})$. There exists a primitive ideal of $U(\mathfrak{k})$ which is generic for K.*

From 4.5.7, the elements of Fract $(U(\mathfrak{k})/K)$ annihilated by the adjoint representation of $\mathfrak{g}$ are algebraic over k. Let $\mathscr{P}$ be the set of the prime ideals P of $U(\mathfrak{k})$ such that $P \supset K$, $P \neq K$, and $[\mathfrak{g},P] \subset P$. From 4.5.5, the intersection $\hat{K}$ of the elements of $\mathscr{P}$ is distinct from K. There exists a primitive ideal J of $U(\mathfrak{k})$ such that $J \supset K$, $J \not\supset \hat{K}$ (3.1.15). Let K' be the largest two sided ideal of $U(\mathfrak{k})$ which is contained in J and such that $[\mathfrak{g},K'] \subset K'$. Then $K \subset K'$. Let us assume that $K \neq K'$. This will lead to a contradiction, which will prove that J is generic for K. Let $P_1, \ldots, P_n$ be the minimal prime ideals of $U(\mathfrak{k})$ containing K'. Since J is primitive and contains K', J contains P_{i_0} for some i_0. Now $K \subset P_{i_0}$, $K \neq P_{i_0}$, and $[\mathfrak{g},P_{i_0}] \subset P_{i_0}$ (3.3.3). Hence $\hat{K} \subset P_{i_0}$, which contradicts $\hat{K} \not\subset J$.

References: [31], [35], [51].

4.6. Heisenberg and Weyl algebras

In this section we shall introduce certain Lie algebras and certain very special algebras. We shall group together all the lemmas concerning these algebras which will be useful in the remainder of the book.

4.6.1. Let n be an integer $\geqq 0$, and $(x_1, \ldots, x_n, y_1, \ldots, y_n, z)$ a basis for a vector space $\mathfrak{n}$. It is immediately seen that we can define a Lie algebra structure on this space by setting $[x_i,y_i] = -[y_i,x_i] = z$, the other brackets of basis elements being zero. The centre of $\mathfrak{n}$ is kz. If $n = 0$, then $\mathfrak{n}$ is one-dimensional. If $n > 0$, then $[\mathfrak{n},\mathfrak{n}] = kz$, and $\mathfrak{n}$ is nilpotent. The Lie algebras of the above type are termed *Heisenberg algebras*.

Let $\mathfrak{g}$ be a Lie algebra with centre $\mathfrak{c}$. If $\mathfrak{c} = [\mathfrak{g},\mathfrak{g}]$ and $\dim \mathfrak{c} = 1$, then $\mathfrak{g}$ is a Heisenberg algebra. For, let $z \in \mathfrak{c} - \{0\}$. There exists an alternating bilinear form B on $\mathfrak{g}$ such that $[x,y] = B(x,y)z$ for all $x,y \in \mathfrak{g}$. The kernel of B is $\mathfrak{c}$, whence our assertion follows from the properties of alternating forms.

4.6.2. Lemma. *Let* $\mathfrak{n}$ *be a nilpotent Lie algebra. Assume that every commutative characteristic ideal of* $\mathfrak{n}$ *is of dimension* $\leqq 1$. *Then* $\mathfrak{n}$ *is either null or a Heisenberg algebra.*

Let $\mathfrak{n}' = [\mathfrak{n},\mathfrak{n}]$. If $\dim \mathfrak{n}' > 1$, there exist ideals $\mathfrak{n}_1$ and $\mathfrak{n}_2$ of $\mathfrak{n}$, of dimensions 1 and 2, respectively, such that $\mathfrak{n}_1 \subset \mathfrak{n}_2 \subset \mathfrak{n}'$; we have

$$[\mathfrak{n}',n_2] \subset [[\mathfrak{n},\mathfrak{n}_2]\mathfrak{n}] + [[n_2,\mathfrak{n}],\mathfrak{n}] \subset [\mathfrak{n}_1,\mathfrak{n}] = 0.$$

Hence the centre of $\mathfrak{n}'$ is of dimension >1. Now this centre is a characteristic ideal of $\mathfrak{n}$, which is contradictory. Hence $\dim \mathfrak{n}' \leqq 1$. If $\mathfrak{n}' = 0$, then $\mathfrak{n}$ is commutative and thus of dimension $\leqq 1$. Let us assume that $\dim \mathfrak{n}' = 1$. Let $\mathfrak{c}$ be the centre of $\mathfrak{n}$. Then $[\mathfrak{n},\mathfrak{n}'] = 0$, hence $\mathfrak{c} \supset \mathfrak{n}'$. Moreover, $\dim \mathfrak{c} \leqq 1$, hence $\mathfrak{c} = \mathfrak{n}'$ and it is sufficient to apply 4.6.1.

4.6.3. Let $n \in N$. We denote the algebra defined by $2n$ generators $p_1,q_1, \ldots, p_n,q_n$, and the relations

$$[p_i,q_i] = 1,$$
$$[p_i,q_j] = [p_i,p_j] = [q_i,q_j] = 0 \quad \text{for } i \neq j.$$

by $A_n(k)$, or simply by A_n.

The algebras A_n are termed *Weyl algebras*. Clearly, the elements $p_1^{i_1}q_1^{j_1} \cdots p_n^{i_n}q_n^{j_n}$ $(i_1,j_1, \ldots, i_n, j_n \in \mathbf{N})$ generate the vector space A_n.

In the vector space $E = k[X_1, \ldots, X_n]$, let P_i be the endomorphism $\partial/\partial X_i$ and Q_i the endomorphism of multiplication by X_i. We have

$$[P_i,Q_i] = 1,$$
$$[P_i,Q_j] = [P_i,P_j] = [Q_i,Q_j] = 0 \quad \text{for } i \neq j,$$

and hence there exists a homomorphism ϱ of A_n into $\operatorname{End}(E)$ such that $\varrho(p_i) = P_i$, $\varrho(q_i) = Q_i$ for all i. It is easily seen that the $P_1^{i_1}Q_1^{j_1} \cdots P_n^{i_n}Q_n^{j_n}$ are linearly independent. It follows at the same time that the $p_1^{i_1}q_1^{j_1} \cdots p_n^{i_n}q_n^{j_n}$ form a basis for the vector space A_n, and that ϱ is injective. From this we deduce that A_n is canonically isomorphic to $A_1 \otimes A_1 \otimes \cdots \otimes A_1$ (n factors). The representation ϱ of A_n in E is termed the *standard representation* of A_n. It follows directly that E is a simple A_n-module (known as the *standard module* over A_n), and that the set of A_n-endomorphisms of E is k.

In the case of A_1, we set $p_1 = p$ and $q_1 = q$.

4.6.4. Let B_m be the set of linear combinations of the

$$p_1^{i_1}q_1^{j_1}\cdots p_n^{i_n}q_n^{j_n}\in A_n \quad \text{such that } i_1+j_1+\cdots+i_n+j_n\leqq m.$$

Then $B_m B_{m'}\subset B_{m+m'}$. It is easily seen that the graded algebra associated with A_n equipped with the filtration $(B_0,B_1,\ldots)$ is the polynomial algebra in $2n$ variables. Consequently, A_n is integral and Noetherian.

4.6.5. The centre of A_n is reduced to k. In fact, it is sufficient to prove this for A_1. Now, if a $\sum_{i,j}\lambda_{ij}p^iq^j$ is a central element of A_1, then

$$0=[p,a]=\sum_{i,j} j\lambda_{ij}p^iq^{j-1},\qquad 0=[q,a]=-\sum_{i,j} i\lambda_{ij}p^{i-1}q^j,$$

whence $\lambda_{ij}=0$ for $i+j>0$.

4.6.6. The algebra A_n is simple. In fact, from 4.6.3 and 4.6.5, it is sufficient to prove it for A_1. Let I be a non-null two-sided ideal of A_1 and $a=\sum_{i,j}\lambda_{ij}p^iq^j\in I-\{0\}$. Then

$$\sum_{i,j} j\lambda_{i,j}p^iq^{j-1}\in I,\qquad \sum_{i,j} i\lambda_{i,j}p^{i-1}q^j\in I.$$

Step by step, we deduce that $1\in I$.

4.6.7. Lemma. (i) *Let A be an algebra, and B and C permutable subalgebras of A which generate A. Assume that C is isomorphic to A_n. Then the canonical homomorphism of $B\otimes C$ into A is an isomorphism.*

(ii) *Let M be an algebra, X and Y indeterminates, and Δ the derivation of $M[Y]$ such that $\Delta(M)=0$, $\Delta(Y)=1$. In $(M[Y])_\Delta[X]$, we have $[X,Y]=1$, which permits us to identify the subalgebra generated by X and Y with A_1. The canonical morphism of $M\otimes A_1$ into $(M[Y])_\Delta[X]$ is an isomorphism.*

From 4.6.5 and 4.6.6, every two-sided ideal of $B\otimes C$ is of the form $J\otimes C$, where J is a two-sided ideal of B (cf. 4.5.1). Let φ be the canonical morphism of $B\otimes C$ into A. Since $\varphi|B$ is injective, we have $\operatorname{Ker}\varphi=0$. Clearly, φ is surjective, whence (i). Assertion (ii) is a special case of (i).

4.6.8. Lemma. *Every derivation of A_n is an inner derivation.*

Let $p_1,q_1,\ldots,p_nq_n$ be the canonical generators of A_n. Every element a of A_n can be written as

$$a=\sum_{i_1,j_1,\ldots,i_nj_n}\alpha_{i_1j_1\ldots i_nj_n}p_1^{i_1}q_1^{j_1}\cdots p_n^{i_n}q_n^{j_n},$$

where the $\alpha_{i_1 j_1 \ldots i_n j_n}$ belong to k. If we identify a with a polynomial in the indeterminates p_i, q_i, then

$$(1) \qquad [p_i,a] = \frac{\partial a}{\partial q_i}, \qquad [q_i,a] = -\frac{\partial a}{\partial p_i}.$$

Let D be a derivation of A_n. It will be convenient to set

$$D(p_i) = a_{q_i}, \qquad D(q_i) = -a_{p_i}.$$

The relations $[p_i,p_j] = [q_i,q_j] = 0$ and $[p_i,q_j] = \delta_{ij}$ imply that

$$[Dp_i,p_j] + [p_i,Dp_j] = [Dq_i,q_j] + [q_i,Dq_j]$$
$$= [Dp_i,q_j] + [p_i,Dq_j] = 0,$$

i.e.,

$$-\frac{\partial a_{q_i}}{\partial q_j} + \frac{\partial a_{q_j}}{\partial q_i} = -\frac{\partial a_{p_i}}{\partial p_j} + \frac{\partial a_{p_j}}{\partial p_i} = \frac{\partial a_{q_i}}{\partial p_j} - \frac{\partial a_{p_j}}{\partial p_i} = 0.$$

Hence there exists $b \in A_n$ such that

$$\frac{\partial b}{\partial p_i} = a_{p_i}, \qquad \frac{\partial b}{\partial q_i} = a_{q_i}.$$

Let D' be the inner derivation of A_n defined by $-b$. Then

$$D(p_i) = \quad a_{q_i} = \quad \frac{\partial b}{\partial q_i} = [p_i,b] = D'(p_i),$$

$$D(q_i) = -a_{p_i} = -\frac{\partial b}{\partial p_i} = [q_i,b] = D'(q_i),$$

and hence $D = D'$.

4.6.9. LEMMA. *Let $p_1,q_1,\ldots,p_n,q_n$ be the canonical generators of A_n. Let us define vector subspaces S,T of A_n in the following way:*

$$S = \sum_{i=1}^{n} (kp_i + kq_i),$$

$$T = \sum_{1 \leq i,j \leq n} (k \cdot \tfrac{1}{2}(p_i q_j + q_j p_i) + kp_i p_j + kq_i q_j).$$

Then:

(i) *$k \oplus S$ and T are Lie subalgebras of A_n; $k \oplus S$ is a Heisenberg algebra.*
(ii) *The set of the $a \in A_n$ such that $[a,S] \subset k \oplus S$ is $k \oplus S \oplus T$.*
(iii) *The set of the $a \in A_n$ such that $[a,S] \subset S$ is $k \oplus T$.*
(iv) *Let B be the non-degenerate alternating bilinear form on S such that*

$[x,y] = B(x,y) \cdot 1$ for $x,y \in S$. Let η be the Lie algebra consisting of the endomorphisms u of the vector space S such that $B(ux,y) + B(x,uy) = 0$ for all $x,y \in S$. For all $a \in T$, let $\delta(a)$ be the restriction of $\mathrm{ad}_{A_n} a$ to S. Then δ is an isomorphism of the Lie algebra T onto the Lie algebra η.

(v) *The set of the $a \in T$ such that $kp_1 + \cdots + kp_n$ is invariant under δ/a is*

$$\sum_{1 \leq i,j \leq n} (k \cdot \tfrac{1}{2}(p_i q_j + q_j p_i) + kp_i p_j).$$

It is obvious that $k \oplus S$ is a Lie subalgebra of A_n, and is a Heisenberg algebra. Let us identify the elements of A_n with polynomials as we did in the proof of 4.6.8. Then $k \oplus S \oplus T$ is the set of polynomials of degree $\leqq 2$. Let $a \in A_n$. From (1), in order that $[a,S] \subset k + S$, it is necessary and sufficient that a is of degree $\leqq 2$, and hence that $a \in k \oplus S \oplus T$. We have

$$[p_i p_j, p_l] = 0, \qquad [p_i p_j, q_l] = \delta_{il} p_j + \delta_{jl} p_i$$

hence $[p_i p_j, S] \subset S$; similarly, it can be seen that $[q_i q_j, S] \subset S$, $[p_i q_j, S] \subset S$ and $[q_j p_i, S] \subset S$. From this we deduce (ii) and (iii). On the other hand,

$$[p_i p_j, \tfrac{1}{2}(p_m q_l + q_l p_m)] = [p_i p_j, p_m q_l] = \delta_{il} p_m p_j + \delta_{jl} p_m p_i,$$

$$[q_i q_j, \tfrac{1}{2}(p_m q_l + q_l p_m)] = [q_i q_j, p_m q_l] = -\delta_{im} q_j q_l - \delta_{jm} q_i q_l,$$

$$[\tfrac{1}{2}(p_i q_j + q_j p_i), \tfrac{1}{2}(p_m q_l + q_l p_m)] = [p_i q_j, p_m q_l]$$

$$= \delta_{il} p_m q_j - \delta_{jm} p_i q_l$$

$$= \delta_{il}(\tfrac{1}{2}(p_m q_j + q_j p_m) + \tfrac{1}{2}\delta_{jm}) - \delta_{jm}(\tfrac{1}{2}(p_i q_l + q_l p_i) + \tfrac{1}{2}\delta_{il})$$

$$= \tfrac{1}{2}\delta_{il}(p_m q_j + q_j p_m) - \tfrac{1}{2}\delta_{jm}(p_i q_l + q_l p_i),$$

$$[p_i p_j, q_m q_l] = \delta_{im} p_j q_l + \delta_{jm} p_i q_l + \delta_{il} q_m p_j + \delta_{jl} q_m p_i.$$

Let us denote the latter element by a. If $l = m$ or if $i = j$, clearly $a \in T$. If $l \neq m$ and $i \neq j$, we may assume that $m < l$ and $i < j$. We distinguish the following cases:

(1) $i = m$, $j = l$; then

$$a = p_j q_j + q_i p_i = \tfrac{1}{2}(p_j q_j + q_j p_j + p_i q_i + q_i p_i);$$

(2) $i = m$, $j \neq l$; then

$$a = p_j q_l = \tfrac{1}{2}(p_j q_l + q_l p_j);$$

(3) $i = l$; then

$$a = q_m p_j = \tfrac{1}{2}(p_j q_m + q_m p_j);$$

(4) $j = m$; then

$$a = p_i q_l = \tfrac{1}{2}(p_i q_l + q_l p_i),$$

(5) $j = l$, $i \neq m$; then

$$a = q_m p_i = \tfrac{1}{2}(p_i q_m + q_m p_i).$$

In the other cases, $a = 0$. Thus we have proved (i).

If $a \in T$, then $\mathrm{ad}_{A_n} a$ is a derivation of A_n under which S is stable, and hence $\delta(a) \in \eta$; if $\delta(a) = 0$, then $\mathrm{ad}_{A_n} a$ is zero on S and hence on A_n, whence $a \in k \cap T = 0$. Clearly, δ is a Lie algebra homomorphism. The fact that δ is surjective can be seen for example by calculating the dimensions. Assertion (v) follows from a simple calculation.

4.6.10. Lemma. *The algebras A_n $(n = 0,1,2,\ldots)$ are pairwise non-isomorphic.*

From 4.6.9 (i), A_n is a quotient of $U(\mathfrak{g})$, where $\mathfrak{g}$ is the Heisenberg algebra of dimension $2n + 1$. Hence the Krull dimension of A_n is finite (3.5.7). Let us identify A_{n+1} with $A_n \otimes A_1$, and let us consider the strictly decreasing infinite sequence

$$A_n \otimes A_1 \supset A_n \otimes A_1 q \supset A_n \otimes A_1 q^2 \supset \cdots$$

of left ideals of A_{n+1}. Among the A_{n+1}-submodules of

$$A_n \otimes (A_1 q^i / A_1 q^{i+1}),$$

are the $M \otimes (A_1 q^i / A_1 q^{i+1})$, where M is a left ideal of A_n. This proves that $\mathrm{Kdim}\,(A_{n+1}) > \mathrm{Kdim}\,(A_n)$.

4.6.11. If R is an algebra (over k), we set $A_n(k) \otimes R = A_n(R)$. [If, moreover, R is a commutative field, the notation $A_n(R)$ is compatible with the notation $A_n(k)$.]

References: [30], [31], [45], [99], [110].

4.7. Centre and core in the nilpotent case

4.7.1. Proposition ($\mathfrak{g}$ nilpotent). *Let I be a two-sided ideal of $U(\mathfrak{g})$.*

(i) *Every non-null ideal of $U(\mathfrak{g})/I$ has a non-null intersection with $Z(\mathfrak{g}; I)$.*

(ii) *If I is prime, then $C(\mathfrak{g}; I)$ is the field of fractions of $Z(\mathfrak{g}; I)$.*

Assertion (i) and (ii) follow from 4.4.1 and 4.4.2 respectively, and 4.3.2.

4.7.2. Corollary ($\mathfrak{g}$ nilpotent). *$C(\mathfrak{g})$ is the field of fractions of $Z(\mathfrak{g})$.*

4.7.3. PROPOSITION ($\mathfrak{g}$ nilpotent). *Let I be a two-sided ideal of $U(\mathfrak{g})$. The following conditions are equivalent:*

(i) *I is prime;*

(ii) *$Z(\mathfrak{g}; I)$ is integral.*

If I is prime, clearly $Z(\mathfrak{g}; I)$ is integral. If I is non-prime, there exist in $U(\mathfrak{g})/I$ two non-null ideals with null product; from 4.7.1 (i), there exist in zero $Z(\mathfrak{g}; I)$ two non-zero elements with product zero.

4.7.4. PROPOSITION ($\mathfrak{g}$ nilpotent). *Let I be a two-sided ideal of $U(\mathfrak{g})$. The following conditions are equivalent:*

(i) *I is maximal;*

(ii) *I is primitive;*

(iii) *$Z(\mathfrak{g}; I)$ is a field.*

(We recall that, from 4.5.7, condition (ii) can be replaced by four other equivalent conditions.)

(i) $\Rightarrow$ (ii). This is obvious (cf. 3.1.6).

(ii) $\Rightarrow$ (iii). If I is primitive, $C(\mathfrak{g}; I)$ is an algebraic extension of k (4.5.7); now every subalgebra of an algebraic extension is a field.

(iii) $\Rightarrow$ (i). This follows from 4.7.1 (i).

4.7.5. LEMMA. *Let A be an algebra, D a locally nilpotent derivation of A, α an element of the centre of A such that $D\alpha = 1$, $\bar{A}$ the algebra $A/A\alpha$, λ the canonical homomorphism of A onto $\bar{A}$, Y an indeterminate, and Δ the derivation of $\bar{A}[Y]$ which is zero on $\bar{A}$ and transforms Y into 1. For all $a \in A$, let us set*

$$\chi(a) = \sum_{n \geqq 0} \frac{1}{n!} \lambda(D^n a) Y^n.$$

Then χ is an isomorphism of the algebra A onto the algebra $\bar{A}[Y]$ such that $D = \chi^{-1} \Delta \chi$. We have $\chi^{-1}(Y) = \alpha$, and, for all $a \in A$,

$$\chi^{-1}(\lambda a) = \sum_{m \geqq 0} \frac{(-1)^m}{m!} (D^m a) \alpha^m.$$

We immediately verify that $\chi(aa') = \chi(a)\chi(a')$ for $a, a' \in A$, and that $\chi \circ D = \Delta \circ \chi$. If $b \in A$, we have

$$\sum_{m \geqq 0} \frac{(-1)^m}{m!} (D^m(b\alpha))\alpha^m =$$

$$= \sum_{m \geqq 0} \frac{(-1)^m}{m!} ((D^m b)\alpha + m(D^{m-1} b))\alpha^m$$

$$= \sum_{m \geqq 0} \frac{(-1)^m}{m!} (D^m b)\alpha^{m+1} + \sum_{n \geqq 0} \frac{(-1)^{n+1}}{n!} (D^n b)\alpha^{n+1} = 0.$$

Hence there exists one and only one linear mapping χ' of $\bar{A}[Y]$ into A such that, for $a \in A$ and $n \in \mathbf{N}$,

$$\chi'((\lambda a) Y^n) = \sum_{m \geqq 0} \frac{(-1)^m}{m!} (D^m a) \alpha^{m+n}.$$

We immediately verify that χ' is a homomorphism of algebras. We have

$$(\chi'\chi)(a) = \sum_{m,n \geqq 0} \frac{1}{n!} \frac{(-1)^m}{m!} (D^{m+n} a) \alpha^{m+n} = a,$$

$$(\chi\chi')(Y) = \chi(\alpha) = Y,$$

$$(\chi\chi')(\lambda a) = \chi\left(\sum_{m \geqq 0} \frac{(-1)^m}{m!} (D^m a) \alpha^m\right)$$

$$= \sum_{m,n \geqq 0} \frac{(-1)^m}{m!n!} \lambda(D^n(D^m a \cdot \alpha^m)) Y^n$$

$$= \sum_{m,p,q \geqq 0} \frac{(-1)^m}{m!p!q!} \lambda((D^{p+m} a)(D^q \alpha^m)) Y^{p+q}$$

$$= \sum_{p,m \geqq 0} \frac{(-1)^m}{m!p!m!} \lambda(D^{p+m} a) m! Y^{p+m} = \lambda a,$$

hence $\chi\chi' = 1$ and $\chi'\chi = 1$.

4.7.6. LEMMA. *Let $A, D, \alpha, \bar{A}, \lambda$ be as in 4.7.5, and let X be an indeterminate. There exists one and only one homomorphism χ of the algebra $A_D[X]$ into the algebra $\bar{A} \otimes A_1$ such that*

$$\chi(X) = 1 \otimes p,$$

$$\chi(a) = \sum_{n \geqq 0} \frac{1}{n!} \lambda(D^n a) \otimes q^n \quad \text{for all } a \in A.$$

This homomorphism is an isomorphism. We have

$$\chi^{-1}(1 \otimes p) = X, \qquad \chi^{-1}(1 \otimes q) = \alpha,$$

$$\chi^{-1}(\lambda a \otimes 1) = \sum_{m \geqq 0} \frac{(-1)^m}{m!} (D^m a) \alpha^m \quad \textit{for all } a \in A.$$

Lemmas 4.7.5 and 4.6.7 (ii) define the isomorphisms

$$A_D[X] \xrightarrow{\chi_1} (\bar{A}[Y])_\Delta[X] \xrightarrow{\chi_2} \bar{A} \otimes A_1.$$

Let χ be the composite isomorphism. Then

$$\chi(X) = \chi_2(X) = 1 \otimes p,$$

$$\chi(\alpha) = \chi_2(Y) = 1 \otimes q$$

and, for $a \in A$,

$$\chi(a) = \chi_2\Big(\sum_{n \geq 0} \frac{1}{n!}\lambda(D^n a)Y^n\Big) = \sum_{n \geq 0} \frac{1}{n!}\lambda(D^n a) \otimes q^n,$$

$$\chi^{-1}(\lambda a \otimes 1) = \chi_1^{-1}(\lambda a) = \sum_{m \geq 0} \frac{(-1)^m}{m!}(D^m a)\alpha^m.$$

4.7.7. Let us assume that $\mathfrak{g}$ is nilpotent. A *reducing quadruple* of $\mathfrak{g}$ is a quadruple $(x,y,z,\mathfrak{h})$ such that:

(1) $x,y,z \in \mathfrak{g}$, where z is central and non-zero in $\mathfrak{g}$ and $[x,y] = z$;

(2) $\mathfrak{h}$ is the centralizer of y in $\mathfrak{g}$;

(3) $\mathfrak{g} = \mathfrak{h} \oplus kx$.

It follows from these conditions that $kx + ky + kz$ is a three-dimensional Lie subalgebra of $\mathfrak{g}$, that $ky + kz$ is a two-dimensional ideal of $\mathfrak{g}$, and that $\mathfrak{h}$ is an ideal of $\mathfrak{g}$.

If the centre $\mathfrak{z}$ of $\mathfrak{g}$ is one-dimensional and $\mathfrak{g} \neq \mathfrak{z}$, there do exist reducing quadruples. Indeed, there exists an ideal $\mathfrak{y}$ of $\mathfrak{g}$ such that $\mathfrak{y} \supset \mathfrak{z}$, $\dim \mathfrak{y} = 2$ and $\mathfrak{y}/\mathfrak{z}$ is central in $\mathfrak{g}/\mathfrak{z}$. Let z be a non-zero element of $\mathfrak{z}$, and y an element of $\mathfrak{y}$ which does not belong to $\mathfrak{z}$. For all $u \in \mathfrak{g}$, we have $[u,y] = \mu(u)z$, where $\mu \in \mathfrak{g}^*$. We have $\mu \neq 0$ since $y \notin \mathfrak{z}$, hence the centralizer $\mathfrak{h}$ of y in $\mathfrak{g}$, namely Ker μ, is of codimension 1 in $\mathfrak{g}$. Let $x \in \mathfrak{g}$ be such that $\mu(x) = 1$. Then $(x,y,z,\mathfrak{h})$ is a reducing quadruple.

4.7.8. LEMMA ($\mathfrak{g}$ nilpotent). *Let $(x,y,z,\mathfrak{h})$ be a reducing quadruple of $\mathfrak{g}$, $\bar{\mathfrak{h}} = \mathfrak{h}/ky$, δ the (locally nilpotent) derivation of $U(\mathfrak{h})$ defined by x, λ the canonical homomorphism of $U(\mathfrak{h})$ onto $U(\bar{\mathfrak{h}})$, and $\bar{z} = \lambda z$.*

(i) *There exists one and only one homomorphism φ of the algebra $U(\mathfrak{g})$ into the algebra $U(\bar{\mathfrak{h}}) \otimes A_1$ such that*

$$\varphi(x) = 1 \otimes p,$$

$$\varphi(u) = \sum_{n \geq 0} \frac{1}{n!}\lambda(\delta^n u) \otimes q^n \quad \textit{for all } u \in U(\mathfrak{h}).$$

The homomorphism φ can be uniquely extended to a homomorphism ψ of the algebra $U(\mathfrak{g})_z$ into the algebra $U(\bar{\mathfrak{h}})_{\bar{z}} \otimes A_1$. The homomorphism ψ is an isomorphism. We have

$$\psi^{-1}(1 \otimes p) = x, \qquad \psi^{-1}(1 \otimes q) = yz^{-1}, \qquad \psi^{-1}(\bar{z} \otimes 1) = z,$$

$$\psi^{-1}(\lambda u \otimes 1) = \sum_{m \geq 0} \frac{(-1)^m}{m!}(\delta^m u)y^m z^{-m} \quad \textit{for all } u \in U(\mathfrak{h}).$$

(ii) *Let I be a two-sided ideal of $U(\mathfrak{g})$, distinct from $U(\mathfrak{g})$, such that $z - 1 \in I$. There exists one and only two-sided ideal J of $U(\bar{\mathfrak{h}})$ distinct from $U(\bar{\mathfrak{h}})$ such* *that* $\bar{z} - 1 \in J$ *and* $\psi(I_z) = J_{\bar{z}} \otimes A_1$. *By passage to the quotients,* ψ *defines an isomorphism of* $U(\mathfrak{g})_z/I_z$ *onto* $(U(\bar{\mathfrak{h}})_{\bar{z}}/J_{\bar{z}}) \otimes A_1$. *The canonical homomorphisms* $U(\mathfrak{g})/I \to U(\mathfrak{g})_z/I_z$ *and* $U(\bar{\mathfrak{h}})/J \to U(\bar{\mathfrak{h}})_{\bar{z}}/J_{\bar{z}}$ *are isomorphisms, whence we have a sequence of isomorphisms*

$$U(\mathfrak{g})/I \to U(\mathfrak{g})_z/J_z \to (U(\bar{\mathfrak{h}})_{\bar{z}}/J_{\bar{z}}) \otimes A_1 \to (U(\bar{\mathfrak{h}})/J) \otimes A_1.$$

Let A be the subalgebra $U(\mathfrak{h})_z$ of $U(\mathfrak{g})_z$. The derivation δ can be uniquely extended to a derivation D of A which is locally nilpotent. Let $\alpha = yz^{-1}$. which is a central element of A, such that $D\alpha = 1$. The kernel of λ is the ideal $U(\mathfrak{h})y$ of $U(\mathfrak{h})$. The homomorphism λ can be uniquely extended to a homomorphism, which we shall also denote by λ, of A onto $\bar{A} = U(\bar{\mathfrak{h}})_{\bar{z}}$. Thus we arrive at the situation of 4.7.6. On the other hand, from 2.1.11, $U(\mathfrak{g})_z$ can be canonically identified with $A_D[X]$ in such a way that x can be identified with X. There thus exists an isomorphism ψ of $U(\mathfrak{g})_z$ onto $U(\bar{\mathfrak{h}})_{\bar{z}} \otimes A_1$ such that

$$\psi(x) = 1 \otimes p,$$

$$\psi(u) = \sum_{n \geq 0} \frac{1}{n!} \lambda(D^n u) \otimes q^n \quad \text{for all } u \in U(\mathfrak{h})_z.$$

This proves the existence of the homomorphism φ of the lemma; its uniqueness is obvious. Since $\varphi(z) = \bar{z} \otimes 1$ and $\bar{z} \otimes 1$ is invertible in $U(\bar{\mathfrak{h}})_{\bar{z}} \otimes A_1$, φ can be uniquely extended to a homomorphism of $U(\mathfrak{g})_z$ into $U(\bar{\mathfrak{h}})_{\bar{z}} \otimes A_1$; necessarily, this homomorphism is ψ. The final equations of (i) follow from 4.7.6.

Let us prove (ii). Since $z - 1 \in I$, z is not a divisor of zero modulo I, and the conditions of 3.6.15 then apply to the consideration of I_z. From 4.5.1, $\psi(I_z)$ is the tensor product of a unique two-sided ideal J' of $U(\bar{\mathfrak{h}})_{\bar{z}}$ and A_1. There exists a unique two-sided ideal J of $U(\bar{\mathfrak{h}})$ such that $J' = J_{\bar{z}}$ (3.6.15). We have $(\bar{z} - 1) \otimes 1 = \psi(z - 1) \in J_{\bar{z}} \otimes A_1$, hence $\bar{z} - 1 \in J_{\bar{z}} \cap U(\bar{\mathfrak{h}}) = J$. Since $z - 1 \in I$, the image of z in $U(\mathfrak{g})/I$ is invertible, and so the canonical homomorphism of $U(\mathfrak{g})/I$ into $U(\mathfrak{g})_z/I_z$ is bijective. The same applies to the canonical homomorphism of $U(\bar{\mathfrak{h}})/J$ into $U(\bar{\mathfrak{h}})_{\bar{z}}/J_{\bar{z}}$.

4.7.9. THEOREM ($\mathfrak{g}$ nilpotent). *Let I be a two-sided ideal of $U(\mathfrak{g})$. The following conditions are equivalent:*

(i) *I is rational;*

(ii) *$Z(\mathfrak{g}; I) = k$;*

(iii) *there exists* $n \in \mathbf{N}$ *such that* $U(\mathfrak{g})/I$ *is isomorphic to* A_n;
(iv) I *is the kernel of an absolutely simple representation of* $U(\mathfrak{g})$.

(i) $\Rightarrow$ (ii). This is obvious.

(ii) $\Rightarrow$ (iii). We assume that $Z(\mathfrak{g};I) = k$ and prove (iii). This is obvious if $\dim \mathfrak{g} \leqq 1$. Let $p = \dim \mathfrak{g} > 1$, and assume that our assertion has been proved for all dimensions $<p$. Let $\mathfrak{z}$ be the centre of $\mathfrak{g}$, and $\mathfrak{z}_0 = I \cap \mathfrak{z}$, which is an ideal of $\mathfrak{g}$. First we assume that $\mathfrak{z}_0 \neq 0$. We have $U(\mathfrak{g}/\mathfrak{z}_0) = U(\mathfrak{g})/U(\mathfrak{g})_{\mathfrak{z}_0}$, and $I \supset U(\mathfrak{g})\mathfrak{z}_0$; let I' be the canonical image of I in $U(\mathfrak{g}/\mathfrak{z}_0)$; then $U(\mathfrak{g})/I = U(\mathfrak{g}')/I'$ and $Z(\mathfrak{g}';I') = k$, and it is sufficient to apply the induction hypothesis. Let us assume that $\mathfrak{z}_0 = 0$. Since $Z(\mathfrak{g};I) = k$, every element of $\mathfrak{z}$ is congruent to k modulo I, hence $\dim \mathfrak{z} = 1$. Thus there exists a reducing quadruple $(x,y,z,\mathfrak{h})$ and we may assume that $z \equiv 1$ (mod I). Let $\bar{\mathfrak{h}}, \bar{z}, J$ be as in 4.7.8. Since $U(\mathfrak{g})/I$ is isomorphic to $(U(\bar{\mathfrak{h}})/J) \otimes A_1$, we have $Z(\bar{\mathfrak{h}};J) = k$, and it is sufficient to apply the induction hypothesis.

(iii) $\Rightarrow$ (iv). This follows from the fact that the standard representation of A_n is absolutely simple.

(iv) $\Rightarrow$ (ii). This follows from 2.6.5.

(ii) $\Rightarrow$ (i). If $Z(\mathfrak{g};I) = k$, then I is prime (4.7.3), and therefore rational from 4.7.1 (ii).

4.7.10. Let us assume that $\mathfrak{g}$ is nilpotent, and let I be a rational ideal of $U(\mathfrak{g})$. The integer n of 4.7.9 (iii) is uniquely determined by I (4.6.10). It is termed the *weight* of I.

4.7.11. Let us assume that $\mathfrak{g}$ is nilpotent. For every extension K of k, we denote the set of rational ideals of $U(\mathfrak{g} \otimes K)$ by $\mathscr{R}(\mathfrak{g},K)$ or $\mathscr{R}(K)$. Let K' be another extension of k, and $f: K \to K'$ a homomorphism of extensions (hence necessarily injective, and reducing to the identity mapping on k). Let $I \in \mathscr{R}(K)$. Then $U(\mathfrak{g} \otimes K \otimes_K K')/I \otimes_K K'$ is isomorphic to $(U(\mathfrak{g} \otimes K)/I) \otimes_K K'$, and so has K' as its centre; consequently, $I \otimes_K K' \in \mathscr{R}(K')$ The mapping $I \mapsto I \otimes_K K'$ of $\mathscr{R}(K)$ into $\mathscr{R}(K')$ is denoted by $\mathscr{R}(f)$. If $f' : K' \to K''$ is a homomorphism of extensions of k, then

$$\mathscr{R}(f' \circ f) = \mathscr{R}(f') \circ \mathscr{R}(f).$$

4.7.12. Let K be an extension of k, and $I \in \mathscr{R}(K)$. Then $I \cap U(\mathfrak{g}) = P$ is a prime ideal of $U(\mathfrak{g})$. The canonical homomorphism φ of $U(\mathfrak{g})/P$ into $U(\mathfrak{g} \otimes K)/I$ is injective, and its image generates $U(\mathfrak{g} \otimes K)/I$ as a K-algebra. Hence $\varphi | Z(\mathfrak{g};P)$ is an injective homomorphism of $Z(\mathfrak{g};P)$ into K which can be uniquely extended to a k-homomorphism ψ of $C(\mathfrak{g};P)$ into K. We say that the prime ideal P and the homomorphism ψ of $C(\mathfrak{g};P)$ into K are *canonically associated with I.*

4.7.13. We still assume that $\mathfrak{g}$ is nilpotent. Let K be an extension of k, P a prime ideal of $U(\mathfrak{g})$, and ψ a k-homomorphism of $C(\mathfrak{g};P)$ into K. Let us consider the commutative diagram

$$\begin{array}{ccc} U(\mathfrak{g}) & \xrightarrow{\quad i \quad} & U(\mathfrak{g}) \otimes K = U(\mathfrak{g} \otimes K) \\ \downarrow p & & \downarrow q \\ (U(\mathfrak{g})/P) \otimes_{Z(\mathfrak{g};p)} C(\mathfrak{g};P) & \xrightarrow{1 \otimes \psi} & (U(\mathfrak{g})/P) \otimes_{Z(\mathfrak{g};p)} K \end{array}$$

where i,p,q are the canonical mappings, and let $I = \operatorname{Ker} q$. Since q is surjective, we have

$$U(\mathfrak{g} \otimes K)/I = (U(\mathfrak{g})/P) \otimes_{Z(\mathfrak{g};P)} K = (U(\mathfrak{g})/P) \otimes_{Z(\mathfrak{g};P)} C(\mathfrak{g};P) \otimes_{C(\mathfrak{g};P)} K.$$

Now the centre of $(U(\mathfrak{g})/P) \otimes_{Z(\mathfrak{g};P)} C(\mathfrak{g};P)$ is $C(\mathfrak{g};P)$ (4.1.5), hence the centre of $U(\mathfrak{g} \otimes K)/I$ is K. Thus I is an element of $\mathscr{R}(K)$, and is said to be *canonically associated* with P and ψ.

4.7.14. PROPOSITION ($\mathfrak{g}$ nilpotent). *Let K be an extension of k. Let* $\mathscr{R}'(K)$ *be the set of pairs* (P,ψ), *where P is a prime ideal of* $U(\mathfrak{g})$ *and* ψ *is a k-homomorphism of* $C(\mathfrak{g};P)$, *into K. Then the mappings* $\mathscr{R}(K) \to \mathscr{R}'(K)$ *and* $\mathscr{R}'(K) \to \mathscr{R}(K)$ *defined in* 4.7.12 *and* 4.7.13 *are mutually reciprocal bijections.*

Adopting the notation of 4.7.13, we have

$$I \cap U(\mathfrak{g}) = \operatorname{Ker}(q \circ i) = \operatorname{Ker}((1 \otimes \psi) \circ p) = \operatorname{Ker} p = P.$$

The homomorphism of $U(\mathfrak{g})/P$ into $U(\mathfrak{g} \otimes K)/I = (U(\mathfrak{g})/P) \otimes_{Z(\mathfrak{g};P)} K$ deduced from i is the restriction of $1 \otimes \psi$ to $U(\mathfrak{g})/P$; its restriction to $Z(\mathfrak{g};P)$ is hence equal to the restriction of ψ. Thus P and ψ are canonically associated with I.

Conversely, as in 4.7.12, let us start with an element I of $\mathscr{R}(K)$. Let P be the prime ideal of $U(\mathfrak{g})$ and ψ the homomorphism of $C(\mathfrak{g};P)$ into K which is canonically associated with I. We form diagram (1) of 4.7.13. Let φ be the canonical homomorphism of $U(\mathfrak{g})/P$ into $(U(\mathfrak{g}) \otimes K)/I$. The homomorphisms φ and ψ have the same restriction to $Z(\mathfrak{g};P)$. Hence there exists one and only one homomorphism φ' of $(U(\mathfrak{g})/P) \otimes_{Z(\mathfrak{g};P)} K$ into $(U(\mathfrak{g}) \otimes K)/I$ which extends φ and is K-linear; the homomorphism $\varphi' \circ q$ and the canonical homomorphism of $U(\mathfrak{g}) \otimes K$ onto $(U(\mathfrak{g} \otimes K)/I$ coincide on $U(\mathfrak{g})$ and hence are equal. Consequently $\operatorname{Ker} q \subset I$, whence $\operatorname{Ker} q = I$

since Ker q is a maximal two-sided ideal. Thus I is canonically associated with P and ψ.

4.7.15. The bijections of 4.7.14 are said to be *canonical.*

4.7.16. Let $f: K \to K'$ be a k-homomorphism of extensions of k. Let $I \in \mathscr{R}(K)$, $I' = I \otimes_K K' \in \mathscr{R}(K')$, and (P, ψ) be the pair associated with I. It follows immediately that the pair associated with I' is $(P, f \circ \psi)$.

4.7.17. We still assume that $\mathfrak{g}$ is nilpotent. Let P be a prime ideal of $U(\mathfrak{g})$, and let $Z = Z(\mathfrak{g}; P)$, $C = C(\mathfrak{g}; P)$. Then P and the identity mapping of C define a rational ideal of $U(\mathfrak{g} \otimes C)$, namely the kernel of the canonical homomorphism

$$U(\mathfrak{g} \otimes C) \to (U(\mathfrak{g})/P) \otimes_Z C$$

(cf. 4.7.13). *This rational ideal is denoted by* $P^\wedge$. From 4.7.9, $(U(\mathfrak{g})/P) \otimes_Z C$ *is isomorphic to* $A_n(C)$, where n is an integer which is uniquely determined by P and termed the *weight* of P. This definition generalises 4.7.10.

4.7.18. PROPOSITION ($\mathfrak{g}$ nilpotent). *There exist a pure extension* D *of finite type of* k *and an integer* n *such that* $K(\mathfrak{g})$ *is isomorphic with the field of fractions of* $A_n(D)$.

This follows from 4.7.17 applied to the case where $P = 0$, from 4.4.8 and from 4.1.4.

Reference: [6], [8], [30], [52], [99].

4.8. Invariant ideals of the symmetric algebra (the nilpotent case)

The results of this section, with the exception of 4.8.11 and 4.8.12, are in fact lemmas only intended for use in Chapter VI. However, for technical reasons it is convenient to insert them here.

4.8.1. We assume that $\mathfrak{g}$ is nilpotent. Let $\mathscr{A}$ be the adjoint group of $\mathfrak{g}$ (CH, p. 181). This group operates in $S(\mathfrak{g})$ by automorphisms, whence we have the notion of an $\mathscr{A}$-invariant (or simply *invariant*) ideal of $S(\mathfrak{g})$; if J is such an ideal, $\mathscr{A}$ operates in $S(\mathfrak{g})/J$ by automorphisms, and we may therefore consider the subalgebra of $\mathscr{A}$-invariant elements of $S(\mathfrak{g})/J$; this subalgebra is denoted by $(S(\mathfrak{g})/J)^{\mathscr{A}}$. The adjoint representation of $\mathfrak{g}$ defines a representation of $\mathfrak{g}$ in $S(\mathfrak{g})$ (1.2.14). The invariant ideals of $S(\mathfrak{g})$ are the ideals which are invariant under this representation (BO, p. 137).

4.8.2. PROPOSITION ($\mathfrak{g}$ nilpotent). *Let $\mathscr{A}$ be the adjoint group of $\mathfrak{g}$. Let J be an invariant ideal of $S(\mathfrak{g})$. Every non-null invariant ideal of $S(\mathfrak{g})/J$ has a non-null intersection with $(S(\mathfrak{g})/J)^{\mathscr{A}}$.*

Let σ be the representation of $\mathfrak{g}$ in $S(\mathfrak{g})/J$ deduced from the adjoint representation. Let K be an non-null invariant ideal of $S(\mathfrak{g})/J$. For all $n \geqq 0$, let

$$S(\mathfrak{g})_n = S(\mathfrak{g})^0 + S(\mathfrak{g})^1 + \cdots + S(\mathfrak{g})^n,$$

and
$$T_n = S(\mathfrak{g})_n / S(\mathfrak{g})_n \cap J.$$

There exists n_0 such that $K \cap T_{n_0} \neq 0$. For all $x \in \mathfrak{g}$, $\sigma(x)|K \cap T_{n_0}$ is nilpotent, hence there exists a non-zero t in $K \cap T_{n_0}$ which is annihilated by $\sigma(\mathfrak{g})$. Then $t \in (S(\mathfrak{g})/J)^{\mathscr{A}}$.

4.8.3. We assume that $\mathfrak{g}$ is nilpotent. The vector space $\mathfrak{g}$ considered as a commutative Lie algebra is denoted by $\mathfrak{g}^{\text{co}}$, and the semi-direct product of $\mathfrak{g}$ with $\mathfrak{g}^{\text{co}}$ corresponding to the adjoint representation is denoted by $\mathfrak{g}^{\sim}$; it is a nilpotent Lie algebra. We identify $S(\mathfrak{g})$ with $U(\mathfrak{g}^{\text{co}}) \subset U(\mathfrak{g}^{\sim})$. From 2.5.3 there exists on $S(\mathfrak{g})$ a unique $\mathfrak{g}^{\sim}$-module structure such that:

(1) if $x \in \mathfrak{g}$, then $x_{S(\mathfrak{g})}$ is the derivation of $S(\mathfrak{g})$ which extends ad $x|\mathfrak{g}^{\text{co}}$;

(2) if $y \in \mathfrak{g}^{\text{co}}$, then $y_{S(\mathfrak{g})}$ is multiplication by y.

The sub-$\mathfrak{g}^{\sim}$-modules of $S(\mathfrak{g})$ are invariant ideals.

4.8.4. PROPOSITION ($\mathfrak{g}$ nilpotent). *Let $\mathscr{A}$ be the adjoint group of $\mathfrak{g}$, and J an invariant ideal of $S(\mathfrak{g})$. The following conditions are equivalent:*

(i) *J is prime;*

(ii) *$(S(\mathfrak{g})/J)^{\mathscr{A}}$ is integral.*

If J is prime, it is clear that $(S(\mathfrak{g})/J)^{\mathscr{A}}$ is integral. Let us assume that J is not prime. The non-null minimal prime ideals of $S(\mathfrak{g})/J$ are $\mathscr{A}$-invariant (3.3.3). Hence there exist two non-null invariant ideals of $S(\mathfrak{g})/J$ whose product is null. From 4.8.2, there exist in $(S(\mathfrak{g})/J)^{\mathscr{A}}$ two non-null elements whose product is null.

4.8.5. PROPOSITION ($\mathfrak{g}$ nilpotent). *Let $\mathscr{A}$ be the adjoint group of $\mathfrak{g}$, and J an invariant ideal of $S(\mathfrak{g})$. The following conditions are equivalent:*

(i) *J is invariant maximal;*

(ii) *$(S(\mathfrak{g})/J)^{\mathscr{A}}$ is a field;*

(iii) *$(S(\mathfrak{g})/J)^{\mathscr{A}}$ is an algebraic extension of finite degree of k.*

(iii) $\Rightarrow$ (ii). This is obvious.

(ii) $\Rightarrow$ (i). This follows from 4.8.2.

(i) $\Rightarrow$ (iii). Let us assume that J is maximal invariant. Then $S(\mathfrak{g})/J$ is a simple $\mathfrak{g}^{\sim}$-module (4.8.3). Let D be the algebra of all endomorphisms of this $\mathfrak{g}^{\sim}$-module; then D is an algebraic extension of k (2.6.4). If $s \in (S(\mathfrak{g})/J)^{\mathscr{A}}$, the multiplication by s in $S(\mathfrak{g})/J$ belongs to D. Then $(S(\mathfrak{g})/J)^{\mathscr{A}}$ can be identified with a subalgebra of D, and hence is an algebraic extension of k. Finally, this extension is contained in the algebra $S(\mathfrak{g})/J$ of finite type, hence is of finite type, and consequently of finite degree.

4.8.6. We assume that $\mathfrak{g}$ is nilpotent. Let $\mathscr{A}$ be the adjoint group of $\mathfrak{g}$. An invariant ideal J of $S(\mathfrak{g})$ such that $(S(\mathfrak{g})/J)^{\mathscr{A}} = k$ is termed a *rational invariant ideal* of $S(\mathfrak{g})$. From 4.8.4 and 4.8.5, such an ideal is prime and is maximal invariant.

4.8.7. PROPOSITION ($\mathfrak{g}$ nilpotent). *Let J be a rational invariant ideal of $S(\mathfrak{g})$. Then $S(\mathfrak{g})/J$ is isomorphic to an algebra of polynomials over k.*

The algebra $A = S(\mathfrak{g})/J$ is the algebra of regular functions on an irreducible affine manifold V. Let $\mathscr{A}$ be the adjoint group of $\mathfrak{g}$. The group $\mathscr{A}$ operates regularly in V, and $A^{\mathscr{A}} = k$. Hence A is an algebra of polynomials over k (11.2.1).

4.8.8. Let us assume that $\mathfrak{g}$ is nilpotent. For every extension K of k, we denote the set of rational invariant ideals of $S(\mathfrak{g} \otimes K)$ by $\mathscr{S}(\mathfrak{g}, K)$ or $\mathscr{S}(K)$. Let us denote the adjoint group of $\mathfrak{g}$ by $\mathscr{A}$, and the adjoint group of $\mathfrak{g} \otimes K$, which can be deduced from $\mathscr{A}$ by extension of scalars, by $\mathscr{A}_K$. Let $f: K \to K'$ be a homomorphism of extensions of k. If $J \in \mathscr{S}(K)$, the algebra $(S(\mathfrak{g} \otimes K \otimes_K K')/J \otimes_K K')^{\mathscr{A}_{K'}}$ is isomorphic to $(S(\mathfrak{g} \otimes K)/J)^{\mathscr{A}_K} \otimes_K K'$, and hence $J \otimes_K K' \ni \mathscr{S}(K')$. The mapping $J \mapsto J \otimes_K K'$ of $\mathscr{S}(K)$ into $\mathscr{S}(K')$ is denoted by $\mathscr{S}(f)$. If $f': K' \to K''$ is a homomorphism of extensions of k, then

$$\mathscr{S}(f' \circ f) = \mathscr{S}(f') \circ \mathscr{S}(f).$$

4.8.9. Let K be an extension of k, and $J \in \mathscr{S}(K)$. Then $J \cap S(\mathfrak{g}) = Q$ is a prime invariant ideal of $S(\mathfrak{g})$. The canonical homomorphism of $S(\mathfrak{g})/Q$ into $S(\mathfrak{g} \otimes K)/J$ is injective; with the notation of 4.8.8, it sends $(S(\mathfrak{g})/Q)^{\mathscr{A}}$ to $(S(\mathfrak{g} \otimes K)/J)^{\mathscr{A}_K} = K$; its restriction to $(S(\mathfrak{g})/Q)^{\mathscr{A}}$ can be extended to a homomorphism of extensions Fract $(S(\mathfrak{g})/Q)^{\mathscr{A}} \to K$. We say that Q and ω are *canonically associated with* J.

Let $\mathscr{S}'(K)$ be the set of pairs (Q, ω), where Q is a prime invariant ideal of $S(\mathfrak{g})$ and ω: Fract$(S(\mathfrak{g})/Q)^{\mathscr{A}} \to K$ a homomorphism of extensions. By a simple analogy with 4.7.13 and 4.7.14, we define, for all $(Q, \omega) \in \mathscr{S}'(K)$, a rational invariant ideal of $S(\mathfrak{g} \otimes K)$ which is said to be *canonically asso-*

ciated with (Q, ω), and we verify that we have thus constructed reciprocal bijections, termed *canonical*, of $\mathscr{S}(K)$ onto $\mathscr{S}'(K)$ and of $\mathscr{S}'(K)$ onto $\mathscr{S}(K)$. The property analogous to 4.7.16 is true.

4.8.10. Let Q a prime invariant ideal of $S(\mathfrak{g})$, $Z = (S(\mathfrak{g})/Q)^{\mathscr{A}}$, and $C = \text{Fract}\, Z$. Then Q and the identity mapping of C define a rational invariant ideal, denoted by $Q^\wedge$, of $S(\mathfrak{g} \otimes C)$, namely the kernel of the canonical homomorphism $S(\mathfrak{g} \otimes C) \to (S(\mathfrak{g})/Q) \otimes_Z C$. From 4.8.7, $(S(\mathfrak{g})/Q) \otimes_Z C$ is isomorphic to an algebra of polynomials over C.

The parallelism between this section and section 4.7 will be expressed much more precicely in section 6.3.

The rest of this section is concerned with a different question.

4.8.11. LEMMA ($\mathfrak{g}$ nilpotent). *Let* $(x, y, z, \mathfrak{h})$ *be a reducing quadruple of* $\mathfrak{g}$. *Then* $Y(\mathfrak{g}) \subset Y(\mathfrak{h})$.

Let $(x_1, \ldots, x_n)$ be a basis for $\mathfrak{g}$ such that $(x_1, \ldots, x_{n-1})$ is a basis for $\mathfrak{h}$ and $x_1 = z$, $x_2 = y$, $x_n = x$. Every element of $S(\mathfrak{g})$ can be uniquely written as $f(x_1, \ldots, x_n)$, where $f \in k[X_1, \ldots, X_n]$. If $f(x_1, \ldots, x_n) \in Y(\mathfrak{g})$, then

$$0 = y \cdot f(x_1, \ldots, x_n) = [y, x_1]f'_{x_1} + \cdots + [y, x_n]f'_{x_n} = -zf'_{x_n},$$

and hence $f'_{x_n} = 0$. Then $f \in S(\mathfrak{h})$, whence obviously $f \in Y(\mathfrak{h})$.

4.8.12. PROPOSITION ($\mathfrak{g}$ nilpotent). *Let* φ *be the canonical mapping of* $S(\mathfrak{g})$ *onto* $U(\mathfrak{g})$. *Then* $\varphi \mid Y(\mathfrak{g})$ *is an isomorphism of the algebra* $Y(\mathfrak{g})$ *onto the algebra* $Z(\mathfrak{g})$.

From 2.4.11, it is sufficient to establish that $\varphi \mid Y(\mathfrak{g})$ is multiplicative. This is clear if $\dim \mathfrak{g} \leq 1$. Let us assume that $\dim \mathfrak{g} > 1$, and let us reason by induction on $\dim \mathfrak{g}$. Let $\mathfrak{z}$ be the centre of $\mathfrak{g}$.

Let us assume that $\dim \mathfrak{z} = 1$. Let $(x, y, z, \mathfrak{h})$ be a reducing quadruple of $\mathfrak{g}$. From the induction hypothesis, $\varphi \mid Y(\mathfrak{h})$ is multiplicative. From 4.8.11, $\varphi \mid Y(\mathfrak{g})$ is multiplicative.

Let us assume that $\dim \mathfrak{z} > 1$. Let δ be a one-dimensional subspace of $\mathfrak{z}$; it is an ideal of $\mathfrak{g}$. Let us consider the commutative diagram

$$\begin{array}{ccc}
U(\mathfrak{g}) & \xrightarrow{\theta} & U(\mathfrak{g}/\mathfrak{d}) \\
\varphi^{-1}\big\downarrow & & \big\downarrow\varphi'^{-1} \\
S(\mathfrak{g})) & \xrightarrow{\eta} & S(\mathfrak{g}/\mathfrak{d})
\end{array}$$

where θ, η, φ' are the canonical mappings. Let $f, g \in Z(\mathfrak{g})$. Then $\theta(f), \theta(g) \in Z(\mathfrak{g}/\mathfrak{d})$. From the induction hypothesis, we have

$$\varphi'^{-1}(\theta(f)\,\theta(g)) = \varphi'^{-1}(\theta(f))\varphi'^{-1}(\theta(g)),$$

i.e.

$$\eta(\varphi^{-1}(fg) - \varphi^{-1}(f)\varphi^{-1}(g)) = 0.$$

Thus the element $\varphi^{-1}(fg) - \varphi^{-1}(f)\varphi^{-1}(g)$ of $S(\mathfrak{g})$ is divisible by the elements of $\mathfrak{d}$ and hence finally by all elements of $\mathfrak{z}$. Since dim $\mathfrak{z} > 1$, this implies that

$$\varphi^{-1}(fg) - \varphi^{-1}(f)\varphi^{-1}(g) = 0.$$

References: [29], [99].

4.9. Supplementary remarks

4.9.1. Proposition 4.1.7 is due to Gabriel. Proposition 4.2.2 is due to Solomon and Verma [120]. Proposition 4.3.5 can be found in [31]. Proposition 4.4.8 is due to Bernat [8] and proposition 4.4.11 to Gabriel [51]. Theorem 4.5.7 can be found in [31] for k algebraically closed, and in [51] for the general case. Theorem 4.7.9 is essentially contained in [30]; the appearance of Weyl algebras in these questions goes back to [28] (th. 2) (at least). The remainder of section 4.7 and section 4.8 are due to Gabriel and Nouazé [99] (except for 4.8.12 which is proved in [29]), and in the first place anticipate chapter VI.

4.9.2. Assume that $\mathfrak{g}$ is semi-simple. Let I be a two-sided ideal of $U(\mathfrak{g})$. Then 4.2.2 cannot be extended from $U(\mathfrak{g})$ to $U(\mathfrak{g})/I$. (Take $\mathfrak{g} = \mathfrak{sl}(2,\mathbf{C})$. Let z be a generator of the algebra $Z(\mathfrak{g})$, $\lambda \in \mathbf{C}$, and I the two-sided ideal of $U(\mathfrak{g})$ generated by $z - \lambda$. The centre of $U(\mathfrak{g})/I$ is $\mathbf{C}$ (4.2.5). But compare 4.9.22.) Cf. problem 8.

4.9.3. (a) Proposition 4.2.5 is no longer true if $\mathfrak{g}$ is nilpotent instead of semi-simple. (Take $\mathfrak{g},x,y,z$ as in 1.14.10 and $I = U(\mathfrak{g}) \cdot z$.)

(b) Assume that $\mathfrak{g}$ is nilpotent. Let $\mathfrak{c}$ be the centre of $\mathfrak{g}$. There exists a finite number of one-dimensional subspaces $\mathfrak{c}_1, \ldots, \mathfrak{c}_n$ of $\mathfrak{c}$ with the following property: if $\mathfrak{c}'$ is a one-dimensional subspace of $\mathfrak{c}$ distinct from $\mathfrak{c}_1, \ldots, \mathfrak{c}_n$, and φ designates the canonical mapping of $U(\mathfrak{g})$ onto $U(\mathfrak{g}/\mathfrak{c}')$, then $\varphi(Z(\mathfrak{g}))$ and $Z(\mathfrak{g}/\mathfrak{c}')$ have the same field of fractions, and $\delta(\mathfrak{g}/\mathfrak{c}') = \delta(\mathfrak{g}) - 1$ (where $\delta(\mathfrak{h})$ designates the degree of transcendence of $Z(\mathfrak{h})$ over k) [29].

4.9.4. If $\mathfrak{g} \neq 0$, the semi-centre of $U(\mathfrak{g})$ is never reduced to k. In particular, if the radical of $\mathfrak{g}$ is nilpotent and $\mathfrak{g} \neq 0$, then $Z(\mathfrak{g}) \neq k$ [33]. Cf. 4.9.8. We may have $Z(\mathfrak{g}) = k$ with $\mathfrak{g}$ unimodular $\neq 0$ [146].

4.9.5. Let S be the set of linear forms on $\mathfrak{g}$ which are distinguished relative to the ideal 0. Then S is a semigroup under addition. This semigroup is not always finitely generated [146].

4.9.6. Let $\mathfrak{g}$ be the completely solvable Lie algebra with basis (x,y,z,t) such that

$$[x,y] = y, \qquad [x,z] = -z, \qquad [y,z] = t, \qquad [\mathfrak{g},t] = 0.$$

(a) The Lie algebra $\mathfrak{g}$ is not nilpotent, but the semi-centre of $U(\mathfrak{g})$ is equal to its centre. (Use 4.3.4 with $I = 0$, $\mathfrak{g}' = ky + kz + kt$.)

(b) Let β be the canonical mapping of $S(\mathfrak{g})$ into $U(\mathfrak{g})$. Then $t \in Y(\mathfrak{g})$, $xt + yz \in Y(\mathfrak{g})$, and

$$(\beta(xt + yz))^2 = \beta((xt + yz)^2) + 1/12\, t^2$$

[29].

4.9.7. We equip $K(\mathfrak{g})$ with the adjoint representation. Let

$$D(\mathfrak{g}) = \sum_{\lambda \in \mathfrak{g}^*} K(\mathfrak{g})_\lambda .$$

(a) $D(\mathfrak{g})$ is a commutative subalgebra of $K(\mathfrak{g})$, equal to $C(\mathfrak{g})$ if the radical of $\mathfrak{g}$ is nilpotent.

(b) Let $\lambda \in \mathfrak{g}^*$ be such that $K(\mathfrak{g})_\lambda \neq 0$, and $\mathfrak{g}' = \operatorname{Ker} \lambda$. Then $K(\mathfrak{g})_\mu \subset K(\mathfrak{g}')$ for all $\mu \in \mathfrak{g}^*$.

(c) Assume that k is algebraically closed. If $u \in K(\mathfrak{g})_\lambda$, there exist $\mu \in \mathfrak{g}^*$, $u_1 \in U(\mathfrak{g})_{\lambda+\mu}$ and $u_2 \in U(\mathfrak{g})_\mu$ such that $u = u_1 u_2^{-1}$ ([15], [111]).

4.9.8. Let $\mathfrak{g}$ be the completely solvable Lie algebra with basis (x,y,z) such that $[x,y] = y$, $[x,z] = z$, $[y,z] = 0$. Then $Z(\mathfrak{g}) = k$, but $C(\mathfrak{g})$ is the subfield of $K(\mathfrak{g})$ generated by yz^{-1}.

4.9.9. Assume that $\mathfrak{g}$ is solvable. Let J be a non-null two-sided ideal of $U(\mathfrak{g})$. There exists $\lambda \in \mathfrak{g}^*$ such that $J \cap U(\mathfrak{g})_\lambda \neq 0$. (Argue as in 4.4.1, using in addition a Galois argument and the fact that $U(\mathfrak{g})$ is integral.) Cf. problem 8.

4.9.10. Assume that k is algebraically closed and $\mathfrak{g}$ is solvable. There exists a subset L of $K(\mathfrak{g})$ with the following properties:

(a) L is a maximal commutative subfield of $K(\mathfrak{g})$;

(b) as a field L is generated, by k and a finite number of elements which are algebraically independent over k;

(c) $[\mathfrak{g},L] \subset L$

[96].

4.9.11. Assume that $k = \mathbf{R}$ and $\mathfrak{g}$ is solvable. Then $C(\mathfrak{g})$ is a purely transcendental extension of $\mathbf{R}$ [8]. Cf. problem 40.

4.9.12. Assume that $\mathfrak{g}$ is solvable. Let I be a prime ideal of $U(\mathfrak{g})$, $C = C(\mathfrak{g};I)$, k' be an extension of k, $\mathfrak{g}' = \mathfrak{g} \otimes k'$, $\mathscr{E}$ be the set of prime ideals I' of $U(\mathfrak{g}')$ such that $I' \cap U(\mathfrak{g}) = I$, and $\mathscr{F}$ be the set of prime ideals of $C \otimes k'$.

(a) There exists a canonical bijection β of $\mathscr{E}$ onto $\mathscr{F}$. If $I' \in \mathscr{E}$, then $C(\mathfrak{g}';I')$ is canonically isomorphic to Fract $(C \otimes k'/\beta(I'))$.

(b) If I is rational, then $\mathscr{E}$ reduces to a single rational element ([15], (51]).

4.9.13. Assume that $\mathfrak{g}$ is solvable. Let I be a two-sided ideal of $U(\mathfrak{g})$. For I to be right primitive (i.e. the annihilator of a simple *right* $U(\mathfrak{g})$-module), it is necessary and sufficient that I is primitive [45]. Cf. problem 19.

4.9.14. Let I be a two-sided ideal of $U(\mathfrak{g})$. The following conditions are equivalent:

(i) I is absolutely primitive;

(ii) there exists an algebraically closed extension k' of k such that $I \otimes k'$ is a primitive ideal of $U(\mathfrak{g} \otimes k')$;

(iii) I is primitive and $C(\mathfrak{g};I) = k$.

If g is completely solvable, these conditions are moreover equivalent to the following:

(iv) I is rational;

(v) I is the kernel of an absolutely simple representation.

(Gabriel, unpublished. Use 4.1.7.)

4.9.15. We have Kdim $A_n = n$ [110]. The global homological dimension of A_n is n [113].

4.9.16. (a) Let p and q be the canonical generators of A_1. Then the $p^i q^j$ form a basis for the vector space A_1, which allows us to define the degree of an element of A_1 in an obvious way. Let ϱ be a simple representation of A_1 in V. For all $v \in V$, the annihilator of v in A_1 is not null. We may thus define the *height* of ϱ as the smallest integer $m > 0$ such that there exists an $a \in A_1$ of degree m with Ker $\varrho(a) \neq 0$.

(b) Let $A_1^\wedge$ be the set of classes of simple representations of A_1. The notion of height has a meaning for the elements of $A_1^\wedge$. The group G of automorphisms of A_1 operates in $A_1^\wedge$ by transport of the structure.

(c) Let u be an element of A_1 of degree 1. There exists one, and up to equivalence only one simple representation ϱ_u of A_1 such that $\operatorname{Ker} \varrho_u(u) \neq 0$. ϱ_u is equivalent to $\varrho_{u'}$ if and only if u and u' are proportional. The representation ϱ_p is equivalent to the standard representation. Let G' be the set of the $g \in G$ such that

$$g(p) = \alpha p + \beta q + \gamma, \qquad g(q) = \alpha' p + \beta' q + \gamma',$$

where $\alpha, \beta, \ldots, \gamma' \in k$, $\alpha\beta' - \beta\alpha' = 1$. The group G' operates transitively in the set of elements of $A_1^\wedge$ with height 1.

(d) Let $n \in \mathbf{N}$. Let g be the element of G such that $g(q) = q$ and $g(p) = p + q^n$. Then g transforms the standard representation of A_1 into a representation with height n.

(e) The group G has infinitely many orbits in $A_1^\wedge$ ([30], [34]).

4.9.17. If $\mathfrak{g}$ is nilpotent, the ring $Z(\mathfrak{g})$ is factorial [29]. Cf. also [221].

4.9.18. Assume that $\mathfrak{g}$ is nilpotent. Let $\mathfrak{g}'$ be an ideal of codimension 1 of $\mathfrak{g}$. Then either $Z(\mathfrak{g})$ strictly contains $Z(\mathfrak{g}')$, in which case $Z(\mathfrak{g}') = Z(\mathfrak{g}) \cap U(\mathfrak{g}')$, or $Z(\mathfrak{g}')$ strictly contains $Z(\mathfrak{g})$ [28].

4.9.19. (a) Assume that k is algebraically closed and $\mathfrak{g}$ is nilpotent. Let I be a prime ideal of $U(\mathfrak{g})$, $A = U(\mathfrak{g})/I$, and $Z = Z(\mathfrak{g}; I)$. For all $z \in Z$, let $\mathscr{E}_z$ and $\mathscr{E}'_z$ be the sets of primitive ideals of A and Z respectively, which do not contain z. There exists $z \in Z$ such that the mappings $J \mapsto J \cap Z$ and $K \mapsto AK$ are reciprocal bijections of $\mathscr{E}_z$ onto $\mathscr{E}'_z$ and of $\mathscr{E}'_z$ onto $\mathscr{E}_z$.

(b) Let $\mathfrak{g}$ be the nilpotent Lie algebra with a basis (x,y,z,t,u) such that

$$[x,y] = t, \qquad [x,z] = u, \qquad [y,z] = 0, \qquad [\mathfrak{g},t] = [\mathfrak{g},u] = 0.$$

Then $Z(\mathfrak{g})$ is the algebra generated by the algebraically independent elements t,u and $yu - zt$. Let K and K' be the ideals of $Z(\mathfrak{g})$ generated by t,u and $t,u,\ yu - zt$ respectively. Then

$$(U(\mathfrak{g})K) \cap Z(\mathfrak{g}) = (U(\mathfrak{g})K') \cap Z(\mathfrak{g}).$$

In particular, $U(\mathfrak{g})$ is not a free module over $Z(\mathfrak{g})$ ([32], [98], [99]).

(c) For an extension of (a) to the solvable case, see [15], [93].

4.9.20. (a) Let $\mathfrak{a}$ and $\mathfrak{b}$ be commutative Lie algebras. We are given a linear mapping of $\mathfrak{a}$ into $\operatorname{End}(\mathfrak{b})$. whence a semi-direct product $\mathfrak{c} = \mathfrak{a} \oplus \mathfrak{b}$. For every $x \in \mathfrak{a}$, let M_x be the image of $\operatorname{ad}_{\mathfrak{c}} x$; then $M_x \subset \mathfrak{b}$. Let $M_x^\perp$ be the orthogonal subspace of M_x in $\mathfrak{b}^*$. Let us make the following hypothesis:

(★) $$b^* - \bigcup_{\substack{x \in a \\ x \neq 0}} M_x^{\perp} \quad \text{is dense in } \mathfrak{b}^*.$$

Then $Y(\mathfrak{c})$ is the set of $\mathfrak{a}$-invariant elements of $S(\mathfrak{b})$.

(b) Let $\mathfrak{a},\mathfrak{b},\mathfrak{c},M_x$ be as in (a). Assume that

(1) $\mathrm{ad}_{\mathfrak{c}} x \neq 0$ for all $x \in \mathfrak{a} - \{0\}$;

(2) there exists a basis $(e_1, \ldots, e_n)$ for $\mathfrak{b}$ such that, for all $x \in \mathfrak{a}$, M_x is generated by some of the e_i. Then condition (★) of (a) is satisfied.

(c) Let

$$a_{11}, a_{12}, \ldots, a_{1,16}$$

$$a_{21}, a_{22}, \ldots, a_{2,16}$$

$$a_{31}, a_{32}, \ldots, a_{3,16}$$

be complex numbers which are algebraically independent over $\mathbf{Q}$. Let $V = \mathbf{C}^{32}$. Let $\mathfrak{g}$ the set of matrices $\begin{pmatrix} 0 & 0 \\ X & 0 \end{pmatrix} \in \mathfrak{gl}(V)$, where X is a diagonal matrix with 16 rows and 16 columns whose diagonal elements $x_1, \ldots, x_{16}$ satisfy $a_{i1}x_1 + \cdots + a_{i,16}x_{16} = 0$ for $i = 1,2,3$. Then $\mathfrak{g}$ operates in $S(V)$ by derivations, and the algebra I of invariants of $S(V)$ is not of finite type. (M. Nagata, *Proc. Int. Congress*, 1958, Cambridge University Press, pp. 459—462.)

Let $\mathfrak{h}$ be the semi-direct product of $\mathfrak{g}$ and V defined by the action of $\mathfrak{g}$ in V (it is a nilpotent Lie algebra of dimension 45.) Then the algebra $Z(\mathfrak{h})$ is not of finite type. (By using 4.8.12 and (b), it can be shown that $Z(\mathfrak{h})$ is isomorphic to I.)

(d) Let $\mathfrak{g}$ be a nilpotent Lie algebra and $a_1, \ldots, a_q$ elements of $Z(\mathfrak{g})$ such that $a_1, \ldots, a_q, k$ generate the field $C(\mathfrak{g})$. There exists a non-zero element a of $k[a_1, \ldots, a_q]$ such that $Z(\mathfrak{g}) \subset k[a_1, \ldots, a_q, a^{-1}]$ [28].

(e) The calculation of $Z(\mathfrak{g})$ for nilpotent Lie algebras of dimension ≤ 5 and for the Lie algebra of strictly lower triangular matrices can be found in *Can. J. Math.* **10** (1958) 321—348 and **11** (1959) 321—344.

4.9.21. (a) If $\mathfrak{g} = \mathfrak{sl}(n,k)$, or $\mathfrak{g} = \mathfrak{gl}(n,k)$, then $K(\mathfrak{g})$ is isomorphic to the field of fractions of $A_p(C(\mathfrak{g}))$ with $p = n(n-1)/2$. Cf. 4.7.18 and problem 3.

(b) We take for $\mathfrak{g}$ the complex solvable algebra with basis (x,y,z) such that $[x,y] = y$, $[x,z] = \alpha z$, $[y,z] = 0$, with $\alpha \in \mathbf{C}$, $\alpha \notin \mathbf{Q}$. Then there does not exist an extension k' of $\mathbf{C}$ and an integer p such that $K(\mathfrak{g})$ is isomorphic to the field of fractions of $A_p(k')$ [52].

4.9.22. Let $\mathfrak{g} = \mathfrak{sl}(2,\mathbf{C})$, and let h,e,f be as in 1.8. Let $Q = 4ef + h^2 - 2h \in Z(\mathfrak{g})$.

(a) For all $\lambda \in \mathbf{C}$, let $I_\lambda = U(\mathfrak{g})(\mathbf{Q} - \lambda)$. Then the mapping $\lambda \mapsto I_\lambda$ is a bijection of $\mathbf{C}$ onto the set of primitive ideals of $U(\mathfrak{g})$ of infinite codimension. If λ is not of the form $n^2 + 2n$ ($n \in \mathbf{N}$), then I_λ is a maximal two-sided ideal of $U(\mathfrak{g})$. If $\lambda = n^2 + 2n$ ($n \in \mathbf{N}$), there exists one and only one two-sided ideal I'_λ of $U(\mathfrak{g})$ such that $I_\lambda \subset I'_\lambda \subset U(\mathfrak{g})$ and $I_\lambda \neq I'_\lambda \neq U(\mathfrak{g})$. This ideal I'_λ is of codimension $(n+1)^2$ and it is the kernel of the $(n+1)$-dimensional simple representation of $U(\mathfrak{g})$. The I'_λ are the primitive ideals of finite codimension of $U(\mathfrak{g})$. The non-null prime ideals of $U(\mathfrak{g})$ are primitive [99].

(b) For $\lambda,\lambda' \in \mathbf{C}$ and $\lambda \neq \lambda'$, the algebras $U(\mathfrak{g})/I_\lambda$ and $U(\mathfrak{g})/I_{\lambda'}$ are not isomorphic [41].

4.9.23. Assume that $\mathfrak{g}$ is completely solvable. Let I be a prime ideal of $U(\mathfrak{g})$, $Z = Z(\mathfrak{g};I)$, and $C = C(\mathfrak{g};I)$. Let E be the union of the $(U(\mathfrak{g})/I_\lambda - \{0\}$ when λ runs through $\mathfrak{g}^*$.

(a) The algebra $(U(\mathfrak{g})/I)_E$ exists (cf. 3.3.9). Similarly, if $e \in E$ and $T = \{1,e,e^2,\ldots\}$, the algebra $(U(\mathfrak{g})/I)_T$, which we shall denote by $(U(\mathfrak{g})/I_e)$, also exists.

(b) The algebra $(U(\mathfrak{g})/I)_E$ is simple and has centre C.

(c) I is primitive if and only if there exists $e \in E$ such that the algebra $(U(\mathfrak{g})/I)_e$ is simple.

(d) Let V be a finite-dimensional vector space, δ an alternating bilinear form on V, and G an additive subgroup of finite type of V^*. Let Φ be the quotient of the tensor algebra of V by the two-sided ideal which is generated by elements of the form

$$x \otimes y - y \otimes x - \delta(x,y) \qquad (x,y \in V).$$

Then $V \subset \Phi$. For all $g \in G$, the mapping $v \mapsto v + g(v)$ of V into Φ can be uniquely extended to an element $\theta(g)$ of $\mathrm{Aut}(\Phi)$, and θ is a homomorphism of G into $\mathrm{Aut}(\Phi)$. The cross product of Φ by G which is defined by θ is denoted by $A(V,\delta,G)$. Let V^δ be the kernel of δ, V^G the orthogonal subspace of G in V, and $V^{\delta G}$ the kernel of $\delta | V^G$. Then $A = A(V,\delta,G)$ is simple if and only if $V^G \cap V^\delta = 0$. The integers $\dim(V)$, $\dim(V^G)$, $\dim(V^{\delta G})$ and $\mathrm{rank}(G)$ only depend on the algebra A and not on its presentation of the form $A(V,\delta,G)$.

(e) The algebra $(U(\mathfrak{g})/I)_E$ is isomorphic to an algebra $A(V,\delta,G)$ over the field C, where, from (d), there are four other integers which intrinsically correspond to I.

(f) If the semi-centre of $U(\mathfrak{g})/I$ is equal to Z, there exists $e \in Z - \{0\}$ such that $(U(\mathfrak{g})/I)_e$ is isomorphic to the tensor product of a Weyl algebra and Z.

(g) If $\mathfrak{g}$ is algebraic, there exist $m,n \in \mathbf{N}$ and $e \in E$ such that $(U(\mathfrak{g})/I)_e$ is isomorphic to $A'_m \otimes A_n \otimes Z'$, where Z' is the centre of $(U(\mathfrak{g})/I)_e$. (In the Weyl algebra A_m, with canonical generators $p_1, q_1, \ldots, p_m, q_m$, let S be the set of monomials with respect to $q_1, \ldots, q_m$; denote the algebra $(A_m)_S$ by A'_m.) The field Fract$(U(\mathfrak{g})/I)$ is isomorphic to Fract $A_{m+n}(C)$. Cf. 4.9.21.

(h) Let Λ be the set of linear forms on $\mathfrak{g}$ which are distinguished relative to I, $\mathfrak{g}'$ the intersection of the Ker λ for $\lambda \in \Lambda$, and $I' = I \cap U(\mathfrak{g}')$. Then the semi-centre of $U(\mathfrak{g})/I$ is contained in $Z(\mathfrak{g}'; I')$, and is equal to $Z(\mathfrak{g}'; I')$ if $\mathfrak{g}$ is algebraic ([15], [72], [93], [94]).

4.9.24. Let $R(\mathfrak{g}) =$ Fract $S(\mathfrak{g})$. The adjoint representation of $\mathfrak{g}$ in $S(\mathfrak{g})$ can be extended to a representation of $\mathfrak{g}$ by derivations of $R(\mathfrak{g})$. Let $Y'(\mathfrak{g})$ be the set of $\mathfrak{g}$-invariant elements of $R(\mathfrak{g})$, r the index of $\mathfrak{g}$, $f \in \mathfrak{g}^*$, be $\mathfrak{z}$ the centre of $\mathfrak{g}^f$, and $Y''(\mathfrak{g})$ the set of those elements of $Y'(\mathfrak{g})$ which, when interpreted as rational functions on $\mathfrak{g}^*$, are defined at f. If $v \in Y''(\mathfrak{g})$, the differential of v at f can be identified with an element $\varphi(v)$ of $\mathfrak{g}$.

(a) We have $\varphi(Y''(\mathfrak{g})) \subset \mathfrak{z}$. The inclusion can be strict even for $\mathfrak{g}$ nilpotent.

(b) Let us assume that $\mathfrak{g}$ is algebraic. Then $Y'(\mathfrak{g})$ has r as its degree of transcendence. If f belongs to a suitable open set in $\mathfrak{g}^*$, there exist $v_1, \ldots, v_r \in Y''(\mathfrak{g})$ whose differentials at f are linearly independent, and $\varphi(Y''(\mathfrak{g})) = \mathfrak{g}^f = \mathfrak{z}$.

(c) If $\mathfrak{g}$ is semi-simple and there exist r elements of $Y(\mathfrak{g})$ whose differentials at f are linearly independent, then f is regular. This does not hold if $\mathfrak{g}$ is nilpotent ([28], [49], [78]).

4.9.25. For explicit generators of $U(\mathfrak{sl}(n,k))$, cf. [141].

CHAPTER 5

INDUCED REPRESENTATIONS

5.1. Induced representations

5.1.1. Let $\mathfrak{h}$ be a Lie subalgebra of $\mathfrak{g}$ and W an $\mathfrak{h}$-module. We may consider $U(\mathfrak{g})$ as a right $U(\mathfrak{h})$-module, and hence can form the left $U(\mathfrak{g})$-module $V = U(\mathfrak{g}) \otimes_{U(\mathfrak{h})} W$. We shall term it the $\mathfrak{g}$-*module induced by* W, and denote it by $\operatorname{ind}(W,\mathfrak{h} \uparrow \mathfrak{g})$ or simply by $\operatorname{ind}(W,\mathfrak{g})$. If the representation of $\mathfrak{h}$ corresponding to W is designated by ϱ and that of $\mathfrak{g}$ corresponding to V by π, we say that π is the *representation of* $\mathfrak{g}$ *induced by* ϱ, and denote it by $\operatorname{ind}(\varrho,\mathfrak{h} \uparrow \mathfrak{g})$, or simply by $\operatorname{ind}(\varrho,\mathfrak{g})$.

5.1.2. Since $U(\mathfrak{g})$ is a free right $U(\mathfrak{h})$-module, the mapping $w \mapsto 1 \otimes w$ of W into V is injective; it is obviously an $\mathfrak{h}$-homomorphism. W can be identified with a sub-$\mathfrak{h}$-module of V under this mapping. The $\mathfrak{g}$-module V is generated by W. This embedding of W in V is universal in the following sense:

5.1.3. PROPOSITION. *Let $\mathfrak{h}$ be a Lie subalgebra of $\mathfrak{g}$, W an $\mathfrak{h}$-module, and V the $\mathfrak{g}$-module induced by W. Let V' be a $\mathfrak{g}$-module, and ψ an $\mathfrak{h}$-homomorphism of W into V'. Then ψ can be uniquely extended to a $\mathfrak{g}$-homomorphism φ of V into V'. The mapping $\psi \mapsto \varphi$ is a bijection of* $\operatorname{Hom}_{\mathfrak{h}}(W,V')$ *onto* $\operatorname{Hom}_{\mathfrak{g}}(V,V')$.

This follows from the general properties of the tensor product.

5.1.4. PROPOSITION. *Let $\mathfrak{h}$ be a Lie subalgebra of $\mathfrak{g}$, let W and W' be $\mathfrak{h}$-modules, let V and V' be the induced $\mathfrak{g}$-modules, and let ψ be an $\mathfrak{h}$-homomorphism of W into W'. Then ψ can be uniquely extended to a $\mathfrak{g}$-homomorphism φ of V into V'. The passage from φ to ψ transforms exact sequences into exact sequences.*

This follows from the general properties of the tensor product, and from the fact $U(\mathfrak{g})$ is a free right $U(\mathfrak{h})$-module.

5.1.5. In particular, if W' is a sub-$\mathfrak{h}$-module of W, then V' can be identified with a sub-$\mathfrak{g}$-module of V. We have $V' \cap W = W'$. The $\mathfrak{g}$-module V/V' can be identified with $\operatorname{ind}(W/W',\mathfrak{g})$.

5.1.6. Let $\mathfrak{h}$ be a Lie subalgebra of $\mathfrak{g}$, W an $\mathfrak{h}$-module, and $(e_1, \ldots, e_n)$ a basis for a complement of $\mathfrak{h}$ in $\mathfrak{g}$. The $e^\nu = e_1^{\nu_1} \cdots e_n^{\nu_n}$, where $\nu = (\nu_1, \ldots, \nu_n) \in \mathbf{N}^n$, form a basis for the right $U(\mathfrak{h})$-module $U(\mathfrak{g})$. If we set $V = \operatorname{ind}(W,\mathfrak{g})$, we thus have

$$V = \bigoplus_{\nu \in \mathbf{N}^n} e^\nu \otimes W.$$

Let π be the representation of $\mathfrak{g}$ corresponding to V. If $w \in W$, then

$$e^\nu \otimes w = \pi(e^\nu)w.$$

Hence $\pi(e^\nu)|W$ is an isomorphism of the vector space W onto the vector space $\pi(e^\nu)(W)$, and V is the direct sum of the $\pi(e^\nu)(W)$ for ν running through $\mathbf{N}^n$.

If $u \in U(\mathfrak{g})$, then $\pi(u)$ can be calculated in the following way. For $\nu \in \mathbf{N}^n$, we write

$$ue^\nu = \sum_{\mu \in \mathbf{N}^n} e^\mu u_{\mu\nu},$$

where the $u_{\mu\nu}$ belong to $U(\mathfrak{h})$; then, if $w \in W$, we have

$$\pi(u)(\pi(e^\nu)w) = \Big(\sum_\mu e^\mu u_{\mu\nu}\Big)(1 \otimes w)$$

$$= \sum_\mu e^\mu \otimes u_{\mu\nu} w = \sum_\mu \pi(e^\mu)(u_{\mu\nu} w).$$

5.1.7. PROPOSITION. *Let $\mathfrak{h}$, ϱ, W, π and V be as in 5.1.1, so that $W \subset V$. Let J be the kernel of ϱ [in $U(\mathfrak{h})$].*

(i) *The annihilator of W in $U(\mathfrak{g})$ is left ideal $U(\mathfrak{g})J$.*

(ii) *The kernel of π [in $U(\mathfrak{g})$] is the largest two-sided ideal of $U(\mathfrak{g})$ contained in $U(\mathfrak{g})J$.*

We use the notation of 5.1.6. Let $u = \sum_{\nu \in N^n} e^\nu u_\nu$ [where the u_ν belong to $U(\mathfrak{h})$] be an element of $U(\mathfrak{g})$. Then

$$u(W) = 0 \Leftrightarrow \sum e^\nu \otimes (u_\nu W) = 0$$

$$\Leftrightarrow u_\nu W = 0 \quad \text{for all } \nu$$

$$\Leftrightarrow u_\nu \in J \quad \text{for all } \nu.$$

On the other hand, $U(\mathfrak{g})J = \bigoplus e^\nu J$, whence (i).

Let $v \in U(\mathfrak{g})$. Then

$$v \in \operatorname{Ker} \pi \Leftrightarrow v(U(\mathfrak{g})W) = 0$$
$$\Leftrightarrow vU(\mathfrak{g}) \subset U(\mathfrak{g})J \quad \text{from (i)}$$
$$\Leftrightarrow U(\mathfrak{g})vU(\mathfrak{g}) \subset U(\mathfrak{g})J,$$

whence (ii).

5.1.8. LEMMA. *Let $f\colon \mathfrak{g} \to k$ be a one-dimensional representation of $\mathfrak{g}$, N the kernel of f in $U(\mathfrak{g})$, and L and R the left and right ideals respectively of $U(\mathfrak{g})$ generated by the $x - f(x)$, where $x \in \mathfrak{g}$. Then $N = L = R$.*

We have $L \subset N$. On the other hand, let $(x_1, \ldots, x_n)$ be a basis for $\mathfrak{g}$ such that $f(x_2) = \cdots = f(x_n) = 0$. Every element $x_1^{\nu_1} \cdots x_n^{\nu_n}$ such that $\nu_2 + \cdots + \nu_n > 0$ belongs to L, and $x_1^{\nu_1} - f(x_1)_1^{\nu_1} \in L$. Hence L is of co-dimension $\leqq 1$ in $U(\mathfrak{g})$, so that $L = N$. Similarly, we see that $R = N$.

5.1.9. PROPOSITION. *Let $\mathfrak{h}, \varrho, W, \pi$ and V be as in 5.1.1, so that $W \subset V$. Let w be a generator of the $U(\mathfrak{h})$-module W, and L the annihilator of w in $U(\mathfrak{h})$.*

(i) *The mapping φ of $U(\mathfrak{g})$ into V defined by $\varphi(u) = uw$ for all $u \in U(\mathfrak{g})$ is surjective and has kernel $U(\mathfrak{g}) \cdot L$.*

(ii) *Let ψ be the mapping of $U(\mathfrak{g})/U(\mathfrak{g}) \cdot L$ into V deduced from φ by passage to the quotient. Then ψ is a $\mathfrak{g}$-module isomorphism ($U(\mathfrak{g})/U(\mathfrak{g}) \cdot L$ being provided with the left regular representation).*

(iii) *If ϱ is one-dimensional and can hence be identified with a linear form on $\mathfrak{h}$, then $U(\mathfrak{g}) \cdot L$ is the left ideal of $U(\mathfrak{g})$ generated by the $x - \varrho(x)$, where x runs through $\mathfrak{h}$.*

We have $U(\mathfrak{h})w = W$, hence $U(\mathfrak{g})w \supset U(\mathfrak{g})W = V$, so that φ is surjective. We use the notation of 5.1.6. Let $u = \bigwedge_{\nu \in N} e^{\nu} u_\nu$ [where the u_ν belong to $U(\mathfrak{h})$] be an element of $U(\mathfrak{g})$. Then

$$uw = 0 \Leftrightarrow \sum e^{\nu} \otimes u_\nu w = 0$$
$$\Leftrightarrow u_\nu w = 0 \quad \text{for all } \nu$$
$$\Leftrightarrow u_\nu \in L \quad \text{for all } \nu$$
$$\Leftrightarrow u \in U(\mathfrak{g}) \cdot L,$$

hence $\operatorname{Ker} \varphi = U(\mathfrak{g}) \cdot L$. Clearly, φ is a $\mathfrak{g}$-homomorphism [$U(\mathfrak{g})$ being provided with the left regular representation], whence (ii). Assertion (iii) follows from 5.1.8.

5.1.10. Proposition. *Let* $\mathfrak{h}$, *W and V be as in* 5.1.1. *If V is simple (absolutely simple); then W is simple (absolutely simple).*

If V is simple, then W is simple from 5.1.5. The rest can be deduced from this by extension of the base field.

5.1.11. Proposition. *Let* $\mathfrak{h}$ *be a Lie subalgebra of* $\mathfrak{g}$, $\mathfrak{k}$ *a Lie subalgebra of* $\mathfrak{h}$, σ *a representation of* $\mathfrak{k}$, $\varrho = \operatorname{ind}(\sigma,\mathfrak{h})$, *and* $\pi = \operatorname{ind}(\varrho,\mathfrak{g})$. *Then* π *is equivalent to* $\operatorname{ind}(\sigma,\mathfrak{g})$.

Indeed, if W is the $\mathfrak{k}$-module corresponding to σ, the $U(\mathfrak{g})$-module $U(\mathfrak{g}) \otimes_{U(\mathfrak{h})} (U(\mathfrak{h}) \otimes_{U(\mathfrak{k})} W)$ is canonically isomorphic to

$$(U(\mathfrak{g}) \otimes_{U(\mathfrak{h})} U(\mathfrak{h})) \otimes_{U(\mathfrak{k})} W = U(\mathfrak{g}) \otimes_{U(\mathfrak{k})} W.$$

5.1.12. Proposition. *Let* $\mathfrak{h}$ *be a Lie subalgebra of* $\mathfrak{g}$, $\mathfrak{n}$ *an ideal of* $\mathfrak{g}$ *contained in* $\mathfrak{h}$, $\mathfrak{g}_1 = \mathfrak{g}/\mathfrak{n}$ *and* $\mathfrak{h}_1 = \mathfrak{h}/\mathfrak{n}$, $\eta : \mathfrak{g} \to \mathfrak{g}_1$ *and* $\zeta : \mathfrak{h} \to \mathfrak{h}_1$ *be the canonical mappings,* ϱ_1 *be a representation of* $\mathfrak{h}$, *and let*

$$\varrho = \varrho_1 \circ \zeta, \qquad \pi_1 = \operatorname{ind}(\varrho_1,\mathfrak{g}_1), \qquad \pi = \operatorname{ind}(\varrho,\mathfrak{g}).$$

Then π *is equivalent to* $\pi_1 \circ \eta$.

Let W be the space of ϱ and ϱ_1. The space of π_1 is $U(\mathfrak{g}_1) \otimes_{U(\mathfrak{h}_1)} W$ and that of π is $U(\mathfrak{g}) \otimes_{U(\mathfrak{h})} W$. The canonical mapping φ of $U(\mathfrak{g})$ onto $U(\mathfrak{g}_1)$ defines a mapping ψ of $U(\mathfrak{g}) \otimes_{U(\mathfrak{h})} W$ onto $U(\mathfrak{g}_1) \otimes_{U(\mathfrak{h}_1)} W$ which is compatible with the left module structures on $U(\mathfrak{g})$ and $U(\mathfrak{g}_1)$. It is then sufficient to prove that ψ is bijective. We use the notation of 5.1.6. Then $(\eta e_1, \ldots, \eta e_n)$ is a basis for a complement of $\mathfrak{h}_1$ in $\mathfrak{g}_1$, $U(\mathfrak{g}) \otimes_{U(\mathfrak{h})} W$ is the direct sum of the $e^\nu \otimes W$, and $U(\mathfrak{g}_1) \otimes_{U(\mathfrak{h}_1)} W$ is the direct sum of the $\varphi(e^\nu) \otimes W$, whence our assertion follows.

5.1.13. Proposition. *Let* $\mathfrak{h}$ *be a Lie subalgebra of* $\mathfrak{g}$, $\mathfrak{n}$ *an ideal of* $\mathfrak{g}$ *contained in* $\mathfrak{h}$, ϱ *a representation of* $\mathfrak{h}$ *such that* $\varrho([\mathfrak{g},\mathfrak{n}]) = 0$ *and* $\pi = \operatorname{ind}(\varrho,\mathfrak{g})$. *Then* $\pi|\mathfrak{n}$ *is a multiple of* $\varrho|\mathfrak{n}$.

Let $n \in \mathfrak{n}$ and $x_1, \ldots, x_p \in \mathfrak{g}$. Then

$$nx_1x_2 \cdots x_p \in x_1x_2 \cdots x_pn + U(\mathfrak{g})\,[\mathfrak{g},\mathfrak{n}]. \tag{1}$$

Indeed, this is obvious for $p = 0$. If (1) is true for p and if $x_0 \in \mathfrak{g}$, then

$$\begin{aligned} nx_0x_1 \cdots x_p &= x_0nx_1 \cdots x_p + [n,x_0]\, x_1 \cdots x_p \\ &\in x_0x_1 \cdots x_pn + x_0U(\mathfrak{g})\,[\mathfrak{g},\mathfrak{n}] + x_1 \cdots x_p[n,x_0] + U(\mathfrak{g})\,[\mathfrak{g},\mathfrak{n}] \\ &\subset x_0x_1 \cdots x_pn + U(\mathfrak{g})[\mathfrak{g},\mathfrak{n}]. \end{aligned}$$

Given this, if $n \in \mathfrak{n}$, $u \in U(\mathfrak{g})$ and w belongs to the space of ϱ, then $nu = un + \sum_i u_i n_i$, where $u_i \in U(\mathfrak{g})$ and $n_i \in [\mathfrak{g},\mathfrak{n}]$ for all i, hence

$$\pi(n)\,(u \otimes w) = nu \otimes w = \Big(un + \sum_i u_i n_i\Big) \otimes w$$

$$= u \otimes \varrho(n)w + \sum_i u_i \otimes \varrho(n_i)w = u \otimes \varrho(n)w,$$

hence $\pi(n) = 1 \otimes \varrho(n)$.

5.1.14. PROPOSITION. *Let* $\mathfrak{h}$ *and* $\mathfrak{k}$ *be Lie subalgebras of* $\mathfrak{g}$ *such that* $\mathfrak{g} = \mathfrak{h} + \mathfrak{k}$. *Let* $\mathfrak{l} = \mathfrak{h} \cap \mathfrak{k}$, ϱ *be a representation of* $\mathfrak{h}$, *and* $\pi = \operatorname{ind}(\varrho,\mathfrak{g})$. *Then* $\pi|\mathfrak{k}$ *is equivalent to* $\operatorname{ind}(\varrho|\mathfrak{l}, \mathfrak{k})$.

Taking 2.2.9 into account, we have the canonical vector space isomorphisms

$$U(\mathfrak{k}) \otimes_{U(\mathfrak{l})} W \xrightarrow{\varphi} U(\mathfrak{k}) \otimes_{U(\mathfrak{l})} U(\mathfrak{h})) \otimes_{U(\mathfrak{h})} W \xrightarrow{\psi} U(\mathfrak{g}) \otimes_{U(\mathfrak{h})} W.$$

If $u \in U(\mathfrak{k})$ and $w \in W$, then

$$\varphi(u \otimes w) = u \otimes 1 \otimes w, \qquad \psi(u \otimes 1 \otimes w) = u \otimes w,$$

hence $\psi \circ \varphi$ is a $\mathfrak{k}$-module isomorphism.

5.1.15. PROPOSITION. *Let* $\mathfrak{h}$ *be a Lie subalgebra of* $\mathfrak{g}$, ϱ *a representation of* $\mathfrak{h}$, $\pi = \operatorname{ind}(\varrho,\mathfrak{g}), f \in \mathfrak{g}^*$ *a one-dimensional representation of* $\mathfrak{g}, f' = f|\mathfrak{h}, \varrho' = \varrho \otimes f'$, *and* $\pi' = \operatorname{ind}(\varrho,'\mathfrak{g})$. *Then* π' *is equivalent to* $\pi \otimes f$.

Let Z,Z' be the $\mathfrak{h}$-modules corresponding to ϱ, ϱ' and W the common underlying vector space of Z and Z'. If $u \in U(\mathfrak{g})$ and $w \in W$, we denote by $u \otimes w$ and $u \otimes' w$ the tensor products of u and w calculated in $U(\mathfrak{g}) \otimes_{U(\mathfrak{h})} Z$ and $U(\mathfrak{g}) \otimes_{U(\mathfrak{h})} Z'$. We use the notation of 2.2.23, and write

$$F(u,w) = \alpha_{-f}(u) \otimes' w.$$

For all $y \in \mathfrak{h}$, we have

$$F(uy,w) = \alpha_{-f}(uy) \otimes' w = \alpha_{-f}(u)\alpha_{-f'}(y) \otimes' w$$

$$= \alpha_{-f}(u) \otimes' \varrho'(\alpha_{-f'}(y))w$$

$$= \alpha_{-f}(u) \otimes' \varrho(y)w = F(u,\varrho(y)w),$$

hence there exists a linear mapping Φ of $U(\mathfrak{g}) \otimes_{U(\mathfrak{h})} Z$ (which is the space of π and of $\pi \otimes f$) into $U(\mathfrak{g}) \otimes_{U(\mathfrak{h})} Z'$ (which is the space of π') such that $\Phi(u \otimes w) = \alpha_{-f}(u) \otimes' w$ for $u \in U(\mathfrak{g})$, $w \in W$. This mapping is bijective from 2.2.7 and 2.2.23. If $x \in \mathfrak{g}$, then

$$\Phi(\pi \otimes f)(x)\,(u \otimes w) =$$

$$= \Phi(\pi(x)\, u \otimes w + f(x)u \otimes w) = \Phi(xu \otimes w + f(x)u \otimes w)$$

$$= \alpha_{-f}((x + f(x))u) \otimes' w = x\alpha_{-f}(u) \otimes' w$$

$$= \pi'(x)\,(\alpha_{-f}(u) \otimes' w) = \pi'(x)\,\Phi(u \otimes w),$$

hence $\Phi \circ (\pi \otimes f)(x) = \pi'(x) \circ \Phi$.

5.1.16. Proposition. *Let* $\mathfrak{h}$ *be a Lie subalgebra of* $\mathfrak{g}$, ϱ *a simple representation of* $\mathfrak{h}$, ξ *a finite-dimensional simple representation of* $\mathfrak{g}$, $m = \operatorname{mtp}(\varrho,\xi)$, $\pi = \operatorname{ind}(\varrho,\mathfrak{g})$, *and* V *be the space of* π.

(i) *Let* $\mathscr{F}$ *be the set of the sub-*$\mathfrak{g}$*-modules* T *of* V *such that* V/T *is isotypic of type* ξ. *Then* $\mathscr{F}$ *has a smallest element* T_0.

(ii) $\operatorname{mtp}(\xi,V/T_0) = m$.

Let Z be the space of ξ, and W the space of ϱ. If T is an element of $\mathscr{F}$, then $\operatorname{Hom}_{\mathfrak{g}}(V/T,Z)$ can be identified with a vector subspace of $\operatorname{Hom}_{\mathfrak{g}}(V,Z)$. We then have, from 5.1.3,

$$\dim \operatorname{Hom}_{\mathfrak{g}}(V/T,Z) \leqq \dim \operatorname{Hom}_{\mathfrak{h}}(W,Z) = m.$$

Consequently, $\dim(V/T) \leqq \dim \xi$.

Since the intersection of two elements of $\mathscr{F}$ is also an element of $\mathscr{F}$, we see that $\mathscr{F}$ has a least element T_0.

If f is a $\mathfrak{g}$-homomorphism of V into Z, then $\operatorname{Ker} f \in \mathscr{F}$. Hence $\operatorname{Hom}_{\mathfrak{g}}(V,Z)$ can be identified with $\operatorname{Hom}_{\mathfrak{g}}(V/T_0,Z)$. The integer $\operatorname{mtp}(\xi,V/T_0)$ is at the same time the dimension of $\operatorname{Hom}_{\mathfrak{g}}(Z,V/T_0)$ and of $\operatorname{Hom}_{\mathfrak{g}}(V/T_0,Z)$, hence of $\operatorname{Hom}_{\mathfrak{g}}(V,Z)$, hence of $\operatorname{Hom}_{\mathfrak{h}}(W,Z)$ from 5.1.3.

References: [30], [67].

5.2. Twisted induced representations

5.2.1. Let E be a finite-dimensional vector space, F a vector subspace of E, and u an endomorphism of E under which F is stable. We denote the scalar $\operatorname{tr} u - \operatorname{tr}(u|F)$ by $\operatorname{tr}_{E/F} u$.

If $\mathfrak{h}$ is a Lie subalgebra of $\mathfrak{g}$, we write, for all $x \in \mathfrak{h}$.

$$\theta_{\mathfrak{g},\mathfrak{h}}(x) = \tfrac{1}{2} \operatorname{tr}_{\mathfrak{g}/\mathfrak{h}} \operatorname{ad}_{\mathfrak{g}} x.$$

Then $\theta_{\mathfrak{g},\mathfrak{h}}$ is a linear form on $\mathfrak{h}$ which is zero on $[\mathfrak{h},\mathfrak{h}]$. If $\mathfrak{h}$ is an ideal of $\mathfrak{g}$, then $\theta_{\mathfrak{g},\mathfrak{h}} = 0$. If $\mathfrak{k}$ is a Lie subalgebra of $\mathfrak{h}$, then

$$\theta_{\mathfrak{g},\mathfrak{k}}(x) = \theta_{\mathfrak{g},\mathfrak{h}}(x) + \theta_{\mathfrak{h},\mathfrak{k}}(x) \quad \text{for all } x \in \mathfrak{k}.$$

5.2.2. Let $\mathfrak{h}$ be a Lie subalgebra of $\mathfrak{g}$, and ϱ a representation of $\mathfrak{h}$ in W. For all $x \in \mathfrak{h}$, we write

$$\varrho^{\sim}(x) = \varrho(x) + \theta_{\mathfrak{g},\mathfrak{h}}(x) \cdot 1.$$

Then $\varrho^{\sim}$ is a representation of $\mathfrak{h}$ in W. The representation $\operatorname{ind}(\varrho^{\sim},\mathfrak{g})$ is termed the *twisted representation of* $\mathfrak{g}$ *induced by* ϱ and is denoted by $\operatorname{ind}^{\sim}(\varrho,\mathfrak{h}\uparrow\mathfrak{g})$ or $\operatorname{ind}^{\sim}(\varrho,\mathfrak{g})$. The corresponding $\mathfrak{g}$-module is termed *the twisted* $\mathfrak{g}$*-module induced by the* $\mathfrak{h}$*-module* W and is denoted by $\operatorname{ind}^{\sim}(W,\mathfrak{h}\uparrow\mathfrak{g})$ or $\operatorname{ind}^{\sim}(W,\mathfrak{g})$. The utility of this notion will become clear for the first time in 6.1.4 and 6.6.2. Some of the properties of twisted induced representations are obvious consequences of those of ordinary induced representations. Others require a certain amount of proof, which we shall indicate.

5.2.3. PROPOSITION. *Let* $\mathfrak{h}$, $\mathfrak{k}$ *and* σ *be as in* 5.1.11, *and*

$$\varrho = \operatorname{ind}^{\sim}(\sigma,\mathfrak{h}), \pi = \operatorname{ind}^{\sim}(\varrho,\mathfrak{g}).$$

Then π *is equivalent to* $\operatorname{ind}^{\sim}(\sigma,\mathfrak{g})$.

Let $\pi' = \operatorname{ind}^{\sim}(\sigma,\mathfrak{g})$, and let V_σ, V_ϱ, V_π and $V_{\pi'}$ be the spaces of σ, ϱ, π and π' respectively. Then $V_\sigma \subset V_\varrho \subset V_\pi$. For $x \in \mathfrak{k}$ and $v \in V_\sigma$, we have

$$\begin{aligned}\pi(x)v = \varrho(x)v + \theta_{\mathfrak{g},\mathfrak{h}}(x)v &= \sigma(x)v + \theta_{\mathfrak{h},\mathfrak{k}}(x)v + \theta_{\mathfrak{g},\mathfrak{h}}(x)v \\ &= (\sigma(x) + \theta_{\mathfrak{g},\mathfrak{k}}(x)\cdot 1)v.\end{aligned}$$

From 5.1.3, there exists a homomorphism ψ of the $\mathfrak{g}$-module $V_{\pi'}$ onto the $\mathfrak{g}$-module V_π which extends the identity mapping of V_σ. By the repeated application of 5.1.6, we see that ψ is injective.

5.2.4. PROPOSITION. *Let* $\mathfrak{h}$, $\mathfrak{n}$, $\mathfrak{g}_1$, $\mathfrak{h}_1$, η, ζ, ϱ_1 *and* ϱ *be as in* 5.1.12, $\pi_1 = \operatorname{ind}^{\sim}(\varrho_1,\mathfrak{g}_1)$, *and* $\pi = \operatorname{ind}^{\sim}(\varrho,\mathfrak{g})$. *Then* π *is equivalent to* $\pi_1 \circ \eta$.

If $x \in \mathfrak{h}$, then

$$\operatorname{tr}\operatorname{ad}_{\mathfrak{g}}x = \operatorname{tr}(\operatorname{ad}_{\mathfrak{g}}x \mid \mathfrak{n}) + \operatorname{tr}\operatorname{ad}_{\mathfrak{g}_1}\zeta(x),$$

and

$$\operatorname{tr}\operatorname{ad}_{\mathfrak{h}}x = \operatorname{tr}(\operatorname{ad}_{\mathfrak{h}}x \mid \mathfrak{n}) + \operatorname{tr}\operatorname{ad}_{\mathfrak{h}_1}\zeta(x),$$

hence

$$\theta_{\mathfrak{g},\mathfrak{h}} = \theta_{\mathfrak{g}_1,\mathfrak{h}_1} \circ \zeta.$$

Let us set

$$\varrho^{\sim}(x) = \varrho(x) + \theta_{\mathfrak{g},\mathfrak{h}}(x) \qquad \text{for } x \in \mathfrak{h},$$

and

$$\varrho_1^{\sim}(y) = \varrho_1(y) + \theta_{\mathfrak{g}_1,\mathfrak{h}_1}(y) \quad \text{for } y \in \mathfrak{h}_1.$$

Then $\varrho^{\sim} = \varrho_1^{\sim} \circ \zeta$, so that 5.2.4 follows from 5.1.12.

5.2.5. Let $\mathfrak{h}$ be a Lie subalgebra of $\mathfrak{g}$. For $u \in U(\mathfrak{g})$ and $x \in \mathfrak{h}$, let us set $u \star x = ux - \theta_{\mathfrak{g},\mathfrak{h}}(x)u$. Then

$$(u \star x) \star x' - (u \star x') \star x = u \star [x,x'] \quad \text{if } x' \in \mathfrak{h}.$$

Hence there exists on $U(\mathfrak{g})$ one and only one right $U(\mathfrak{h})$-module structure such that $u \cdot x = u \star x$ for $u \in U(\mathfrak{g})$ and $x \in \mathfrak{h}$. If $(x_1, \ldots, x_p)$ is a basis for a complement of $\mathfrak{h}$ in $\mathfrak{g}$, the $x_1^{\nu_1} \cdots x_p^{\nu_p}$, where $\nu_1, \ldots, \nu_p \in \mathbf{N}$, form a basis for this right $U(\mathfrak{h})$-module, as may be easily seen.

Let ϱ and W be as in 5.2.2. Let $V = U(\mathfrak{g}) \otimes_{U(\mathfrak{h})} W$, where we use on $U(\mathfrak{h})$ the above right $U(\mathfrak{h})$-module structure. We thus obtain a left $U(\mathfrak{g})$-module. With the above notation, V is the direct sum of the $x_1^{\nu_1} \cdots x_n^{\nu_n} \otimes W$. For $x \in \mathfrak{h}$, $w \in W$, we have

$$x \cdot (1 \otimes w) = (1 \star (x + \theta_{\mathfrak{g},\mathfrak{h}}(x))) \otimes w = 1 \otimes (x + \theta_{\mathfrak{g},\mathfrak{h}}(x))w = 1 \otimes \varrho^{\sim}(x)w,$$

so that the $\mathfrak{g}$-module V is isomorphic to $\text{ind}^{\sim}(W,\mathfrak{h})$.

5.2.6. Given this, we may repeat the proof of 5.1.7 in order to obtain the following result:

PROPOSITION. *Let* $\mathfrak{h}$ *and* ϱ *be as in* 5.2.2, *and let* $J = \text{Ker}\,\varrho$, $I = \text{Ker ind}^{\sim}(\varrho,\mathfrak{g})$. *Then* I *is the largest two-sided ideal of* $U(\mathfrak{g})$ *contained in* $U(\mathfrak{g}) \star J$ (*cf.* 5.2.5).

References: [26], [31].

5.3. A criterion for the simplicity of induced representations

5.3.1. Let $\mathfrak{k}$ be an ideal of $\mathfrak{g}$, V a vector space, and σ a representation of $\mathfrak{k}$ in V. The set of the $y \in \mathfrak{g}$ such that there exists $s \in \text{End}(V)$ satisfying

$$\sigma([y,x]) = [s,\sigma(x)]$$

for all $x \in \mathfrak{k}$ [and consequently for all $x \in U(\mathfrak{k})$] is termed the *stabilizer of* σ in $\mathfrak{g}$ and is denoted by $\mathfrak{st}(\sigma,\mathfrak{g})$. It is easily verified that $\mathfrak{st}(\sigma,\mathfrak{g})$ is a Lie subalgebra of $\mathfrak{g}$ containing $\mathfrak{k}$.

Let $f \in \mathfrak{g}^*$ be such that $f([\mathfrak{k},\mathfrak{k}]) = 0$. Then $f|\mathfrak{k}$ is a one-dimensional representation of $\mathfrak{k}$, and we have

$$\mathfrak{k}^f = \mathfrak{st}(f|\mathfrak{k},\mathfrak{g}).$$

(cf. 1.11.2).

5.3.2. Let $\mathfrak{k}$ be an ideal of $\mathfrak{g}$, and K a two-sided ideal of $U(\mathfrak{k})$. The set of the $y \in \mathfrak{g}$ such that $[y,K] \subset K$ is termed the *stabilizer of K in $\mathfrak{g}$* and is denoted by $\mathfrak{st}(K,\mathfrak{g})$; it is a Lie subalgebra of $\mathfrak{g}$ containing $\mathfrak{k}$.

5.3.3. Proposition. *Let $\mathfrak{k}$ be an ideal of $\mathfrak{g}$, σ a representation of $\mathfrak{k}$, and K its kernel. Then:*

(i) $\mathfrak{st}(\sigma,\mathfrak{g}) \subset \mathfrak{st}(K,\mathfrak{g})$.

(ii) *Let us assume that $\mathfrak{k}$ is nilpotent and σ is absolutely simple. For all $y \in \mathfrak{st}(K,\mathfrak{g})$, there exists $u_0 \in U(\mathfrak{k})$ such that $\sigma([y,x]) = [\sigma(u_0),\sigma(x)]$ for all $x \in \mathfrak{k}$. We have $\mathfrak{st}(\sigma,\mathfrak{g}) = \mathfrak{st}(K,\mathfrak{g})$.*

Assertion (i) is obvious. Let $y \in \mathfrak{st}(K,\mathfrak{g})$. Then y defines, by the adjoint representation, a derivation δ of $U(\mathfrak{k})/K$. If $\mathfrak{k}$ is nilpotent and σ is absolutely simple, $U(\mathfrak{k})/K$ is isomorphic to an algebra $A_n(k)$ (4.7.9). Hence δ is an inner derivation (4.6.8). Thus there exists $u_0 \in U(\mathfrak{k})$ such that $[y,x] \equiv [u_0,x]$ (mod K) for all $x \in \mathfrak{k}$, whence $\sigma([y,x]) = [\sigma(u_0),\sigma(x)]$ and $y \in \mathfrak{st}(\sigma,\mathfrak{g})$.

5.3.4. Lemma. *Let $\mathfrak{k}$ be an ideal of $\mathfrak{g}$, σ a representation of $\mathfrak{k}$, and k' an algebraic extension of k. Let $\mathfrak{g}'$, $\mathfrak{k}'$ and σ' be the objects deduced from $\mathfrak{g}$,$\mathfrak{k}$ and σ by extension of the scalar field from k to k'. Then $\mathfrak{st}(\sigma',\mathfrak{g}') = \mathfrak{st}(\sigma,\mathfrak{g}) \otimes k'$.*

(a) Obviously, $\mathfrak{st}(\sigma,\mathfrak{g}) \subset \mathfrak{st}(\sigma',\mathfrak{g}')$, and hence $\mathfrak{st}(\sigma,\mathfrak{g}) \otimes k' \subset \mathfrak{st}(\sigma',\mathfrak{g}')$. Let $y \in \mathfrak{st}(\sigma',\mathfrak{g}')$. Let V be the space of σ, and $V' = V \otimes k'$. There exists $s \in \mathrm{End}_{k'}(V')$ such that $\sigma'([y,x]) = [s,\sigma'(x)]$ for all $x \in \mathfrak{k}$.

(b) Let us assume that $[k' : k] < +\infty$. Let $(\lambda_1, \ldots, \lambda_n)$ be a basis for k' over k. Then $y = \lambda_1 y_1 + \cdots + \lambda_n y_n$ and $s = \lambda_1 s_1 + \cdots + \lambda_n s_n$, with $y_1, \ldots, y_n \in \mathfrak{g}$ and $s_1, \ldots, s_n \in \mathrm{End}(V)$. For all $x \in \mathfrak{k}$, we may write

$$\begin{aligned}\lambda_1\sigma([y_1,x]) + \cdots + \lambda_n\sigma([y_n,x]) &= \sigma'([y,x])\\ &= [\lambda_1 s_1 + \cdots + \lambda_n s_n,\sigma'(x)]\\ &= \lambda_1[s_1,\sigma(x)] + \cdots + \lambda_n[s_n,\sigma(x)].\end{aligned}$$

For $i = 1, \ldots, n$, we deduce from it that $\sigma([y_i,x]) = [s_i,\sigma(x)]$, whence y_i $y_i \in \mathfrak{st}(\sigma,\mathfrak{g})$ and $y \in \mathfrak{st}(\sigma,g) \otimes k'$.

(c) We now pass to the general case. There exists a subextension of finite degree $k'' \subset k'$ such that $y \in \mathfrak{g} \otimes k''$. Let $V'' = V \otimes k''$, let η be a

k''-linear projection of k' onto k'', let ζ be the projection $1 \otimes \eta$ of V' onto V'', and let

$$t = (\zeta \circ s) \mid V'' : V'' \to V''.$$

For all $x \in \mathfrak{k}$, we have

$$\sigma'([y,x]) \mid V'' = \zeta\sigma'([y,x]) \mid V'' = \zeta s\sigma'(x) \mid V'' - \zeta\sigma'(x)s \mid V''$$
$$= \zeta s\sigma'(x) \mid V'' - \sigma'(x)\zeta s \mid V'' = [t,\sigma'(x) \mid V''].$$

By (b), we deduce from it that $y \in \mathfrak{st}(\sigma,\mathfrak{g}) \otimes k''$.

5.3.5. LEMMA. *Let $\mathfrak{k}$ be an ideal of $\mathfrak{g}$,σ an absolutely simple representation of $\mathfrak{k}$, $\mathfrak{h} = \mathfrak{st}(\sigma,\mathfrak{g})$, ϱ a representation of $\mathfrak{h}$ such that $\varrho|\mathfrak{k}$ is a multiple of σ, and $\pi = \mathrm{ind}(\varrho,\mathfrak{g})$ or $\mathrm{ind}^{\sim}(\mathfrak{b},\mathfrak{g})$. Let V,W be the spaces of π,ϱ so that $V = U(\mathfrak{g}) \otimes_{U(\mathfrak{h})} W$. For all $n \in \mathbf{N}$, let V_n be the set of linear combinations of the $u \otimes w$, where $u \in U_n(\mathfrak{g})$ and $w \in W$. Let p be an integer >0 and $t \in V_p - \{0\}$. Then there exists $z \in U(\mathfrak{k})$ such that $zt \in V_{p-1} - \{0\}$.*

Since $\theta_{\mathfrak{g},\mathfrak{h}}|\mathfrak{k} = 0$, it follows from 2.2.22 that the restrictions of $\mathrm{ind}(\varrho,\mathfrak{g})$ and of $\mathrm{ind}^{\sim}(\varrho,\mathfrak{g})$ to $\mathfrak{k}$ are the same. For instance, let us assume that $\pi = \mathrm{ind}(\varrho,\mathfrak{g})$.

(a) Let $(x_1, \ldots, x_n)$ be the basis for a complement of $\mathfrak{h}$ in $\mathfrak{g}$. For $\nu = (\nu_1, \ldots, \nu_n) \in \mathbf{N}^n$, we set $x^\nu = x_1^{\nu_1} \cdots x_n^{\nu_n}$. The element t can be uniquely written in the form

$$t = \sum_{|\nu| \leq p} x^\nu \otimes w_\nu \quad (w_\nu \in W \text{ for all } \nu).$$

If $|\nu| = p$ implies that $w_\nu = 0$, the lemma is obvious. We shall hence assume that the w_ν, where $|\nu| = p$, are not all zero.

(b) By hypothesis, W can be written as $\oplus_{\lambda \in \Lambda} W_\lambda$, where the W_λ are $\mathfrak{k}$-modules isomorphic to an absolutely simple $\mathfrak{k}$-module C. For all $\lambda \in \Lambda$, let ζ_λ be a $\mathfrak{k}$-homomorphism of W onto C with kernel $\oplus_{\lambda' \neq \lambda} W_{\lambda'}$. The $\zeta_\lambda(w_\nu)$ are zero except for a finite number of pairs (λ,ν). Let us choose a non-zero element c of C. From 2.6.5, there exist $z \in U(\mathfrak{k})$ and $\xi_{\lambda,\nu} \in k$ which are not all zero such that $z(\zeta_\lambda w_\nu) = \xi_{\lambda,\nu} c$ for all λ and ν with $|\nu| = p$. Now, from 2.2.22, we have

$$zt \equiv \sum_{|\nu| = p} x^\nu \otimes zw_\nu \quad (\mathrm{mod}\ V_{p-1}).$$

Substituting zt for t, we may thus assume that, for $|\nu| = p$ and $\lambda \in \Lambda$, we have $\zeta_\lambda w_\nu = \xi_{\lambda,\nu} c$, where $\xi_{\lambda,\nu} \in k$, and that there exist ν^0,λ_0 with $\xi_{\lambda_0,\nu^0} \neq 0$.

(c) Let us set $\nu^0 = (\nu_1^0, \ldots, \nu_n^0)$, and let $i \in \{1, \ldots, n\}$ be such that $\nu_i^0 > 0$. For $j = 1, \ldots, n$, let $\varepsilon_j = (0, \ldots, 0,1,0, \ldots, 0)$, where 1 appears in the jth place. Let $\mu = \nu^0 - \varepsilon_i$. For all $z \in U(\mathfrak{k})$, we have, from 2.2.22,

$$(1) \qquad zt = \sum_{|\nu| \leq p} (zx^\nu) \otimes w_\nu$$

$$\equiv \sum_{|\nu|=p} x^\nu \otimes zw_\nu + \sum_{|\nu|=p} \sum_{j=1}^{n} \nu_j x^{\nu-\varepsilon_j} \otimes [z,x_j]w_\nu$$

$$+ \sum_{|\nu|=p-1} x^\nu \otimes zw_\nu \quad (\text{mod } W_{p-2}).$$

The term in x^μ in the expansion of zt is hence

$$(2) \qquad x^\mu \otimes zw_\mu + \sum_{j=1}^{n} (\mu_j + 1)x^\mu \otimes [z,x_j]w_{\mu+\varepsilon_j}.$$

(d) Let us assume that $U(\mathfrak{k})t \cap V_{p-1} = 0$. Let $z \in U(\mathfrak{k})$ such that $zc = 0$. Then $zw_\nu = 0$ for $|\nu| = p$. From (1), we have $zt \in V_{p-1}$, hence $zt = 0$ and consequently, from (2),

$$(3) \qquad zw_\mu + \sum_{j=1}^{n} (\mu_j + 1)\,[z,x_j]w_{\mu+\varepsilon_j} = 0.$$

Applying ζ_{λ_0} to (3), we obtain

$$0 = z(\zeta_{\lambda_0} w_\mu) + \sum_{j=1}^{n} (\mu_j + 1)\,[z,x_j]\xi_{\mu+\varepsilon_j,\lambda_0} c$$

$$= z(\zeta_{\lambda_0} w_\mu) + \left[z, \sum_{j=1}^{n} (\mu_j + 1)\xi_{\mu+\varepsilon_j,\lambda_0} x_j\right] c.$$

Let us set

$$c' = \zeta_{\lambda_0} w_\mu \in C, \qquad y = \sum_{j=1}^{n} (\mu_j + 1)\xi_{\mu+\varepsilon_j,\lambda_0} x_j \in \mathfrak{g}.$$

Since $\xi_{\mu+\varepsilon_i,\lambda_0} = \xi_{\nu^0,\lambda_0} \neq 0$, we have $y \notin \mathfrak{h}$. On the other hand, the above proves the existence of an $s \in \text{End}(C)$ such that

$$s(zc) = zc' + [z,y]c \quad \text{for all } z \in U(\mathfrak{k}).$$

For $x \in \mathfrak{k}$ and $z \in U(\mathfrak{k})$, we then have

$$[s,x_C](zc) = s(xzc) - xs(zc)$$

$$= xzc' + [xz,y]c - xzc' - x[z,y]c = [x,y]zc,$$

whence $[y,x]_C = [-s,x_C]$. This proves that $y \in \mathfrak{h}$, which leads to a contradiction, which establishes the lemma.

5.3.6. THEOREM. *Let $\mathfrak{k}$ be an ideal of $\mathfrak{g}$, σ an absolutely simple representation of $\mathfrak{k}$, $\mathfrak{h} = \mathfrak{st}(\sigma,\mathfrak{g})$, ϱ a representation of $\mathfrak{h}$ such that $\varrho|\mathfrak{k}$ is a multiple of σ, and $\pi = \mathrm{ind}(\varrho,\mathfrak{g})$ or $\mathrm{ind}^{\sim}(\varrho,\mathfrak{g})$. If ϱ is simple (absolutely simple), then π is simple (absolutely simple).*

Let V,W be the spaces of π, ϱ, so that $W \subset V$. Let T be a non-null sub-$\mathfrak{g}$-module of V. From 5.3.5, we deduce in a stepwise fashion that $T \cap W \neq 0$. Let us assume that ϱ is simple. Then $T \supset W$ and hence $T = V$, which proves that π is simple. Let us assume that ϱ is absolutely simple. Let k' be an algebraic closure of k. Let $\mathfrak{g}'$, $\mathfrak{k}'$, σ', $\mathfrak{h}'$, ϱ' and π' be the objects deduced from $\mathfrak{g}$, $\mathfrak{k}$, σ, $\mathfrak{h}$, ϱ and π by extension of the scalar field from k to k'. Taking the above and 5.3.4 into account, π' is simple. Hence π is absolutely simple (2.6.5).

5.3.7. THEOREM. *Let $\mathfrak{k}$ be an ideal of $\mathfrak{g}$, σ an absolutely simple representation of $\mathfrak{k}$, $\mathfrak{h} = \mathfrak{st}(\sigma,\mathfrak{g})$, and ϱ_1 and ϱ_2 representations of $\mathfrak{h}$ in W_1 and W_2 such that $\varrho_1|\mathfrak{k}$ and $\varrho_2|\mathfrak{k}$ are multiples of σ. For $i = 1,2$, let $\pi_i = \mathrm{ind}(\varrho_i,\mathfrak{g})$ [or $\pi_i = \mathrm{ind}^{\sim}(\varrho_i,\mathfrak{g})$], and let V_i be the space of π_i. For all $u \in \mathrm{Hom}_{\mathfrak{h}}(W_1,W_2)$ let $\varphi(u)$ be the unique element of $\mathrm{Hom}_{\mathfrak{g}}(V_1,V_2)$ which extends u. Then φ is a bijection of $\mathrm{Hom}_{\mathfrak{h}}(W_1,W_2)$ onto $\mathrm{Hom}_{\mathfrak{g}}(V_1,V_2)$.*

The case of $\mathrm{ind}^{\sim}(\varrho_i,\mathfrak{g})$ can easily be deduced from the case of $\mathrm{ind}(\varrho_i,\mathfrak{g})$. We shall assume that $\pi_i = \mathrm{ind}(\varrho_i,\mathfrak{g})$.

Clearly, φ is injective. Let $v \in \mathrm{Hom}_{\mathfrak{g}}(V_1,V_2)$. We must prove that $v(W_1) \subset W_2$. Let C be a simple sub-$\mathfrak{k}$-module of W_1. The proof will be complete if we establish that $v(C) \subset W_2$.

The $\mathfrak{g}$-module $V_1 \times V_2$ is induced by the $\mathfrak{h}$-module $W_1 \times W_2$. Let T be the set of the (x,vx), where $x \in V_1$; it is a sub-$\mathfrak{g}$-module of $V_1 \times V_2$. Let c be a non-zero element of C, and $t = (c,vc) \in T$. From 5.3.5, there exists $z \in U(\mathfrak{k})$ such that $(zc,vzc) = (zc,zvc) = zt$ is a non-zero element of $W_1 \times W_2$. Then $zc \in C - \{0\}$ and

$$v(C) = v(U(\mathfrak{k})zc) = U(\mathfrak{k})v(zc) \subset U(\mathfrak{k})W_2 \subset W_2.$$

References: [13], [35], [45].

5.4. The construction of primitive ideals by induction

5.4.1. PROPOSITION. *Let π be a simple (absolutely simple) representation of $\mathfrak{g}$, and $\mathfrak{k}$ an ideal of $\mathfrak{g}$. We assume that $\pi|\mathfrak{k}$ possesses an absolutely simple* *subrepresentation* σ. *Let* $\mathfrak{h} = \mathfrak{st}(\sigma,\mathfrak{g})$. *Then there exists a simple (absolutely simple) representation* ϱ *of* $\mathfrak{h}$ *such that* $\varrho|\mathfrak{k}$ *is a multiple of* σ *and* $\operatorname{ind}(\varrho,\mathfrak{g})$ *is equivalent to* π.

Let V,X be the spaces of π,σ. Let W be the sum of the sub-$\mathfrak{k}$-modules of V which are isomorphic to the $\mathfrak{k}$-module X. Let $y \in \mathfrak{h}$. There exists $s \in \operatorname{End}(X)$ such that $\sigma([y,x]) = [s,\sigma x]$ for all $x \in \mathfrak{k}$. Let $u: X \to V$ be the restriction of $\pi(y)$ to X. We shall consider $u - s$ as a mapping of X into V. For all $x \in \mathfrak{k}$ and all $z \in X$, we have

$$\begin{aligned}\pi(x)\,(u - s) &= \pi(x)\pi(y)z - \sigma(x)sz\\ &= \pi(x)\pi(y)z + \pi([y,x])z - s\sigma(x)z\\ &= \pi(y)\pi(x)z - s\sigma(x)z = (u - s)\sigma(x)z,\end{aligned}$$

hence $u - s$ is a $\mathfrak{k}$-module homomorphism. Consequently,

$$u(X) \subset (u - s)(X) + s(X) \subset W + X = W.$$

This proves that $\pi(y)(W) \subset W$. Thus there exists a representation ϱ of $\mathfrak{h}$ in W such that ϱ is a subrepresentation of $\pi|\mathfrak{h}$. Clearly, $\varrho|\mathfrak{k}$ is a multiple of σ. Let $\pi' = \operatorname{ind}(\varrho,\mathfrak{g})$, and V' be the space of π'. From 5.1.3 there exists a $\mathfrak{g}$-homomorphism φ of V' into V which reduces to the identity mapping on W. Since π is simple, φ is surjective. Let $T = \operatorname{Ker}\varphi$, which is a sub-$\mathfrak{g}$-module of V'. If $T \neq 0$, we have $T \cap W \neq 0$ from 5.3.5. This is impossible since $\varphi|W = \mathrm{id}_W$. Hence $T = 0$ and φ is an isomorphism. Thus π is equivalent to $\operatorname{ind}(\varrho,\mathfrak{g})$. Consequently, $\operatorname{ind}(\varrho,\mathfrak{g})$ is simple (absolutely simple) and hence ϱ is simple (absolutely simple) from 5.1.10.

5.4.2. LEMMA. *Let* $\mathfrak{k}$ *be an ideal of* $\mathfrak{g}$, $(x_1, \ldots, x_n)$ *a basis for a complement of* $\mathfrak{k}$ *in* $\mathfrak{g}$, I *a two-sided ideal of* $U(\mathfrak{g})$, $K = I \cap U(\mathfrak{k})$, $M = U(\mathfrak{g})/I$ *and* $N = U(\mathfrak{k})/K$. *For all* $\nu = (\nu_1, \ldots, \nu_n) \in \mathbf{N}^n$, *we denote by* x^ν *the class of* $x_1^{\nu_1} \cdots x_n^{\nu_n}$ *modulo* I. *We order* $\mathbf{N}^n$ *as in* 2.6.1 *and write*

$$M_\nu = \sum_{\nu' \leqq \nu} x^{\nu'} N, \qquad M_\nu^- = \sum_{\nu' < \nu} x^{\nu} N.$$

Then:

(i) M_ν *and* M_ν^- *are left and right sub-N-modules of* M; *the union of the* M_ν *is* M.

(ii) *The annihilator* A_ν *of* M_ν/M_ν^- *in* N *is the same, whether we consider* M_ν/M_ν^- *as a left or as a right N-module.*

(iii) A_ν *is a two-sided ideal of* N.

(iv) *Let* $\nu = (\nu_1, \ldots, \nu_n) \in \mathbf{N}^n$ *and* $\nu' = (\nu'_1, \ldots, \nu'_n) \in \mathbf{N}^n$. *If* $\nu_1 \leqq \nu'_1, \ldots, \nu_n \leqq \nu'_n$, then $A_\nu \subset A_{\nu'}$.

(v) *The set of non-null* A_ν, *has a finite number of minimal elements; let* A *be their intersection. Then* $A \subset A_\nu$ *for all* ν *such that* $A_\nu \neq 0$.

(vi) *If* L *is a left ideal of* N *such that* $ML = M$, *then* L *contains a power of* A.

It is clear that M_ν and M_ν^- are right sub-N-modules of M and that the union of the M_ν is M. If $n \in \mathbf{N}$, we have, from 2.2.22.

$$x^\nu n \in nx^\nu + \sum_{|\nu'|<|\nu|} x^{\nu'} N \tag{1}$$

hence by recurrence

$$M_\nu = \sum_{\nu' \leqq \nu} Nx^{\nu'}, \qquad M_\nu^- = \sum_{\nu' < \nu} Nx^{\nu'}. \tag{2}$$

Thus M_ν and M_ν^- are left sub-N-modules of M. Let $n \in N$. From (1) and (2), we have

$$nM_\nu \subset M_\nu^- \Leftrightarrow nx^\nu \in M_\nu^- \Leftrightarrow x^\nu n \in M_\nu^- \Leftrightarrow M_\nu n \subset M_\nu^- .$$

This proves (ii) and (iii).

Let us prove (iv). Let ε_i be the element of $\mathbf{N}^n$ all of whose co-ordinates are zero except for the ith, which is equal to 1. It is sufficient to prove that for all $\nu \in \mathbf{N}^n$ we have $A_\nu \subset A_{\nu+\varepsilon_i}$. Since

$$x_j x_{j'} \in x_{j'} x_j + kx_1 + \cdots + kx_n + \mathfrak{k},$$

it is easily seen that

$$x^{\nu+\varepsilon_i} \in x^{\varepsilon_i} x^\nu + \sum_{|\nu'| \leqq |\nu|} x^{\nu'} N.$$

Let $y \in A_\nu$. Then $x^\nu y \in M_\nu^-$, hence

$$\begin{aligned} x^{\nu+\varepsilon_i} y &\in x^{\varepsilon_i} x^\nu y + \sum_{|\nu'| \leqq |\nu|} x^{\nu'} N \\ &\subset x^{\varepsilon_i} \sum_{\nu' < \nu} x^{\nu'} N + \sum_{|\nu'| \leqq |\nu|} x^{\nu'} N \\ &\subset \sum_{\nu' < \nu} x^{\nu'+\varepsilon_i} N + \sum_{|\nu'| \leqq |\nu|} x^{\nu'} N. \end{aligned}$$

From 2.6.1 (c) this implies that

$$x^{\nu+\varepsilon_i} y \in \sum_{\nu'' < \nu+\varepsilon_i} x^{\nu''} N + \sum_{|\nu'| < |\nu+\varepsilon_i|} x^{\nu'} N = M_{\nu+\varepsilon_i}^- .$$

Thus $y \in A_{\nu+\varepsilon_i}$, and the proof of (iv) is complete.

From 2.6.2, the set P of the $\nu \in \mathbf{N}^n$ such that $A_\nu \neq 0$, ordered by the product ordering on $\mathbf{N}^n$, has a finite number of minimal elements $\mu_1, \ldots, \mu_t$, and every element of P is greater than one of the μ_i. From (iv), for all $\nu \in P$, A_ν contains one of the A_{μ_i}. This proves (v).

Let us set

$$B_\nu = A_\nu \quad \text{if } A_\nu \neq 0,$$

and

$$B_\nu = N \quad \text{if } A_\nu = 0.$$

Let L be a left ideal of N. Then

$$ML = \sum_\nu x^\nu NL = \sum_\nu x^\nu L.$$

Let $\nu_1, \nu_2 \in \mathbf{N}^n$ with $0 < \nu_1 < \nu_2$, and assume that

$$\prod_{\nu_1 < \nu \leq \nu_2} B_\nu \subset \sum_{\nu \leq \nu_1} x^\nu L. \tag{3}$$

From this we shall deduce that

$$\prod_{\nu_1 \leq \nu \leq \nu_2} B_\nu \subset \sum_{\nu < \nu_1} x^\nu L. \tag{4}$$

First, we assume that $A_{\nu_1} = 0$. Then $\prod_{\nu_1 \leq \nu \leq \nu_2} B_\nu = \prod_{\nu_1 < \nu \leq \nu_2} B_\nu$. Let $y \in \prod_{\nu_1 < \nu \leq \nu_2} B_\nu$. From (3), we have $y = \sum_{\nu \leq \nu_1} x^\nu y_\nu$ with $y_\nu \in L$ for all ν. Then

$$\begin{aligned} M_{\nu_1} y_{\nu_1} &\subset N x^{\nu_1} y_{\nu_1} + M_{\nu_1}^- \subset Ny + \sum_{\nu \neq \nu_1} N x^\nu y_\nu + M_{\nu_1}^- \\ &\subset NN + M_{\nu_1}^- = M_{\nu_1}^-, \end{aligned}$$

hence $y_{\nu_1} \in A_{\nu_1} = 0$ and $y \in \sum_{\nu < \nu_1} x^\nu L$, which proves (4) for this case.

Let us now assume that $A_{\nu_1} \neq 0$. Then $B_{\nu_1} = A_{\nu_1}$, hence

$$\prod_{\nu_1 \leq \nu \leq \nu_2} B_\nu = A_{\nu_1} \Big(\prod_{\nu_1 < \nu \leq \nu_2} B_\nu \Big) \subset A_{\nu_1} M_{\nu_1} L \subset M_{\nu_1}^- L = \sum_{\nu < \nu_1} x^\nu L.$$

Let us assume that $ML = M$. There exists a family $(y_\nu)_{0 \leq \nu \leq \nu_1}$ of elements of L such that $1 = \sum_{\nu \leq \nu_1} x^\nu y_\nu$. If $\nu_1 = 0$, then $L = N \supset A$. Let us assume that $\nu_1 > 0$. Then $x^{\nu_1} y_{\nu_1} \in M_{\nu_1}^-$, hence $y_{\nu_1} \in A_{\nu_1}$. If $A_{\nu_1} = 0$, we see that $1 = \sum_{\nu < \nu_1} x^\nu y_\nu$. Thus we are brought back in steps to the case where $A_{\nu_1} \neq 0$. Then

$$B_{\nu_1} = A_{\nu_1} \subset \sum_{\nu \leq \nu_1} A_{\nu_1} x^\nu L \subset M_{\nu_1}^- L \subset \sum_{\nu < \nu_1} x^\nu L.$$

Using the implication (3) $\Rightarrow$ (4), we deduce in a stepwise fashion that $\prod_{0 \leq \nu \leq \nu_1} B_\nu \subset L$. *A fortiori*, L contains a power of A.

5.4.3. Proposition. *Let* $\mathfrak{k}$ *be an ideal of* $\mathfrak{g}$, I *a maximal two-sided ideal of* $U(\mathfrak{g})$ *and* $K = I \cap U(\mathfrak{k})$.

(i) *There exists a primitive ideal* J *of* $U(\mathfrak{k})$ *which is generic for* K.

(ii) *If* k *is algebraically closed, there exists a primitive ideal* J *of* $U(\mathfrak{k})$ *which has the following properties*:

(a) J *is generic for* K;

(b) *for every simple representation* σ *of* $\mathfrak{k}$ *with kernel* J, *there exists a simple representation* ϱ *of* $\mathfrak{st}(\sigma,\mathfrak{g})$ *such that* $\varrho|\mathfrak{k}$ *is a multiple of* σ *and* $\operatorname{ind}(\varrho,\mathfrak{g})$ *is simple with kernel* I.

We introduce the notation of 5.4.2. From 3.3.4, K is prime. Hence $A \neq 0$. From 3.1.15, there exists a primitive ideal J of $U(\mathfrak{k})$ containing K and such that $A \not\subset J/K$. Let σ be a simple representation of $\mathfrak{k}$ with kernel J, and L the annihilator in $U(\mathfrak{k})$ of a fixed non-zero element in the space of σ. Then J is the largest two-sided ideal of $U(\mathfrak{k})$ contained in L. For all integers $r > 0$, we have $A^r \not\subset J/K$ (since J is prime), hence $A^r \not\subset L/K$. From 5.4.2 (vi), we have $I + U(\mathfrak{g})L \neq U(\mathfrak{g})$. Let L' be a maximal left ideal of $U(\mathfrak{g})$ such that $L' \supset I + U(\mathfrak{g})L$. Let J_1 be the largest two-sided ideal of $U(\mathfrak{k})$ contained in J and such that $[\mathfrak{g},J_1] \subset J_1$. Then $K \subset J_1 \subset J$. On the other hand, $U(\mathfrak{g})J_1$ is a two-sided ideal of $U(\mathfrak{g})$, and

$$I \subset I + U(\mathfrak{g})J_1 \subset I + U(\mathfrak{g})L \subset L' \neq U(\mathfrak{g}),$$

whence $I + U(\mathfrak{g})J_1 = I$ (since I is maximal), and

$$U(\mathfrak{g})J_1 \subset I, \qquad J_1 \subset I \cap U(\mathfrak{k}) = K, \qquad J_1 = K.$$

This proves that J is generic for K.

Let π be the simple representation of $U(\mathfrak{g})$ defined by L'. Its kernel contains I, and hence is equal to I. Since $L' \cap U(\mathfrak{k}) \supset L$ and $1 \notin L' \cap U(\mathfrak{k})$, we have $L' \cap U(\mathfrak{k}) = L$, hence σ is a subrepresentation of $\pi|\mathfrak{k}$. Let us assume that k is algebraically closed. From 5.4.1, there exists a simple representation ϱ of $\mathfrak{st}(\sigma,\mathfrak{g})$ such that $\varrho|\mathfrak{k}$ is a multiple of σ and $\operatorname{ind}(\varrho,\mathfrak{g})$ is equivalent to π. Then $\operatorname{ind}(\varrho,\mathfrak{g})$ is simple with kernel I.

5.4.4. Proposition (k algebraically closed, $\mathfrak{g}$ solvable). *Let* $\mathfrak{k}$ *be a nilpotent ideal of* $\mathfrak{g}$, I *a primitive ideal of* $U(\mathfrak{g})$, *and* $K = I \cap U(\mathfrak{k})$. *There exist a primitive ideal* J *of* $U(\mathfrak{k})$ *which is generic for* K, *a simple representation* σ *of* $\mathfrak{k}$ *with kernel* J *and a simple representation* ϱ *of* $\mathfrak{st}(\sigma,\mathfrak{g})$ *such that* $\varrho|\mathfrak{k}$ *is a multiple of* σ *and* $\operatorname{ind}(\varrho,\mathfrak{g})$ *is simple with kernel* I.

The proposition is obvious if $\dim \mathfrak{g} = 0$. We shall assume that it has been established for Lie algebras of dimension $<\dim \mathfrak{g}$. If I is a maximal two-sided ideal of $U(\mathfrak{g})$, the proposition follows from 5.4.3. Henceforth we shall therefore assume that there exists a two-sided ideal I_1 of $U(\mathfrak{g})$ such that $I \subset I_1$ and $I \neq I_1 \neq U(\mathfrak{g})$. From 4.4.1, there exist $u \in I_1$ and $\lambda \in \mathfrak{g}^*$ such that $u \notin I$ and $[x,u] \in \lambda(x)u + I$ for all $x \in \mathfrak{g}$. If $\lambda = 0$, then $u \bmod I$ belongs to the centre of $U(\mathfrak{g})/I$, and hence $u \in k \cdot 1 + I$ since I is primitive (4.5.7). Since $u \in I_1$, we conclude that $I_1 = U(\mathfrak{g})$, which is a contradiction. Hence $\lambda \neq 0$. Let $\mathfrak{g}' = \operatorname{Ker} \lambda$; it is an ideal of codimension 1 in $\mathfrak{g}$. Let $I' = I \cap U(\mathfrak{g}')$. From 3.3.8, I is the two-sided ideal of $U(\mathfrak{g})$ generated by I'. The centre of $\operatorname{Fract}(U(\mathfrak{g})/I)$ is k (4.5.7); then, from 4.3.4, we have $u \bmod I \in U(\mathfrak{g}')/I'$. By adding to u a suitable element of I, we may thus assume that $u \in U(\mathfrak{g}')$ and $u \notin I'$. There exists a primitive ideal J' of $U(\mathfrak{g}')$ which is generic for I' relative to $\mathfrak{g}$ (4.5.9). Let us assume that $u \in J'$. For all $x \in \mathfrak{g}$, we have

$$[x,u] \in (\lambda(x)u + I) \cap U(\mathfrak{g}') \subset ku + I';$$

then the two-sided ideal of $U(\mathfrak{g}')$ generated by $ku + I'$ is $\mathfrak{g}$-stable and contained in J', which is a contradiction; hence $u \notin J'$. On the other hand, since $\lambda(\mathfrak{g}') = 0$, $u \bmod I'$ is central in $U(\mathfrak{g}')/I'$.

Since $\mathfrak{k}$ is nilpotent, we have $\mathfrak{k} \subset \mathfrak{g}'$. From the induction hypothesis, there exist a primitive ideal J of $U(\mathfrak{k})$ which is generic for $J' \cap U(\mathfrak{k})$ relative to $\mathfrak{g}'$, a simple representation θ of $\mathfrak{k}$ with kernel J and a simple representation s_0 of $\mathfrak{h}_0 = \mathfrak{st}(\sigma,\mathfrak{g}')$ such that $\varrho_0|\mathfrak{k}$ is a multiple of σ and $\theta = \operatorname{ind}(\varrho_0,\mathfrak{g}')$ is ϱimple with kernel J'.

Let us show that J is generic for K relative to $\mathfrak{g}$. Firstly, K is a two-sided ideal of $U(\mathfrak{k})$ contained in J and such that $[\mathfrak{g},K] \subset K$. Let K_1 be a two-sided ideal of $U(\mathfrak{k})$ such that $K_1 \subset J$ and $[\mathfrak{g},K_1] \subset K_1$. Then $[\mathfrak{g}',K_1] \subset K_1$, and hence $K_1 \subset J' \cap U(\mathfrak{k})$. Let K_1' be the two-sided ideal of $U(\mathfrak{g}')$ generated by K_1. Then $K_1 \subset J'$, hence $K_1' \subset J'$, and

$$\begin{aligned}[\mathfrak{g},K_1'] &= [\mathfrak{g},U(\mathfrak{g}')K_1U(\mathfrak{g}')] = [\mathfrak{g},U(\mathfrak{g}')K_1] \\ &\subset [\mathfrak{g},U(\mathfrak{g}')]K_1 + U(\mathfrak{g}')[\mathfrak{g},K_1] \subset U(\mathfrak{g}')K_1 \subset K_1'.\end{aligned}$$

Hence $K_1' \subset I'$, and consequently

$$K_1 \subset I' \cap U(\mathfrak{k}) = I \cap U(\mathfrak{k}) = K.$$

This proves our assertion.

We have $\theta(u) \neq 0$ since $u \notin J'$. If $x \in \mathfrak{g}$ is such that $x \notin \mathfrak{g}'$, then $\theta([x,u]) = \lambda(x)\theta(u)$ and $\lambda(x) \neq 0$; on the other hand, $\theta(u)$ is scalar since $u \bmod I'$ is central in $U(\mathfrak{g}')/I'$. Hence $x \notin \mathfrak{st}(\theta,\mathfrak{g})$. Thus $\mathfrak{st}(\theta,\mathfrak{g}) = \mathfrak{g}'$. From

5.3.6, $\operatorname{ind}(\theta,\mathfrak{g})$ is simple. On the other hand, $\operatorname{ind}(\theta,\mathfrak{g})$ can be identified with $\pi = \operatorname{ind}(\varrho_0,\mathfrak{g})$ (5.1.11). The representation $\pi|\mathfrak{h}_0$ has a subrepresentation equivalent to ϱ_0, hence $\pi|\mathfrak{k}$ has a subrepresentation equivalent to σ. From 5.4.1, there exists a simple representation ϱ of $\mathfrak{st}(\sigma,\mathfrak{g})$ such that $\varrho|\mathfrak{k}$ is a multiple of σ and $\operatorname{ind}(\varrho,\mathfrak{g})$ is equivalent to π, and hence simple.

Let N be the kernel of $\operatorname{ind}(\theta,\mathfrak{g})$. It remains for us to prove that $N = I$. Since $I = U(\mathfrak{g})I' \subset U(\mathfrak{g})J'$, we have $I \subset N$ (5.1.7). Let $x \in \mathfrak{g}$ be such that $x \notin \mathfrak{g}'$. Let

$$u = x^n u_n + x^{n-1} u_{n-1} + \cdots + u_0 \in N,$$

with $u_n, \ldots, u_0 \in U(\mathfrak{g}')$. Then, for all $p \geqq 0$ we have

$$x^n \cdot (\operatorname{ad} x)^p u^n + x^{n-1} \cdot (\operatorname{ad} x)^p u_{n-1} + \cdots + (\operatorname{ad} x)^p u_0 = (\operatorname{ad} x)^p u$$
$$\in N \subset U(\mathfrak{g})J' = \bigoplus_{i \geqq 0} x^i J',$$

hence $(\operatorname{ad} x)^p u_i \in J'$ for all i and all p, whence $u_i \in I'$ for all i, and $u \in I$.

References: [35], [48].

5.5. Co-induced representations

5.5.1. Let $\mathfrak{h}$ be a Lie subalgebra of $\mathfrak{g}$, and W an $\mathfrak{h}$-module. We may consider $U(\mathfrak{g})$ as a left $U(\mathfrak{h})$-module, and hence we set

$$V = \operatorname{Hom}_{U(\mathfrak{h})}(U(\mathfrak{g}), W).$$

If $f \in V$ and $u \in U(\mathfrak{g})$, let us define the mapping $u \cdot f$ of $U(\mathfrak{g})$ into W by $(u \cdot f)(v) = f(vu)$ for all $v \in U(\mathfrak{g})$. It follows directly that $u \cdot f \in V$ and that V is thus equipped with the structure of a left $U(\mathfrak{g})$-module. This module V is termed the *$\mathfrak{g}$-module co-induced by W*. If ϱ designates the representation of $\mathfrak{h}$ corresponding to W and π the representation of $\mathfrak{g}$ corresponding to V, then π is said to be the *representation of $\mathfrak{g}$ co-induced by ϱ* and is denoted by $\operatorname{coind}(\varrho,\mathfrak{g})$.

5.5.2. The mapping $f \mapsto f(1)$ of V into W is a surjective $\mathfrak{h}$-homomorphism. We identify W with a quotient $\mathfrak{h}$-module of V by means of this mapping, which is termed *canonical*.

5.5.3. PROPOSITION. *Let $\mathfrak{h}$ be a Lie subalgebra of $\mathfrak{g}$, W an $\mathfrak{h}$-module, V the $\mathfrak{g}$-module co-induced by W, and ε the canonical mapping of V onto W. Let V' be a $\mathfrak{g}$-module, and ψ an $\mathfrak{h}$-homomorphism of V' into W. Then there exists one and only one $\mathfrak{g}$-homomorphism φ of V' into V such that $\psi = \varepsilon \circ \varphi$. The mapping $\psi \mapsto \varphi$ is a bijection of* $\operatorname{Hom}_{\mathfrak{h}}(V',W)$ *onto* $\operatorname{Hom}_{\mathfrak{g}}(V',V)$.

This follows from the general properties of the functor Hom (or can easily be proved directly).

5.5.4. PROPOSITION. *Let* $\mathfrak{h}$ *be a Lie subalgebra of* $\mathfrak{g}$, W *an* $\mathfrak{h}$*-module,* W^* *the dual* $\mathfrak{h}$*-module,* $V = U(\mathfrak{g}) \otimes_{U(\mathfrak{h})} W$ *the* $\mathfrak{g}$*-module induced by* W, *and* $V' = \mathrm{Hom}_{U(\mathfrak{h})}(U(\mathfrak{g}), W^*)$ *the* $\mathfrak{g}$*-module co-induced by* W^*. *For* $\psi \in V^*$ *and* $u \in U(\mathfrak{g})$, *let* $\hat{\psi}(u)$ *be the linear form* $w \mapsto \langle \psi, u^\mathsf{T} \otimes w \rangle$ *on* W. *Then* $\hat{\psi} \in V'$, *and the mapping* $\psi \mapsto \hat{\psi}$ *is an isomorphism of the* $\mathfrak{g}$*-module* V^* *onto the* $\mathfrak{g}$*-module* V'.

Let $\psi \in V^*$, $u \in U(\mathfrak{g})$, $V \in U(\mathfrak{h})$ and $w \in W$. Then

$$\begin{aligned}\langle \hat{\psi}(vu), w \rangle &= \langle \psi, u^\mathsf{T} v^\mathsf{T} \otimes w \rangle = \langle \psi, u^\mathsf{T} \otimes v^\mathsf{T} w \rangle \\ &= \langle \hat{\psi}(u), v^\mathsf{T} w \rangle = \langle v \cdot \hat{\psi}(u), w \rangle,\end{aligned}$$

hence $\hat{\psi} \in V'$. If $u' \in U(\mathfrak{g})$, then

$$\begin{aligned}\langle (u' \cdot \hat{\psi})(u), w \rangle &= \langle \hat{\psi}(uu'), w \rangle = \langle \psi, u'^\mathsf{T} u^\mathsf{T} \otimes w \rangle \\ &= \langle u' \cdot \psi, u^\mathsf{T} \otimes w \rangle = \langle (u' \cdot \psi)^\wedge(u), w \rangle,\end{aligned}$$

hence $\psi \mapsto \hat{\psi}$ is a $\mathfrak{g}$-homomorphism. If $\hat{\psi} = 0$, then clearly $\psi = 0$. Finally, let $f \in V'$. The mapping $(u,w) \mapsto \langle f(u^\mathsf{T}), w \rangle$ of $U(\mathfrak{g}) \times W$ into k is bilinear, and for $v \in U(\mathfrak{h})$ we have

$$\langle f((uv)^\mathsf{T}), w \rangle = \langle f(v^\mathsf{T} u^\mathsf{T}), w \rangle = \langle v^\mathsf{T} f(u^\mathsf{T}), w \rangle = \langle f(u^\mathsf{T}), vw \rangle.$$

Hence there exists $\psi \in V^*$ such that

$$\langle \psi, u \otimes w \rangle = \langle f(u^\mathsf{T}), w \rangle \tag{1}$$

for $u \in U(\mathfrak{g})$ and $w \in W$. Clearly, $\hat{\psi} = f$.

5.5.5. We can identify V' with V^* under the isomorphism of 5.5.4. Thus V and V' are dual, so that, from (1),

$$\langle f, u \otimes w \rangle = \langle f(u^\mathsf{T}), w \rangle$$

for $u \in U(\mathfrak{g})$, $w \in W$, $f \in V'$.

5.5.6. COROLLARY. *Let* $\xi \in \mathfrak{g}^\wedge$ *and* $m \in \mathbf{N}$. *Then, with the notation of* 5.5.4, *the following conditions are equivalent:*

(i) $\mathrm{mtp}(\xi^*, V') = m$;

(ii) *let* $\mathscr{F}$ *be the set of sub-*$\mathfrak{g}$*-modules* T *of* V *such that* V/T *is isotopic of type* ξ; *then* $\mathscr{F}$ *has a least element* T_0, *and* $\mathrm{mtp}(\xi, V/T_0) = m$.

Let S be a finite-dimensional vector subspace of V^*. It is orthogonal to a vector subspace T of finite codimension of V. To say that S is isotypic of type ξ^* is equivalent to saying that V/T is isotypic of type ξ; and, if so, we have $\text{mtp}(\xi^*,S) = \text{mtp}(\xi,V/T)$. Given, this the corollary follows from 5.5.4.

5.5.7. PROPOSITION. *Let* $\mathfrak{h}$ *be a Lie subalgebra of* $\mathfrak{g}$ *which is reductive in* $\mathfrak{g}$, ϱ *a finite-dimensional simple representation of* $\mathfrak{h}$, ξ *a finite-dimensional simple representation of* $\mathfrak{g}$, *and* $\nu = \text{coind}(\varrho,\mathfrak{g})$. *Then* $\text{mtp}(\xi,\nu) = \text{mtp}(\varrho,\xi)$.

This follows from 5.5.3.

5.5.8. PROPOSITION. *Let* $\mathfrak{h}$ *and* $\mathfrak{k}$ *be Lie subalgebras of* $\mathfrak{g}$ *such that* $\mathfrak{g} = \mathfrak{h} + \mathfrak{k}$. *Let* $\mathfrak{l} = \mathfrak{h} \cap \mathfrak{k}$, ϱ *be a representation of* $\mathfrak{h}$, *and* $\pi = \text{coind}(\varrho,\mathfrak{g})$. *Then* $\pi|\mathfrak{k}$ *is equivalent to* $\text{coind}(\varrho|\mathfrak{l},\mathfrak{k})$.

Let W be the space of ϱ, and $\varphi: \text{Hom}_{U(\mathfrak{h})}(U(\mathfrak{g}),W) \to \text{Hom}_{U(\mathfrak{l})}(U(\mathfrak{k}),W)$ the restriction mapping. Clearly φ is a $\mathfrak{k}$-homomorphism. Since $U(\mathfrak{g}) = U(\mathfrak{h}) \cdot U(\mathfrak{k})$, φ is injective. Let $f \in \text{Hom}_{U(\mathfrak{l})}(U(\mathfrak{k}),W)$. There exists a $U(\mathfrak{h})$-homomorphism g of $U(\mathfrak{h}) \otimes_{U(\mathfrak{l})} U(\mathfrak{k})$ into W such that $g(u \otimes v) = uf(v)$ for $u \in U(\mathfrak{h})$ and $v \in U(\mathfrak{k})$. From 2.2.9, there then exists an $h \in \text{Hom}_{U(\mathfrak{h})}(U(\mathfrak{g}),W)$ such that $h(uv) = uf(v)$ for $u \in U(\mathfrak{h})$ and $v \in U(\mathfrak{k})$. Then $h|U(\mathfrak{k}) = f$, hence φ is surjective.

Reference: [67].

5.6. Supplementary remarks

5.6.1. The notions of induced and co-induced representations of algebras are certainly of long standing since they are concerned with the extension of scalars in a module. In any case, these notions are explicit in [67] (with the terminology "induced", "produced"). When the co-induced representations of $U(\mathfrak{g})$ are not finite-dimensional, they are of non-denumerable dimension, and consequently cannot be simple; hence we shall above all operate with induced representations (cf., however, the construction of the principle series in 9.3).

Let G be a real Lie group, and $\mathfrak{g}$ its Lie algebra. To each (for example) unitary representation of G in a complex Hilbert space H there corresponds a representation of $U(\mathfrak{g} \otimes \mathbf{C})$ in the space H_∞ of indefinitely differentiable vectors of H and a representation of $U(\mathfrak{g} \otimes \mathbf{C})$ in the space H^∞ of distribution vectors of H. To the induction of unitary representations corresponds, from the point of view of Lie algebras either the co-induction if we use H_∞, or the induction if we use H^∞. Further information on this may be found in [13] and [45].

Theorem 5.3.6 is due to Blattner [13]. Proposition 5.4.3 can be found in [35] and [48], and proposition 5.4.4 in [35].

5.6.2. Assume that k is algebraically closed.

(a) Let $\mathfrak{h}$ be a Lie subalgebra of $\mathfrak{g}$, and ϱ a simple representation of $\mathfrak{h}$. There exists a simple representation π of $\mathfrak{g}$ such that $\pi|\mathfrak{h}$ has a simple subrepresentation which is equivalent to ϱ.

(b) Let $\mathfrak{k}$ be an ideal of $\mathfrak{g}$,σ a simple representation of $\mathfrak{k}$, and $\mathfrak{h}$ a Lie subalgebra of $\mathfrak{g}$ containing $\mathfrak{k}$. The following conditions are equivalent:

(i) $\mathfrak{h} \subset \mathfrak{st}(\sigma,\mathfrak{g})$;

(ii) there exists a representation ϱ of $\mathfrak{h}$ such that $\varrho|\mathfrak{k}$ is a multiple of σ;

(iii) there exists a simple representation ϱ of $\mathfrak{h}$ such that $\varrho|\mathfrak{k}$ is a multiple of σ.

(c) Let $\mathfrak{k}$ be an ideal of $\mathfrak{g}$, K a primitive ideal of $U(\mathfrak{k})$, and $\mathfrak{h}$ a Lie subalgebra of $\mathfrak{g}$ containing $\mathfrak{k}$. The following conditions are equivalent:

(i) $\mathfrak{h} \subset \mathfrak{st}(K,\mathfrak{g})$;

(ii) there exists a two-sided ideal I of $U(\mathfrak{h})$ such that $K = I \cap U(\mathfrak{k})$;

(iii) there exists a primitive ideal I of $U(\mathfrak{h})$ such that $K = I \cap U(\mathfrak{k})$.

(d) Let $\mathfrak{k}$ be an ideal of $\mathfrak{g}$, and σ a simple representation of $\mathfrak{k}$. It is possible that $\mathfrak{st}(\sigma,\mathfrak{g}) \neq \mathfrak{st}(\mathrm{Ker}\,\sigma,\mathfrak{g})$.

(e) Let $\mathfrak{k}$ be an ideal of $\mathfrak{g}$, and π a simple representation of $\mathfrak{g}$ such that $\pi|\mathfrak{k}$ has a simple subrepresentation σ. Let $L = \mathrm{Ker}\,\sigma$ and $\mathfrak{h} = \mathfrak{st}(L,\mathfrak{g})$. There exists a simple representation ϱ of $\mathfrak{h}$ such that $\mathrm{Ker}\,\varphi \cap U(\mathfrak{k}) = L$ and $\mathrm{ind}(\varrho,\mathfrak{g})$ is equivalent to π [35].

5.6.3. If $\mathfrak{g}'$ is a Lie subalgebra of $\mathfrak{g}$ and I' a two-sided ideal of $U(\mathfrak{g}')$, we denote by $\mathrm{ind}(I',\mathfrak{g})$ the largest two-sided ideal of $U(\mathfrak{g})$ contained in $U(\mathfrak{g})I'$. This notation is justified by 5.1.7.

(a) We adopt the notation $\mathfrak{g},x,y,z$ of 1.14.10. Let $\mathfrak{g}' = ky + kz$, and I' be the ideal of $U(\mathfrak{g}')$ generated by $z - 1$. Then $\mathrm{ind}(I',\mathfrak{g})$ is equal to $U(\mathfrak{g})(z - 1)$, and hence is primitive although I' is not primitiv

(b) Assume that k is algebraically closed. Let $\mathfrak{k}$ be an ideal of $\mathfrak{g}$, I a maximal two-sided ideal of $U(\mathfrak{g})$, and $K = I \cap U(\mathfrak{k})$. There exists a primitive ideal L of $U(\mathfrak{k})$ which is generic for K and a maximal two-sided ideal I' of $U(\mathfrak{st}(L,\mathfrak{g}))$ such that $I = \mathrm{ind}(I',\mathfrak{g})$ [35].

(c) If I' is completely prime, then I is completely prime [24].

5.6.4. Assume that k is algebraically closed and $\mathfrak{g}$ solvable. Let $\mathfrak{k}$ be an ideal of $\mathfrak{g}$, and I a primitive ideal of $U(\mathfrak{g})$. There exist a simple representation σ of $\mathfrak{k}$ and a simple representation τ of $\mathfrak{st}(\sigma,\mathfrak{g})$ such that $\tau|\mathfrak{k}$ is a multiple of σ and $\mathrm{ind}(\tau,\mathfrak{g})$ is simple with kernel I [27].

5.6.5. Assume that k is algebraically closed. Let $\mathfrak{k}$ be a solvable ideal of $\mathfrak{g}$, I a primitive ideal of $U(\mathfrak{g})$, and $\mathscr{A}$ the algebraic adjoint group of $\mathfrak{g}$.

(a) There exists a primitive ideal of $U(\mathfrak{k})$ which is generic for $I \cap U(\mathfrak{k})$ [39].

(b) The primitive ideals of $U(\mathfrak{k})$ which are generic for $I \cap U(\mathfrak{k})$ form an $\mathscr{A}$-orbit in Prim $U(\mathfrak{k})$ [15].

(c) Assume that $\mathfrak{k}$ is nilpotent. For every primitive ideal L of $U(\mathfrak{k})$ which is generic for $I \cap U(\mathfrak{k})$, there exists a simple representation σ of $\mathfrak{k}$ with kernel L and a simple representation ϱ of $\mathfrak{st}(\sigma,\mathfrak{g})$ such that $\varrho|\mathfrak{k}$ is a multiple of σ and ind$(\varrho,\mathfrak{g})$ is simple with kernel I [22].

5.6.6. Let $\mathfrak{g},x,y,z$ be as in 1.14.10, and $\mathfrak{g}' = ky + kz$. Let ϱ be the representation of $\mathfrak{g}$ in $k[X]$ such that $\varrho(x)P = \mathrm{d}P/\mathrm{d}X$, $\varrho(y)P = XP$, $\varrho(z)P = P$ for all $P \in k[X]$. Then $\varrho|\mathfrak{g}'$ does not have any simple subrepresentation.

5.6.7. Let $\mathfrak{h}$ be a Lie subalgebra of $\mathfrak{g}$.

(a) Let W be an $\mathfrak{h}$-module, and $X = \mathrm{coind}(W,\mathfrak{g})$. For all $n \in \mathbf{N}$, let X_n be the set of the $x \in X$ such that $x(U_n(\mathfrak{g})) = 0$. Then (X_n) is a decreasing filtration of X with intersection 0 such that $U_n(\mathfrak{g})X_{m+n} \subset X_m$.

(b) Let W_1 and W_2 be $\mathfrak{h}$-modules, let

$$V_1 = \mathrm{coind}(W_1,\mathfrak{g}), \qquad V_2 = \mathrm{coind}(W_2,\mathfrak{g}), \qquad V_3 = \mathrm{coind}(W_1 \otimes W_2,\mathfrak{g}),$$

and let c be the coproduct of $U(\mathfrak{g})$. If $v_1 \in V_1$ and $v_2 \in V_2$, let $v_1 \times v_2$ be the element of

$$\mathrm{Hom}_{U(\mathfrak{h})\otimes U(\mathfrak{h})}(U(\mathfrak{g}) \otimes U(\mathfrak{g}), W_1 \otimes W_2)$$

defined by

$$(v_1 \times v_2)(u \otimes u') = v_1(u) \otimes v_2(u').$$

Then we define $v_1v_2 \in V_3$ by the formula $(v_1v_2)(u) = (v_1 \times v_2)(cu)$. The mapping $(v_1,v_2) \mapsto v_1v_2$ of $V_1 \times V_2$ into V_3 is bilinear, associative in an obvious sense, and $(V_1)_p \cdot (V_2)_q \subset (V_3)_{p+q}$. If $x \in \mathfrak{g}$, then $x(v_1v_2) = (xv_1)v_2 + v_1(xv_2)$. In particular, by considering k as a trivial $\mathfrak{h}$-module, $F = \mathrm{coind}(k,\mathfrak{g})$ is an algebra equipped with a decreasing filtration in which $\mathfrak{g}$ operates by derivations. If $q = \dim(\mathfrak{g}/\mathfrak{h})$, this algebra is isomorphic to the algebra of formal series in q indeterminates on k. This generalises 2.7.5.

(c) Let Y be a $\mathfrak{g}$-module. An F-module structure (cf. (b)) on Y such that $x(fy) = (xf)y + f(xy)$ for $x \in \mathfrak{g}$, $f \in F$, $y \in Y$ is termed a *transitive system of imprimitivity based on* $\mathfrak{g}/\mathfrak{h}$ *for* Y. Let W be an $\mathfrak{h}$-module and $V = \mathrm{coind}(W,\mathfrak{g})$. Then (b) defines an F-module structure on V which is a transitive system of imprimitivity based on $\mathfrak{g}/\mathfrak{h}$.

(d) Let Y be a $\mathfrak{g}$-module equipped with a transitive system of imprimitivity based on $\mathfrak{g}/\mathfrak{h}$. Let $W = Y/F_1 Y$ which is in a natural way an $\mathfrak{h}$-module. Let us assume that $\bigcap_{n \geqq 0} F_n Y = 0$ and $\dim W < +\infty$. For all $y \in Y$, let θy be the element of coind$(W, \mathfrak{g})$ which transforms each $u \in U(\mathfrak{g})$ into the canonical image of uy in W. Then θ is an isomorphism of Y onto coind$(W, \mathfrak{g})$ for the $\mathfrak{g}$-module and the F-module structures [13].

CHAPTER 6

PRIMITIVE IDEALS (THE SOLVABLE CASE)

This chapter is one of the central parts of the book. In it we succeed in completing, for k algebraically closed and $\mathfrak{g}$ solvable, the programme which was described in the introduction, namely the determination of all the primitive ideals of $U(\mathfrak{g})$.

Some results can be directly established in the solvable case, while others first require a thorough knowledge of the nilpotent case. Moreover, the theorems are sometimes more complete in the nilpotent case. As a result the logical progression of the chapter is rather tortuous.

6.1. The ideals $I(f)$

6.1.1. THEOREM. *Let* $s = (\mathfrak{g}_0, \mathfrak{g}_1, \ldots, \mathfrak{g}_n)$ *be an increasing sequence of ideals of* $\mathfrak{g}$ *such that* $\dim \mathfrak{g}_i = i$ *and* $\mathfrak{g}_n = \mathfrak{g}$. *Let* $f \in \mathfrak{g}^*$, *and* $\mathfrak{p} = \mathfrak{p}(f,s) \in P(f)$ (*cf.* 1.12.11). *Then* $\operatorname{ind}(f|\mathfrak{p},\mathfrak{g})$ *and* $\operatorname{ind}^{\sim}(f|\mathfrak{p},\mathfrak{g})$ *are absolutely simple.*

We reason by induction on $\dim \mathfrak{g}$.

Let us assume that there exists a non-null ideal $\mathfrak{a}$ of $\mathfrak{g}$ such that $f(\mathfrak{a}) = 0$. Let $\mathfrak{g}' = \mathfrak{g}/\mathfrak{a}$, let π be the canonical mapping of $\mathfrak{g}$ onto $\mathfrak{g}'$, and let f' be the element of $\mathfrak{g}'^*$ such that $f = f' \circ \pi$, $\mathfrak{g}'_i = \pi(\mathfrak{g}_i)$, $f'_i = f'|\mathfrak{g}'_i$ and $f_i = f|\mathfrak{g}_i$. Then $\pi(\mathfrak{g}_i^{f_i}) = \mathfrak{g}_i'^{f'_i}$. Let s' be the sequence deduced from $(\mathfrak{g}'_0, \ldots, \mathfrak{g}'_n)$ after the elimination of repeated terms. Then $\pi(\mathfrak{p}) = \mathfrak{p}(f',s')$ and $\mathfrak{a} \subset \mathfrak{p}$. It is sufficient to apply the induction hypothesis to $\mathfrak{g}'$, s', f', and 5.1.12, 5.2.4.

Let us henceforth assume that $\operatorname{Ker} f$ does not contain a non-null ideal of $\mathfrak{g}$. Let $\mathfrak{z}$ be the centre of $\mathfrak{g}$. Then $(\operatorname{Ker} f) \cap \mathfrak{z} = 0$, and hence $\dim \mathfrak{z} \leq 1$.

Let $\mathfrak{b} = \mathfrak{g}_{i_0}$ be the smallest non-central ideal in the sequence s. Then $i_0 = 1$ or 2, and $\mathfrak{g}_{i_0}$ is commutative. Let $\mathfrak{g}'$ be the subalgebra $\mathfrak{b}^f = \mathfrak{st}(f|\mathfrak{b},\mathfrak{g})$ of $\mathfrak{g}$. Since $[\mathfrak{g},\mathfrak{b}]$ is a non-null ideal of $\mathfrak{g}$, we have $f([\mathfrak{g},\mathfrak{b}]) \neq 0$, and hence $\mathfrak{g} \neq \mathfrak{g}$. Let $f' = f|\mathfrak{g}'$, $\mathfrak{g}'_i = \mathfrak{g}_i \cap \mathfrak{g}'$, and $f'_i = f|\mathfrak{g}'_i$. Let s' be the sequence deduced from $(\mathfrak{g}'_0, \ldots, \mathfrak{g}'_n)$ after the elimination of repeated terms. Let $x \in \mathfrak{g}_i^{f_i}$; if $i_0 \leq i$, then $\mathfrak{g}_{i_0} \subset \mathfrak{g}_i$, hence $x \in \mathfrak{g}'$ and then $x \in \mathfrak{g}_i'^{f'_i}$; if $i_0 > i$, then x is central in $\mathfrak{g}$, hence $x \in \mathfrak{g}_i'^{f'_i}$. This proves that $\mathfrak{p} \subset \mathfrak{p}(f',s') \subset \mathfrak{g}'$. Since $\mathfrak{p} \in P(f)$ and $\mathfrak{p}(f',s') \in P(f')$, we have $\mathfrak{p} = \mathfrak{p}(f',s')$. From the induction hypothesis, $\varrho = \operatorname{ind}(f'|\mathfrak{p},\mathfrak{g}')$ and $\varrho^\sim = \operatorname{ind}^\sim(f'|\mathfrak{p},\mathfrak{g}')$ are absolutely simple. We have $\mathfrak{b} \subset \mathfrak{g}'^{f'} \subset \mathfrak{p}$. The representation $\operatorname{ind}(f|\mathfrak{p},\mathfrak{g})$ is equivalent to $\operatorname{ind}(\mathfrak{p},\mathfrak{g})$, and the representation $\operatorname{ind}^\sim(f|\mathfrak{p},\mathfrak{g})$ is equivalent to $\operatorname{ind}^\sim(\varrho^\sim,\mathfrak{g})$ (5.1.11 and 5.2.3). From 5.1.13, $\varrho|\mathfrak{b}$ and $\varrho^\sim|\mathfrak{b}$ are multiples of $f|\mathfrak{b}$. The theorem then follows from 5.3.6.

6.1.2. LEMMA. (i) *Let* $\mathfrak{g}_1$ *be a Lie algebra with basis* (x,y) *such that* $[x,y] = y$, (x^*,y^*) *is the dual basis for* $\mathfrak{g}_1^*$, $\alpha \in k$, $f = \alpha x^* + y^*$, $\mathfrak{h}_1 = kx$, $\mathfrak{h}_2 = ky$, $\varrho_1 = \operatorname{ind}^\sim(f|\mathfrak{h}_1,\mathfrak{g}_1)$ *and* $\varrho_2 = \operatorname{ind}^\sim(f|\mathfrak{h}_2,\mathfrak{g}_1)$. *Then* ϱ_1 *and* ϱ_2 *have the same kernel (namely null).*

(ii) *Let* $\mathfrak{g}_2$ *be a Lie algebra with basis* (x,y,z) *such that* $[x,y] = y$, $[x,z] = [y,z] = 0$, (x^*,y^*,z^*) *is the dual basis for* $\mathfrak{g}_2^*$, $\alpha \in k$, $f = \alpha x^* + y^* + z^*$, $\mathfrak{h}_1 = kx + kz$, $\mathfrak{h}_2 = ky + kz$, $\varrho_1 = \operatorname{ind}^\sim(f|\mathfrak{h}_1,\mathfrak{g}_2)$ *and* $\varrho_2 = \operatorname{ind}^\sim(f|\mathfrak{h}_2,\mathfrak{g}_2)$. *Then* ϱ_1 *and* ϱ_2 *have the same kernel, namely* $(z-1)U(\mathfrak{g}_2)$.

(iii) *Let* $\mathfrak{g}_3$ *be a Lie algebra with basis* (x,y,z) *such that* $[x,y] = z$, $[x,z] = [y,z] = 0$, (x^*,y^*,z^*) *is the dual basis for* $\mathfrak{g}_3^*$, $\alpha,\beta \in k$, $f = \alpha x^* + y^* + z^*$, $\mathfrak{h}_1 = kx + kz$, $\mathfrak{h}_2 = ky + kz$, $\varrho_1 = \operatorname{ind}^\sim(f|\mathfrak{h}_1,\mathfrak{g}_3)$ *and* $\varrho_2 = \operatorname{ind}^\sim(f|\mathfrak{h}_2,\mathfrak{g}_3)$. *Then* ϱ_1 *and* ϱ_2 *have the same kernel, namely* $(z-1)U(\mathfrak{g}_3)$.

(iv) *Let* $\mathfrak{g}_4$ *be a Lie algebra with basis* (x,y,z,t) *such that* $[x,y] = y$, $[x,t] = -t$, $[t,y] = z$, $[\mathfrak{g},z] = 0$, (x^*,y^*,z^*,t^*) *is the dual basis for* $\mathfrak{g}_4$, $f = z^*$, $\mathfrak{h}_1 = kx + ky + kz$, $\mathfrak{h}_2 = kx + kt + kz$, $\varrho_1 = \operatorname{ind}^\sim(f|\mathfrak{h}_1,\mathfrak{g}_4)$ *and* $\varrho_2 = \operatorname{ind}^\sim(f|\mathfrak{h}_2,\mathfrak{g}_4)$. *Then* ϱ_1 *and* ϱ_2 *have the same kernel.*

Let us adopt the notation of (i). We have

$$[x,x^m y^n] = n x^m y^n,$$

$$[y,x^m y^n] = (-m x^{m-1} + \tfrac{1}{2}m(m-1)\, x^{m-2} - \cdots + (-1)^m) y^{n+1}$$

from 3.3.7. From this we deduce that the $U(\mathfrak{g}_1)_\lambda$ (cf. 4.3.1) are the vector subspaces k, ky, $ky^2, \ldots$ If J is a non-null two-sided ideal of $U(\mathfrak{g}_1)$, then J contains some y^i (4.4.1). From 5.1.6, the space V_1 of ϱ_1 has a basis $(e_0,e_1,\ldots)$ such that $\varrho_1(y)e_n = e_{n+1}$, and the space V_2 of ϱ_2 has a basis $(e'_0,e'_1,\ldots)$ such that

$$\begin{aligned}\varrho_2(y)e'_n &= \varrho_2(y)\varrho_2(x)^n e'_0\\ &= \varrho_2(x-1)^n \varrho_2(y)e'_0 = e'_n - n e'_{n-1} + \cdots + (-1)^n e'_0.\end{aligned}$$

It can be seen that $\varrho_1(y^i) \neq 0$ and $\varrho_2(y^i) \neq 0$ for all i, and hence

$$\operatorname{Ker} \varrho_1 = \operatorname{Ker} \varrho_2 = 0.$$

Let us adopt the notation of (ii). From 5.1.6, the representations ϱ_1, ϱ_2 operate in the same spaces as in (i) and x,y operate in the same way. Moreover, $\varrho_1(z) = 1$ and $\varrho_2(z) = 1$. Let $\sum_{\alpha\beta} x^\alpha y^\beta p_{\alpha\beta}(z)$ be an element of $U(\mathfrak{g}_2)$, where the $p_{\alpha\beta}$ are polynomials. Then

$$\begin{aligned}
&\sum_{\alpha,\beta} x^\alpha y^\beta p_{\alpha\beta}(z) \in \operatorname{Ker} \varrho_1\\
\Leftrightarrow &\sum_{\alpha,\beta} p_{\alpha\beta}(1) x^\alpha y^\beta \in \operatorname{Ker} \varrho_1\\
\Leftrightarrow &\, p_{\alpha\beta}(1) = 0 \quad \text{for all } \alpha,\beta \text{ (from (i))}\\
\Leftrightarrow &\, \varrho_{\alpha\beta}(z) \quad \text{is divisible by } z - 1 \text{ for all } \alpha,\beta.
\end{aligned}$$

Hence $\operatorname{Ker} \varrho_1 = (z-1)U(\mathfrak{g}_2)$ and similarly $\operatorname{Ker} \varrho_2 = (z-1)U(\mathfrak{g}_2)$.

Let us adopt the notation of (iii). We have $\varrho_1(z) = 1$ and $\varrho_2(z) = 1$. Now the classes x', y' of x,y modulo $(z-1)U(\mathfrak{g}_3)$ satisfy $[x',y'] = 1$. Hence $U(\mathfrak{g}_3)/(z-1)U(\mathfrak{g}_3)$ is isomorphic to the Weyl algebra A_1, which is simple (4.6.6). Consequently,

$$\operatorname{Ker} \varrho_1 = (z-1)U(\mathfrak{g}_3) = \operatorname{Ker} \varrho_2.$$

Let us adopt the notation of (iv). We have $\varrho_1(z) = 1$ and $\varrho_2(z) = 1$. It is easily verified that $xz - ty \in Z(\mathfrak{g}_4)$. The space V_1 of ϱ_1 has a basis $(e_0, e_1, \ldots)$ such that $\varrho_1(t)e_n = e_{n+1}$, and

$$\begin{aligned}
\varrho_1(xz) - ty)e_n &= \varrho_1((xz - ty)t^n)e_0 = \varrho_1(t^n(xz - ty))e_0\\
&= \varrho_1(t)^n(f(x) - \tfrac{1}{2})f(z)e_0 - \varrho_1(t)^{n+1}f(y)e_0 = -\tfrac{1}{2}\, e_n.
\end{aligned}$$

Similarly, the space V_2 of ϱ_2 has a basis $(e_0', e_1', \ldots)$ such that $\varrho_2(y)e_n' = e_{n+1}'$, and

$$\begin{aligned}
\varrho_2(xz - ty)e_n' &= \varrho_2((xz - ty)y^n)e_0' = \varrho_2(y^n(xz - ty))e_0'\\
&= \varrho_2(y^n(xz - yt - z))e_0'\\
&= \varrho_2(y)^n(f(x) + \tfrac{1}{2})f(z)e_0' - \varrho_2(y)^{n+1}f(t)e_0' - \varrho_2(y)^n f(z)e_0'\\
&= -\tfrac{1}{2}\, e_n'.
\end{aligned}$$

Let J be the two-sided ideal of $U(\mathfrak{g}_4)$ generated by $z - 1$ and $x - ty + \frac{1}{2}$. Then the algebra $U(\mathfrak{g}_4)/J$ is generated by the canonical images y', t' of y, t, and $[y',t'] = -1$, hence this algebra is isomorphic to A_1 and consequently simple. Hence $\operatorname{Ker} \varrho_1 = J = \operatorname{Ker} \varrho_2$.

6.1.3. LEMMA. *Let* $\mathfrak{a}$ *be a commutative ideal of* $\mathfrak{g}$, $\mathfrak{a}'$ *the centralizer of* $\mathfrak{a}$ *in* $\mathfrak{g}$, $f \in \mathfrak{g}^*$, $\mathfrak{h}$ *a Lie subalgebra of* $\mathfrak{g}$ *subordinate to* f, *and* $\mathfrak{h}' = (\mathfrak{h} \cap \mathfrak{a}^f) + \mathfrak{a}$. *Then* $\mathfrak{h}'$ *is a Lie subalgebra of* $\mathfrak{g}$ *subordinate to* f, *and* $\mathfrak{h} \cap \mathfrak{a}' \cap (\operatorname{Ker} f)$ *is an ideal of the Lie algebra* $\mathfrak{h} + \mathfrak{a}$.

Since $\mathfrak{a}^f$ is a Lie subalgebra of $\mathfrak{g}$, clearly $\mathfrak{h}'$ is a Lie subalgebra of $\mathfrak{g}$. We have $[\mathfrak{h}',\mathfrak{h}'] \subset [\mathfrak{h},\mathfrak{h}] + [\mathfrak{a}^f,\mathfrak{a}]$, and hence $\mathfrak{h}'$ is subordinate to f. Finally

$$[\mathfrak{h} + \mathfrak{a}, \mathfrak{h} \cap \mathfrak{a}' \cap \operatorname{Ker} f] = [\mathfrak{h},\mathfrak{h} \cap \mathfrak{a}' \cap \operatorname{Ker} f] \subset [\mathfrak{h},\mathfrak{h}] \cap \mathfrak{a}'$$
$$\subset \mathfrak{h} \cap \mathfrak{a}' \cap \operatorname{Ker} f.$$

6.1.4. THEOREM ($\mathfrak{g}$ completely solvable). *Let* $f \in \mathfrak{g}^*$, $\mathfrak{h}_1 \in P(f)$ $\mathfrak{h}_2 \in P(f)$, $\varrho_1 = \operatorname{ind}^\sim(f|\mathfrak{h}_1,\mathfrak{g})$, *and* $\varrho_2 = \operatorname{ind}^\sim(f|\mathfrak{h}_2,\mathfrak{g})$. *Then* $\operatorname{Ker} \varrho_1 = \operatorname{Ker} \varrho_2$ (*cf.* 6.6.2).

We assume that $\dim \mathfrak{g} \geqq 1$, and that the theorem has been proved for dimensions $<\dim \mathfrak{g}$.

(a) Let us assume that there exists a non-null ideal $\mathfrak{a}$ of $\mathfrak{g}$ such that $f(\mathfrak{a}) = 0$. Then $\mathfrak{h}_1 \supset \mathfrak{a}$, $\mathfrak{h}_2 \supset \mathfrak{a}$, and we easily return to the study of $\mathfrak{g}/\mathfrak{a}$ to which we can apply the induction hypothesis. Henceforth we assume that $\operatorname{Ker} f$ does not contain a non-null ideal of $\mathfrak{g}$. It follows from this that the centre $\mathfrak{z}$ of $\mathfrak{g}$ is of dimension 0 or 1.

(b) Let us assume that there exists a one-dimensional non-central ideal $\mathfrak{a} = ka$ in $\mathfrak{g}$. We may assume that $f(a) = 1$. There exists $\lambda \in \mathfrak{g}^*$ such that $[x,a] = \lambda(x)a$ for all $x \in \mathfrak{g}$. Then $\mathfrak{a}^f = \operatorname{Ker} \lambda$ is the centralizer $\mathfrak{a}'$ of $\mathfrak{a}$ in $\mathfrak{g}$.

We shall now establish the following intermediate assertion:

($\bigstar$) Let $\mathfrak{h} \in P(f)$ and $\varrho = \operatorname{ind}^\sim(f|\mathfrak{h},\mathfrak{g})$. There exists $\mathfrak{h}' \in P(f)$ such that $\mathfrak{h}' \subset \mathfrak{a}^f$ and such that, if $\varrho' = \operatorname{ind}^\sim(f|\mathfrak{h}',\mathfrak{g})$, we have $\operatorname{Ker} \varrho = \operatorname{Ker} \varrho'$.

We may assume that $\mathfrak{h} \not\subset \mathfrak{a}^f$. There then exists $x \in \mathfrak{h}$ such that $[x,a] = a$. Since $f(a) = 1$ and $f([\mathfrak{h},\mathfrak{h}]) = 0$, we have $a \notin \mathfrak{h}$. Clearly, $\mathfrak{h} = (\mathfrak{h} \cap \mathfrak{a}^f) \oplus kx$. Let us set $\mathfrak{h}' = (\mathfrak{h} \cap \mathfrak{a}^f) \oplus ka$. From 6.1.3, $\mathfrak{h}'$ is a Lie subalgebra of $\mathfrak{g}$ subordinate to f; we have $\dim \mathfrak{h}' = \dim \mathfrak{h}$, hence $\mathfrak{h}' \in P(f)$, and $\mathfrak{h}' \subset \mathfrak{a}^f$. Let ϱ and ϱ' be as in ($\bigstar$), and let us prove that $\operatorname{Ker} \varrho = \operatorname{Ker} \varrho'$. Let $\mathfrak{k}$ be the Lie algebra $\mathfrak{h} + \mathfrak{a}$, $\sigma = \operatorname{ind}^\sim(f|\mathfrak{h},\mathfrak{k})$, and $\sigma' = \operatorname{ind}^\sim(f|\mathfrak{h}',\mathfrak{k})$. Then ϱ is equivalent to $\operatorname{ind}^\sim(\sigma,\mathfrak{g})$ and ϱ' is equivalent to $\operatorname{ind}^\sim(\sigma,\mathfrak{g})$ (5.2.3). From 5.2.6, it is sufficient to prove that $\operatorname{Ker} \sigma = \operatorname{Ker} \sigma'$. Clearly, $\mathfrak{h},\mathfrak{h}' \in P(f|\mathfrak{k})$. Let $\mathfrak{b} = \mathfrak{h} \cap \mathfrak{a}' \cap \operatorname{Ker} f$, which is an ideal of $\mathfrak{k}$ (6.1.3). If $\mathfrak{b} \neq 0$, then $\operatorname{Ker} \sigma = \operatorname{Ker} \sigma'$ from part (a) of the proof. Let us assume that $\mathfrak{b} = 0$.

If $\mathfrak{h} \cap \mathfrak{a}^f = 0$, then

$$\mathfrak{h} = kx, \qquad \mathfrak{h}' = ka, \qquad \mathfrak{k} = kx + ka,$$

$$[x,a] = a, \qquad f(a) = 1,$$

and it is sufficient to apply 6.1.2 (i). Otherwise we have $\dim(\mathfrak{h} \cap \mathfrak{a}^f) = 1$. Let z be a non-zero element of $\mathfrak{h} \cap \mathfrak{a}^f$. Then

$$\mathfrak{h} = kz \oplus kx, \qquad \mathfrak{h}' = kz \oplus ka, \qquad \mathfrak{k} = kz \oplus kx \oplus ka,$$

$$[x,a] = a, \qquad [z,a] \in [\mathfrak{a}',a] = 0.$$

Since $\mathfrak{b} = 0$, we have $f(z) \neq 0$, and hence we may assume that $f(z) = 1$; now $[x,z] \in \mathfrak{h} \cap \mathfrak{a}^f$, hence $[x,z]$ is proportional to z, and

$$f([x,z]) \in f([\mathfrak{h},\mathfrak{h}]) = 0$$

hence $[x,z] = 0$. It is then sufficient to apply 6.1.2 (ii).

We have thus established (★), which allows us to assume that $\mathfrak{h}_1 \subset \mathfrak{a}^f$ and $\mathfrak{h}_2 \subset \mathfrak{a}^f$. From the induction hypothesis, $\mathrm{ind}^{\sim}(f|\mathfrak{h}_1,\mathfrak{a}^f)$ and $\mathrm{ind}^{\sim}(f|\mathfrak{h}_2,\mathfrak{a}^f)$ have the same kernel, and it is sufficient to apply 5.2.3 and 5.2.6.

(c) Let us assume that there does not exist a one-dimensional non-central ideal in $\mathfrak{g}$. Then $\dim \mathfrak{z} = 1$, and $f(\mathfrak{z}) \neq 0$. Let $z \in \mathfrak{z}$ such that $f(z) = 1$. There exists an ideal $\mathfrak{a}$ of $\mathfrak{g}$ which contains $\mathfrak{z}$ and is of dimension 2 (at least if $\dim \mathfrak{g} > 1$, which we may assume to be the case); this ideal is commutative; let $\mathfrak{a}'$ be its centralizer in $\mathfrak{g}$. There exists $a \in \mathfrak{a}$ such that $a \neq 0$ and $f(a) = 0$, and then $\mathfrak{a} = ka \oplus kz$. There exist $\lambda,\mu \in \mathfrak{g}^*$ such that $[t,a] = \lambda(t)a + \mu(t)z$ for all $t \in \mathfrak{g}$. If $\mu = 0$, then ka is a one-dimensional non-central ideal, which is a contradiction; hence $\mu \neq 0$. We have $\mathfrak{a}^f = \mathrm{Ker}\mu$. We shall now establish (★), after which the proof is concluded as in (b). We may assume that $\mathfrak{h} \not\subset \mathfrak{a}^f$; let us choose $x \in \mathfrak{h}$ such that $\mu(x) = 1$. Then $\mathfrak{z} \subset \mathfrak{h}$. We set

$$\mathfrak{h}' = (\mathfrak{h} \cap \mathfrak{a}^f) + \mathfrak{a}, \qquad \mathfrak{b} = \mathfrak{h} \cap \mathfrak{a}' \cap \mathrm{Ker} f, \qquad \mathfrak{k} = \mathfrak{h} + \mathfrak{a}.$$

Since

$$f([\mathfrak{h},\mathfrak{h}]) = 0, \qquad f([x,a]) = \mu(x)f(z) = 1,$$

we have $a \notin \mathfrak{h}$. Hence $\dim \mathfrak{h}' = \dim \mathfrak{h}$ and $\mathfrak{h}' \in P(f)$. As in (b), we return to proving that $\mathrm{ind}^{\sim}(f|\mathfrak{h},\mathfrak{k})$ and $\mathrm{ind}^{\sim}(f|\mathfrak{h}',\mathfrak{k})$ have the same kernel, and we may assume that $\mathfrak{b} = 0$. Then $\dim \mathfrak{h} \cap \mathfrak{a}' \leq 1$, and since $\mathfrak{z} \subset \mathfrak{h} \cap \mathfrak{a}'$, we have $\mathfrak{h} \cap \mathfrak{a}' = \mathfrak{z}$.

Let us assume that $\lambda = 0$. Then $\mathfrak{a}' = \operatorname{Ker} \mu$ and $[x,a] = z$. Clearly,

$$\mathfrak{h} = (\mathfrak{h} \cap \mathfrak{a}') \oplus kx = kx \oplus kz,$$
$$\mathfrak{h}' = (\mathfrak{h} \cap \mathfrak{a}') \oplus ka = ka \oplus kz,$$
$$\mathfrak{k} = kx \oplus ka \oplus kz.$$

Since $f(z) = 1$, it is sufficient to apply 6.1.2 (iii).

Let us assume that $\lambda = \alpha\mu$, with $\alpha \in k - \{0\}$. Then, for all $t \in \mathfrak{g}$, we have

$$[t, a + \alpha^{-1}z] = \alpha\mu(t)a + \mu(t)z = \alpha\mu(t)\,(a + \alpha^{-1}z),$$

and hence $k(a + \alpha^{-1}z)$ is a one-dimensional non-central ideal, which is a contradiction.

Finally, let us assume that λ and μ are linearly independent. Then

$$\mathfrak{a}' = \operatorname{Ker} \lambda \cap \operatorname{Ker} \mu$$

is of codimension 1 in $\mathfrak{a}^f$. Hence $\mathfrak{z} = \mathfrak{h} \cap \mathfrak{a}'$ is of codimension $\leqq 1$ in $\mathfrak{h} \cap \mathfrak{a}^f$. If $\mathfrak{z} = \mathfrak{h} \cap \mathfrak{a}^f$, then

$$\mathfrak{h} = kz \oplus kx, \qquad \mathfrak{h}' = kz \oplus ka,$$
$$[x,a] = \lambda(x)a + z, \quad \mathfrak{k} = kx \oplus ka \oplus kz,$$

According to whether $\lambda(x) \neq 0$ or $\lambda(x) = 0$, it is sufficient to apply 6.1.2 (ii) or 6.1.2 (iii). Let us assume that $\dim(\mathfrak{h} \cap \mathfrak{a}^f) = 2$. Then $\mathfrak{h}/\mathfrak{z} = \mathfrak{h}/\mathfrak{h} \cap \mathfrak{a}'$ is of dimension 2, and hence $\lambda|\mathfrak{h}$ and $\mu|\mathfrak{h}$ are linearly independent. By adding an element of $\mathfrak{h} \cap \mathfrak{a}^f$ to x, we may assume that $\lambda(x) = 0$. On the other hand, there exists $y \in \mathfrak{h} \cap \mathfrak{a}^f$ such that $\lambda(y) = 1$ and $\mu(y) = 0$. By adding elements of $\mathfrak{z}$ to x and y, we may assume that $f(x) = f(y) = 0$. Then

$$\mathfrak{h} = kx \oplus ky \oplus kz, \qquad \mathfrak{h}' = ky \oplus ka \oplus kz, \qquad \mathfrak{k} = kx \oplus ky \oplus ka \oplus kz,$$
$$[x,a] = z, \qquad [y,a] = a, \qquad [\mathfrak{k},z] = 0$$

and, since $[x,y] \in [\mathfrak{h},\mathfrak{h}] \subset \mathfrak{h} \cap \operatorname{Ker} f$, we have

$$[x,y] = \alpha x + \beta y, \quad \text{with } \alpha,\beta \in k.$$

Moreover,

$$\begin{aligned} 0 &= [a,[x,y]] + [x,[y,a]] + [y,[a,x]] \\ &= [a, \alpha x + \beta y] + [x,a] - [y,z] \\ &= -\alpha z - \beta a + z, \end{aligned}$$

whence $\beta = 0$, $\alpha = 1$, and $[x,y] = x$. Since $f(x) = f(y) = f(a) = 0$ and $f(z) = 1$, it is sufficient to apply 6.1.2 (iv).

6.1.5. Let us assume that $\mathfrak{g}$ is completely solvable. Let $f \in \mathfrak{g}^*$. *A two-sided ideal $I(f)$ of $U(\mathfrak{g})$ is associated with f* in the following way: choose $\mathfrak{h} \in P(f)$, and set $I(f) = \operatorname{Ker} \operatorname{ind}^{\sim}(f|\mathfrak{h},\mathfrak{g})$; from 6.1.4, $I(f)$ only depends on f. From 6.1.1, *$I(f)$ is primitive*, and is even the kernel of an absolutely simple representation. Let $\mathscr{A}$ be the algebraic adjoint group of $\mathfrak{g}$. It operates in $\mathfrak{g}^*$ and in $U(\mathfrak{g})$. If $f \in \mathfrak{g}^*$ and $a \in \mathscr{A}$, then $I(af) = a(I(f))$ by transport of structure, and hence, from 2.4.17, $I(af) = I(f)$. The mapping I of $\mathfrak{g}^*$ into $\operatorname{Prim} U(\mathfrak{g})$ thus defines a mapping, denoted by $\bar{I}$, of $\mathfrak{g}^*/\mathscr{A}$ into $\operatorname{Prim} U(\mathfrak{g})$. If k is algebraically closed, $\bar{I}$ is bijective; we shall prove this in 6.2.4 for $\mathfrak{g}$ nilpotent and in 6.5.12 for $\mathfrak{g}$ solvable.

6.1.6. LEMMA. *Let $f \in \mathfrak{g}^*$, let $\mathfrak{h}$ be a Lie subalgebra of $\mathfrak{g}$ subordinate to f, and let $\varrho = \operatorname{ind}(f|\mathfrak{h},\mathfrak{g})$. We assume that ϱ is simple. Then $\mathfrak{h}$ contains all one-dimensional ideals of $\mathfrak{g}$.*

Let $y \in \mathfrak{g}$ such that ky is an ideal of $\mathfrak{g}$ and $y \notin \mathfrak{h}$. Then $\mathfrak{g}' = \mathfrak{h} + ky$ is a Lie subalgebra of $\mathfrak{g}$. Let $\varrho' = \operatorname{ind}(f|\mathfrak{h},\mathfrak{g}')$, and let V' be the space of ϱ'. From 5.1.10, ϱ' is simple. From 5.1.6, there exists a basis $(e_0,e_1, \ldots)$ for V' such that $\varrho'(y)e_i = e_{i+1}$ for all i. In particular, $W' = \varrho'(y)(V')$ is distinct from V'. Now there exists $\lambda \in \mathfrak{g}^*$ such that $[x,y] = \lambda(x)y$ for all $x \in \mathfrak{g}$. In particular, $[\varrho'(x),\varrho'(y)] = \lambda(x)\varrho'(y)$ for all $x \in \mathfrak{g}'$. This implies that $\varrho'(y)(V')$ is stable under $\varrho'(\mathfrak{g}')$, which leads to a contradiction, which establishes the lemma.

6.1.7. THEOREM (k algebraically closed, $\mathfrak{g}$ solvable). *Let I be a primitive ideal of $U(\mathfrak{g})$. There exists $f \in \mathfrak{g}^*$ such that $I = I(f)$.*

This is obvious if $\dim \mathfrak{g} \leq 1$. Let us assume that $\dim \mathfrak{g} > 1$ and let us reason by induction on $\dim \mathfrak{g}$.

If $I \cap \mathfrak{g} \neq 0$, it is sufficient to apply the induction hypothesis to the ideal of $U(\mathfrak{g}/(I \cap \mathfrak{g}))$ deduced from I by passage to the quotient. Let us henceforth assume that $I \cap \mathfrak{g} = 0$. Since the centre of $U(\mathfrak{g})/I$ is k (2.6.5), the centre $\mathfrak{z}$ of $\mathfrak{g}$ is of dimension ≤ 1. If $\mathfrak{z} = 0$, we denote a one-dimensional ideal of $\mathfrak{g}$ by $\mathfrak{a}$; if $\dim \mathfrak{z} = 1$, we denote a two-dimensional ideal of $\mathfrak{g}$ containing $\mathfrak{z}$ by $\mathfrak{a}$. In both cases, $\mathfrak{a}$ is commutative and $[\mathfrak{g},\mathfrak{a}] \neq 0$.

From 5.4.4, there exists a primitive ideal J of $U(\mathfrak{a})$ which is generic for $I \cap U(\mathfrak{a})$, a simple representation σ of $\mathfrak{a}$ with kernel J, and a simple representation ϱ of $\mathfrak{g}' = \mathfrak{st}(\sigma,\mathfrak{g})$, such that $\varrho|\mathfrak{a}$ is a multiple of σ and $\operatorname{ind}^{\sim}(\varrho,\mathfrak{g})$ is simple with kernel I. Since σ is one-dimensional, $\mathfrak{g}'$ is the set of the $x \in \mathfrak{g}$

such that $\sigma([x,\mathfrak{a}]) = 0$; hence $\varrho([\mathfrak{g}',\mathfrak{a}]) = 0$, if $\mathfrak{g}' = \mathfrak{g}$, we deduce from this that $I \supset [\mathfrak{g},\mathfrak{a}]$, which is a contradiction. Hence $\mathfrak{g}' \neq \mathfrak{g}$.

From the induction hypothesis, there exist $f' \in \mathfrak{g}'^*$ and $\mathfrak{h} \in P(f')$ such that, if $\varrho' = \text{ind}^{\sim}(f'|\mathfrak{h},\mathfrak{g}')$, we have $\text{Ker}\,\varrho = \text{Ker}\,\varrho'$. From 6.1.1. and 6.1.4, we may even assume that ϱ' is simple. From 5.2.6, I is the kernel of $\text{ind}^{\sim}(f'|\mathfrak{h},\mathfrak{g})$. Let $f \in \mathfrak{g}^*$ be an extension of f'. We shall now prove that $\mathfrak{h} \in P(f)$, and the proof will be concluded.

Firstly, $f([\mathfrak{h},\mathfrak{h}]) = f'([\mathfrak{h},\mathfrak{h}]) = 0$. Let $x \in \mathfrak{g}$ such that $f([x,\mathfrak{h}]) = 0$, and let us show that $x \in \mathfrak{h}$. If $\mathfrak{a} = ky$ with some $y \in \mathfrak{g} - \{0\}$, then $\mathfrak{a} \subset \mathfrak{h}$ and $y - f'(y) \in \text{Ker}\,\varrho'$ (since y is in the centre of $\mathfrak{g}'$), and $y - \sigma(y) \in \text{Ker}\,\varrho$ (since $\varrho|\mathfrak{a}$ is a multiple of σ). Since $\text{Ker}\,\varrho = \text{Ker}\,\varrho'$, it can be seen that $f(y) = \sigma(y)$. We have $f([x,\mathfrak{a}]) = 0$, hence $\sigma([x,\mathfrak{a}]) = 0$, whence $x \in \mathfrak{g}'$ and then $x \in \mathfrak{h}$. Let us henceforth assume that $\dim \mathfrak{a} = 2$. Then $\mathfrak{a} = ky \oplus kz$ with $z \in \mathfrak{z}$. We may assume that $z - 1 \in I$, and *a fortiori* $z - 1 \in \text{Ker}\,\sigma$, whence $\sigma(z) = 1$. On the other hand, $z - f'(z) \in \text{Ker}\,\varrho' = \text{Ker}\,\varrho$, and $\varrho|\mathfrak{a}$ is a multiple of σ, whence $f'(z) = \sigma(z) = 1$. By changing y, we may assume that $\sigma(y) = 0$, whence $\varrho(y) = 0$ and $\varrho'(y) = 0$. There exist $\lambda,\mu \in \mathfrak{g}^*$ such that $[u,y] = \lambda(u)y + \mu(u)z$ for all $u \in \mathfrak{g}$, whence $\sigma([u,y]) = \mu(u)$ and $\mathfrak{g}' = \text{Ker}\,\mu$. It can then be seen that ky is a one-dimensional ideal of $\mathfrak{g}'$, whence $y \in \mathfrak{h}$ (6.1.6), and the equality $\varrho'(y) = 0$ implies that $f'(y) = 0$. Thus, $f|\mathfrak{a} = \sigma|\mathfrak{a}$. Since $f([x,\mathfrak{h}]) = 0$, it can be seen that $\sigma([x,y]) = 0$, whence $\sigma([x,\mathfrak{a}]) = 0$, $x \in \mathfrak{g}'$ and thus $x \in \mathfrak{h}$.

6.1.8. PROPOSITION ($\mathfrak{g}$ completely solvable). *Let $f,h \in \mathfrak{g}^*$. We assume that $h([\mathfrak{g},\mathfrak{g}]) = 0$; let α be the automorphism of $U(\mathfrak{g})$ such that*

$$\alpha(x) = x - h(x) \quad \textit{for all } x \in \mathfrak{g}$$

(2.2.23). *Then $I(f + h) = \alpha(I(f))$.*

Let $\mathfrak{h} \in P(f)$. Since $B_f = B_{f+h}$, we have $\mathfrak{h} \in P(f + h)$. Let L and L' be the left ideals of $U(\mathfrak{g})$ generated by the $x - f(x) - \theta_{\mathfrak{g},\mathfrak{h}}(x)$ where $x \in \mathfrak{h}$ and by the $x - f(x) - h(x) - \theta_{\mathfrak{g},\mathfrak{h}}(x)$ where $x \in \mathfrak{h}$, respectively. Then $I(f)$ and $I(f + h)$, are the largest two-sided ideals of $U(\mathfrak{g})$ contained in L and L', respectively (5.1.7, 5.1.9). Now $\alpha(L) = L'$, and hence $\alpha(I(f)) = I(f + h)$.

References: [9], [27], [31], [45], [124].

6.2. Rational ideals in the nilpotent case

In the nilpotent case, the results of section 6.1 allow of the improvements 6.2.2, 6.2.3 and 6.2.9 set out below.

6.2.1. LEMMA ($\mathfrak{g}$ nilpotent). *Let* $(x,y,z,\mathfrak{h})$ *be a reducing quadruple of* $\mathfrak{g}$. *Let* $\bar{\mathfrak{h}}, \bar{z}$ *and* ψ *be as in* 4.7.8. *Let* $f \in \mathfrak{g}^*$ *such that* $f(y) = 0$ *and* $f(z) = 1$. *Let* $g = f|\mathfrak{h}$ *and let* $\bar{g}$ *be the linear form on* $\bar{\mathfrak{h}}$ *deduced from* g *by passage to the quotient. Then* $\psi(I(f)_z) = I(\bar{g})_{\bar{z}} \otimes A_1$. *If* $\mathfrak{k} \in P(g)$, *then* $\mathfrak{k} \in P(f)$.

Let $\mathfrak{k} \in P(g)$. Then $y \in \mathfrak{h}^g \subset \mathfrak{k}$, and $f([x,y]) = 1$, hence $y \notin \mathfrak{g}^f$ and consequently $\mathfrak{k} \in P(f)$ (1.12.2). Let L and L_z be the left ideals of $U(\mathfrak{g})$ and $U(\mathfrak{g})_z$ respectively, generated by the $k - f(k)$ where k runs through $\mathfrak{k}$. Since $\psi(yz^{-1}) = 1 \otimes q$, the formulae of 4.7.8 prove that $\psi(L_z)$ is the left ideal of $U(\bar{\mathfrak{h}})_{\bar{z}} \otimes A_1$ generated by $1 \otimes q$ and the $(\bar{k} - \bar{g}(\bar{k})) \otimes 1$ where k runs through the image $\bar{\mathfrak{k}}$ of $\mathfrak{k}$ in $\bar{\mathfrak{h}}$. Since $\mathfrak{k} \in P(f)$, $I(f)_z$ is the largest two-sided ideal of $U(\mathfrak{g})_z$ contained in L_z (5.1.7, 5.1.9). Since $\mathfrak{k} \in P(g)$, $I(\bar{g})_{\bar{z}}$ is contained in the left ideal generated by the $\bar{k} - \bar{g}(\bar{k})$, where $\bar{k} \in \bar{\mathfrak{k}}$; hence

$$I(\bar{g})_{\bar{z}} \otimes A_1 = (1 \otimes A_1)\,(I(\bar{g})_{\bar{z}} \otimes 1) \subset \psi(L_z),$$

and hence $I(\bar{g})_{\bar{z}} \otimes A_1$ is contained in $\psi(I(f)_z)$; since $U(\bar{\mathfrak{h}})_{\bar{z}} \otimes A_1/I(\bar{g})_{\bar{z}} \otimes A_1$ is simple (4.7.8, 4.7.9), we have

$$\psi(I(f)_z) = I(\bar{g})_{\bar{z}} \otimes A_1.$$

6.2.2. PROPOSITION ($\mathfrak{g}$ nilpotent). *Let* I *be a rational ideal with weight* r *of* $U(\mathfrak{g})$. *There exists a linear form* f *on* $\mathfrak{g}$ *which possesses the following properties:*

(i) B_f *is of rank* $2r$;

(ii) $I = I(f)$.

This is obvious if $\dim \mathfrak{g} \leq 1$. Let us assume that $\dim \mathfrak{g} > 1$ and let us reason by induction on $\dim \mathfrak{g}$.

Let $\mathfrak{z}$ be the centre of $\mathfrak{g}$. If $I \cap \mathfrak{g} \neq 0$, we return to $\mathfrak{g}/(I \cap \mathfrak{z})$. Let us assume that $I \cap \mathfrak{z} = 0$, and hence that $\dim \mathfrak{z} = 1$. Let $(x,y,z,\mathfrak{h})$ be a reducing quadruple. We may assume that $z - 1 \in I$. Let $\bar{\mathfrak{h}}, \bar{z}$ and ψ be as in 4.7.8. There exists a two-sided ideal J of $U(\bar{\mathfrak{h}})$ such that

$$\bar{z} - 1 \in J, \qquad \psi(I_z) = J_z \otimes A_1, \qquad U(\mathfrak{g})_z/I_z = U(\mathfrak{g})/I,$$

$$U(\bar{\mathfrak{h}})_{\bar{z}}/J_{\bar{z}} = U(\bar{\mathfrak{h}})/J, \qquad U(\mathfrak{g})_z/I_z = (U(\bar{\mathfrak{h}})_{\bar{z}}/J_{\bar{z}}) \otimes A_1.$$

Then J is rational, and hence there exists $g \in \mathfrak{h}^*$ such that:

(1) $g(y) = 0$;

(2) $J = I(\bar{g})$, where $\bar{g}$ designates the linear form on $\bar{\mathfrak{h}}$ deduced from g by passage to the quotient;

(3) if $2s$ is the rank of $B_{\bar{g}}$, then $U(\bar{\mathfrak{h}})/J$ is isomorphic to A_s.
Let f be a linear form on $\mathfrak{g}$ which extends g. Then

$$f(y) = g(y) = 0, \qquad f(z) = g(z) = \bar{g}(\bar{z}) = 1$$

since $J = I(\bar{g})$ and $\bar{z} - 1 \in J$. From 6.2.1, we have

$$\psi(I(f)_z) = I(\bar{g})_{\bar{z}} \otimes A_1 = J_{\bar{z}} \otimes A_1 = \psi(I_z),$$

whence $I(f)_z = I_z$ and $I(f) = I$. The rank of B_f is $2s + 2$ from 6.2.1 and 1.12.2, and $U(\mathfrak{g})/I$ is isomorphic to $(U(\bar{\mathfrak{h}})/J) \otimes A_1$, and hence to A_{s+1}.

6.2.3. PROPOSITION ($\mathfrak{g}$ nilpotent). *Let $\mathscr{A}$ be the adjoint group of $\mathfrak{g}$, considered as operating in $\mathfrak{g}^*$. Let $f, f' \in \mathfrak{g}^*$. In order that $I(f) = I(f')$, it is necessary and sufficient that $f' \in \mathscr{A}f$.*

Let $a \in \mathscr{A}$, and let a_U be the automorphism of $U(\mathfrak{g})$ defined by a. If $a(f) = f'$, then $a(I(f)) = I(f')$ by transport of structure, and hence, from 2.4.17, $I(f) = I(f')$.

Let us assume that $I(f) = I(f')$ and prove that $f' \in \mathscr{A}f$ by induction on $\dim \mathfrak{g}$. We may assume that $\dim \mathfrak{g} > 1$. Let $\mathfrak{z}$ be the centre of $\mathfrak{g}$, and

$$\mathfrak{z}_0 = I(f) \cap \mathfrak{z} = I(f') \cap \mathfrak{z};$$

then $f(\mathfrak{z}_0) = f'(\mathfrak{z}_0) = 0$. If $\mathfrak{z}_0 \neq 0$, we return to $\mathfrak{g}/\mathfrak{z}_0$. Let us assume that $\mathfrak{z}_0 = 0$, and hence $\dim \mathfrak{z} = 1$. Let $(x,y,z,\mathfrak{h})$ be a reducing quadruple.

Let $\bar{\mathfrak{h}}, \bar{z}$ and ψ be as in 4.7.8. We may assume that $f(z) = f'(z) = 1$ and $f(y) = 0$. For $\lambda \in k$, we have

$$((\exp \operatorname{ad} \lambda x)f')(y) = f'(y - \lambda[x,y]) = f'(y) - \lambda);$$

replacing f' by $(\exp \operatorname{ad} \lambda x)f'$ where necessary, we may hence assume that $f'(y) = 0$. Let $g = f|\mathfrak{h}$, $g' = f'|\mathfrak{h}$, and let $\bar{g}$ and $\bar{g}'$ be the linear forms over $\bar{\mathfrak{h}}$ deduced from g, g' by passage to the quotient. From 6.2.1, we have

$$I(\bar{g})_{\bar{z}} \otimes A_1 = \psi(I(f)_z) = \psi(I(f')_z) = I(\bar{g})_{\bar{z}} \otimes A_1,$$

whence $I(\bar{g}) = I(\bar{g}')$. From the induction hypothesis, we return to the case where $\bar{g} = \bar{g}'$, whence $f|\mathfrak{h} = f'|\mathfrak{h}$. For all $\mu \in k$, we have

$$(\exp \operatorname{ad} \mu y)f \,|\, \mathfrak{h} = f \,|\, \mathfrak{h}, \qquad (\exp \operatorname{ad} \mu y)f' \,|\, \mathfrak{h} = f' \,|\, \mathfrak{h}.$$

Finally,

$$(\exp \operatorname{ad} \mu y)f)(x) = f(x - \mu[y,x]) = f(x) + \mu,$$

whence $(\exp \operatorname{ad} \mu y)f = f'$ for a suitable choice of μ.

6.2.4. THEOREM ($\mathfrak{g}$ nilpotent). *Let $\mathscr{A}$ be the adjoint group of $\mathfrak{g}$. The mapping $f \mapsto I(f)$ defines by passage to the quotient a bijection of $\mathfrak{g}^*/\mathscr{A}$ onto the set of rational ideals of $U(\mathfrak{g})$ (i.e.* Prim $U(\mathfrak{g})$ *if k is algebraically closed).*

This follows from 6.1.5, 6.2.2 and 6.2.3.

6.2.5 The bijection of 6.2.4 and its converse are termed *canonical.* If ω is an $\mathscr{A}$-orbit in $\mathfrak{g}^*$, the rank $2r$ of B_f is obviously constant when f runs through ω, and r is the weight of the rational ideal associated with ω (6.2.2).

6.2.6. LEMMA ($\mathfrak{g}$ completely solvable). *Let $\mathfrak{n}$ be a nilpotent ideal of $\mathfrak{g}$, x an element of $\mathfrak{g}$ such that $\mathfrak{g} = \mathfrak{n} \oplus kx$, D the derivation of $U(\mathfrak{n})$ defined by x, g a linear form on $\mathfrak{g}$ such that $g(x) = 0$ and $g([x,\mathfrak{n}]) = 0$, $f = g|\mathfrak{n}$, and $A = U(\mathfrak{n})/I(f)$. Then:*

(i) $I(g) \cap U(\mathfrak{n}) = I(f)$.

(ii) *The canonical homomorphism of $U(\mathfrak{n})$ into $U(\mathfrak{g})/I(g)$ defines by passage to the quotient an isomorphism ψ of A onto $U(\mathfrak{g})/I(g)$.*

(iii) $\mathfrak{st}(I(f),\mathfrak{g}) = \mathfrak{g}$.

(iv) *Let D' be the derivation of A deduced from D by passage to the quotient. Let a be the element of A such that $\psi(a)$ is the class of x. Then D' is the inner derivation of A defined by a.*

(v) *Let $\mathfrak{h}$ be an element of $P(f)$ such that $[x,\mathfrak{h}] \subset \mathfrak{h}$ (we recall that, from* 1.12.10, *such an element exists). Let $\sigma_{\mathfrak{h}} = \operatorname{ind}(f|\mathfrak{h},\mathfrak{n})$, $\bar{\sigma}_{\mathfrak{h}}$ be the representation of A deduced from $\sigma_{\mathfrak{h}}$ by passage to the quotient, and $\lambda = \operatorname{tr}_{\mathfrak{n}/\mathfrak{h}} D$. Then*

$$\bar{\sigma}_{\mathfrak{h}}(a)(u \otimes 1) = (D + \tfrac{1}{2}\lambda)\, u \otimes 1 \quad \textit{for all } u \in U(\mathfrak{n}).$$

(vi) *Let $\mathfrak{h}' = \mathfrak{h} \oplus kx \in P(g)$ and $\tau_{\mathfrak{h}} = \operatorname{ind}^{\sim}(g|\mathfrak{h}',\mathfrak{g})$. Then the space of $\tau_{\mathfrak{h}}$ can be identified with that of $\sigma_{\mathfrak{h}}$, so that $\tau_{\mathfrak{h}}|\mathfrak{n} = \sigma_{\mathfrak{h}}$ and $\tau_{\mathfrak{h}}(x) = \bar{\sigma}(a)$.*

Let $\mathfrak{h} \in P(f)$ such that $[x,\mathfrak{h}] \subset \mathfrak{h}$. Let us introduce $\sigma_{\mathfrak{h}}$, $\bar{\sigma}_{\mathfrak{h}}$ and λ as in (v). Let L be the left ideal of $U(\mathfrak{n})$ generated by the $h - f(h)$, where $h \in \mathfrak{h}$; let $V = U(\mathfrak{n})/L = U(\mathfrak{n}) \otimes_{U(\mathfrak{h})} k$ be the space of $\sigma_{\mathfrak{h}}$ (5.1.9). If $h \in \mathfrak{h}$, then $f(D\mathfrak{h}) = 0$, hence $D(h - f(h)) = Dh - f(Dh) \in L$; this proves that $D(L) \subset L$. Hence D defines by passage to the quotient an endomorphism $\omega_{\mathfrak{h}}$ of the space V. For all $u \in U(\mathfrak{n})$, the class of u modulo L can be identified with $u \otimes 1$ under the canonical isomorphism $U(\mathfrak{n})/L \to U(\mathfrak{n}) \otimes_{U(\mathfrak{h})} k$; hence $\omega_{\mathfrak{h}}(u \otimes 1) = Du \otimes 1$. On the other hand, the largest two-sided ideal of $U(\mathfrak{n})$ contained in L, that is, $I(f)$, is stable under D. This proves (iii).

Let $\mathfrak{h}'$ be as in (vi). Then

$$g([\mathfrak{h}',\mathfrak{h}']) = f([\mathfrak{h},\mathfrak{h}]) + f(D\mathfrak{h}) = 0.$$

Since the dimension of the maximal totally isotropic subspaces for B_g exceeds that of the maximal totally isotropic sub-spaces for B_f by at most 1, we have $\mathfrak{h}' \in P(g)$. From 5.1.6, the space of $\tau_{\mathfrak{h}} = \operatorname{ind}^{\sim}(g|\mathfrak{h}',\mathfrak{g})$ can be identified with that of $\sigma_{\mathfrak{h}}$, so that $\tau_{\mathfrak{h}}|\mathfrak{n} = \sigma_{\mathfrak{h}}$. This proves that $I(g) \cap U(\mathfrak{n}) = I(f)$ (whence (i)), and that the canonical homomorphism $U(\mathfrak{n}) \to U(\mathfrak{g})/I(\mathfrak{g})$ defines by passage to the quotient an injective homomorphism of A into $U(\mathfrak{g})/I(\mathfrak{g})$. By virtue of this homomorphism we identify A with a subalgebra of $U(\mathfrak{g})/I(g)$. If $u \in U(\mathfrak{n})$, then

$$\begin{aligned}\tau_{\mathfrak{h}}(x)(u \otimes 1) &= xu \otimes 1 = ux \otimes 1 + Du \otimes 1 \\ &= u \otimes x \cdot 1 + Du \otimes 1 = \tfrac{1}{2}\lambda u \otimes 1 + \omega_{\mathfrak{h}}(u \otimes 1),\end{aligned}$$

hence

$$\tau_{\mathfrak{h}}(x) = \omega_{\mathfrak{h}} + \tfrac{1}{2}\lambda.$$

From 6.1.5, 4.7.9 and 4.6.8, there exists $a \in A$ such that D' is the inner derivation of A defined by a. Let $\bar{x}$ be the class of x in $U(\mathfrak{g})/I(g)$. Then $\bar{x} - a$ commutes with A; in particular, $\bar{x} - a$ commutes with a, hence with $\bar{x}$, and, finally, $\bar{x} - a$ is central in $U(\mathfrak{g})/I(g)$, and hence is scalar. Consequently, $\bar{x} \in A$ and $U(\mathfrak{g})/I(g) = A$. We have thus proved (ii). By adding a suitable scalar to a, we have $\bar{x} = a$ and (iv) is established. Then

$$\bar{\sigma}_{\mathfrak{h}}(a) = \tau_{\mathfrak{h}}(x) = \omega_{\mathfrak{h}} + \tfrac{1}{2}\lambda$$

(whence (vi)), hence $\bar{\sigma}_{\mathfrak{h}}(a)\,(u \otimes 1) = (D + \tfrac{1}{2}\lambda)\,u \otimes 1$ for all $u \in U(\mathfrak{n})$, which proves (v).

6.2.7. Lemma. *Let* $\mathfrak{n}$ *be a nilpotent ideal of* $\mathfrak{g}$, $f \in \mathfrak{n}^*$, $\mathfrak{g}'$ *and* $\mathfrak{g}'$ *be set of the* $x \in \mathfrak{g}$ *such that* $f([x,\mathfrak{n}]) = 0$. *Then* $\mathfrak{st}(I(f),\mathfrak{g}) \subset \mathfrak{g}' + \mathfrak{n}$.

The group $\operatorname{Aut}(\mathfrak{n})$ operates in $U(\mathfrak{n})$ by automorphisms. Let $\mathscr{B}$ be the set of those $b \in \operatorname{Aut}(\mathfrak{n})$ such that $b(I(f)) = I(f)$; it is an algebraic subgroup of $\operatorname{Aut}(\mathfrak{n})$ (2.4.16); let $\mathfrak{b}$ be its Lie algebra. Let $\mathscr{A}$ be the adjoint group of $\mathfrak{n}$. If $b \in \mathscr{B}$, then $b \cdot f \in \mathscr{A} \cdot f$ (6.2.3). Let $x \in \mathfrak{st}(I(f),\mathfrak{g})$. Then $\operatorname{ad}_{\mathfrak{n}} x \in \mathfrak{b}$ (2.4.16), hence there exists $y \in \mathfrak{n}$ such that $(\operatorname{ad}_{\mathfrak{n}} x) \cdot f = (\operatorname{ad}_{\mathfrak{n}} y) \cdot f$, that is, $x - y \in \mathfrak{g}'$.

6.2.8. Proposition (k algebraically closed *or* $\mathfrak{g}$ completely solvable). *Let* $\mathfrak{n}$ *be a nilpotent ideal of* $\mathfrak{g}$, $f \in \mathfrak{n}^*$, *and* $\mathfrak{g}'$ *be the set of those* $x \in \mathfrak{g}$ *such that* $f([x,\mathfrak{n}]) = 0$. *Then* $\mathfrak{st}(I(f),\mathfrak{g}) = \mathfrak{g}' + \mathfrak{n}$.

Obviously, $\mathfrak{n} \subset \mathfrak{st}(I(f),\mathfrak{g})$. From 6.2.7, it is sufficient to prove that $\mathfrak{g}' \subset \mathfrak{st}(I(f),\mathfrak{g})$. Let $x \in \mathfrak{g}'$. If $\mathfrak{g}$ is completely solvable, or *if* k is algebraically closed, $\mathfrak{n} \oplus kx$ is completely solvable, hence $x \in \mathfrak{st}(I(f),\mathfrak{g})$ from 6.2.6 (iii).

6.2.9. PROPOSITION ($\mathfrak{g}$ nilpotent). *Let* $f \in \mathfrak{g}^*$ *and* $\mathfrak{h} \in P(f)$. *Then* $\operatorname{ind}(f|\mathfrak{h},\mathfrak{g})$ *is absolutely simple.*

Let $(\mathfrak{g}_0,\mathfrak{g}_1,\ldots,\mathfrak{g}_s)$ be an increasing sequence of Lie subalgebras of $\mathfrak{g}$ such that $\mathfrak{g}_0 = \mathfrak{h}$, $\mathfrak{g}_s = \mathfrak{g}$, and $\mathfrak{g}_i$ is an ideal of codimension 1 in $\mathfrak{g}_{i+1}$. Let $f_i = f|\mathfrak{g}_i$. Clearly, $\mathfrak{h} \in P(f_i)$. If $x \in \mathfrak{g}_i$ is such that $f([x,\mathfrak{g}_{i-1}]) = 0$ and $x \notin \mathfrak{g}_{i-1}$, then $f([x,\mathfrak{g}_i]) = 0$, hence $x \in \mathfrak{g}_i^{f_i} \subset \mathfrak{h}$, whence we have a contradiction. From 6.2.8 and 5.3.3, the stabilizer of $\operatorname{ind}(f|\mathfrak{h},\mathfrak{g}_{i-1})$ in $\mathfrak{g}_i$ is hence $\mathfrak{g}_{i-1}$ if $\operatorname{ind}(f|\mathfrak{h},\mathfrak{g}_{i-1})$ is absolutely simple; we then see that $\operatorname{ind}(f|\mathfrak{h},\mathfrak{g}_i)$ is absolutely simple (5.3.6). Proceeding in a stepwise fashion, this proves the proposition.

References: [30], [45], [99].

6.3. Prime ideals of the enveloping algebra and invariant prime ideals of the symmetric algebra (the nilpotent case)

6.3.1. If $f \in \mathfrak{g}^*$, we denote the set of elements of $S(\mathfrak{g})$ which are zero on the orbit of f for the algebraic adjoint group by $J(f)$.

6.3.2. PROPOSITION ($\mathfrak{g}$ nilpotent). *Let* $\mathscr{A}$ *be the adjoint group of* $\mathfrak{g}$.

(i) *Let* $f \in \mathfrak{g}^*$, *and let* $2r$ *be the rank of* B_f. *Then* $J(f)$ *is a rational invariant ideal of* $S(\mathfrak{g})$, *and the algebra* $S(\mathfrak{g})/J(f)$ *is isomorphic to* $k[\xi_1,\ldots,\xi_{2r}]$ (*where* $\xi_1,\ldots,\xi_{2r}$ *are indeterminates*).

(ii) *The mapping* $f \mapsto J(f)$ *defines by passage to the quotient a bijection of* $\mathfrak{g}^*/\mathscr{A}$ *onto the set of rational invariant ideals of* $S(\mathfrak{g})$.

Clearly, $J(f)$ is invariant. Let $p \in S(\mathfrak{g})$ be such that the image of p in $S(\mathfrak{g})/J(f)$ belongs to $(S(\mathfrak{g})/J(f))^{\mathscr{A}}$. For all $a \in \mathscr{A}$, $p - ap$ is zero on $\mathscr{A}f$, hence $p|\mathscr{A}f$ is constant, which proves that $J(f)$ is rational. From 11.2.4, $\mathscr{A}f$ is a closed submanifold of $\mathfrak{g}^*$, and $S(\mathfrak{g})/J(f)$ is an algebra of polynomials over k; since $\dim(\mathscr{A}f)$ is equal to $\operatorname{rank}(B_f)$ from 1.11.3, we have proved (i).

Let J be a rational invariant ideal of $S(\mathfrak{g})$. From 4.8.7, there exists a homomorphism of $S(\mathfrak{g})$ into k whose kernel contains J. This homomorphism is defined by some $f \in \mathfrak{g}^*$. Then $J \subset J(f)$. But J is maximal invariant (4.8.6), hence $J = J(f)$. The mapping considered in (ii) is hence surjective. Finally, let $f_1, f_2 \in \mathfrak{g}^*$ and assume that $J(f_1) = J(f_2)$. Then $\mathscr{A}f_1$ and $\mathscr{A}f_2$ have the same closure; but $\mathscr{A}f_1$ and $\mathscr{A}f_2$ are closed (11.2.4), hence $\mathscr{A}f_1 = \mathscr{A}f_2$. This proves that the mapping considered in (ii) is injective.

6.3.3. By composition with the bijection of 6.2.4, we obtain a bijection of the set of rational ideals of $U(\mathfrak{g})$ onto the set of rational invariant ideals of $S(\mathfrak{g})$. All these bijections and their inverses are said to be *canonical.*

6.3.4. Let $f: K' \to K$ be a homomorphism of extensions of k. The diagram

$$\begin{array}{ccc} \mathscr{R}(K') & \xrightarrow{\mathscr{R}(f)} & \mathscr{R}(K) \\ \downarrow & & \downarrow \\ \mathscr{S}(K') & \xrightarrow{\mathscr{S}(f)} & \mathscr{S}(K) \end{array}$$

where the vertical arrows are the canonical bijections (6.3.3), is commutative. Indeed, we may assume that $K' = k$. Let $I \in \mathscr{R}(k)$ and $J \in \mathscr{S}(k)$ be corresponding elements; we shall prove that $I \otimes K$ and $J \otimes K$ correspond to each other. Let $g \in \mathfrak{g}^*$ and $\mathfrak{h} \in P(g)$ such that $I = \operatorname{Ker} \operatorname{ind}(g|\mathfrak{h},\mathfrak{g})$. It follows immediately that $\mathfrak{h} \otimes K$ is a polarization of $\mathfrak{g} \otimes K$ in $g \otimes 1$ and that

$$I \otimes K = \operatorname{Ker} \operatorname{ind}(g \otimes 1 \mid \mathfrak{h} \otimes K, \mathfrak{g} \otimes K).$$

On the other hand, J is the ideal of the elements of $S(\mathfrak{g})$ which are zero on the orbit of g, hence $J \otimes K$ is the ideal of the elements of $S(\mathfrak{g} \otimes K)$ which are zero on the orbit of $g \otimes 1$, whence our assertion.

6.3.5. PROPOSITION ($\mathfrak{g}$ nilpotent). *Let $\mathscr{P}$ be the set of prime ideals of $U(\mathfrak{g})$, and $\mathscr{Q}$ the set of invariant prime ideals of $S(\mathfrak{g})$.*

(i) *Let $P \in \mathscr{P}$, let K be an extension of k, ψ an extension homomorphism of $C(\mathfrak{g};P)$ into K, I the element of $\mathscr{R}(K)$ associated with (P,ψ), J the corresponding element of $\mathscr{S}(K)$, and (Q,ω) the pair associated with J. Then Q only depends on P and not on the choice of K and ψ.*

(ii) *The mapping $P \mapsto Q$ is a bijection β of $\mathscr{P}$ onto $\mathscr{Q}$. If P is rational, then $\beta(P)$ is the rational invariant ideal corresponding to P.*

(a) Let ψ_0 be the identity isomorphism of $K_0 = C(\mathfrak{g};P)$, I_0 the element of $\mathscr{R}(K_0)$ associated with (P,ψ_0), J_0 the corresponding element of $\mathscr{S}(K_0)$, and (Q_0,ω_0) the pair associated with J_0. From 4.7.16, we have $I = \mathscr{R}(\psi)(I_0)$, and hence, from 6.3.4, $J = \mathscr{S}(\psi)(J_0)$; consequently, the pair associated with J is $(Q_0,\psi \circ \omega_0)$ (4.8.9). Hence $Q = Q_0$, which proves (i). We denote the mapping of $\mathscr{P}$ into $\mathscr{Q}$ which is defined in this way by β.

(b) Let $P_1,P_2 \in \mathscr{P}$, let K be an extension of k such that there exist extension homomorphisms $\psi_1 : C(\mathfrak{g};P_1) \to K$ and $\psi_2 : C(\mathfrak{g};P_2) \to K$. For all $P \in \mathscr{P}$, let $\mathscr{R}_P$ be the set of elements of $\mathscr{R}(K)$ which are associated with pairs (P,ψ). Then $\mathscr{R}(K)$ is the disjoint union of the $\mathscr{R}_P$ from 4.7.14. For all $Q \in \mathscr{Q}$, let $\mathscr{S}_Q$ be the set of elements of $\mathscr{S}(K)$ which are associated with pairs (Q,ω). Then $\mathscr{S}(K)$ is the disjoint union of the $\mathscr{S}_Q$ from 4.8.9. The canonical bijection of $\mathscr{R}(K)$ onto $\mathscr{S}(K)$ sends $\mathscr{R}_P$ into $\mathscr{S}_{\beta(P)}$ from (i). Consequently, if $P_1 \neq P_2$, then $\beta(P_1) \neq \beta(P_2)$, which proves that β is injective.

(c) Let $Q \in \mathcal{Q}$. There exist an extension K of k, and a $J \in \mathcal{S}(K)$ associated with a pair (Q,ω) (4.8.10). Let I be the element of $\mathcal{R}(K)$ corresponding to J. Then I is associated with a pair (P,ψ), and $Q = \beta(P)$.

(d) Thus $\beta : \mathcal{P} \to \mathcal{Q}$ is bijective. The last assertion of the proposition is obvious.

6.3.6. The bijections $\mathcal{P} \to \mathcal{Q}$ and $\mathcal{Q} \to \mathcal{P}$ defined in 6.3.5 are termed *canonical.*

6.3.7. PROPOSITION ($\mathfrak{g}$ nilpotent). *Let $\mathcal{P}$ and $\mathcal{Q}$ be as in* 6.3.5, *$\mathcal{A}$ the adjoint group of $\mathfrak{g}$, and β the canonical bijection of $\mathcal{P}$ onto $\mathcal{Q}$. Let $P \in \mathcal{P}$ and $Q = \beta(P) \in \mathcal{Q}$. Then there exists one and only one isomorphism $\beta_P : C(\mathfrak{g};P) \to \mathrm{Fract}((S(\mathfrak{g})/Q)^{\mathcal{A}})$ of extensions of k such that, for every extension homomorphism $\psi : C(\mathfrak{g};P) \to K$ (with which an $I \in \mathcal{R}(K)$, hence a $J \in \mathcal{S}(K)$ and hence a pair of the form (Q,ω) are canonically associated), we have $\psi = \omega \circ \beta_P$.*

In the proof of 6.3.5 (a), starting with the identity automorphism of $C(\mathfrak{g};P)$ we defined $I_0, J_0, Q_0 = Q$, and an extension homomorphism $\omega_0 : \mathrm{Fract}((S(\mathfrak{g})/Q)^{\mathcal{A}}) \to C(\mathfrak{g};P)$; and, from the above proof, we have $\omega = \psi \circ \omega_0$ with the notation of 6.3.7. Similarly, starting with the identity automorphism of $\mathrm{Fract}((S(\mathfrak{g})/Q)^{\mathcal{A}})$, we define an extension homomorphism $\varphi_0 : C(\mathfrak{g};P) \to \mathrm{Fract}((S(\mathfrak{g})/Q)^{\mathcal{A}})$, and we have $\psi = \omega \circ \varphi_0$. In particular, $\varphi_0 \circ \omega_0 = 1$ and $\omega_0 \circ \varphi_0 = 1$, hence ω_0 and φ_0 are mutually inverse isomorphisms. This proves the existence of β_P. If β'_P has the properties of the proposition, then $1 = \omega_0 \circ \beta'_P$, whence $\beta'_P = \beta_P$.

6.3.8. With the notation of 6.3.7, the isomorphism β_P is termed *canonical.* Let us identify $C = C(\mathfrak{g};P)$ with $\mathrm{Fract}((S(\mathfrak{g})/Q)^{\mathcal{A}})$ by means of β_P. Then the rational ideal $P^\wedge$ of $U(\mathfrak{g} \otimes C)$ associated with P and the rational invariant ideal $Q^\wedge$ of $S(\mathfrak{g} \otimes C)$ associated with Q correspond to one and the same orbit ω in $(\mathfrak{g} \otimes C)^*$. Let r be the weight of P (4.7.17); then:

(1) if we set $Z = Z(\mathfrak{g};P)$, then $(U(\mathfrak{g})/P) \otimes_Z C = U(\mathfrak{g} \otimes C)/P^\wedge$ as a C-algebra is isomorphic to $A_r(C)$ (4.7.17);

(2) B_f has rank $2r$ for every $f \in \omega$ (6.2.2);

(3) if we set $Z' = (S(\mathfrak{g})/Q)^{\mathcal{A}}$, then $(S(\mathfrak{g})/Q) \otimes_{Z'} C = S(\mathfrak{g} \otimes C)/Q^\wedge$ as a C-algebra is isomorphic to $C[\xi_1, \ldots, \xi_{2r}]$ where $\xi_1, \ldots, \xi_{2r}$ are indeterminates (6.3.2).

Reference: [99].

6.4. The Jacobson topology

6.4.1. LEMMA. *Let d be an integer $\geqq 0$, and T_d the set of the*

$$(f,\mathfrak{h}) \in \mathfrak{g}^* \times \mathrm{Gr}(\mathfrak{g},d)$$

such that $\mathfrak{h}$ is a Lie subalgebra of $\mathfrak{g}$ subordinate to f. For $t = (f,\mathfrak{h}) \in T_d$, let $\varrho(t) = \mathrm{ind}^{\sim}(f|\mathfrak{h},\mathfrak{g})$. *Let $u \in U(\mathfrak{g})$.*

(i) *The set of the $t \in T_d$ such that $\varrho(t)(u)$ is scalar is a closed subset $F_{d,u}$ of T_d. Let $u(t)$ be this scalar.*

(ii) *The mapping $t \mapsto u(t)$ of $F_{d,u}$ into k is a rational function defined everywhere on $F_{d,u}$.*

(iii) *The set $E_{d,u}$ of the $t \in T_d$ such that $\varrho(t)(u) = 0$ is closed in T_d.*

Let $b = (e_1, \ldots, e_n)$ be a basis for $\mathfrak{g}$. Let $O_b \subset \mathrm{Gr}(\mathfrak{g},d)$ be the set of complements of $ke_{d+1} + \cdots + ke_n$ in $\mathfrak{g}$. For $t = (f,\mathfrak{h}) \in T_d \cap (\mathfrak{g}^* \times O_b)$, let us define the following objects:

(1) f_t is the linear form $x \mapsto f(x) + \frac{1}{2}\,\mathrm{tr}\,\mathrm{ad}_{\mathfrak{g}/\mathfrak{h}}(x)$ on $\mathfrak{h}$;

(2) k_t is the vector space k considered as an $\mathfrak{h}$-module by virtue of f_t;

(3) the scalars $\alpha_{ij}(t)$ ($i = 1, \ldots, d$; $j = d+1, \ldots, n$) are defined by

$$h_i = e_i - \alpha_{i,d+1}e_{d+1} - \cdots - \alpha_{i,n}e_n \in \mathfrak{h} \qquad (i = 1, \ldots, d).$$

For $\nu = (\nu_{d+1}, \ldots, \nu_n) \in \mathbf{N}^{n-d}$, we set $e^\nu = e_{d+1}^{\nu_{d+1}} \cdots e_n^{\nu_n} \in U(\mathfrak{g})$. The $e^\nu \oplus 1$ constitute a basis for the space $U(\mathfrak{g}) \oplus_{U(\mathfrak{h})} k_t$ of $\varrho(t)$. In $U(\mathfrak{g})$ we have

$$ue^\nu = \sum_{\xi,m_1,\ldots,m_d} \beta_{\nu,\xi,m_1,\ldots,m_d}(\alpha_{ij}(t))e^\xi h_1^{m_1} \cdots h_d^{m_d},$$

where the $\beta_{\nu,\xi,m_1,\ldots,m_d}$ are polynomial functions of $d(n-d)$ variables. Then

$$\varrho(t)(u)(e^\nu \otimes 1) = \\ = \sum_{\xi,m_1,\ldots,m_d} \beta_{\nu,\xi,m_1,\ldots,m_d}(\alpha_{ij}(t))f_t(h_1)^{m_1} \cdots f_t(h_d)^{m_d}(e^\xi \otimes 1).$$

With respect to the basis $(e^\nu \otimes 1)$, the coefficients $c_{\nu,\xi}(t)$ of the matrix of $\varrho(t)(u)$ are hence polynomial functions of the $\alpha_{ij}(t)$ and $f(e_1), \ldots, f(e_n)$. Now $F_{d,u} \cap (\mathfrak{g}^* \cap O_b)$ is the set of the $t \in T_d \cap (\mathfrak{g}^* \times O_b)$ such that, for $\nu,\xi \in N^{n-d}$,

$$c_{\nu,\xi}(t) = 0 \quad \text{si} \quad \nu \neq \xi,$$

$$c_{\nu,\nu}(t) = c_{0,0}(t).$$

Thus $F_{d,u} \cap (\mathfrak{g}^* \times O_b)$ is closed in $T_d \cap (\mathfrak{g}^* \times O_b)$. Since this is true for every basis b, we deduce (i) from it.

If $t \in F_{d,u}$, then $u(t) = c_{0,0}(t)$, which proves (ii). Assertion (iii) follows from (ii).

6.4.2. Lemma ($\mathfrak{g}$ completely solvable). *Let $f \in \mathfrak{g}^*$, and let $\mathfrak{h}_1$ and $\mathfrak{h}_2$ be Lie subalgebras of $\mathfrak{g}$ subordinate to f and such that $\mathfrak{h}_1 \subset \mathfrak{h}_2$. Then*

$$\text{Ker ind}^{\sim}(f|\mathfrak{h}_1,\mathfrak{g}) \subset \text{Ker ind}^{\sim}(f|\mathfrak{h}_2,\mathfrak{g}).$$

(a) From 5.2.3 and 5.2.6, it is sufficient to consider the case where $\mathfrak{h}_2 = \mathfrak{g}$. We then have $f([\mathfrak{g},\mathfrak{g}]) = 0$, and $I(f)$ is the ideal of codimension 1 in $U(\mathfrak{g})$ which is the kernel of the representation f. We set $\mathfrak{h}_1 = \mathfrak{h}$, and we must prove that $\text{Ker ind}^{\sim}(f|\mathfrak{h},\mathfrak{g}) \subset I(f)$. From 5.2.3 and 5.2.6, we may assume that $\dim \mathfrak{g}/\mathfrak{h} = 1$.

(b) If $\text{tr ad}_{\mathfrak{g}/\mathfrak{h}} x = 0$ for all $x \in \mathfrak{h}$, the assertion follows from 5.1.9 (iii).

(c) The assertion is obnvious if $\dim \mathfrak{g} \leqq 1$. We hence assume that $\dim \mathfrak{g} \geqq 2$ and we reason by induction on $\dim \mathfrak{g}$. If $\mathfrak{h} \cap \text{Ker} f$ contains a non-null ideal $\mathfrak{b}$ of $\mathfrak{g}$, it is sufficient to apply the induction hypothesis to $\mathfrak{g}/\mathfrak{b}$ and $\mathfrak{h}/\mathfrak{b}$. Henceforth we assume that $\mathfrak{h} \cap \text{Ker} f$ does not contain any non-null ideal of $\mathfrak{g}$. The ideal $\text{Ker} f$ contains a one-dimensional ideal $\mathfrak{a}$ of $\mathfrak{g}$. We have $\mathfrak{g} = \mathfrak{h} \oplus \mathfrak{a}$. Let $a \in \mathfrak{a} - \{0\}$. There exists $\lambda \in \mathfrak{g}^*$ such that $[x,a] = \lambda(x)a$ for all $x \in \mathfrak{g}$. If $\text{Ker}\,\lambda \supset \mathfrak{h}$, the assertion follows from (b). Let us assume that $\text{Ker}\,\lambda \not\supset \mathfrak{h}$. The set $\mathfrak{h} \cap \text{Ker}\,\lambda \cap \text{Ker} f$ is an ideal of $\mathfrak{h}$ which commutes with $\mathfrak{a}$, hence is an ideal of $\mathfrak{g}$, and hence is null. If $f \in k\lambda$, then $\dim \mathfrak{h} = 1$, and $\text{Ker ind}^{\sim}(f|\mathfrak{h},\mathfrak{g}) = 0$ from 6.1.2 (i). If $f \notin k\lambda$, there exists a basis (x,z) for $\mathfrak{h}$ such that

$$\lambda(z) = f(x) = 0, \qquad \lambda(x) = f(z) = 1;$$

we have $[x,a] = a$, and $[z,a] = 0$, and

$$[x,z] = 0 \qquad \text{because } [\mathfrak{h},\mathfrak{h}] \subset \mathfrak{h} \cap \text{Ker}\,\lambda \cap \text{Ker} f = 0;$$

to calculate $\text{ind}^{\sim}(f|\mathfrak{h},\mathfrak{g})$, the value of $f(a)$ has no significance, hence 6.1.2 (ii) proves that

$$\text{Ker ind}^{\sim}(f|\mathfrak{h},\mathfrak{g}) = (z-1)U(\mathfrak{g}) \subset I(f).$$

6.4.3. Lemma ($\mathfrak{g}$ completely solvable). *Let $f \in \mathfrak{g}^*$, and let $\mathfrak{h}$ be a Lie subalgebra of $\mathfrak{g}$ subordinate to f. Then* $\text{Ker ind}^{\sim}(f|\mathfrak{h},\mathfrak{g}) \subset I(f)$.

By virtue of 6.4.2, we may assume that $\mathfrak{h}$ is maximal among the Lie subalgebras of $\mathfrak{g}$ subordinate to f, and hence contains the centre $\mathfrak{z}$ of $\mathfrak{g}$. We reason by induction on $\dim \mathfrak{g}$.

(a) Let us assume that f is zero on a non-null ideal $\mathfrak{a}$ of $\mathfrak{g}$. Then $\mathfrak{a} + \mathfrak{h}$ is subordinate to f, hence $\mathfrak{a} \subset \mathfrak{h}$ and it is sufficient to apply the induction hypothesis to $\mathfrak{g}/\mathfrak{a}$ and $\mathfrak{h}/\mathfrak{a}$. Henceforth we shall assume that $\operatorname{Ker} f$ does not contain a non-null ideal. It follows from this that $\mathfrak{z}$ has dimension 0 or 1.

(b) Let us assume that there exists a one-dimensional non-central ideal $\mathfrak{a} = ka$ in $\mathfrak{g}$. We may assume that $f(a) = 1$. There exists $\lambda \in \mathfrak{g}^*$ such that $[x,a] = \lambda(x)a$ for all $x \in \mathfrak{g}$. Then $\mathfrak{g}' = \operatorname{Ker} \lambda$ is the centralizer of $\mathfrak{a}$ in $\mathfrak{g}$. Let $f' = f|\mathfrak{g}'$. From assertion ($\bigstar$) of 6.1.4 (b), there exists a representation σ' of $\mathfrak{g}'$ such that $\operatorname{Ker} \sigma' = I(f')$ and $\operatorname{Ker} \operatorname{ind}^{\sim}(\sigma',\mathfrak{g}) = I(f)$.

Let us assume that $\mathfrak{h} \subset \mathfrak{g}'$. Then $\operatorname{Ker} \operatorname{ind}^{\sim}(f'|\mathfrak{h},\mathfrak{g}') \subset I(f')$ from the induction hypothesis, and hence, from 5.2.6 and the above, $\operatorname{Ker} \operatorname{ind}^{\sim}(f|\mathfrak{h},\mathfrak{g}) \subset I(f)$.

Let us assume that $\mathfrak{h} \not\subset \mathfrak{g}'$. From 6.1.3, $\mathfrak{h}' = (\mathfrak{h} \cap \mathfrak{g}') + \mathfrak{a}$ is a Lie subalgebra of $\mathfrak{g}$ subordinate to f. Taking the foregoing into account, it is sufficient to prove that $\operatorname{Ker} \operatorname{ind}^{\sim}(f|\mathfrak{h},\mathfrak{g}) = \operatorname{Ker} \operatorname{ind}^{\sim}(f|\mathfrak{h}',\mathfrak{g})$. Now $[\mathfrak{h},\mathfrak{a}] = \mathfrak{a}$ and $f(\mathfrak{a}) \neq 0$, hence $\mathfrak{h} + \mathfrak{a}$ is not subordinate to f. Since $\mathfrak{h}$ and $\mathfrak{h}'$ have codimension 1 in $\mathfrak{h} + \mathfrak{a}$, we have $\mathfrak{h},\mathfrak{h}' \in P(f/\mathfrak{h} + \mathfrak{a})$, and hence

$$\operatorname{Ker} \operatorname{ind}^{\sim}(f|\mathfrak{h}, \mathfrak{h} + \mathfrak{a}) = \operatorname{Ker} \operatorname{ind}^{\sim}(f|\mathfrak{h}', \mathfrak{h} + \mathfrak{a})$$

(6.1.4), whence, from 5.2.3 and 5.2.6, our assertion follows.

(c) Let us assume that $\mathfrak{g}$ contains no one-dimensional non-central ideal. Then $\dim \mathfrak{z} = 1$ and $f(\mathfrak{z}) \neq 0$. Let $z \in \mathfrak{z}$ be such that $f(z) = 1$. There exists an ideal $\mathfrak{a}$ of $\mathfrak{g}$ containing $\mathfrak{z}$ and of dimension 2. There exists a basis (z,a) for $\mathfrak{a}$ such that $f(a) = 0$, and $\lambda,\mu \in \mathfrak{g}^*$ such that $[x,a] = \lambda(x)a + \mu(x)z$ for all $x \in \mathfrak{g}$. If $\mu = 0$, then ka is a one-dimensional non-central ideal, which is a contradiction; hence $\mu \neq 0$. Let $\mathfrak{g}' = \operatorname{Ker} \mu$, which is a Lie subalgebra of codimension 1 of $\mathfrak{g}$. Let $f' = f|\mathfrak{g}'$. From assertion ($\bigstar$) of 6.1.4 (c), there exists a representation σ' of $\mathfrak{g}'$ such that $\operatorname{Ker} \sigma' = I(f')$ and $\operatorname{Ker} \operatorname{ind}^{\sim}(\sigma',\mathfrak{g}) = I(f)$.

Let us assume that $\mathfrak{h} \subset \mathfrak{g}'$. Then the proof may be concluded as in (b).

Let us assume that $\mathfrak{h} \not\subset \mathfrak{g}'$. There exists $x \in \mathfrak{h}$ such that $\mu(x) = 1$. Let $\alpha,\beta \in k$. Then

$$f([x,\alpha a + \beta z]) = \alpha f([x,a]) = \alpha f(\lambda(x)a + z) = \alpha.$$

Hence $\alpha = 0$ if $\alpha a + \beta z \in \mathfrak{h}$, whence $\mathfrak{h} \cap \mathfrak{a} = \mathfrak{z}$. From 6.1.3, $\mathfrak{h}' = (\mathfrak{h} \cap \mathfrak{g}') + \mathfrak{a}$ is a Lie subalgebra of $\mathfrak{g}$ subordinate to f; and $\mathfrak{h},\mathfrak{h}'$ have codimension 1 in $\mathfrak{h} + \mathfrak{a}$. We have $f([x,a]) = 1$, hence $\mathfrak{h} + \mathfrak{a}$ is not subordinate to f, and the proof may be concluded as in (b).

6.4.4. THEOREM ($\mathfrak{g}$ completely solvable). *The mapping* $f \mapsto I(f)$ *of* $\mathfrak{g}^*$ *into*

Prim $U(\mathfrak{g})$ *is continuous* (*with respect to the Zariski and the Jacobson topology*).

Let $u \in U(\mathfrak{g})$, and let F be the set of the $f \in \mathfrak{g}^*$ such that $u \in I(f)$. We must show that F is closed. Let T_d and $E_{d,u}$ be as in 6.4.1, and let p_d be the canonical projection of $\mathfrak{g}^* \times \mathrm{Gr}(\mathfrak{g},d)$ onto $\mathfrak{g}^*$. Since $\mathrm{Gr}(\mathfrak{g},d)$ is a complete manifold and $E_{d,u}$ is closed, $p(E_{d,u})$ is closed in $\mathfrak{g}^*$. Let us show that $F = \bigcup_d p(E_{d,u})$. If $f \in F$, let $\mathfrak{h} \in P(f)$ and $d_0 = \dim \mathfrak{h}$; we have $u \in \mathrm{Ker}\,\mathrm{ind}^{\sim}(f|\mathfrak{h},\mathfrak{g})$, and hence $f \in p(E_{d_0},u)$. Let $d \in \mathbf{N}$ and $g \in p(E_{d,u})$. There exists a Lie subalgebra $\mathfrak{h}$ of $\mathfrak{g}$ subordinate to g such that $u \in \mathrm{Ker}\,\mathrm{ind}^{\sim}(g|\mathfrak{h},\mathfrak{g})$. Hence $u \in I(g)$ from 6.4.3, and $g \in F$.

6.4.5. LEMMA ($\mathfrak{g}$ nilpotent). *Let $\mathscr{A}$ be the adjoint group of $\mathfrak{g}$, P a prime ideal of $U(\mathfrak{g})$, P' the prime invariant ideal of $S(\mathfrak{g})$ associated with P, r the weight of P,*

$$Z = Z(\mathfrak{g};P), \qquad C = C(\mathfrak{g};P), \qquad Z' = (S(\mathfrak{g})/P')^{\mathscr{A}}, \qquad C' = \mathrm{Fract}\,Z',$$

and ε the canonical isomorphism of C onto C'. There exist $z \in Z - \{0\}$ and $z' \in Z' - \{0\}$ having the following properties:

(i) *Z_z and Z'_z are algebras of finite type* (*over k*);

(ii) $\varepsilon(Z_z) = Z'_{z'}$;

(iii) *the isomorphisms*

$$\theta : (U(\mathfrak{g})/P) \otimes_Z C \to A_r(C), \qquad \theta' : (S(\mathfrak{g})/P') \otimes_{Z'} C' \to C'[\xi_1, \ldots, \xi_{2r}]$$

define by restriction the isomorphisms

$$(U(\mathfrak{g})/P) \otimes_Z Z_z \to A_r(Z_z),$$

$$(S(\mathfrak{g})/P') \otimes_{Z'} Z'_{z'} \to Z'_{z'}[\xi_1, \ldots, \xi_{2r}].$$

Since $U(\mathfrak{g})/P$ is an algebra of finite type, there exists $z_1 \in Z - \{0\}$ such that

$$\theta(U(\mathfrak{g})/P) \subset A_r(Z_{z_1}).$$

Hence there exists $z_2 \in Z - \{0\}$ such that $\theta((U(\mathfrak{g})/P) \otimes_Z Z_{z_2})$ contains $p_1,q_1, \ldots, p_r,q_r$. Let $z_3 = z_1 z_2$. Then

$$\theta((U(\mathfrak{g})/P) \otimes_Z Z_{z_3}) \subset A_r(Z_{z_1}) \cdot Z_{z_3} = A_r(Z_{z_3}),$$

$$\{p_1,q_1, \ldots, p_r,q_r\} \cup Z_{z_3} \subset \theta((U(\mathfrak{g})/P) \otimes_Z Z_{z_3}),$$

hence $\theta((U(\mathfrak{g})/P) \otimes_Z Z_{z_3}) = A_r(Z_{z_3})$. The algebra $(U(\mathfrak{g})/P) \otimes_Z Z_{z_3}$ is generated (over k) by $U(\mathfrak{g})/P$ and z_3^{-1}, hence $A_r(Z_{z_3})$ is an algebra of finite type. By filtering $A_r(Z_{z_3})$ in an obvious way, the associated gradied algebra (which is of finite type over k) is an algebra of polynomials over Z_{z_3}. Hence Z_{z_3} is an algebra of finite type.

A similar reasoning shows that there exists $z_3' \in Z' - \{0\}$ such that $Z'_{z_3'}$ is an algebra of finite type and that

$$\theta'((S(\mathfrak{g})/P') \otimes_{Z'} Z'_{z_3'} = Z'_{z_3'}[\xi_1, \ldots, \xi_{2r}].$$

Let $z_4 \in Z$ and $z_4' \in Z'$ such that $\varepsilon(Z_{z_3}) \subset Z'_{z_3' z_4'} \subset \varepsilon(Z_{z_3 z_4})$. Let $z_5' \in Z'$ and $l \in \mathbf{N}$ such that $\varepsilon(z_3 z_4) = z_5'(z_3' z_4')^{-1}$. We write $z = z_3 z_4$ and $z' = z_3' z_4' z_5'$. Then $\varepsilon(Z_z) \subset Z'_{z'} \subset \varepsilon(Z_z)$, whence $\varepsilon(Z_z) = Z'_{z'}$. Properties (i) and (iii) of the lemma follow immediately.

6.4.6. THEOREM (k algebraically closed, $\mathfrak{g}$ nilpotent). *Let $\mathscr{A}$ be the adjoint group of $\mathfrak{g}$, P a prime ideal of $U(\mathfrak{g})$, W the irreducible closed subset of* Prim $U(\mathfrak{g})$ *corresponding to P, P' the invariant prime ideal of $S(\mathfrak{g})$ associated with P, and W' the $\mathscr{A}$-invariant irreducible closed subset of $\mathfrak{g}^*$ corresponding to P'. There exists a non-empty open subset W_1 of W and an $\mathscr{A}$-stable non-empty open subset W_1' of W' such that $\bar{I}|(W_1'/\mathscr{A})$ is a homeomorphism of $W_1'/\mathscr{A}$ onto W_1.*

(a) Let $r, Z, C, Z', C', \varepsilon, \theta, \theta'$ be as in 6.4.5. We consider the homomorphisms

$$S(\mathfrak{g} \otimes C') \to S(\mathfrak{g} \otimes C')/P'^{\wedge} \to (S(\mathfrak{g})/P') \otimes_{Z'} C' \xrightarrow{\theta'} C'[\xi_1, \ldots, \xi_{2r}] \to C',$$

where the last homomorphism is the identity mapping on C' and transforms $\xi_1, \ldots, \xi_{2r}$ into 0. The composite homomorphism defines a C'-linear form f_0 on $\mathfrak{g} \otimes C'$, and all elements of $P'^{\wedge}$ are zero at f_0. The rank of B_{f_0} is hence $2r$. Let $\mathfrak{h}_0$ be a polarization of $\mathfrak{g} \otimes C'$ at f_0, and $P^{\vee}$ the rational ideal Ker ind$(f_0|\mathfrak{h}_0, \mathfrak{g} \otimes C')$. By definition we have $P = P^{\vee} \cap U(\mathfrak{g})$; the canonical mapping of $U(\mathfrak{g})/P$ into $U(\mathfrak{g} \otimes C')/P^{\vee}$ sends Z into C', and its restriction to Z is none other than $\varepsilon|Z$.

There exists a basis $(x_1, \ldots, x_n)$ for $\mathfrak{g}$ such that $(x_1, \ldots, x_r)$ is a basis for a complement of $\mathfrak{h}_0$ in $\mathfrak{g} \otimes C'$; by changing the numbering of $x_{r+1}, \ldots, x_n$ where necessary, we may assume that

$$\delta = \det (f_0([x_i, x_j]))_{1 \leq i,j \leq 2r}$$

is non-null. Let us write

$$x_i = h_i + \sum_{j=1}^{r} \alpha_{ij} x_j \quad (i = r+1, \ldots, n),$$

where $h_i \in h_0$, $\alpha_{ij} \in C'$.

(b) We choose z and z' with the properties of 6.4.5 and, moreover, such that δ, δ^{-1} and the α_{ij} are in $Z'_{z'}$. Let us set

$$\mathfrak{h}_0 \cap (\mathfrak{g} \otimes Z'_{z'}) = \mathfrak{h}_1, \qquad f_0 \,|\, \mathfrak{g} \otimes Z'_{z'} = f_1.$$

Then f_1 is a $Z'_{z'}$-linear form on $\mathfrak{g} \otimes Z'_{z'}$. On the other hand, $x_1, \ldots, x_r$, $h_{r+1}, \ldots, h_n$ belong to $\mathfrak{g} \otimes Z'_{z'}$, are linearly independent over C' hence over $Z'_{z'}$, and each element of $\mathfrak{g} \otimes Z'_{z'}$ is a $Z'_{z'}$-linear combination of $x_1, \ldots, x_n$ and hence of $x_1, \ldots, x_r, h_{r+1}, \ldots, h_n$. In brief, $\mathfrak{g} \otimes Z'_{z'}$ has the basis $(x_1, \ldots, x_r, h_{r+1}, \ldots, h_n)$. Now $\mathfrak{h}_1$ contains $h_{r+1}, \ldots, h_n$, and its intersection with $C'x_1 + \cdots + C'x_r$ is null. Hence $(h_{r+1}, \ldots, h_n)$ is a basis for $\mathfrak{h}_1$ over $Z'_{z'}$.

(c) Let W_1 be the set of primitive ideals R of $U(\mathfrak{g})$ containing P and such that $z \notin R/P$; it is a non-empty open subset of W. Let W'_1 be the set of the $f \in W'$ such that $z'(f) \neq 0$; it is an $\mathscr{A}$-stable non-empty open subset of W'.

Let $\mathscr{M}$ be the space of maximal ideals of Z_z. The mapping $S \mapsto A_r(k) \otimes S$ is a bijection of the set of ideals of Z_z onto the set of two-sided ideals of $A_r(k) \otimes Z_z = (U(\mathfrak{g})/P)_z$; this bijection preserves the inclusion, transforms prime ideals into prime ideals, and maximal ideals into two-sided maximal ideals (4.5.1). On the other hand, the mapping $R \mapsto R_z$ is a bijection of the set of prime ideals of $U(\mathfrak{g})/P$ not containing z onto the set of prime ideals of $(U(\mathfrak{g})/P)_z$, and $R = R_z \cap (U(\mathfrak{g})/P)$ (3.6.17); if R is a maximal two-sided ideal of $U(\mathfrak{g})/P$ not containing z, we therefore see that R_z is a maximal two-sided ideal of $(U(\mathfrak{g})/P)_z$; if R_z is maximal, z is congruent to a scalar modulo R_z and hence modulo R, hence $(U(\mathfrak{g})/P)_z/R_z = U(\mathfrak{g})/R$ and R is maximal. From all this it follows that if to every $Q \in W_1$ we associate the element $\chi(Q)$ of $\mathscr{M}$ such that $(Q/P)_z = A_r(k) \otimes \chi(Q)$, then χ is a homeomorphism of W_1 onto $\mathscr{M}$.

Let $\mathscr{M}'$ be the space of maximal ideals of $Z'_{z'}$; for $u \in Z'_{z'}$ and $M' \in \mathscr{M}'$, we denote by $u(M')$ the value of u at M'. The space W'_1 can be identified with the space of maximal ideals of $(S(\mathfrak{g})/P')_{z'} = k[\xi_1, \ldots, \xi_{2r}] \otimes Z'_{z'}$, i.e. to $k^{2r} \times \mathscr{M}'$. For every $R \in W'_1$, let $\pi(R) = R \cap Z'_{z'} \in \mathscr{M}'$. Then π can be identified with the projection of $k^{2r} \times \mathscr{M}'$ onto $\mathscr{M}'$, hence π is continuous and open.

Two $\mathscr{A}$-conjugate elements of W'_1 have the same image under π. For every $M' \in \mathscr{M}'$, let $f(M')$ be the element of W'_1 which can be identified with the ideal

$$k[\xi_1, \ldots, \xi_{2r}] \otimes M'.$$

Then $\pi \circ f = \mathrm{id}_{\mathscr{M}'}$. If $x \in \mathfrak{g}$, the definitions imply that

$$f(M')(x) = f_1(x)(M').$$

The restriction of ε to Z_z defines a homeomorphism ψ of $\mathscr{M}'$ onto $\mathscr{M}$. Hence we have the following diagram:

$$\begin{array}{ccc} W'_1 & & W_1 \\ \pi \downarrow \uparrow f & & \downarrow \chi \\ \mathscr{M}' & \xrightarrow{\psi} & \mathscr{M} \end{array} \tag{1}$$

(d) For all $M' \in \mathscr{M}'$, $\pi^{-1}(M')$ is a manifold isomorphic to k^{2r}. On the other hand, the alternating bilinear form $B_{f(M')}$ has rank $2r$ since δ is invertible in $Z'_{z'}$. The $\mathscr{A}$-orbit of $f(M')$ thus has dimension $2r$. It is closed (11.2.4) and contained in $\pi^{-1}(M')$, and therefore equal to $\pi^{-1}(M')$.

(e) For all $M' \in \mathscr{M}'$, we consider the linear mapping of $\mathfrak{g} \otimes Z'_{z'}$ onto $\mathfrak{g}$ which transforms $x \otimes u$ into $u(M')x$ ($x \in \mathfrak{g}$, $u \in Z'_{z'}$). Let $\mathfrak{h}(M')$ be the image of $\mathfrak{h}_1$ under this mapping. Then $\mathfrak{h}(M')$ is a Lie subalgebra of $\mathfrak{g}$ subordinate to $f(M')$, and has the family of the

$$x_i - \sum_{j=1}^{r} \alpha_{ij}(M')x_j \qquad (i = r+1, \ldots, n)$$

as a basis. Taking (d) into account, we have $\mathfrak{h}(M') \in P(f(M'))$.

(f) Let $M' \in \mathscr{M}'$. Let us consider the homomorphisms

$$k \leftarrow Z'_{z'} \rightarrow C'$$

(that on the left being defined by M' and that on the right being the canonical injection). They define the homomorphisms (cf. 2.1.12)

$$U(\mathfrak{g}) \leftarrow U(\mathfrak{g} \otimes Z'_{z'}) \rightarrow U(\mathfrak{g} \otimes C'),$$

$$U(\mathfrak{h}(M') \leftarrow U(\mathfrak{h}_1) \rightarrow U(\mathfrak{h}_0),$$

$$U(\mathfrak{g}) \otimes_{U(\mathfrak{h}(M'))} k \leftarrow U(\mathfrak{g} \otimes Z'_{z'}) \otimes_{U(\mathfrak{h}_1)} Z'_{z'} \xrightarrow{\gamma} U(\mathfrak{g} \otimes C') \otimes_{U(\mathfrak{h}_0)} C'$$

(in the latter tensor products, $\mathfrak{h}(M')$ operates in k by virtue of $f(M')$, $\mathfrak{h}_1$ operates in $Z'_{z'}$ by virtue of f_1, $\mathfrak{h}_0$ operates in C' by virtue of f_0). Since $x_1, \ldots, x_r$ is a basis (over $Z'_{z'}$) for a complement of $\mathfrak{h}_1$ in $\mathfrak{g} \otimes Z'_{z'}$, the ordered monomials in $x_1, \ldots, x_r$ form a basis over $Z'_{z'}$ for $U(\mathfrak{g} \otimes Z'_{z'}) \otimes_{U(\mathfrak{h}_1)} Z'_{z'}$ (2.2.8); their transforms under γ form a basis over C' for $U(\mathfrak{g} \otimes C') \otimes_{U(\mathfrak{h}_0)} C'$; hence γ is injective. Every element of P annihilates the module $U(\mathfrak{g} \otimes C') \otimes_{U(\mathfrak{h}_0)} C'$, hence the module $U(\mathfrak{g} \otimes Z'_{z'}) \otimes_{U(\mathfrak{h}_1)} Z'_{z'}$, hence the module $U(\mathfrak{g}) \otimes_{U(\mathfrak{h}(M'))} k$. This proves that $P \subset I(f(M'))$.

(g) Let $u \in Z$. By definition of $P^{\vee}$ and of ε, we have for every element w of $U(\mathfrak{g} \otimes C') \otimes_{U(\mathfrak{h}_0)} C'$,

$$u \cdot w = \varepsilon(u) \cdot w.$$

In particular, this equality holds for $w \in U(\mathfrak{g} \otimes Z'_{z'}) \otimes_{U(\mathfrak{h}_1)} Z'_{z'}$. We deduce that, for every $w' \in U(\mathfrak{g}) \otimes_{U(\mathfrak{h}(M'))} k$ and every $M' \in \mathscr{M}'$,

$$u \cdot w' = \varepsilon(u)(M') \cdot w'.$$

Now u operates in $U(\mathfrak{g}) \otimes_{U(\mathfrak{h}(M'))} k$ according to the homothety with ratio $u \bmod (I(f(M'))/P)$. Hence

$$u(\psi(M')) = \varepsilon(u)(M') = u \bmod (I(f(M'))/P. \tag{2}$$

Since z is invertible in Z_z, $\varepsilon(z)$ is invertible in $Z'_{z'}$; for every $M' \in \mathscr{M}'$ we therefore have

$$z \bmod (I(f(M'))/P) = \varepsilon(z)(M') \neq 0,$$

hence

$$I(f(M')) \in W_1.$$

Taking (d) into account we see that $I(W'_1) \subset W_1$ and diagram (1) gives us the following diagram:

$$\begin{array}{ccc} W'_1 & \xrightarrow{I|W'_1} & W_1 \\ \downarrow{\scriptstyle \pi} & & \downarrow{\scriptstyle \chi} \\ \mathscr{M}' & \xrightarrow{\psi} & \mathscr{M} \end{array} \tag{3}$$

Diagram (3) is commutative. For, let $w'_1 \in W'_1$; we show that

$$\chi(I(w'_1)) = \psi(\pi(w'_1)).$$

Taking (d) into account, it is sufficient to consider the case where w'_1 is of the form $f(M')$, where $M' \in \mathscr{M}'$. We then have, for every $u \in Z$,

$$u(\chi(I(w'_1))) = u \bmod (I(f(M'))/P),$$

$$u(\psi(\pi(w'_1))) = u(\psi(M')),$$

so that it is sufficient to use (2).

Since χ and ψ are homeomorphisms, the theorem follows from what was said in (c) concerning π.

6.4.7. REMARK. In fact, we have proved a more precise result than what was announced: let $\mathscr{A}, P, W, P', W'$ be as in 6.4.6, r the weight of P, Z, C, Z', C',

ε as in 6.4.5. There exist $z \in Z - \{0\}$, $z' \in Z' - \{0\}$ having the following properties :

(i), (ii), (iii) are as in 6.4.5;

(iv) let W_1 be the set of primitive ideals R of $U(\mathfrak{g})$ containing P and such that $z \notin R/P$; let W_1' be the set of the $f \in W'$ such that $z'(f) \neq 0$; then $\bar{I}$ defines a homeomorphism of $W_1'/\mathscr{A}$ onto W_1;

(v) if $u \in Z$ and $f \in W_1'$, then

$$u \bmod (I(f)/P) = \varepsilon(u)(f).$$

6.4.8. COROLLARY (k algebraically closed, $\mathfrak{g}$ nilpotent). *Let P be a prime ideal of $U(\mathfrak{g})$, P' the invariant prime ideal of $S(\mathfrak{g})$ associated with P and $W' \subset \mathfrak{g}^*$ the set of zeros of P'. Then $P = \bigcap_{f \in W'} I(f)$.*

With the notation of 6.4.6, we have $I(f) \supset P$ for $f \in W_1'$, hence for $f \in W'$ (6.4.4). On the other hand, since W_1 is dense in W, we have

$$\bigcap_{f \in W_1'} I(f) = \bigcap_{Q \in W_1} Q = \bigcap_{Q \in W} Q = P.$$

References: [20], [24], [107].

6.5. The injectivity of the mapping $\bar{I}$

6.5.1. LEMMA ($\mathfrak{g}$ completely solvable). *Let $\mathscr{A}$ be the algebraic adjoint group of $\mathfrak{g}$, $\mathfrak{g}'$ an ideal of $\mathfrak{g}$, $f \in \mathfrak{g}^*$ and $f' = f|\mathfrak{g}'$. Then*

$$I(f) \cap U(\mathfrak{g}') = \bigcap_{\alpha \in \mathscr{A}} \alpha(I(f')).$$

From 1.12.10, there exists $\mathfrak{h} \in P(f)$ such that $\mathfrak{h}' = \mathfrak{h} \cap \mathfrak{g}' \in P(f)$. Let L (or L') be the left ideal of $U(\mathfrak{g})$ (or $U(\mathfrak{g}')$) generated by the $h - f(h) - \theta_{\mathfrak{g},\mathfrak{h}}(h)$ where $h \in \mathfrak{h}$ (or by the $h - f'(h) - \theta_{\mathfrak{g}',\mathfrak{h}'}(h)$ where $h \in \mathfrak{h}'$). From 5.1.7 and 5.1.9 we have $I(f) = \bigcap_{\alpha \in \mathscr{A}} \alpha(L)$. Since $\theta_{\mathfrak{g},\mathfrak{h}}|\mathfrak{h}' = \theta_{\mathfrak{g}',\mathfrak{h}'}$ (because $\mathfrak{g}'$ is an ideal of $\mathfrak{g}$), the Poincaré–Birkhoff–Witt theorem implies that $L \cap U(\mathfrak{g}') = L'$. Then, by denoting the adjoint group of $\mathfrak{g}'$ by $\mathscr{A}'$, we have

$$I(f) \cap U(\mathfrak{g}') = \bigcap_{\alpha \in \mathscr{A}} \alpha(L) \cap U(\mathfrak{g}') = \bigcap_{\alpha \in \mathscr{A}} \alpha(L')$$

$$= \bigcap_{\alpha \in \mathscr{A}} \bigcap_{\alpha' \in \mathscr{A}'} \alpha\alpha'(L') = \bigcap_{\alpha \in \mathscr{A}} \alpha(I(f')).$$

6.5.2. LEMMA (k algebraically closed). *Let $\mathfrak{n}$ be an ideal of $\mathfrak{g}$ containing*

$[\mathfrak{g},\mathfrak{g}]$, $\mathfrak{n}^{\perp}$ *the orthogonal of* $\mathfrak{n}$ *in* $\mathfrak{g}^*$, $\pi : \mathfrak{g}^* \to \mathfrak{n}^*$ *the restriction mapping and* $\mathscr{A}$ *the adjoint algebraic group of* $\mathfrak{g}$.

(i) *If* $f \in \mathfrak{n}^{\perp}$, *then* $\mathscr{A}f = f$.

(ii) *If* $\alpha \in \mathscr{A}$, $f \in \mathfrak{n}^{\perp}$ *and* $f' \in \mathfrak{g}^*$, *then* $\alpha(f+f') = f + \alpha(f')$.

(iii) *If* $\alpha \in \mathscr{A}$, $f \in \mathfrak{g}^*$ *and* $g = \pi(f)$, *then* $\pi(\alpha f) = \alpha\pi(f)$.

(iv) *Let* ω *be an* $\mathscr{A}$*-orbit in* $\mathfrak{n}^*$. *The* $\mathscr{A}$*-orbits contained in* $\pi^{-1}(\omega)$ *are closed in* $\pi^{-1}(\omega)$.

(v) *For all* $f \in \mathfrak{g}^*$, *let* Λ_f *be the set of the* $h \in \mathfrak{n}^{\perp}$ *such that* $\mathscr{A}(f+h) = \mathscr{A}f$ *Then the* Λ_f *are vector subspaces of* $\mathfrak{n}^{\perp}$. *If* $f, f' \in \mathfrak{g}^*$ *are such that* $\pi(\mathscr{A}f) = \pi(\mathscr{A}f')$, *then* $\Lambda_f = \Lambda_{f'}$.

If $f \in \mathfrak{n}^{\perp}$ and $x \in \mathfrak{g}$, then $f([x,\mathfrak{g}]) \subset f(\mathfrak{n}) = 0$, hence $x \cdot f = 0$. This proves (i), and (i) implies (ii). Assertion (iii) is obvious (note that $\mathscr{A}\mathfrak{n} \subset \mathfrak{n}$). From (ii) and (iii), the $\mathscr{A}$-orbits contained in $\pi^{-1}(\omega)$ all have the same dimension; hence they are closed in $\pi^{-1}(\omega)$ (BO, p. 98). It is clear that Λ_f is a subgroup of $\mathfrak{n}^{\perp}$. It is also, from (iv), the set of the $h \in \mathfrak{n}^{\perp}$ such that $f + h \in (\mathscr{A}f)^-$; hence it is an algebraic subgroup of $\mathfrak{n}^{\perp}$, i.e. a vector subspace of $\mathfrak{n}^{\perp}$. The final assertion follows from (ii).

6.5.3. LEMMA ($\mathfrak{g}$ completely solvable). *Let* $\mathfrak{n}$ *be a nilpotent ideal of* $\mathfrak{g}$ *containing* $[\mathfrak{g},\mathfrak{g}]$, $\mathfrak{n}^{\perp}$ *the orthogonal of* $\mathfrak{n}$ *in* $\mathfrak{g}^*$, $\pi : \mathfrak{g}^* \to \mathfrak{n}^*$ *the restriction mapping,* $\mathscr{A}$ *the algebraic adjoint group of* $\mathfrak{g}$, $\mathscr{N}$ *the irreducible subgroup of* $\mathscr{A}$ *with Lie algebra* $\mathrm{ad}_{\mathfrak{g}}\mathfrak{n}$, $f \in \mathfrak{g}^*$, *and* $g = \pi(f) \in \mathfrak{n}^*$. *The following properties are equivalent:*

(i) $(\mathscr{N}f + h) \cap (\mathscr{N}f) = \emptyset$ *for all* $h \in \mathfrak{n}^{\perp} - \{0\}$;

(ii) $\pi|\mathscr{N}f$ *is a bijection of* $\mathscr{N}f$ *onto* $\mathscr{N}g$;

(iii) $\mathfrak{n}^g \subset \mathfrak{g}^f$;

(iv) $\mathfrak{n}^g = \mathfrak{n} \cap \mathfrak{g}^f$;

(v) $\mathfrak{g} = \mathfrak{n} + \mathfrak{n}^f$.

If these conditions are satisfied, then

$$\mathscr{N}g = \mathscr{A}g, \qquad J(f) \cap S(\mathfrak{n}) = J(g), \qquad I(f) \cap U(\mathfrak{n}) = I(g).$$

The equivalence (i) $\Leftrightarrow$ (ii) is clear. On the other hand, we always have

$$\mathfrak{n} \cap \mathfrak{n}^f = \mathfrak{n}^g \supset \mathfrak{n} \cap \mathfrak{g}^f \tag{1}$$

and $\mathfrak{n} \subset (\mathfrak{n}^f)^f$, hence

$$\mathfrak{n}^g \subset \mathfrak{n}^f \cap (\mathfrak{n}^f)^f = (\mathfrak{n} + \mathfrak{n}^f)^f. \tag{2}$$

Given this, (v) $\Rightarrow$ (iii) follows from (2), (iii) $\Rightarrow$ (iv) follows from (1), and (iv) implies that $\mathfrak{n} \cap \mathfrak{n}^f \subset \mathfrak{g}^f$ (from (1)), whence $\mathfrak{g} \subset \mathfrak{n}^f + (\mathfrak{n}^f)^f = \mathfrak{n}^f + \mathfrak{n}$, i.e. (v).

Let us assume that (i) is satisfied. Let $y \in \mathfrak{n}^g$. Then $f([y,\mathfrak{n}]) = 0$. Hence, for every $\lambda \in k$, $(\exp \operatorname{ad} \lambda y) \cdot f - f \in \mathfrak{n}^\perp$. Assumption (i) implies that

$$(\exp \operatorname{ad} \lambda y) \cdot f - f = 0,$$

whence

$$y \cdot f = 0,$$

i.e. $y \in \mathfrak{g}^f$. Thus (i) $\Rightarrow$ (iii).

Let us assume that condition (iv) is satisfied. The stabilizers $\mathcal{N}_f$ and $\mathcal{N}_g$ of f and g in $\mathcal{N}$ are algebraic subgroups of $\mathcal{N}$, and hence are irreducible (CH′, p. 124). Now their Lie algebras are $\operatorname{ad}_\mathfrak{g}(\mathfrak{n} \cap \mathfrak{g}^f)$ and $\operatorname{ad}_\mathfrak{g}(\mathfrak{n}^\mathfrak{g})$, whence $\mathcal{N}_f = \mathcal{N}_g$. Since the canonical mappings $\mathcal{N}/\mathcal{N}_f \to \mathcal{N}f$ and $\mathcal{N}/\mathcal{N}_g \to \mathcal{N}_g$ are bijective, we deduce condition (ii). We have thus proved the equivalence of conditions (i) to (v).

Let us assume that conditions (i) to (v) are satisfied. Let $\chi_\mathfrak{g}$ be the homomorphism of $S(\mathfrak{n})$ in k defined by g. Then $\operatorname{Ker} \chi_\mathfrak{g}$ is stable under $\mathfrak{n}^f$. Consequently, $J(g)$, which is the largest ideal of $S(\mathfrak{n})$ contained in $\operatorname{Ker} \chi_\mathfrak{g}$ and stable under $\mathfrak{n}$, is stable under $\mathfrak{n}^f + \mathfrak{n} = \mathfrak{g}$ and hence under $\mathcal{A}$. This proves that $\mathcal{N}g = \mathcal{A}g$. The set of elements of $S(\mathfrak{n})$ which are zero on $\mathcal{A}g$ is also the set of the elements of $S(\mathfrak{n})$ which are zero on $\mathcal{A}f$, whence $S(\mathfrak{n}) \cap J(f) = J(g)$. From 6.2.8, we have $\mathfrak{st}(I(g),\mathfrak{g}) = \mathfrak{n} + \mathfrak{n}^f = \mathfrak{g}$. Then, from 6.5.1,

$$I(f) \cap U(\mathfrak{n}) = \bigcap_{\alpha \in \mathcal{A}} \alpha(I(g)) = I(g).$$

6.5.4. Lemma (k algebraically closed, $\mathfrak{g}$ solvable). *With the notation of* 6.5.3, *the following properties are equivalent:*

(i) $(\mathcal{A}f + h) \cap (\mathcal{A}f) = \emptyset$ *for all* $h \in \mathfrak{n}^\perp - \{0\}$;

(ii) $\pi|\mathcal{A}f$ *is a bijection of* $\mathcal{A}f$ *onto* $\mathcal{A}g$:

(iii) B_f *and* B_g *have the same rank;*

(iv) $\mathfrak{g} = \mathfrak{n} + \mathfrak{g}^f$.

If these conditions are satisfied, then

$$\mathcal{N}g = \mathcal{A}g, \qquad \mathcal{N}f = \mathcal{A}f,$$

$$J(f) \cap S(\mathfrak{n}) = J(g), \qquad I(f) \cap U(\mathfrak{n}) = I(g),$$

and the canonical homomorphisms

$$S(\mathfrak{n})/J(g) \to S(\mathfrak{g})/J(f), \qquad U(\mathfrak{n})/I(g) \to U(\mathfrak{g})/I(f)$$

are isomorphisms.

(i) $\Leftrightarrow$ (ii). This is obvious.

(ii) ⇒ (iii). Let $\mathfrak{a}$ be the Lie algebra of $\mathscr{A}$ and $\mathfrak{s}$ the set of the $x \in \mathfrak{a}$ such that $xf = 0$. Then

$$\dim(\mathscr{A}f) = \dim \mathfrak{a} - \dim \mathfrak{s} \geqq \operatorname{rank}(B_f).$$

On the other hand,

$$\operatorname{rank}(B_f) \geqq \operatorname{rank}(B_g) = \dim \mathscr{N}g.$$

If condition (ii) is satisfied, $\pi|\mathscr{A}f$ is an isomorphism of the manifold $\mathscr{A}f$ onto the normal manifold $\mathscr{A}g$; on the other hand, condition (ii) of 6.5.3 is satisfied, hence $\mathscr{N}g = \mathscr{A}g$, whence

$$\dim(\mathscr{A}f) = \dim(\mathscr{A}g) = \dim(\mathscr{N}g).$$

We conclude that

$$\operatorname{rank}(B_f) = \operatorname{rank}(B_g).$$

(iii) ⇒ (iv). The canonical mapping of $\mathfrak{n}/\mathfrak{n} \cap \mathfrak{g}^f$ into $\mathfrak{g}/\mathfrak{g}^f$ (or into $\mathfrak{n}/\mathfrak{n}^g$) is injective (or surjective), whence

$$\dim(\mathfrak{n}/\mathfrak{n}^g) \leqq \dim(\mathfrak{n}/\mathfrak{n} \cap \mathfrak{g}^f) \leqq \dim(\mathfrak{g}/\mathfrak{g}^f).$$

Condition (iii) implies that these inequalities are equalities, whence

$$\mathfrak{g} = \mathfrak{n} + \mathfrak{g}^f.$$

(iv) ⇒ (ii). Let χ_f be the homomorphism of $S(\mathfrak{g})$ into k defined by f. Then $\operatorname{Ker} \chi_f$ is stable under $\mathfrak{g}^f$. Let I be the ideal of $S(\mathfrak{g})$ defined by $\mathscr{N}f$. It is the largest $\mathfrak{n}$-stable ideal of $S(\mathfrak{g})$ contained in $\operatorname{Ker} \chi_f$. If condition (iv) is satisfied, I is stable under $\mathfrak{g}$. Since $\mathscr{N}f$ is closed (11.2.4), $\mathscr{N}f$ is stable under $\mathscr{A}$, whence $\mathscr{A}f = \mathscr{N}f$. On the other hand, the conditions of 6.5.3 are satisfied, and hence $\mathscr{A}g = \mathscr{N}g$ and $\pi|\mathscr{N}f$ is a bijection of $\mathscr{N}f$ onto $\mathscr{N}g$. We have thus proved (ii).

Conditions (i) to (iv) are thus equivalent. Henceforth let us assume that they are satisfied. In passing we saw that $\mathscr{N}g = \mathscr{A}g$ and $\mathscr{N}f = \mathscr{A}f$. We know from 6.5.3 that $J(f) \cap S(\mathfrak{n}) = J(g)$ and $I(f) \cap U(\mathfrak{n}) = I(g)$. Since π defines an isomorphism from $\mathscr{A}f = \mathscr{N}f$ onto $\mathscr{N}g$ and that these orbits are closed, the canonical homomorphism $S(\mathfrak{n})/J(g) \to S(\mathfrak{g})/J(f)$ is an isomorphism.

Let

$$0 = \mathfrak{g}_0 \subset \mathfrak{g}_1 \subset \cdots \subset \mathfrak{g}_l = \mathfrak{n} \subset \cdots \cap \mathfrak{g}_n = \mathfrak{g}$$

be a sequence of ideals of $\mathfrak{g}$ such that $\dim \mathfrak{g}_i = i$. Let

$$f_i = f|\mathfrak{g}_i, \qquad \mathfrak{h} = \sum_{i=1}^{n} \mathfrak{g}_i^{f_i}, \qquad \mathfrak{h}' = \sum_{i=1}^{l} \mathfrak{g}_i^{f_i}.$$

Then $\mathfrak{h} \in P(f)$, $\mathfrak{h}' \in P(g)$ and $\mathfrak{h}' = \mathfrak{h} \cap \mathfrak{n}$ (1.12.10). Let L (or L') be the left ideal of $U(\mathfrak{g})$ (or $U(\mathfrak{n})$) generated by the $h - f(h) - \theta_{\mathfrak{g},\mathfrak{h}}(h)$ where $h \in \mathfrak{h}$ (or the $h - g(h) - \theta_{\mathfrak{n},\mathfrak{h}'}(h)$ where $h \in \mathfrak{h}'$). Then $\mathfrak{g} = \mathfrak{n} + \mathfrak{g}^f = \mathfrak{n} + \mathfrak{h}$; from 2.1.11, $U(\mathfrak{g})/L$ and $U(\mathfrak{n})/L'$ are canonically isomorphic as $U(\mathfrak{n})$-modules. Let u be the image in $U(\mathfrak{g})/I(f)$ of an element of $\mathfrak{g}$. From 4.6.8, there exists $u' \in U(\mathfrak{n})/I(g)$ such that $[u,v] = [u',v]$ for all $v \in U(\mathfrak{n})/I(g)$. Hence the left multiplication by $u - u'$ defines a $U(\mathfrak{n})$-endomorphism of the module

$$U(\mathfrak{g})/L = U(\mathfrak{n})/L',$$

which is simple (6.1.1). Consequently, $u - u' \in k$, whence $u \in U(\mathfrak{n})/I(g)$. Hence the canonical homomorphism of $U(\mathfrak{n})/I(g)$ into $U(\mathfrak{g})/I(f)$ is surjective.

6.5.5. Let us retain the notation of 6.5.3 and 6.5.4. Since every element of $\mathfrak{n}^{\perp}$ is zero on $[\mathfrak{g},\mathfrak{g}]$, the equivalent properties of 6.5.3, or of 6.5.4, only depend on g and not on the choice of f in $\pi^{-1}(g)$.

6.5.6. LEMMA ($\mathfrak{g}$ completely solvable). *We adopt the notation of* 6.5.3, *and assume that B_f and B_g have the same rank. Then*

$$J(f) \cap S(\mathfrak{n}) = J(g), \qquad I(f) \cap U(\mathfrak{n}) = I(g),$$

and the canonical isomorphisms

$$S(\mathfrak{n})/J(g) \to S(\mathfrak{g})/J(f), \qquad U(\mathfrak{n})/I(g) \to U(\mathfrak{g})/I(f)$$

are isomorphisms.

This follows from 6.5.4 by extension of the scalar field to an algebraic closure of k.

6.5.7. With the hypotheses of 6.5.6, we thus have homomorphisms, which are termed *canonical*, of $S(\mathfrak{g})$ onto $S(\mathfrak{n})J(g)$ and of $U(\mathfrak{g})$ onto $U(\mathfrak{n})/I(g)$. They commute with the adjoint action of $\mathfrak{g}$, and their restrictions to $S(\mathfrak{n})$ and $U(\mathfrak{n})$ are quotient homomorphisms. Let $s \in S(\mathfrak{g})$ and $t \in S(\mathfrak{n})$; if the classes of s and t correspond to each other under the isomorphism of 6.5.6, then we have $s(f) = t(g) = t(f)$ by construction.

6.5.8. LEMMA (k algebraically closed). *Let $\mathfrak{n}$ be a nilpotent ideal of $\mathfrak{g}$ containing $[\mathfrak{g},\mathfrak{g}]$. Let $\mathcal{N}$ be the adjoint group of $\mathfrak{n}$, Q a prime invariant ideal of $S(\mathfrak{n})$, $Z = (S(\mathfrak{g})/Q)^{\mathcal{N}}$, $C =$ Fract Z, $Q^{\wedge}$ the rational invariant ideal of $S(\mathfrak{n} \otimes C)$ associated with Q, $V(Q)$ the set of zeros of Q in $\mathfrak{n}^*$, and $V(Q^{\wedge})$ the set of zeros of $Q^{\wedge}$ in $(\mathfrak{n} \otimes C)^*$.*

(i) *There exist* $r, r_1 \in \mathbf{N}$ *and an open subset* V_1 *of* $V(Q)$ *having the following properties:*

(a) *for all* $g \in V_1$ *and* $G \in V(Q^\wedge)$, *we have* $\operatorname{rank}(B_g) = \operatorname{rank}(B_G) = 2r$;

(b) *for every* $f \in \mathfrak{g}^*$ *such that* $f|\mathfrak{n} \in V_1$ *and every* $F \in (\mathfrak{g} \otimes C)^*$ *such that* $F|\mathfrak{n} \otimes C \in V(Q^\wedge)$, *we have* $\operatorname{rank}(B_f) = \operatorname{rank}(B_F) = 2r_1$.

(ii) *Let us assume the existence of a dense subset* X *of* $V(Q)$ *such that* $\operatorname{rank}(B_g) = \operatorname{rank}(B_f)$ *for all* $g \in X$ *and* $f \in \mathfrak{g}^*$ *such that* $f|\mathfrak{n} = g$. *Then* $\operatorname{rank}(B_G) = \operatorname{rank}(B_F)$ *for all* $G \in V(Q^\wedge)$ *and* $F \in (\mathfrak{g} \otimes C)^*$ *such that* $F|\mathfrak{n} \otimes C = G$.

Assertion (ii) follows directly from (i). Let us prove (i). From 6.3.2 we know that

$$(S(\mathfrak{n})/Q) \otimes_Z C = S(\mathfrak{n} \otimes C)/Q^\wedge$$

can be identified with a polynomial algebra $C[\xi_1, \ldots, \xi_{2r}]$, that $V(Q^\wedge)$ is a $2r$-dimensional orbit for the adjoint group of $\mathfrak{n} \otimes C$, and that B_G has rank $2r$ for every $G \in V(Q^\wedge)$.

From 11.2.4, there exists a $t \in Z - \{0\}$ having the following properties:

(a) $(S(\mathfrak{n})/Q)_t = Z_t[\xi_1, \ldots, \xi_{2r}] = Z_t \otimes k[\xi_1, \ldots, \xi_{2r}]$;

(b) Z_t separates the $\mathcal{N}$-orbits in V_1, where V_1 denotes the set of elements of $V(Q)$ at which t is not zero. We note that Z_t is of finite type since it is the quotient of $(S(\mathfrak{n})/Q)_t$. Let V_2 be the irreducible affine manifold whose algebra of regular functions is Z_t. Then V_1 can be identified with the manifold $V_2 \times k^{2r}$, and the $\mathcal{N}$-orbits in V_1 are the $\{w\} \times k^{2r}$, where $w \in V_2$. For every $w \in V_2$ let

$$g_w = (w, 0) \in V_1.$$

Let us choose $G \in V(Q^\wedge)$ in such a way that $G(\xi_1) = \cdots = G(\xi_{2r}) = 0$ [we identify G with a homomorphism of $S(\mathfrak{n} \otimes C)/Q^\wedge$ into C]. Then $G(\mathfrak{n}) \subset Z_t$, and hence G defines a Z_t-linear form G_1 on $\mathfrak{n} \otimes Z_t$. For all $x \in \mathfrak{n}$ and all $w \in V_2$, we have $g_w(x) = (G_1(x))(w)$.

Let $(x_1, \ldots, x_l)$ be a basis for $\mathfrak{n}$ (or $\mathfrak{g}$). We may assume that t has been chosen in such a way that all non-zero cofactors of the matrix $(G([x_i, x_j]))_{1 \leq i,j \leq 1}$ are invertible in Z_t. Then, for all $w \in V_2$, the matrix $(g_w(x_i, x_j))_{1 \leq i,j \leq l}$ has the same rank as the matrix $(G([x_i, x_j]))_{1 \leq i,j \leq 1}$, whence the property (a) (or (b)).

6.5.9. LEMMA (k algebraically closed). *Let* $\mathfrak{n}$ *be a nilpotent ideal of* $\mathfrak{g}$ *containing* $[\mathfrak{g}, \mathfrak{g}]$, $\mathfrak{n}^\perp$ *the orthogonal subspace of* $\mathfrak{n}$ *in* $\mathfrak{g}^*$, $\pi : \mathfrak{g}^* \to \mathfrak{n}^*$ *the restriction mapping,* $\mathcal{A}$ *the algebraic adjoint group of* $\mathfrak{g}$, X *an* $\mathcal{A}$-*stable irreducible subset of* $\mathfrak{n}^*$ *such that* $(\mathcal{A}f + h) \cap \mathcal{A}f = \varnothing$ *for all* $f \in \pi^{-1}(X)$ *and* $h \in \mathfrak{n}^\perp - \{0\}$,

W' the closure of X in $\mathfrak{n}^*$, P' the $\mathscr{A}$-invariant prime ideal of $S(\mathfrak{n})$ associated with W', P the prime ideal of $U(\mathfrak{n})$ associated with P', and W the set of the primitive ideals of $U(\mathfrak{g})$ containing $PU(\mathfrak{g})$. Then there exists an $\mathscr{A}$-stable non-empty open subset W_1' of W' and an open subset W_1 of W such that I defines a homeomorphism of $\pi^{-1}(W_1')/\mathscr{A}$ onto W_1.

(a) Let $\mathscr{N}$ be the algebraic subgroup of $\mathscr{A}$ with Lie algebra $\mathrm{ad}_{\mathfrak{g}}\mathfrak{n}$, r the weight of P, $Z = Z(\mathfrak{n};P)$, $C = C(\mathfrak{g};P)$, $Z' = (S(\mathfrak{n})/P')^{\mathscr{N}}$, $C' = \mathrm{Fract}\, Z'$ and ε the canonical isomorphism of C onto C'. From 6.5.4, we have $\mathscr{A}g = \mathscr{N}g$ for all $g \in X$, hence every element of C' is $\mathscr{A}$-invariant; by transport of structure, every element of C is $\mathscr{A}$-invariant. From 6.4.7, 6.5.8 and 11.2.4, there exist $z \in Z - \{0\}$ and $z' \in Z' - \{0\}$ with the following properties:

(i) Z_z and $Z'_{z'}$ are algebras of finite type;

(ii) $\varepsilon(Z_z) = Z'_{z'}$;

(iii) $(U(n)/P)_z = A_r(Z_z)$;

(iv) $(S(n)/P')_{z'} = Z'_{z'}[\xi_1, \ldots, \xi_{2r}]$;

(v) if we denote the set of the elements of W' where z' is non-zero by W_1', then $I(g) \supset P$ for all $g \in W_1'$;

(vi) for all $u \in Z$ and $g \in W_1'$, we have $\varepsilon(u)(g) = u \bmod (I((h)/P)$;

(vii) $Z'_{z'}$ separates the $\mathscr{N}$-orbits of W_1';

(viii) for all $f \in \pi^{-1}(W_1')$, $\mathscr{A}f = \mathscr{N}f$ and $\pi(\mathscr{A}f) = \pi(\mathscr{N}f)$ are 2r-dimensional.

(b) By means of the same process as in 6.4.6 (a), let us introduce a C'-linear form on $\mathfrak{n} \otimes C'$, identified with a homomorphism of $S(\mathfrak{n} \otimes C')$ into C'; this form will be denoted here by G'. Then $G'(\mathfrak{n}) \subset Z'_{z'}$, and G' belongs to the orbit in $(\mathfrak{n} \otimes C')^*$ which corresponds to $P'^{\wedge}$. Let F' be a C'-linear form on $\mathfrak{g} \otimes C'$ which extends G' and which is such that $F'(\mathfrak{g}) \subset Z'_{z'}$.

Let F and G be linear forms on $\mathfrak{g} \otimes C$ and $\mathfrak{n} \otimes C$ deduced from F' and G' by virtue of the isomorphism $\varepsilon : C \to C'$. From 6.5.8, we have $\mathrm{rank}(B_{G'}) = \mathrm{rank}(B_{F'})$. We may thus (6.5.7) consider the canonical homomorphisms

$$\theta' : S(\mathfrak{g} \otimes C') \to (S(\mathfrak{n})/P') \otimes_{Z'} C' = S(\mathfrak{n} \otimes C')/J(G'),$$

$$\theta : U(\mathfrak{g} \otimes C) \to (U(\mathfrak{n})/P) \otimes_Z C = U(\mathfrak{n} \otimes C)/I(G)$$

which commute with the adjoint action of $\mathfrak{g}$ and coincide on $S(\mathfrak{n} \otimes C')$ and $U(\mathfrak{n} \otimes C)$ with the quotient homomorphisms. In particular, if $x, y \in \mathfrak{g}$, then

$$y \cdot \theta'(x) = \theta'([y,x]) = [y,x]^-,$$

$$[\bar{y}, \theta(x)] = \theta([y,x]) = [y,x]^-,$$

where we denote the canonical homomorphisms

$$S(\mathfrak{g} \otimes C') \to S(\mathfrak{g} \otimes C')/J(G')S(\mathfrak{g} \otimes C') = (S(\mathfrak{g})/P'S(\mathfrak{g})) \otimes_{Z'} C',$$

$$U(\mathfrak{g} \otimes C) \to U(\mathfrak{g} \otimes C)/I(G)U(\mathfrak{g} \otimes C) = (U(\mathfrak{g})/PU(\mathfrak{g})) \otimes_Z C.$$

by $a \mapsto \bar{a}$. This implies that, for $x \in \mathfrak{g}$, $\bar{x} - \theta(x)$ and $\bar{x} - \theta'(x)$ are $\mathfrak{g}$-invariant. Let $(x_1, \ldots, x_l)$ be a basis for a complement of $\mathfrak{n}$ in $\mathfrak{g}$. By changing z and z' where necessary, we may assume that $\theta(x_i) \in (U(\mathfrak{n})/P)_z$ and $\theta'(x_i) \in (S(\mathfrak{n})/P')_{z'}$ for $i = 1, \ldots, l$. We set

$$S_i = \bar{x}_i - \theta(x_i) \in (U(\mathfrak{g})/PU(\mathfrak{g}))_z,$$

$$S_i' = \bar{x}_i - \theta'(x_i) \in (S(\mathfrak{g})/P'S(\mathfrak{g}))_{z'}.$$

The affine algebra of $\pi^{-1}(W_1')$ is

$$(S(\mathfrak{g})/P'S(\mathfrak{g}))_{z'} = Z_{z'}[\xi_1, \ldots, \xi_{2r}]\,[S_1', \ldots, S_l'].$$

Since the S_i' are $\mathfrak{g}$-invariant, $Z_{z'}'[S_1', \ldots, S_l']$ is the set of the $\mathfrak{g}$-invariants of this affine algebra. Since $Z_{z'}'$ separates the $\mathcal{N}$-orbits of W_1', $Z_{z'}'[S_1', \ldots, S_l']$ separates the $\mathcal{N}$-orbits of $\pi^{-1}(W_1')$ which are also $\mathcal{A}$-orbits and are $2r$-dimensional. The algebra $(U(\mathfrak{g})/PU(\mathfrak{g}))_z$ is isomorphic to $A_r \otimes Z_z[S_1, \ldots, S_l]$.

(c) Throughout the remainder of this proof, we shall denote the space of maximal ideals of every algebra Ξ of finite type by Specm Ξ. The space W_1' can be identified with

$$\operatorname{Specm}((S(\mathfrak{n})/P'))_{z'} = \operatorname{Specm}(Z_{z'}'[\xi_1, \ldots, \xi_{2r}]) = k^{2r} \times \operatorname{Specm}(Z_{z'}').$$

For all $R \in W_1'$, let $\varrho(R) = R \cap Z_{z'}' \in \operatorname{Specm}(Z_{z'}')$. Then ϱ can be identified with the projection of $k^{2r} \times \operatorname{Specm}(Z_{z'}')$ onto $\operatorname{Specm}(Z_{z'}')$:

$$\pi^{-1}(W_1') \underset{\gamma}{\overset{\pi}{\rightleftarrows}} W_1' \underset{\sigma}{\overset{\varrho}{\rightleftarrows}} \operatorname{Specm}(Z_{z'}').$$

Since

$$S(\mathfrak{g} \otimes C')/J(G')S(\mathfrak{g} \otimes C') = (S(\mathfrak{n} \otimes C')/J(G'))[S_1', \ldots, S_l'],$$

we can consider the homomorphism of $S(\mathfrak{g} \otimes C')/J(G')S(\mathfrak{g} \otimes C')$ onto $S(\mathfrak{n} \otimes C')/J(G')$, which is the identity mapping on $S(\mathfrak{n} \otimes C')/J(G')$ and maps $S_1', \ldots, S_l'$ onto 0. Since the S_i' are $\mathcal{A}$-invariant, this homomorphism commutes with the action of $\mathcal{A}$. By restriction, it defines a homomorphism of $(S(\mathfrak{g})P'S(\mathfrak{g}))_{z'}$ onto $(S(\mathfrak{n})/P')_{z'}$, whence we have a mapping $\delta: W_1' \to \pi^{-1}(W_1')$ such that $\pi \circ \delta = \mathrm{id}_{W_1'}$. Similarly, we can consider the homomorphism of $(S(\mathfrak{n})/P') \otimes_{Z'} C' = C'[\xi_1, \ldots, \xi_{2r}]$ onto C' which is the identity mapping on C' and which maps $\xi_1, \ldots, \xi_{2r}$ onto 0; by restriction, it defines a homomorphism of $Z'_{z'}[\xi_1, \ldots, \xi_{2r}]$ onto $Z'_{z'}$, whence we have a mapping $\gamma : \text{Specm}(Z'_{z'}) \to W'_1$ such that $\varrho \circ \gamma = \text{id}_{\text{Specm}(Z'_{z'})}$. For every $M' \in \text{Specm}(Z'_{z'})$, $\gamma(M')$ is, by construction, the linear form $g_{M'} : x \mapsto G'(x)(M')$ on $\mathfrak{n}$, and $(\delta \circ \gamma)(M')$ is the linear form $f_{M'} : x \mapsto F'(x)(M')$ on $\mathfrak{g}$ (cf. 6.5.7).

(d) Let W_1 be the set of the $R \in W$ such that $z \notin R/PU(\mathfrak{g})$. Then W_1 is open in its closure W. The mapping χ of W_1 into Specm $Z_z[S_1, \ldots, S_l]$ which transforms every $R \in W_1$ into

$$(R/PU(\mathfrak{g}))_z \cap Z_z[S_1, \ldots, S_l]$$

is a homeomorphism (this can be seen by a reasoning similar to that in 6.4.6 (c)). The isomorphism of $Z_z[S_1, \ldots, S_l]$ onto $Z'_{z'}[S'_1, \ldots, S'_l]$ which coincides with ε on Z_z and which transforms $S_1, \ldots, S_l$ into $S'_1, \ldots, S'_l$, respectively, defines a homeomorphism ψ of Specm $Z'_{z'}[S'_1, \ldots, S'_l]$ onto Specm $Z_z[S_1, \ldots, S_l]$. The canonical injection of $Z'_{z'}[S'_1, \ldots, S'_l]$ into

$$Z'_{z'}[\xi_1, \ldots, \xi_{2r}]\,[S'_1, \ldots, S'_l] = (S(\mathfrak{g})/P'S(\mathfrak{g}))_{z'}$$

defines a projection σ of $\pi^{-1}(W'_1)$ onto Specm $Z'_{z'}[S'_1, \ldots, S'_l]$. We prove that $I(\pi^{-1}(W'_1)) \subset W_1$ and that the diagram

$$\begin{array}{ccc} \pi^{-1}(W'_1) & \xrightarrow{I|\pi^{-1}(W'_1)} & W_1 \\ \downarrow{\scriptstyle\sigma} & & \downarrow{\scriptstyle\chi} \\ \text{Specm } Z'_{z'}[S'_1, \ldots, S'_l] & \xrightarrow{\psi} & \text{Specm } Z_z[S_1, \ldots, S_l] \end{array}$$

is commutative. This will complete the proof.

Let $h \in \mathfrak{n}^{\perp}$. The mapping $x \mapsto x + h(x)$, where x takes all values in $\mathfrak{g}$, defines automorphisms of $S(\mathfrak{g})$ and of $U(\mathfrak{g})$ and hence automorphisms of $Z_z[S_1, \ldots, S_l]$ and $Z'_{z'}[S'_1, \ldots, S'_l]$ which are trivial on Z_z and $Z'_{z'}$, respectively, and which map S_i and S'_i onto $S_i + h(x_i)$ and $S'_i + h(x_i)$, respectively. The group $\mathfrak{n}^{\perp}$ thus operates in Prim $U(\mathfrak{g})$, W_1, Specm $Z_z[S_1, \ldots, S_l]$, $\pi^{-1}(W'_1)$ and Specm $Z'_{z'}[S'_1, \ldots, S'_l]$. The mappings I, χ, ψ, σ commute with the action of $\mathfrak{n}^{\perp}$ (6.1.8). Hence it is sufficient to prove that, for all $M' \in \text{Specm } Z'_{z'}$, we have $I(f_{M'}) \in W_1$ and that the diagram

$$\begin{array}{ccc} \text{Specm } Z'_{z'} & \xrightarrow{M' \to I(f_{M'})} & W_1 \\ \downarrow{\scriptstyle\tau} & & \downarrow{\scriptstyle\chi} \\ \text{Specm } Z'_{z'}[S'_1, \ldots, S'_l] & \xrightarrow{\psi} & \text{Specm } Z_z[S_1, \ldots, S_l] \end{array}$$

is commutative (τ corresponds with the homomorphism of $Z'_{z'}[S'_1, \ldots, S'_l]$ onto $Z'_{z'}$ which is the identity mapping on $Z'_{z'}$ and which maps $S'_1, \ldots, S'_l$ onto 0). For every $M' \in \operatorname{Specm} Z'_{z'}$, we have $\dim \mathscr{A} f_{M'} = 2r = \dim \mathscr{N} g_{M'}$, hence

$$I(g_{M'}) = I(f_{M'}) \cap U(\mathfrak{n})$$

(6.5.4). From (a) (v) we have $I(g_{M'}) \supset P$, whence $I(f_{M'}) \supset PU(\mathfrak{g})$; from (a)(vi), we have $\varepsilon(u)(M') = u \bmod I(g_{M'})/P$ for every $u \in Z$. Since $\varepsilon(z)$ is invertible in $Z'_{z'}$, we see that $z \notin I(g_{M'})/P$, whence $I(f_{M'}) \in W_1$.

Since $I(g_{M'}) = I(f_{M'}) \cap U(\mathfrak{n})$, we have

$$(I(g_{M'})/P)_z \cap Z_z \subset (I(f_{M'})/PU(\mathfrak{g}))_z \cap Z_z[S_1, \ldots, S_l] = \chi(I(f_{M'})),$$

hence, for all $u \in Z$,

$$u(\chi(I(f_{M'}))) = u \bmod I(g_{M'})/P = \varepsilon(u)(M').$$

To prove that the above diagram is commutative, it is hence sufficient to prove that $S_1, \ldots, S_l$ take the same values at $\psi\tau(M')$ and $\chi I(f_{M'})$. Now $S_1, \ldots, S_l$ are zero at $\psi\tau(M')$. It therefore remains to be shown that $S_i(\chi I(f_{M'})) = 0$, i.e. that $S_i \in (I(f_{M'})/P)_z$ (i and M' being henceforth fixed).

Let $H \in P(F)$; then $H \subset \mathfrak{g} \otimes C$. Let

$$H_1 = H \cap (\mathfrak{g} \otimes Z_z), \qquad F_1 = F \mid \mathfrak{g} \otimes Z_z,$$

and $U_i \in U(\mathfrak{n} \otimes Z_z)$ be a representative of $\theta(x_i)$. Then

$$x_i - U_i \in \operatorname{Ker} \operatorname{ind}^{\sim}(F \mid H, \mathfrak{g} \otimes C).$$

Changing z and z' where necessary we can assume that H_1 is a direct factor in $\mathfrak{g} \otimes Z_z$ and that for all $\alpha \in \mathscr{A}$, $\alpha(x_i - U_i)$ belongs to the left ideal of $U(\mathfrak{g} \otimes Z_z)$ generated by the $x - F_1(x) - \frac{1}{2} \operatorname{tr} \operatorname{ad}_{\mathfrak{g} \otimes Z_z/H_1} x$, where x runs through H_1. Let $\mu_{M'}$ be the mapping from $U(\mathfrak{g} \otimes Z_z)$ onto $U(\mathfrak{g})$ which transforms $u \otimes z$ ($u \in U(\mathfrak{g})$, $z \in Z_z$) into $u \cdot \varepsilon(z)(M')$. Let $\mathfrak{h}(M')$ be the image of H_1 under this mapping. Since

$$\operatorname{rank}(B_{f_{M'}}) = \operatorname{rank}(B_F),$$

we have $\mathfrak{h}(M') \in P(f_{M'})$. Let θ_1 be the restriction of θ to $U(\mathfrak{g} \otimes Z_z)$; since $\theta_1(1 \otimes u) = u$ and $u \bmod I(g_{M'}) = \varepsilon(u)(M')$ for all $u \in Z$, there exists a homomorphism $\theta_{M'}$ which makes the following diagram commutative:

$$\begin{array}{ccc} U(\mathfrak{g} \otimes Z_z) & \xrightarrow{\theta_1} & (U(\mathfrak{n})/P)_z \\ \downarrow{\scriptstyle \mu_{M'}} & & \downarrow \\ U(\mathfrak{g}) & \xrightarrow{\theta_{M'}} & U(\mathfrak{n})/I(g_{M'}) \end{array}$$

where the right-hand arrow is the quotient mapping. We note that $\theta_{M'}$ is the homomorphism ζ defined by $f_{M'}$; indeed, $\theta_{M'}$ and ζ coincide on $U(\mathfrak{n})$, and, for all $\alpha \in \mathscr{A}$, $\alpha(x_i - \mu_{M'}(U_i))$ belongs to the left ideal of $U(\mathfrak{g})$ defined by $f_{M'}$ and $\mathfrak{h}(M')$, whence

$$x_i - \mu_{M'}(U_i) \in I(f_{M'}),$$

and

$$\zeta(x_i - \mu_{M'}(U_i) = 0 = \theta_{M'}(x_i - \mu_{M'}(U_i)).$$

Hence $\operatorname{Ker} \theta_{M'} = I(f_{M'})$. Now θ_1 can be factorized as

$$U(\mathfrak{g}) \otimes Z_z \to (U(\mathfrak{g})/PU(\mathfrak{g}))_z = (U(\mathfrak{n})/P)_z[S_1, \ldots, S_l] \to (U(\mathfrak{n})/P_z,$$

where the first homomorphism transforms $u \otimes u'$ into uu' ($u \in U(\mathfrak{g})$, $u' \in Z$), and the second one maps $S_1, \ldots, S_l$ onto 0. The above diagram implies that we have the commutative diagram

$$\begin{array}{ccc} (U(\mathfrak{g})/PU(\mathfrak{g}))_z = (U(\mathfrak{n})/P_z[S_1, \ldots, S_l] & \longrightarrow & (U(\mathfrak{n})/P)_z \\ \uparrow & & \downarrow \\ U(\mathfrak{g}) & \xrightarrow{\theta_{M'}} & U(\mathfrak{n})/I(g_{M'}) \end{array}$$

where the left-hand arrow is the canonical mapping. Consequently, the kernel of the upper arrow is $(I(f_{M'})/PU(\mathfrak{g}))_z$. We then have $S_i \in (I(f_{M'})/PU(\mathfrak{g}))_z$.

6.5.10. LEMMA (k algebraically closed). *Let* $\mathfrak{n}, \mathfrak{n}^\perp, \pi, \mathscr{A}, f$, *and* g *be as in* 6.5.3. *Let* $\mathscr{H}$ *be an irreducible algebraic group of automorphisms of* $\mathfrak{g}$ *containing* $\mathscr{A}$, *such that* $\mathscr{H}h = \{h\}$ *for all* $h \in \mathfrak{n}^\perp$. *We assume that* $(\mathscr{H}f + h) \cap \mathscr{H}f = \emptyset$ *for all* $h \in \mathfrak{n}^\perp - \{0\}$. *Then* $I(\pi^{-1}(\mathscr{H}g))$ *is open in its closure, and* I *defines a homeomorphism of* $\pi^{-1}(\mathscr{H}g)/\mathscr{A}$ *onto* $I(\pi^{-1}(\mathscr{H}g))$.

The hypotheses of 6.5.9 are satisfied if $X = \mathscr{H}g$. Hence there exists an $\mathscr{A}$-stable non-empty open subset W_1' of $\mathscr{H}g$ such that $I(\pi^{-1}(W_1'))$ is open in its closure and I defines a homeomorphism of $\pi^{-1}(W_1')/\mathscr{A}$ onto $I(\pi^{-1}(W_1'))$. Since I is continuous (6.4.4), $I(\pi^{-1}(W_1'))$ and $I(\pi^{-1}(\mathscr{H}g))$ have the same closure. Hence

$$I(\pi^{-1}(\mathscr{H}g)) = \bigcup_{\gamma \in \mathscr{H}} I(\pi^{-1}(\gamma W_1'))$$

is open in its closure, and it is sufficient to observe that I defines an injective mapping of $\pi^{-1}(\mathscr{H}g)/\mathscr{A}$ into $\operatorname{Prim} U(\mathfrak{g})$. Let $\gamma_1, \gamma_2 \in \mathscr{H}$, $f_1 \in \pi^{-1}(\gamma_1 g)$ and $f_2 \in \pi^{-1}(\gamma_2 g)$ such that $I(f_1) = I(f_2)$. Let U be the set of the $\gamma \in \mathscr{H}$ such that $\gamma g \in W_1'$; it is a non-empty open subset of $\mathscr{H}$. Let $\gamma_0 \in U\gamma_1^{-1} \cap U\gamma_2^{-1}$.

Then $\gamma_0 f_1 \in \pi^{-1}(W_1')$ and $\gamma_0 f_2 \in \pi^{-1}(W_1')$. Since $I(\gamma_0 f_1) = I(\gamma_0 f_2)$, we see that $\gamma_0 f_2 \in \mathscr{A}\gamma_0 f_1$, whence $f_2 \in \mathscr{A} f_1$ because $\mathscr{A}$ is a distinguished subgroup of $\operatorname{Aut}(\mathfrak{g})$.

6.5.11. PROPOSITION (k algebraically closed, $\mathfrak{g}$ nilpotent). *Let $\mathscr{H}$ be an algebraic group of automorphisms of $\mathfrak{g}$. The orbits of $\mathscr{H}$ in* Prim $U(\mathfrak{g})$ *are open in their closure.*

We are immediately led to the case where $\mathscr{H}$ is irreducible, and we can assume that $\mathscr{H}$ contains the adjoint group of $\mathfrak{g}$. We then apply 6.5.10 with $\mathfrak{n} = \mathfrak{g}$.

6.5.12. THEOREM (k algebraically closed, $\mathfrak{g}$ solvable). *Let $\mathscr{A}$ be the algebraic adjoint group of $\mathfrak{g}$. The mapping $\bar{I}$ of $\mathfrak{g}^*/\mathscr{A}$ into* Prim $U(\mathfrak{g})$ *is bijective.*

(a) Let $f_1, f_2 \in \mathfrak{g}^*$ such that $I(f_1) = I(f_2)$; we must prove that $\mathscr{A} f_1 = \mathscr{A} f_2$. Let $\mathfrak{n} = [\mathfrak{g}, \mathfrak{g}]$, $f_1' = f_1|\mathfrak{n}$ and $f_2' = f_2|\mathfrak{n}$. From 6.5.1, we have $\bigcap_{\alpha\in\mathscr{A}} \alpha(I(f_1')) = \bigcap_{\alpha\in\mathscr{A}} \alpha(I(f_2'))$, hence the orbits $\mathscr{A}\cdot I(f_1')$ and $\mathscr{A}\cdot I(f_2')$ have the same closure; from 6.5.11, they coincide. From 6.2.3 there exists $\gamma \in \mathscr{A}$ such that $\gamma f_1' = f_2'$. The problem thus reduces to the case where f_1 and f_2 have a common restriction to $\mathfrak{n}$, which we shall denote by g.

(b) Let $\pi : \mathfrak{g}^* \to \mathfrak{n}^*$ be the restriction mapping. With the notation of 6.5.2, let Λ be the common value of the Λ_f for $f \in \pi^{-1}(\mathscr{A}g)$. Let $\mathfrak{g}_0$ be the orthogonal subspace of Λ in $\mathfrak{g}$; it is an ideal of $\mathfrak{g}$ containing $\mathfrak{n}$. We denote the restriction mappings by $\pi' : \mathfrak{g}^* \to \mathfrak{g}_0^*$ and $\pi_0 : \mathfrak{g}_0^* \to \mathfrak{n}^*$. If $f \in \pi^{-1}(\mathscr{A}g)$ and $f_0 = f|\mathfrak{g}_0$, then $\mathscr{A}f + \Lambda = \mathscr{A}f$, whence $\mathscr{A}f = \pi'^{-1}(\mathscr{A}f_0)$. If $l \in \pi_0^{-1}(\mathscr{A}g)$ and $h \in \mathfrak{g}_0^* - \{0\}$ is zero on $\mathfrak{n}$, then $(Al + h) \cap \mathscr{A}l = \emptyset$; for, if $\alpha \in \mathscr{A}$ is such that $\alpha l = l + h$ and $f \in \mathfrak{g}^*$ is an extension of l, then $\alpha f - f \in \mathfrak{n}^\perp$, hence $\alpha f - f \in \Lambda = \operatorname{Ker} \pi'$, and $h = \alpha l - l = 0$.

(c) Let $f_1^0 = f_1|\mathfrak{g}_0$ and $f_2^0 = f_2|\mathfrak{g}_0$, let W_0 be the canonical image of $\pi_0^{-1}(\mathscr{A}g)$ in Prim $U(\mathfrak{g}_0)$, and Ω_1 and Ω_2 the $\mathscr{A}$-orbits of $I(f_1^0)$ and $I(f_2^0)$ in Prim $U(\mathfrak{g}_0)$. The assumption $I(f_1) = I(f_2)$ and 6.5.1 imply that Ω_1 and Ω_2 have the same closure in Prim $U(\mathfrak{g}_0)$. Let $\mathscr{A}_0$ be the algebraic adjoint group of $\mathfrak{g}_0$. From 6.5.10 (where we replace $\mathfrak{g}$ by $\mathfrak{g}_0$), I defines a homeomorphism of $\pi_0^{-1}(\mathscr{A}g)f\mathscr{A}_0$ onto W_0; now the $\mathscr{A}$-orbits contained in $\pi_0^{-1}(\mathscr{A}g)$ are closed in $\pi_0^{-1}(\mathscr{A}g)$ (because, being conjugate under translations, they have the same dimension); hence Ω_1 and Ω_2 are closed in W_0 and consequently equal; then $\mathscr{A}f_1^0 = \mathscr{A}f_2^0$, whence

$$\mathscr{A}f_1 = \pi'^{-1}(\mathscr{A} f_1^0) = \pi'^{-1}(\mathscr{A} f_2^0) = \mathscr{A} f_2 .$$

References: [22], [109].

6.6. Supplementary remarks

6.6.1. The construction of the $I(f)$ and the results 6.2.4 and 6.2.9 are in [30] and [31]. (The proof of 6.2.9 given here was pointed out orally to me by Rentschler.) The result 6.1.7 is due to Conze, Duflo and Vergne ([27], [45]), 6.2.8 to Duflo [45], 6.3.5 and 6.3.7 to Gabriel and Nouazé [99], 6.4.4 to Conze and Duflo [26] (the proof has been completed — *cf.* 6.4.2 — in accordance with the indications of these authors), 6.4.6 and 6.5.12 to Rentschler ([107], [109]). Better proofs of 6.4.6 and 6.5.12 may be found in [15]; I learnt of them too late to take them into account in this book. Examples of explicit calculations of $\mathfrak{g}^*/\mathscr{A}$, and hence of Prim $U(\mathfrak{g})$, for three-dimensional solvable Lie algebras $\mathfrak{g}$ may be found in [15] (§ 12).

For $\mathfrak{g}$ solvable and k algebraically closed, we do not know if the mapping $\bar{I}$ is a homeomorphism of $\mathfrak{g}^*/\mathscr{A}$ onto Prim $U(\mathfrak{g})$ [cf., however, 6.6.14 (d)]. This has just been established by N. Conze [23] for $\mathfrak{g}$ nilpotent, by imitating a method by which I. Brown proved the analogous result for nilpotent Lie groups (a result conjectured by Kirillov [75]).

6.6.2. Let $\mathfrak{g}_4$, x, y, z, t, f, $\mathfrak{h}_1$ and $\mathfrak{h}_2$ be as in 6.1.2 (iv), $\sigma_1 = \mathrm{ind}(f|\mathfrak{h}_1,\mathfrak{g}_4)$, and $\sigma_2 = \mathrm{ind}(f|\mathfrak{h}_2,\mathfrak{g}_4)$. Then $\mathfrak{h}_1,\mathfrak{h}_2 \in P(f)$, and σ_1 and σ_2 are simple, but

$$\mathrm{Ker}\,\sigma_1 \neq \mathrm{Ker}\,\sigma_2$$

[$\sigma_1(xz - ty)$ and $\sigma_2\,(xz - ty)$ are distinct scalars]. By comparison with 6.1.4, we see the importance of twisted induced representations.

6.6.3. Let $\mathfrak{g}_1,x,y$ be as in 6.1.2 (i) with k algebraically closed. The one-dimensional representations of $\mathfrak{g} = \mathfrak{g}_1$ are the linear forms f_λ such that $f_\lambda(x) = \lambda$ and $f_\lambda(y) = 0$ ($\lambda \in k$). Let I_λ be the kernel of the representation f_λ of $U(\mathfrak{g})$; it is a two-sided ideal of codimension 1. The primitive ideals of $U(\mathfrak{g})$ are: (1) the I_λ; (2) the ideal 0. The latter is hence primitive and not maximal. In Prim $U(\mathfrak{g})$, $\{0\}$ is everywhere dense. All this is an easy consequence of 6.1.7.

We deduce that, if k is algebraically closed and $\mathfrak{h}$ is a non-nilpotent Lie algebra, then $U(\mathfrak{h})$ has a non-maximal primitive ideal [15].

6.6.4. Proposition 6.2.9 fails for $\mathfrak{g}$ solvable and k algebraically closed; this can be seen by taking $\mathfrak{g}$ as in 6.6.3 and $\mathfrak{h} = kx$. This is linked with the "Pukanszky condition"; cf. [9], chap. IV.

6.6.5. Assume that $\mathfrak{g}$ is nilpotent. Let $f \in \mathfrak{g}^*$. Then $(I(f))^{\mathsf{T}} = I(-f)$ [45].

6.6.6. Assume that k is algebraically closed and that $\mathfrak{g}$ is solvable. Let $\mathfrak{n}$ be a nilpotent ideal of $\mathfrak{g}$ such that $\mathfrak{g}/\mathfrak{n}$ is nilpotent. Let $g \in \mathfrak{g}^*$, $f = g|\mathfrak{n}$, and let $\mathfrak{l}$ be a Lie subalgebra of $\mathfrak{g}$ subordinate to g such that $\mathfrak{l} \cap \mathfrak{n} \in P(f)$. Then $\text{ind}^{\sim}(g|\mathfrak{l},\mathfrak{g})$ is simple if and only if $\mathfrak{l} \in P(\mathfrak{g})$ [45].

6.6.7. Assume that $\mathfrak{g}$ is nilpotent. Let $f \in \mathfrak{g}^*$, let $\mathfrak{h}$ be a Lie subalgebra of $\mathfrak{g}$ subordinate to f, and let $\varrho = \text{ind}(f|\mathfrak{h},\mathfrak{g})$. If ϱ is simple, or if the commutant of $\varrho(\mathfrak{g})$ is k, then $\mathfrak{h} \in P(f)$ ([30], p. 503, where, in addition, it is incorrectly asserted that if $\varrho(U(\mathfrak{g}))$ has k as its centre, then $\mathfrak{h} \in P(f)$).

6.6.8. Assume that $\mathfrak{g}$ is nilpotent. Let $f \in \mathfrak{g}^*$, $\mathfrak{h} \in P(f)$, and $\varrho = \text{ind}(f|\mathfrak{h},\mathfrak{g})$. Let us identify $U(\mathfrak{g})/\text{Ker}\ \varrho$ with A_n. Let $\bar{\varrho}$ be the representation of A_n deduced from ϱ by passage to the quotient. Then ϱ can be deduced from the standard representation of A_n by an automorphism of A_n [21].

6.6.9. (a) Let V be a finite-dimensional vector space. Every element f of V defines one and only one derivation $D(f)$ of the algebra $S(V)$ of polynomial functions on V^* such that $D(f)v = \langle f,v \rangle$ for $v \in V$. The mapping $f \mapsto D(f)$ can be uniquely extended to a homomorphism $p \mapsto D(p)$ of the algebra $S(V^*)$ into End $S(V)$. If $p \in S^n(V^*)$, then $D(p)$ is a differential operator with constant coefficients and of order n on V^*, and annihilates every element of $S(V)$ of degree $<n$. Let $S^{\wedge}(V^*)$ be the completed algebra $\prod_{n \geq 0} S^n(V^*)$. From the above, every element p of $S^{\wedge}(V^*)$ also defines an endomorphism $D(p)$ of the vector space $S(V)$ in an obvious way. The transposed mapping of $D(p)$ is multiplication by p in $S^{\wedge}(V^*)$.

(b) Let $x \in \mathfrak{g}$. Let us consider the formal series

$$\frac{\text{sh}\ (\frac{1}{2}\ \text{ad}\ x)}{(\frac{1}{2})\ \text{ad}\ x} = \sum_{n \geq 0} \frac{1}{2^{2n}(2n+1)!} (\text{ad}\ x)^{2n}.$$

The element

$$\det\left(\sum_{n=0}^{m} \frac{1}{2^{2n}(2n+1)!} (\text{ad}\ x)^{2n}\right)$$

of $S(\mathfrak{g}^*)$ tends, when $m \to +\infty$, to an element p of $S^{\wedge}(\mathfrak{g}^*)$ whose constant term is 1. There hence exists one and only one element q of $S^{\wedge}(\mathfrak{g}^*)$ with constant term 1 such that $q^2 = p$, and q is invertible in $S^{\wedge}(\mathfrak{g}^*)$. If $\mathfrak{g}$ is nilpotent, then $q = 1$.

(c) Assume that $\mathfrak{g}$ is nilpotent, or that k is algebraically closed and $\mathfrak{g}$ is solvable. For every $f \in \mathfrak{g}^*$, $I(f) \cap Z(\mathfrak{g})$ is the kernel of a unique homomorphism χ_f of $Z(\mathfrak{g})$ into k. Let β be the canonical mapping of $S(\mathfrak{g})$ into $U(\mathfrak{g})$. Then for all $u \in Y(\mathfrak{g})$ and $f \in \mathfrak{g}^*$, we have

$$\chi_f(\beta u) = (D(q^{-1})u)(f).$$

(d) Assume that $\mathfrak{g}$ is solvable. Let β be as in (c). The mapping $u \mapsto \beta(D(q)u)$ of $Y(\mathfrak{g})$ into $U(\mathfrak{g})$ is an isomorphism of the algebra $Y(\mathfrak{g})$ onto the algebra $Z(\mathfrak{g})$ ([26], [43]). (It would be a good idea to algebraicize the proof of [43]).

6.6.10. Let $\mathfrak{g}$, x, y, z and t be as in 1.14.6, let f be a linear form on $\mathfrak{g}$ such that $f(t) \neq 0$ and $f(z)^2 - 2f(y)f(t) = 0$, and let β be the canonical mapping of $S(\mathfrak{g})$ into $U(\mathfrak{g})$. Then $\beta(J(f)) \neq I(f)$ and $\beta^{-1}(I(f))$ is not even an ideal of $S(\mathfrak{g})$; use the fact that $\beta(x^2(z^2 - 2yt)) = x^2(z^2 - 2yt) - \frac{1}{6}t^2$. (N. Conze, unpublished.)

6.6.11. Assume that $\mathfrak{g}$ is nilpotent. Let P be a prime ideal of $U(\mathfrak{g})$, and Q the corresponding prime invariant ideal of $S(\mathfrak{g})$.

(a) $P \cap Z(\mathfrak{g})$ and $Q \cap Y(\mathfrak{g})$ correspond to each other by means of the isomorphism of 4.8.12 ([98], [99]).

(b) Let $\mathscr{A}$ be the adjoint group of $\mathfrak{g}$, and ε the canonical isomorphism of $C(\mathfrak{g};P)$ onto $\mathrm{Fract}((S(\mathfrak{g})/Q)^{\mathscr{A}})$. Every element of $\varepsilon(Z(\mathfrak{g};P))$ is integral on $S(\mathfrak{g})/Q$ [105]. Cf. problem 25.

6.6.12. Let $\mathfrak{g}$ and $\mathfrak{g}'$ be nilpotent Lie algebras. Let $\mathscr{P}$, $\mathscr{Q}$, $\mathscr{A}$ and β be as in 6.3.7. Let $\mathscr{P}'$, $\mathscr{Q}'$, $\mathscr{A}'$ and β' be defined in a similar way for $\mathfrak{g}'$. Let φ be a homomorphism of $\mathfrak{g}$ into $\mathfrak{g}'$, with which the mappings $\varphi_1 : \mathscr{P}' \to \mathscr{P}$ and $\varphi_2 : \mathscr{Q}' \to \mathscr{Q}$ are associated in an obvious way. Then $\beta \circ \varphi_2 = \varphi_1 \circ \beta'$ [106].

6.6.13. Assume that k is algebraically closed and $\mathfrak{g}$ is nilpotent. Let Γ be an irreducible algebraic group of unipotent automorphisms of $\mathfrak{g}$. Then the Γ-orbits in Prim $U(\mathfrak{g})$ are closed [22].

6.6.14. Assume that k is algebraically closed and $\mathfrak{g}$ is solvable. Let $\mathscr{A}$ be the algebraic adjoint group of $\mathfrak{g}$, $\mathscr{P}$ the set of prime ideals of $U(\mathfrak{g})$, and $\mathscr{Q}$ the set of prime $\mathscr{A}$-invariant ideals of $S(\mathfrak{g})$. Then Prim $U(\mathfrak{g}) \subset \mathscr{P}$. The rational invariant ideals of $S(\mathfrak{g})$ are defined as in 4.8.6; the set of these ideals can be identified with $\mathfrak{g}^*/\mathscr{A}$. Recall (3.8.5) that $\mathscr{P}$ is canonically provided with a topology; the same holds for the set of prime ideals of $S(\mathfrak{g})$ and consequently for $\mathscr{Q}$.

(a) The continuous bijection $\bar{I}$ of $\mathfrak{g}^*/\mathscr{A}$ onto Prim $U(\mathfrak{g})$ can be uniquely extended to a continuous bijection ψ of $\mathscr{Q}$ onto $\mathscr{P}$. If $Q \in \mathscr{Q}$ and V denotes the set of zeros of Q in $\mathfrak{g}^*$, then $\psi(Q) = \bigcap_{f \in V} I(f)$.

(b) Let $P \in \mathscr{P}$. Let K be an algebraically closed extension of $C(\mathfrak{g};P)$. Let P' be the kernel of the canonical mapping

$$U(\mathfrak{g} \otimes K) \to (U(\mathfrak{g})/P) \otimes_{Z(\mathfrak{g};P)} K.$$

This is a primitive ideal of $U(\mathfrak{g} \otimes K)$; let Q' be the corresponding rational invariant ideal in $S(\mathfrak{g} \otimes K)$. Then $Q' \cap S(\mathfrak{g})$ does not depend on the choice of K; it is equal to $\psi^{-1}(P)$. An isomorphism of $C(\mathfrak{g}; P)$ onto $\mathrm{Fract}((S(\mathfrak{g})/Q)^{\mathscr{A}})$ is defined as in 6.3.7.

(c) Let $Q \in \mathscr{Q}$, let A be the closure of $\{0\}$ in $\mathscr{Q}$, and let B be the closure of $\{\psi(Q)\}$ in $\mathscr{P}$. There exists a non-empty subset V which is relatively open in A such that $\psi(V)$ is relatively open in B and $\psi \mid V$ is a homeomorphism of V onto $\psi(V)$.

(d) There exists a partition of $\mathscr{Q}$ into subsets $M_1, \ldots, M_p$ which are locally closed, such that $\psi(M_1) \ldots, \psi(M_p)$ are locally closed in $\mathscr{P}$ and $\psi | M_i$ is a homeomorphism of M_i onto $\psi(M_i)$ for $i = 1, \ldots, p$.

(e) Let Γ be an algebraic subgroup of $\mathrm{Aut}(\mathfrak{g})$. The Γ-orbits in $\mathrm{Prim}\, U(\mathfrak{g})$ are locally closed [15].

CHAPTER 7

VERMA MODULES

Now and in the next three chapters, we turn to the semi-simple case.

The first difficult results, concerning the modules which are the subject of the present chapter, were obtained by D. N. Verma [126] around 1966. Slightly later, I. I. Bernstein, I. M. Gelfand and S. I. Gelfand obtained deep theorems concerning them ([10], [11]). The name "Bernstein—Gelfand—Verma modules", although scientifically justified, seemed less practical to me.

7.0. Notation

In this chapter we denote a splitting semi-simple Lie algebra by $(\mathfrak{g},\mathfrak{h})$, its root system by R, its Weyl group by W, a basis for R by B, the sets of positive and negative roots relative to B by R_+ and R_- respectively, the set of weights of R by P, the set of radicial weights of R by Q, the set of linear combinations of elements of B with coefficients in $\mathbf{N}$ by Q_+, and the set of dominant weights by P_{++}. We denote the ordering relation $\lambda - \mu \in Q_+$ in $\mathfrak{h}^*$ by $\mu \leqq \lambda$ (this ordering relation is different from the one considered in 11.1.7). If $w \in W$, we denote the length of w relative to B by $l(w)$, and the determinant of w, which is equal to 1 or -1 according to whether $l(w)$ is even or odd, by $\varepsilon(w)$. We set

$$\delta = \frac{1}{2} \sum_{\alpha \in R_+} \alpha,$$

$$\mathfrak{n}_+ = \sum_{\alpha \in R_+} \mathfrak{g}^\alpha, \qquad \mathfrak{n}_- = \sum_{\alpha \in R_-} \mathfrak{g}^\alpha,$$

$$\mathfrak{b}_+ = \mathfrak{h} + \mathfrak{n}_+, \quad \mathfrak{b}_- = \mathfrak{h} + \mathfrak{n}_-.$$

If $\nu \in \mathfrak{h}^*$, we denote the number of families $(n_\alpha)_{\alpha \in R_+}$, where the n_α are nonnegative integers. such that $\nu = \sum_{\alpha \in R_+} n_\alpha \alpha$ by $\mathfrak{P}(\nu)$; we have $\mathfrak{P}(\nu) > 0 \Leftrightarrow \nu \in Q_+$.

7.1. The modules $L(\lambda)$ and $M(\lambda)$

7.1.1. Let V be a vector spsce, and π a representation of $\mathfrak{g}$ in V. For all $\mu \in \mathfrak{h}^*$, we defined in 1.2.13 the vector subspace V_μ, which is the set of the $v \in V$ such that $\pi(x)v = \mu(x)v$ for all $x \in \mathfrak{h}$. If $V_\mu \neq 0$, μ is said to be a *weight of* π, and dim V_μ is termed the *multiplicity of the weight* μ (if $V_\mu = 0$, μ is sometimes said to be a weight of multiplicity 0); this notion is a special case of that of 1.2.8. An element of V_μ is said to be *of weight* μ. We recall that the sum of the V_μ is direct (1.2.13).

7.1.2. PROPOSITION (notation as in 7.0). (i) *Let* $\alpha,\mu \in \mathfrak{h}^*$. *Then* $\pi(\mathfrak{g}^\alpha)V_\mu \subset V_{\mu+\alpha}$
(ii) *The sum* $\oplus_{\mu\in\mathfrak{h}^*}V_\mu$ *is stable under* π.

If $h \in \mathfrak{h}$, $x \in \mathfrak{g}^\alpha$ and $v \in V_\mu$, then

$$\begin{aligned} hxv &= xhv + [h,x]v \\ &= x\mu(h)v + \alpha(h)xv = (\mu - \alpha)(h)xv, \end{aligned}$$

whence (i), and (i) implies (ii).

7.1.3. The representation π may have no weight at all. In this chapter we shall study representations π for which, on the contrary, we have $V \oplus_{\mu\in\mathfrak{h}^*} V_\mu$. We shall see (7.2.1) that all finite-dimensional representations of $\mathfrak{g}$ are of this type.

7.1.4. Let $\lambda \in \mathfrak{h}^*$. We define a one-dimensional representation τ_λ of $\mathfrak{b}_+$ by setting $\tau_\lambda(h + n) = \lambda(h)$ for $h \in \mathfrak{h}$ and $n \in \mathfrak{n}$. Let

$$\sigma = \operatorname{ind}^\sim(\tau_\lambda,\mathfrak{g}).$$

If $h \in \mathfrak{h}$, then $\mathfrak{n}_-$ is stable under $\operatorname{ad}_\mathfrak{g} h$, and $\operatorname{tr}(\operatorname{ad}_\mathfrak{g} h|\mathfrak{n}_-) = -2\delta(h)$; on the other hand, if $n \in \mathfrak{n}_+$, then $\operatorname{ad}_\mathfrak{g} n$ is nilpotent. Consequently,

$$\sigma = \operatorname{ind}(\tau_{\lambda-\delta},\mathfrak{g}).$$

We denote the $\mathfrak{g}$-module corresponding to σ by $M(\lambda)$. We term it the *Verma module* associated with $\mathfrak{g}$, $\mathfrak{h}$, B, λ. Hence we have

$$M(\lambda) = U(\mathfrak{g}) \otimes_{U(\mathfrak{g}_+)} k,$$

where k is provided with the $\mathfrak{b}_+$-module structure defined by $\lambda - \delta$.

7.1.5. From 5.1.6, the mapping $u \mapsto u \otimes 1$ of $U(\mathfrak{n}_-)$ into $M(\lambda)$ is an $\mathfrak{n}_-$-module isomorphism if we provide $U(\mathfrak{n}_-)$ with the left regular representation. We sometimes identify $M(\lambda)$ and $U(\mathfrak{n}_-)$ by means of this isomorphism.

7.1.6. PROPOSITION (notation as in 7.0). *Let* $\lambda \in \mathfrak{h}^*$. *For all* $\alpha \in R$, *let* $X_\alpha \in \mathfrak{g}^\alpha - \{0\}$. *Let* $\alpha_1, \ldots, \alpha_n$ *be the (pairwise distinct) elements of* R_+. *Then:*
(i) $M(\lambda) = \oplus_{\mu\in\mathfrak{h}^*} M(\lambda)_\mu$.

(ii) *The weights of* $M(\lambda)$ *have the form* $\lambda - \delta - \sum_{\alpha \in B} n_\alpha \alpha$, *where the* n_α *are non-negative integers. For all* $\mu \in \mathfrak{h}^*$,

$$\dim M(\lambda)_\mu = \mathfrak{P}(\lambda - \delta - \mu).$$

(iii) *For all* $\mu \in \mathfrak{h}^*$,

$$M(\lambda)_\mu = \sum_{\substack{p_1,\ldots,p_n \in \mathbf{N}, \\ \lambda - \delta - p_1\alpha_1 - \cdots - p_n\alpha_n = \mu}} X^{p_1}_{-\alpha_1} \cdots X^{p_n}_{-\alpha n} \otimes k.$$

(iv) *We have*

$$M(\lambda)_{\lambda-\delta} = 1 \otimes k, \qquad M(\lambda) = U(\mathfrak{n}_-)\, M(\lambda)_{\lambda-\delta},$$
$$U(\mathfrak{n}_+)\, M(\lambda)_{\lambda-\delta} = 0.$$

If $h \in \mathfrak{h}$, then

$$\begin{aligned} h \cdot (X^{p_1}_{-\alpha_1} \cdots X^{p_n}_{-\alpha_n} \otimes 1) &= \\ &= [h, X^{p_1}_{-\alpha_1} \cdots X^{p_n}_{-\alpha_n}] \otimes 1 + X^{p_1}_{-\alpha_1} \cdots X^{p_n}_{-\alpha_n} h \otimes 1 \\ &= (-p_1\alpha_1 - \cdots - p_n\alpha_n)(h) X^{p_1}_{-\alpha_1} \cdots X^{p_n}_{-\alpha_n} \otimes 1 \\ &\quad + X^{p_1}_{-\alpha_1} \cdots X^{p_n}_{-\alpha_n} \otimes (\lambda - \delta)(h) \\ &= (\lambda - \delta - p_1\alpha_1 - \cdots - p_n\alpha_n)(h)\,(X^{p_1}_{-\alpha_1} \cdots X^{p_n}_{-\alpha_n} \otimes 1). \end{aligned}$$

This proves (iii). The rest is obvious or follows from (iii).

7.1.7. The element $1 \otimes 1$ of $M(\lambda)_{\lambda-\delta}$ is termed the *canonical generator* of $M(\lambda)$.

7.1.8. PROPOSITION (notation as in 7.0). *Let* V *be a* $\mathfrak{g}$-*module*, $\lambda \in \mathfrak{h}^*$, *and* v *an element of* V_λ *annihilated by* $\mathfrak{n}_+$. *We assume that* V *is generated by* v *as a* $\mathfrak{g}$-*module. Then:*

(i) *There exists one and only one* $\mathfrak{g}$-*homomorphism* φ *of* $M(\lambda + \delta)$ *into* V *such that* $\varphi(1 \otimes 1) = v$. *This homomorphism is surjective.*

(ii) $V = \oplus_{\mu \in \mathfrak{h}^*} V_\mu$. *Each* V_μ *is finite dimensional, and* $\dim V_\lambda = 1$ *if* $V \neq 0$. *Every weight of* V *belongs to* $\lambda - Q_+$.

(iii) $V = U(\mathfrak{n}_-) \cdot v$.

(iv) *Every endomorphism of the* $\mathfrak{g}$-*module* V *is scalar.*

(v) *The module* V *has a central character.*

(vi) *The homomorphism* φ *of* (i) *is bijective if and only if* $V \neq 0$ *and* u_v *is injective for all* $u \in U(\mathfrak{n}_-) - \{0\}$.

Assertion (i) follows from 5.1.3 and the fact that v generates the $\mathfrak{g}$-module V. Assertions (ii) and (iii) follow from (i) and 7.1.6. Let c be a $\mathfrak{g}$-endomorphism of V. For all $h \in \mathfrak{h}$, we have

$$h_V c(v) = c h_V(v) = \lambda(h) c(v),$$

hence

$$c(v) \in V_\lambda.$$

Thus there exists $\xi \in k$ such that $c(v) = \xi v$. Then, for all $u \in U(\mathfrak{g})$, we have $cu_V(v) = u_V c(v) = \xi u_V(v)$, so that $c = \xi \cdot 1$. This proves (iv), and (iv) implies (v). If φ is bijective and $u \in U(\mathfrak{n}_-) - \{0\}$, then u_V is injective from

7.1.5. If φ is not bijective, there exists $u \in U(\mathfrak{n}_-)$ such that $u \neq 0$ and $\varphi(u \otimes 1) = 0$; then

$$u_V v = u_V \varphi(1 \otimes 1) = \varphi(u_V \cdot 1 \otimes 1) = \varphi(u \otimes 1) = 0,$$

hence u_V is not injective if $V \neq 0$.

7.1.9. We denote the central character of $M(\lambda)$ by χ_λ.

7.1.10. We can see that the study of $\mathfrak{g}$-modules of the type considered in 7.1.8 amounts to the study of the sub-$\mathfrak{g}$-modules of $M(\lambda)$. We shall give some elementary results for this problem (which will be studied more deeply in 7.6).

7.1.11. PROPOSITION (notation as in 7.0). *Let* $\lambda \in \mathfrak{h}^*$.

(i) *Let* $M(\lambda)_+ = \sum_{\mu \neq \lambda - \delta} M(\lambda)_\mu$. *Then every sub-*$\mathfrak{g}$*-module of* $M(\lambda)$ *distinct from* $M(\lambda)$ *is contained in* $M(\lambda)_+$.

(ii) *There exists a largest sub-*$\mathfrak{g}$*-module* K *of* $M(\lambda)$ *distinct from* $M(\lambda)$. *The* $\mathfrak{g}$*-module* $M(\lambda)/K$ *is absolutely simple.*

Let F be a sub-$\mathfrak{h}$-module of $M(\lambda)$. Then $F = \sum_{\mu \in \mathfrak{h}^*} F \cap M(\lambda)_\mu$. If F is a sub-$\mathfrak{g}$-module distinct from $M(\lambda)$, then $F \cap M(\lambda)_{\lambda-\delta} = 0$ [since $M(\lambda)_{\lambda-\delta}$ is one-dimensional and generates the $\mathfrak{g}$-module $M(\lambda)$], hence $F \subset M(\lambda)_+$. The sum K of the sub-$\mathfrak{g}$-modules of $M(\lambda)$ distinct from $M(\lambda)$ is hence contained in $M(\lambda)_+$, and is consequently distinct from $M(\lambda)$. The $\mathfrak{g}$-module $M(\lambda)/K$ is simple. It is absolutely simple from 7.1.8 (iv) and 2.6.5.

7.1.12. With the notation of 7.1.11, we denote the $\mathfrak{g}$-module $M(\lambda)/K$ by $L(\lambda)$. We can apply 7.1.8 to it. The image in $L(\lambda)$ of the canonical generator of $M(\lambda)$ is termed the *canonical generator of* $L(\lambda)$.

7.1.13. PROPOSITION (notation as in 7.0). *Let V be a simple $\mathfrak{g}$-module and $\lambda \in \mathfrak{h}^*$. We assume that there exists a non-zero element of $V_{\lambda-\delta}$ annihilated by $\mathfrak{n}_+$. Then V is isomorphic to $L(\lambda)$.*

From 7.1.8 (i) there exists a $\mathfrak{g}$-homomorphism φ of $M(\lambda)$ onto V. Since V is simple, we have $\operatorname{Ker}\varphi = K$ with the notation of 7.1.11, whence the proposition follows.

7.1.14. LEMMA. *Let A be an associative algebra, let x, h, y be elements of A such that $[h,y] = -2y$, $[x,y] = h$, and let m be a non-negative integer. Then, in A,*

$$[x,y^m] = m(h + m - 1)y^{m-1} = my^{m-1}(h - m + 1).$$

This is obvious if $m = 0$ or 1; and, if it is true for m, then

$$\begin{aligned}[x,y^{m+1}] &= hy^m + ym(h + m - 1)y^{m-1} \\ &= hy^m + m(h + m - 1)y^m + m2y^m \\ &= (m + 1)(h + m)y^m,\end{aligned}$$

whence we have the first equality of the lemma. On the other hand,

$$[h,y^{m-1}] = (m - 1)(-2)y^{m-1},$$

whence the second equality.

7.1.15. The advantage of having introduced δ in the definition of the $M(\lambda)$ starts to become apparent in the following proposition:

PROPOSITION (notation as in 7.0). *Let $\lambda \in \mathfrak{h}^*$, $\alpha \in B$, and $m = \lambda(H_\alpha)$. We assume that $m \in \mathbf{N}$. Let v be the canonical generator of $M(\lambda)$, $X_{-\alpha}$ a non-zero element of $\mathfrak{g}^{-\alpha}$, $v' = X^m_{-\alpha}$ and V the sub-$\mathfrak{g}$-module of $M(\lambda)$ generated by v'. Then V is isomorphic to $M(s_\alpha\lambda)$.*

We have $v' \neq 0$ (7.1.5). On the other hand, $s_\alpha\lambda = \lambda - m\alpha$, hence $v' \in M(\lambda)_{s\alpha\lambda-\delta}$ from 7.1.2. For $\beta \in B$ and $\beta \neq \alpha$, we have $[\mathfrak{g}^{-\alpha},\mathfrak{g}^\beta] = 0$ (1.10.15) and $\mathfrak{g}^\beta v = 0$, hence $\mathfrak{g}^\beta v' = 0$. If $X_\alpha \in \mathfrak{g}^\alpha$ is sucht hat $[X_\alpha,X_{-\alpha}] = H_\alpha$, then lemma 7.1.14 proves that

$$\begin{aligned}X_\alpha v' &= X_\alpha X^m_{-\alpha} v \\ &= X^m_{-\alpha} X_\alpha v + mX^{m-1}_{-\alpha}(H_\alpha - m + 1)v \\ &= X^m_{-\alpha} \cdot 0 + mX^{m-1}_{-\alpha}(\lambda(H_\alpha) - \delta(H_\alpha) - m + 1)v \\ &= 0,\end{aligned}$$

taking 11.1.13 into account. From 1.10.15 (ii), we have $\mathfrak{n}_+ v' = 0$. Finally, for all $u \in U(\mathfrak{n}_-) - \{0\}$, u_V is a restriction of $u_{M(\lambda)}$ and so is injective. From 7.1.8 (vi), V is isomorphic to $M(s_\alpha \lambda)$.

References: [16], [63], [66], [71], [118], [126].

7.2. Finite-dimensional representations

7.2.1. Proposition (notation as in 7.0). *Let V be a finite-dimensional $\mathfrak{g}$-module. Then:*

(i) $V = \oplus_{\mu \in \mathfrak{h}^*} V_\mu$.

(ii) *Every weight of V belongs to P.*

(iii) *If μ is a weight of V and $w \in W$, then $w\mu$ is a weight of V with the sane multiplicity as μ.*

To prove (i), we can assume that V is simple, and from 7.1.2 (ii) it is sufficient to prove the existence of a weight. Let k' be an algebraic closure of k, and let $\mathfrak{g}' = \mathfrak{g} \otimes k'$, $\mathfrak{h}' = \mathfrak{h} \otimes k'$, $V' = V \otimes k'$. From 1.3.12, there exists a weight μ of V' relative to $\mathfrak{h}$,' From 1.8.5 applied to $\mathfrak{g}'^{\alpha} + \mathfrak{g}'^{-\alpha} + k'H_\alpha$, we have $\mu(H_\alpha) \in \mathbf{Z}$ for all $\alpha \in R$; hence $\mu(\mathfrak{h}) \subset k$, so that V'_μ is of the form $W'' \otimes k'$, where W'' is a non-null vector subspace of V. Thus $\mu|\mathfrak{h}$ is a weight of V, which proves (i). Moreover it can be seen that if ν is a weight of V, then $\nu(H_\alpha) \in \mathbf{Z}$ for all $\alpha \in R$, whence (ii).

Let ν be a weight of V, m_ν its multiplicity, $\alpha \in R$, and $n = \nu(H_\alpha) \in \mathbf{Z}$. Let $X_\alpha \in \mathfrak{g}^\alpha$ and $X_{-\alpha} \in \mathfrak{g}^{-\alpha}$ be such that $[X_\alpha, X_{-\alpha}] = H_\alpha$. Let $v \in V_\nu - \{0\}$. If $n \geqq 0$, then $X^n_{-\alpha} v \neq 0$ from 1.8.5, and $X^n_{-\alpha} v \in V_{\nu - n\alpha}$ from 7.1.2; hence $(X_{-\alpha})^n_V$ maps V_ν injectively into $V_{\nu - n\alpha}$. If $n \leqq 0$, then $X_\nu^{-n} v \neq 0$ from 1.8.5, and $X_\alpha^{-n} v \in V_{\nu - n\alpha}$ from 7.1.2; hence $(X_\alpha)_V^{-n}$ maps V_ν injectively into $V_{\nu - n\alpha}$. In both cases, we have proved that $s_\alpha \nu = \nu - n\alpha$ is a weight of V with multiplicity $\geqq m$. This proves (iii).

7.2.2. Proposition (notation as in 7.0). *Let V be a simple $\mathfrak{g}$-module of finite dimension. Then:*

(i) *There exists one and only one $\lambda \in \mathfrak{h}^*$ such that V is isomorphic to $L(\lambda + \delta)$. In particular, V is absolutely simple.*

(ii) *$\lambda \in P_{++}$, and λ is a weight with multiplicity* 1.

(iii) *If μ is a weight of V, then $\mu \leqq \lambda$ and $\langle \mu, \mu \rangle \leqq \langle \lambda \lambda \rangle$.*

Since V is finite-dimensional, there exists a weight λ of V such that, for all $\alpha \in R_+$, $\lambda + \alpha$ is not a weight of V. Then $\mathfrak{n}_+ V_\lambda = 0$ from 7.1.2, and (i) follows from 7.1.13. Let $\alpha \in R_+$, $X_\alpha \propto \mathfrak{g}^\alpha$ and $X_{-\alpha} \in \mathfrak{g}^{-\alpha}$ be such that $[X_\alpha, X_{-\alpha}] = H_\alpha$. If $v \in V_\lambda$, then $X_\alpha v = 0$, and hence, from 1.8.5, $\lambda(H_\alpha) \in \mathbf{N}$.

Consequently, $\lambda \in P_{++}$. Let μ be a weight of V. Then $\mu \leqq \lambda$ (7.1.8). Let us prove that $\langle\mu,\mu\rangle \leqq \langle\lambda,\lambda\rangle$. Since $W\mu$ intersects P_{++} (7.2.1 (ii), 11.1.5 and 11.1.12), we may assume that $\mu \in P_{++}$ (7.2.1 (iii)). Then the relations $\lambda + \mu \in P_{++}$ and $\lambda - \mu \in Q_+$ imply that

$$0 \leqq \langle\lambda + \mu, \lambda - \mu\rangle = \langle\lambda,\lambda\rangle - \langle\mu,\mu\rangle .$$

7.2.3. We retain the notation of 7.2.2. Let w_0 be the element of W such that $w_0(B) = -B$. Since λ is the largest weight of V, $w_0\lambda$ is the smallest weight of V; it has multiplicity 1 (7.2.1 (iii)).

7.2.4. LEMMA (notation as in 7.0). *Let V,λ,v satisfy the hypotheses of* 7.1.8. *For all $\alpha \in R$, let X_α be a non-zero element of $\mathfrak{g}^\alpha$. We assume that, for all $\beta \in B$, $X^m_{-\beta}v$ is zero for m sufficiently large. Then V is simple and finite-dimensional.*

Let Π be the set of weights of V, and let $\mu \in \Pi$, $\beta \in B$. Let us set $v^i = X^i_{-\beta}v$ for $i = 0,1,\ldots$ Let m be the largest integer such that $v_m \neq 0$. From 1.8.3 (ii), $kv_0 + \cdots + kv_m$ is stable under $\mathfrak{s}_\beta = \mathfrak{g}^\beta + \mathfrak{g}^{-\beta} + kH_\beta$. The sum of the finite-dimensional simple sub-$\mathfrak{s}_\beta$-modules of V is stable under $\mathfrak{g}$ (1.7.9), and contains v from the above, hence is equal to V. Consequently, V is the direct sum of a family $(T_i)_{i\in I}$ of finite-dimensional simple sub-$\mathfrak{s}_\beta$-modules. Let $t \in V_\mu - \{0\}$, and $t = \sum_{i\in I} t_i$, where $t_i \in T_i$ for all i. Then $H_\beta t_i = \mu(H_\beta)t_i$ for all i. Let $n = \mu(H_\beta)$. Then $n \in \mathbf{Z}$ since at least one of the t_i is non-zero. If $n \geqq 0$, then $X^n_{-\beta}t_i \neq 0$ for all i such that $t_i \neq 0$ (1.8.5); hence $X^n_{-\beta}t \neq 0$ and $s_\beta\mu = \mu - n\beta \in \Pi$. We reach the same conclusion if $n \leqq 0$ by using $X^n_{-\beta}$. Thus we have proved that $\Pi \subset P$ and $W\Pi \subset \Pi$. Now every trajectory of W in P intersects P_{++}. From 7.1.8 (ii), $\Pi \cap P_{++}$ is finite. Hence Π is finite. Since each V_μ is finite-dimensional, V is finite-dimensional. Let $V = S_1 \oplus \cdots \oplus S_p$, where $S_1, \ldots, S_p$ are simple sub-$\mathfrak{g}$-modules. Since $\dim V_\lambda = 1$, v belongs to S_j for a certain j, whence $V = S_j$ and V is simple.

7.2.5. LEMMA (notation as in 7.0). *Let $\lambda \in P_{++}$, let K be the largest sub-$\mathfrak{g}$-module of $M(\lambda + \delta)$ which is distinct from $M(\lambda + \delta)$, and let v be the canonical generator of $M(\lambda + \delta)$. For all $\beta \in B$, let $X_{-\beta}$ be a non-zero element of $\mathfrak{g}^{-\beta}$, and let $m_\beta = \lambda(H_\beta) + 1$. Then*

$$K = \sum_{\beta\in B} U(\mathfrak{g})X^{m_\beta}_{-\beta}v = \sum_{\beta\in B} U(\mathfrak{n}^-)X^{m_\beta}_{-\beta}v,$$

and K has finite codimension in $M(\lambda + \delta)$.

Let $\beta \in B$. Since $(\lambda + \delta)(H_\beta) = m_\beta$ is a positive integer, the sub-$\mathfrak{g}$-module Y_β of $M(\lambda + \delta)$ generated by $X_{-\beta}^{m_\beta} v$ is $U(\mathfrak{n}_-)X_{-\beta}^{m_\beta} v$ (7.1.14) and is distinct from $M(\lambda + \delta)$. Hence $\sum_{\beta \in B} Y_\beta \subset K$. From 7.2.4, $M(\lambda + \delta)/\sum_{\beta \in B} Y_\beta$ is simple and finite-dimensional. Hence $\sum_{\beta \in B} Y_\beta = K$ and K has finite co-dimension in $M(\lambda + \delta)$.

7.2.6. THEOREM (notation as in 7.0). *The mapping* $\lambda \mapsto [L(\lambda + \delta)]$ *is a bijection of* P_{++} *onto the set of classes of finite-dimensional simple* $\mathfrak{g}$*-modules.*

For $\lambda \in P_{++}$, $L(\lambda + \delta)$ is simple and finite-dimensional (7.2.5). The mapping in the theorem is surjective from 7.2.2, and injective since λ is the largest weight of $L(\lambda + \delta)$.

7.2.7. PROPOSITION (notation as in 7.0). *Let* $\lambda \in \mathfrak{h}^*$, *and let* v *and* v' *be the canonical generators of* $M(\lambda + \delta)$ *and* $L(\lambda + \delta)$, *respectively. Let* I *and* I' *be the annihilators of* v *and* v' *in* $U(\mathfrak{g})$, *respectively. Then:*

(i) $I = U(\mathfrak{g})\mathfrak{n}_+ + \sum_{h \in \mathfrak{h}} U(\mathfrak{g})(h - \lambda(h))$.

(ii) *I' is the largest left ideal of $U(\mathfrak{g})$ which is distinct from $U(\mathfrak{g})$ and which contains I.*

(iii) *For all $\beta \in B$, let $m_\beta = \lambda(H_\beta) + 1$, and let $X_{-\beta}$ be a non-zero element of $\mathfrak{g}^{-\beta}$. If $\lambda \in P_{++}$, then*

$$I' = I + \sum_{\beta \in B} U(\mathfrak{g})X_{-\beta}^{m_\beta} = I + \sum_{\beta \in B} U(\mathfrak{n}_-)X_{-\beta}^{m_\beta}.$$

Assertion (i) follows from 5.1.9 (iii). For every sub-$\mathfrak{g}$-module M of $M(\lambda + \delta)$, let I_M be the set of the $u \in U(\mathfrak{g})$ such that $uv \in M$. Then $M \mapsto I_M$ is a bijection of the set of sub-$\mathfrak{g}$-modules of $M(\lambda + \delta)$ onto the set of left ideals of $U(\mathfrak{g})$ containing I. Let K be as in 7.1.11. Then $I' = I_K$, whence (ii). Let us assume that $\lambda \in P_{++}$, and let us use the notation of (iii). Let $u \in U(\mathfrak{g})$. From 7.2.5, we have

$$\begin{aligned} u \in I' &\Leftrightarrow uv \in K \\ &\Leftrightarrow \text{there exists } u_1 \in \sum_{\beta \in B} U(\mathfrak{g})X_{-\beta}^{m_\beta} \text{ such that } uv = u_1 v \\ &\Leftrightarrow u \in I + \sum_{\beta \in B} U(\mathfrak{g})X_{-\beta}^{m_\beta}, \end{aligned}$$

whence we have the first equality of (iii); the second equality can be established similarly.

7.2.8. PROPOSITION (notation as in 7.0). *Let* V *be a finite-dimensional* $\mathfrak{g}$-*module, and* w_0 *the element of* W *which transforms* B *into* $-B$.

(i) *For all* $\lambda \in \mathfrak{h}^*$, *the orthogonal subspace of* V_λ *in* V^* *is* $\sum_{\lambda' \neq -\lambda} V^*_{\lambda'}$, *so that* $V^*_{-\lambda}$ *can be identified with the dual of* V_λ.

(ii) *Let us assume that* V *is simple with largest weight* λ. *Then* V^* *is simple with largest weight* $-w_0\lambda$.

We have $V = \oplus_{\lambda \in \mathfrak{h}^*} V_\lambda$, hence the vextor space V^* can be identified with $\oplus_{\lambda \in \mathfrak{h}^*}(V_\lambda)^*$; for $h \in \mathfrak{h}$, h_{V^*} is equal to $-{}^t(h_V)$, hence its restriction to $(V_\lambda)^*$ is the homothety with ratio $-\lambda(h)$. This proves (i). If V is simple, then clearly V^* is simple. From (i), the largest weight λ^* of V^* is $-\lambda'$, where λ' is the smallest weight of V; hence $\lambda^* = -w_0\lambda$ (7.2.3).

7.2.9. Let $\mathfrak{c}$ be a commutative Lie algebra. Then $\mathfrak{g} \times \mathfrak{c}$ is reductive, and $\mathfrak{h} \times \mathfrak{c}$ is a Cartan subalgebra of $\mathfrak{g} \times \mathfrak{c}$. If π is a representation of $\mathfrak{g} \times \mathfrak{c}$ in V, then the V_μ (where μ runs through $(\mathfrak{h} \times \mathfrak{c})^*$), the weights of π and their multiplicity are defined as in 7.1.1.

Proposition 7.2.1 (i) and (iii) can be extended to the finite-dimensional $(\mathfrak{g} \times \mathfrak{c})$-modules which are $\mathfrak{c}$-diagonalizable; proposition 7.2.1 (ii) becomes: the restriction to $\mathfrak{h}$ of every weight of V belongs to P. For a simple representation π of $\mathfrak{g} \times \mathfrak{c}$ whose restriction to $\mathfrak{c}$ is diagonalizable, we term the weight of π whose restriction to $\mathfrak{h}$ is the largest (or smallest) weight of $\pi|\mathfrak{g}$ the *largest* (or *smallest*) *weight of* π.

References: [16], [63], [66], [71], [118].

7.3. Invariants in the symmetric algebra

7.3.1. LEMMA (notation as in 7.0). *Let* $f \in S(\mathfrak{g}^*)$ *be a polynomial function on* $\mathfrak{g}$. *The following conditions are equivalent:*

(i) f *is an invariant element of the* $\mathfrak{g}$*-modules* $S(\mathfrak{g}^*)$ (*in other words,* $f \in S(\mathfrak{g}^*)^{\mathfrak{g}}$);

(ii) *for every elementary automorphism* θ *of* $\mathfrak{g}$, *we have* $f \circ \theta = f$.

We may assume that f is homogeneous of degree n. Let g be the symmetric n-linear form on $\mathfrak{g}$ such that $f(x) = g(x,x, \ldots, x)$ for all $x \in \mathfrak{g}$. Let us consider the following conditions:

(iii) for all $x,x_1, \ldots, x_n \in \mathfrak{g}$, we have

$$(1) \qquad g([x,x_1],x_2, \ldots, x_n) + \cdots + g(x_1x_2, \ldots, [x,x_n]) = 0;$$

(iv) for all $\theta \in \mathrm{Aut}_e(\mathfrak{g})$ and $x_1, \ldots, x_n \in \mathfrak{g}$, we have

$$g(\theta x_1, \ldots, \theta x_n) = g(x_1, \ldots, x_n);$$

(v) for all $x_1, \ldots, x_n \in \mathfrak{g}$ and every nilpotent element of $\mathfrak{g}$, we have the equality (1);

(vi) for all $x_1, \ldots, x_n \in \mathfrak{g}$, $\tau \in k$ and every nilpotent element x of $\mathfrak{g}$, we have

$$(2) \qquad g((\exp \operatorname{ad} \tau x)x_1, \ldots, (\exp \operatorname{ad} \tau x)x_n) = g(x_1, \ldots, x_n).$$

We have (i) $\Leftrightarrow$ (iii) and (ii) $\Leftrightarrow$ (iv). On the other hand, (iv) $\Leftrightarrow$ (vi) by definition of $\mathrm{Aut}_e(\mathfrak{g})$, and (iii) $\Leftrightarrow$ (v) because $\mathfrak{g}$ is generated, as a Lie algebra, by its nilpotent elements. Finally, the left-hand side of (2) is a polynomial function of τ whose derivative at $\tau = 0$ is

$$g([x,x_1],x_2, \ldots, x_n) + \cdots + g(x_1,x_2, \ldots, [x,x_n])$$

whence (v) $\Leftrightarrow$ (vi).

7.3.2. A polynomial function on $\mathfrak{g}$ which satisfies the equivalent conditions of 7.3.1 is termed an *invariant polynomial function on* $\mathfrak{g}$.

On the other hand, the set of polynomial functions f on $\mathfrak{h}$ such that $f \circ w = f$ for all $w \in W$ is denoted by $S(\mathfrak{h}^*)^W$, and the set of elements of $S(\mathfrak{h}^*)^W$ which are homogeneous of degree m is denoted by $S^m(\mathfrak{h}^*)^W$.

7.3.3. LEMMA (notation as in 7.0). *Let π be a finite-dimensional representation of $\mathfrak{g}$, and let $m \in \mathbf{N}$. The function $x \mapsto \operatorname{tr}(\pi(x)^m)$ on $\mathfrak{g}$ belongs to $S(\mathfrak{g}^*)^{\mathfrak{g}}$.*

This function belongs to $S^m(\mathfrak{g}^*)$. The corresponding symmetric m-linear form g is given by

$$g(x_1, \ldots, x_m) = \frac{1}{m!} \sum_{\sigma \in \mathfrak{S}_m} \operatorname{tr}(\pi(x_{\sigma 1})\pi(x_{\sigma 2}) \cdots \pi(x_{\sigma m})).$$

Now, for all $x \in \mathfrak{g}$, we have

$$m!(xg)(x_1, \ldots, x_m) =$$

$$= \sum_{\sigma \in \mathfrak{S}_m} \sum_{i=1}^{m} \operatorname{tr}(\pi(x_{\sigma 1}) \cdots [\pi(x_{\sigma i}),\pi(x_{\sigma i})] \cdots \pi(x_{\sigma m}))$$

$$= \sum_{\sigma \in \mathfrak{S}_m} (\operatorname{tr} \pi(x)\pi(x_{\sigma 1}) \cdots \pi(x_{\sigma m}) + \operatorname{tr} \pi(x_{\sigma(1)}) \cdots \pi(x_{\sigma m})\pi(x))$$

$$= 0,$$

whence $xg = 0$.

7.3.4. LEMMA (notation as in 7.0). *Let $m \in \mathbf{N}$. Every element of $S^m(\mathfrak{h}^*)^W$ is a linear combination of polynomial functions on $\mathfrak{h}$ of the form $x \mapsto \operatorname{tr}(\pi(x))^m$, where π is a finite-dimensional representation of $\mathfrak{g}$.*

The functions λ^m on $\mathfrak{h}$, where $\lambda \in P$, generate, the vector space $S^m(\mathfrak{h}^*)$. For all $f \in S(\mathfrak{h}^*)$, let $\sigma f = \sum_{w \in W} f \circ w$; since every orbit of W in P intersects

P_{++}, the $\sigma(\lambda^m)$, where $\lambda \in P_{++}$, generate the vector space $S^m(\mathfrak{h}^*)^W$. We prove that, for all $\lambda \in P_{++}$, $\sigma(\lambda^m)$ is a linear combination of functions on $\mathfrak{h}$ of the type specified in the lemma. Let E_λ be the set of the $\mu \in P_{++}$ such that $\mu < \lambda$. Let π be a finite-dimensional simple representation of $\mathfrak{g}$ with largest weight λ. From 7.2.1, the function $x \mapsto g(x) = \operatorname{tr} \pi(x)^m$ on $\mathfrak{h}$ is a linear combination with non-zero coefficients of the $\sigma(\mu^m)$ for the weights μ of π which belong to P_{++}; hence $\sigma(\lambda^m)$ is a linear combinaton of g and the $\sigma(\mu^m)$ with $\mu \in E_\lambda$. Hence it is sufficient to reason by induction on $\operatorname{Card}(E_\lambda)$.

7.3.5. THEOREM (notation as in 7.0). *Let $i: S(\mathfrak{g}^*) \to S(\mathfrak{h}^*)$ be the restriction homomorphism.*

(i) *The mapping $i|S(\mathfrak{g}^*)^\mathfrak{g}$ is an isomorphism of the algebra $S(\mathfrak{g}^*)^\mathfrak{g}$ onto the algebra $S(\mathfrak{h}^*)^W$.*

(ii) *Let $m \in \mathbf{N}$. Then $S^m(\mathfrak{g}^*)^\mathfrak{g}$ is the set of linear combinations of functions of the form $x \to \operatorname{tr} \pi(x)^m$, where π is a finite-dimensional representation of $\mathfrak{g}$.*

(a) From 1.10.19, we have $i(S(\mathfrak{g}^*)^\mathfrak{g}) \subset S(\mathfrak{h}^*)^W$.

(b) Let $f \in S(\mathfrak{g}^*)^\mathfrak{g}$ such that $f|\mathfrak{h} = 0$; we prove that $f = 0$. By extension of the base field, we may assume that k is algebraically closed. Let G be the set of generic elements of $\mathfrak{g}$. From 1.9.9 and 1.9.11, if $x \in G$, there exists $\theta \in \operatorname{Aut}_e(\mathfrak{g})$ such that $\theta x \in \mathfrak{h}$; then $f(x) = f(\theta x) = 0$; hence $f|G = 0$ and consequently $f = 0$.

(c) Let L_m be the set of linear combinations considered in (ii). From 7.3.3 and 7.3.4, we have $L_m \subset S^m(\mathfrak{g}^*)^\mathfrak{g}$ and $i(L_m) \supset S^m(\mathfrak{h}^*)^W$. Taking (a) and (b) into account, this proves (i) and (ii) simultaneously.

7.3.6. LEMMA (notation as in 7.0). *Let α be the Killing isomorphism of $S(\mathfrak{g})$ onto $S(\mathfrak{g}^*)$ (1.5.14). Let $\beta: S(\mathfrak{h}) \to S(\mathfrak{h}^*)$ be the isomorphism defined by the restriction of the Killing form to $\mathfrak{h}$. Let J be the ideal of $S(\mathfrak{g})$ generated by $\mathfrak{n}_+ \cup \mathfrak{n}_-$. Then:*

(i) *β is a W-module isomorphism.*

(ii) *$S(\mathfrak{g}) = S(\mathfrak{h}) \oplus J$. Let j be the homomorphism of the algebra $S(\mathfrak{g})$ onto the algebra $S(\mathfrak{h})$ defined by this decomposition.*

(iii) *Let $i: S(\mathfrak{g}^*) \to S(\mathfrak{h}^*)$ be the restriction homomorphism. If $f \in S(\mathfrak{g})$, then $i(\alpha(f)) = \beta(i(f))$.*

The restriction of the Killing form to $\mathfrak{h}$ is W-invariant, whence (i). Assertion (ii) is obvious. Since i, j, α, β are algebra homomorphisms, it is sufficient to prove (iii) by assuming that $f \in \mathfrak{g}$; in this case, for all $h \in \mathfrak{h}$, we have

$$\begin{aligned}\langle i(\alpha(f)),h\rangle &= \langle \alpha(f),h\rangle = \langle f,h\rangle \\ &= \langle j(f),h\rangle \qquad \text{since } \mathfrak{n}_+ \oplus \mathfrak{n}_- \text{ is orthogonal to } \mathfrak{h} \\ &= \langle \beta(j(f)),h\rangle\end{aligned}$$

whence $i(\alpha(f)) = \beta(j(f))$.

7.3.7. THEOREM (notation as in 7.0). *Let J be the ideal of $S(\mathfrak{g})$ generated by $\mathfrak{n}_+ \cup \mathfrak{n}_-$. Then $S(\mathfrak{g}) = S(\mathfrak{h}) \oplus J$; let j be the homomorphism of the algebra $S(\mathfrak{g})$ onto the algebra $S(\mathfrak{h})$ defined by this decomposition. Let $S(\mathfrak{h})^W$ be the set of invariant elements of the W-module $S(\mathfrak{h})$. Then $j|Y(\mathfrak{g})$ is an isomorphism of the algebra $Y(\mathfrak{g})$ onto the algebra $S(\mathfrak{h})^W$.*

With the notation of 7.3.6, α and β are $\mathfrak{g}$-module and W-module isomorphisms respectively. It is then sufficient to apply 7.3.5 (i) and 7.3.6 (iii).

7.3.8. THEOREM. *Let $\mathfrak{a}$ be a semi-simple Lie algebra, and l its rank.*

(i) *There exist l algebraically independent homogeneous elements $f_1, \ldots, f_l$ of $Y(\mathfrak{a})$ which generate the algebra $Y(\mathfrak{a})$. The degrees $\nu_1, \ldots, \nu_1$ of $f_1, \ldots, f_l$ are independent of the choice of the $(f_1, \ldots, f_l)$. We have*

$$\nu_1 + \cdots + \nu_l = \tfrac{1}{2}(l + \dim \mathfrak{a}).$$

(ii) *The algebra $Z(\mathfrak{a})$ is isomorphic to the algebra of polynomials in l indeterminates.*

For k algebraically closed, (i) follows from 7.3.7 and 11.1.14. The general case may be deduced from it by virtue of 11.2.2. Let us consider the filtration $(U_n(\mathfrak{a}) \cap Z(\mathfrak{a}))$ on $Z(\mathfrak{a})$. From (i), 2.3.6 and 2.4.11, the graded algebra associated with the filtered algebra $Z(\mathfrak{a})$ is generated by l algebraically independent homogeneous elements. This implies (ii) from AC III, p. 39.

References: [16], [118].

7.4. The Harish-Chandra homomorphism

7.4.1. Let $\alpha_1, \ldots, \alpha_n$ be the (pairwise distinct) elements of R^+. For all $\alpha \in R$, let $X_\alpha \in \mathfrak{g}^\alpha - \{0\}$. Let $(H_1, \ldots, H_l)$ be a basis for $\mathfrak{h}$. From 2.1.11, the elements

$$u((q_i),(m_i),(p_i)) = X_{-\alpha_1}^{q_1} \cdots X_{-\alpha n}^{q_n} H_1^{m_1} \cdots H_l^{m_l} X_{\alpha_1}^{p_1} \cdots X_{\alpha_n}^{p_n}$$

(where $q_i,m_i,p_i \in \mathbf{N}$) form a basis for the vector space $U(\mathfrak{g})$. For all $h \in \mathfrak{h}$, we have

$$[h,u((q_i),(m_i),(p_i))] =$$
$$= ((p_1 - q_1)\alpha_1 + \cdots + (p_n - q_n)\alpha_n)(h)u((q_i),(m_i),(p_i)).$$

If we consider $U(\mathfrak{g})$ as a $\mathfrak{g}$-module by virtue of the adjoint representation, we then see that

$$U(\mathfrak{g}) = \oplus_{\lambda\in Q}U(\mathfrak{g})_\lambda.$$

Since $\mathrm{ad}_{U(\mathfrak{g})}h$ is a derivation of $U(\mathfrak{g})$ for all $h \in \mathfrak{h}$, we see moreover that $U(\mathfrak{g})_\lambda U(\mathfrak{g})_\mu \subset U(\mathfrak{g})_{\lambda+\mu}$ for $\lambda,\mu \in Q$, so that the family $(U(\mathfrak{g})_\lambda)_{\lambda\in Q}$ is a graduation of the algebra $U(\mathfrak{g})$. The subspace $U(\mathfrak{g})_0$ is the commutant of $\mathfrak{h}$ in $U(\mathfrak{g})$; it is a sub-algebra of $U(\mathfrak{g})$.

7.4.2. LEMMA (notation as in 7.0). *Let* $L = U(\mathfrak{g})\mathfrak{n}_+ \cap U(\mathfrak{g})_0$. *Then:*
(i) $L = \mathfrak{n}_-U(\mathfrak{g}) \cap U(\mathfrak{g})_0$, *and L is a two-sided ideal of* $U(\mathfrak{g})_0$.
(ii) $U(\mathfrak{g})_0 = U(\mathfrak{h}) \oplus L$.

With the notation of 7.4.1, $\mathfrak{n}_-U(\mathfrak{g})$ and $U(\mathfrak{g})\mathfrak{n}_+$ are the sets of linear combinations of the $u((q_i),(m_i),(p_i))$ such that $\sum q_i > 0$ and $\sum p_i > 0$ respectively. On the other hand,

$$u((q_i),(m_i),(p_i)) \in U(\mathfrak{g})_0 \Leftrightarrow p_1\alpha_1 + \cdots + p_n\alpha_n = q_1\beta\alpha_1 + \cdots + q_n\alpha_n.$$

This implies that

$$\mathfrak{n}_-U(\mathfrak{g}) \cap U(\mathfrak{g})_0 = U(\mathfrak{g})\mathfrak{n}_+ \cap U(\mathfrak{g})_0.$$

Finally, $\mathfrak{n}_-U(\mathfrak{g}) \cap U(\mathfrak{g})_0$ (or $U(\mathfrak{g})\mathfrak{n}_+ \cap U(\mathfrak{g})_0$) is a right (or left) ideal of $U(\mathfrak{g})_0$, whence (i). On the other hand, an element $u((q_i),(m_i),(p_i))$ which is in $U(\mathfrak{g})_0$ belongs to $U(\mathfrak{h})$ (or L) if and only if $p_1 = \cdots = p_n = q_1 = \cdots = q_n = 0$ (or $p_1 + \cdots + p_n + q_1 + \cdots + q_n > 0$), whence (ii).

7.4.3. From 7.4.2, the projection of $U(\mathfrak{g})_0$ onto $U(\mathfrak{h})$ with kernel L is an algebra homomorphism. It is termed the *Harish-Chandra homomorphism* of $U(\mathfrak{g})_0$ onto $U(\mathfrak{h})$ (relative to B).

7.4.4. We recall that $U(\mathfrak{h})$ can be canonically identified with the symmetric algebra of $\mathfrak{h}$, and also with the algebra of polynomial functions on $\mathfrak{h}^*$. We shall use this identification in the following proposition:

PROPOSITION (notation as in 7.0). *Let V be a $\mathfrak{g}$-module, $\lambda \in \mathfrak{h}^*$, v a non-zero element of V annihilated by $\mathfrak{n}_+$; we assume that V is generated by v as a $\mathfrak{g}$-module. Let χ be the central character of V* (7.1.8). *Let φ be the Harish-*

Chandra homomorphism of $U(\mathfrak{g})_0$ *onto* $U(\mathfrak{h})$. *For all* $z \in Z(\mathfrak{g})$, *we have* $\chi(z) = (\varphi(z))(\lambda)$.

There exist $u_1, \ldots, u_p \in U(\mathfrak{g})$ and $n_1, \ldots, n_p \in \mathfrak{n}_+$ such that

$$z = \varphi(z) + u_1 n_1 + \cdots + u_p n_p.$$

Then

$$\chi(z)v = zv = \varphi(z)v + u_1 n_1 v + \cdots + u_p n_p v$$
$$= \varphi(z)v = (\varphi(z))(\lambda)v.$$

7.4.5. Theorem (notation as in 7.0). *Let* γ *be the automorphism of the* algebra $S(\mathfrak{h})$ *which transforms the polynomial function* p *on* $\mathfrak{h}^*$ *into the function* $\lambda \mapsto p(\lambda - \delta)$. *Let* $U(\mathfrak{g})_0$ *be the commutant of* $\mathfrak{h}$ *in* $U(\mathfrak{g})$, φ *the Harish-Chandra homomorphism of* $U(\mathfrak{g})_0$ *onto* $U(\mathfrak{h}) = S(\mathfrak{h})$, *and* $S(\mathfrak{h})^W$ *the set of* W*-invariant elements of* $S(\mathfrak{h})$. *Then* $\gamma \circ \varphi | Z(\mathfrak{g})$ *is an isomorphism of the algebra* $Z(\mathfrak{g})$ *onto the algebra* $S(\mathfrak{h})^W$, *independent of the choice of* B.

(a) Let $\alpha \in B$, $\lambda \in P_{++}$ and $\mu = s_\alpha \lambda$. Then $M(\mu)$ is isomorphic to a sub-$\mathfrak{g}$-module of $M(\lambda)$ (7.1.15), hence $\chi_\lambda = \chi_\mu$. From 7.4.4, we have, for all $z \in Z(\mathfrak{g})$,

$$((\gamma \circ \varphi)(z))(\lambda) = (\varphi(z))(\lambda - \delta) = \chi_\lambda(z)$$
$$= \chi_\mu(z) = ((\gamma \circ \varphi)(z))(\mu).$$

Thus the polynomial functions $(\gamma \circ \varphi)(z)$ and $(\gamma \circ \varphi)(z) \circ s_\alpha$ coincide on P_{++} and are consequently equal. Since the family $(s_\alpha)_{\alpha \in B}$ generates W, this proves that $(\gamma \circ \varphi)(Z(\mathfrak{g})) \subset S(\mathfrak{h})^W$.

(b) Let η be the isomorphism of $Y(\mathfrak{g})$ onto $S(\mathfrak{h})^W$ defined in 7.3.7. We consider the canonical isomorphism of the $\mathfrak{g}$-module $U(\mathfrak{g})$ onto the $\mathfrak{g}$-module $S(\mathfrak{g})$, and let θ be its restriction to $Z(\mathfrak{g})$, which is an isomorphism of the vector space $Z(\mathfrak{g})$ onto the vector space $Y(\mathfrak{g})$ (2.4.11). Let $f \in \mathbf{N}$, and $z \in Z(\mathfrak{g}) \cap U_f(\mathfrak{g})$. With the notation of 7.4.1, we set

$$z = \sum_{q_1 + \cdots + q_n + m_1 + \cdots + m_l + p_1 + \cdots + p_n \leq f} \lambda_{(q_i),(m_i),(p_i)} u((q_i),(m_i),(p_i)).$$

Let $v((q_i),(m_i,(p_i))$ be the monomial $X_{-\alpha_1}^{q_1} \cdots X_{-\alpha_n}^{q_n} H_l^{m_1} \cdots H_l^m X_{\alpha_1}^{p_1} \cdots X_{\alpha_n}^{p_n}$ calculated in $S(\mathfrak{g})$. By setting

$$S^0(\mathfrak{g}) + S^1(\mathfrak{g}) + \cdots + S^{f-1}(\mathfrak{g}) = S_{f-1}(\mathfrak{g}),$$

we have, from 2.1.5 and 2.4.5,

$$\theta(z) \equiv \sum_{q_1 + \cdots + p_n \leq f} \lambda_{(q_i),(m_i),(p_i)} v((q_i),(m_i)(p_i)) \pmod{S_{f-1}(\mathfrak{g})},$$

whence

$$\eta(\theta(z)) \equiv \sum_{m_1+\cdots+m_l=f} \lambda_{(0),(m_i),(0)} v((0),(m_i),(0)) \pmod{S_{f-1}(\mathfrak{g})},$$

and consequently

$$\eta(\theta(z)) \equiv \varphi(z) \quad [\bmod S_{f-1}(\mathfrak{g})].$$

(c) The canonical filtrations on $U(\mathfrak{g})$ and $S(\mathfrak{g})$ induce filtrations on $Z(\mathfrak{g})$, $Y(\mathfrak{g})$ and $S(\mathfrak{h})^W$, whence we have the associated graded vector spaces $\operatorname{gr}(Z(\mathfrak{g}))$, $\operatorname{gr}(Y(\mathfrak{g}))$ and $\operatorname{gr}(S(\mathfrak{h})^W)$. The isomorphisms θ,η are compatible with these filtrations, hence $\eta \circ \theta$ defines an isomorphism $\operatorname{gr}(\eta \circ \theta)$ of $\operatorname{gr}(Z(\mathfrak{g}))$ onto $\operatorname{gr}(S(\mathfrak{h})^W)$. From (b), we have $\operatorname{gr}(\eta \circ \theta) = \operatorname{gr}(\varphi|Z(\mathfrak{g}))$, and it is clear that $\operatorname{gr}(\gamma)$ is the identity mapping. Hence $\operatorname{gr}(\gamma \circ \varphi|Z(\mathfrak{g}))$ is a bijective mapping of $\operatorname{gr}(Z(\mathfrak{g}))$ onto $\operatorname{gr}(S(\mathfrak{h})^W)$. We deduce that $\gamma \circ \varphi|Z(\mathfrak{g})\colon Z(\mathfrak{g}) \to S(\mathfrak{h})^W$ is bijective and consequently is an algebra isomorphism.

(d) Let $\lambda \in P_{++}$, let V be a finite-dimensional simple $\mathfrak{g}$-module with largest weight λ, and χ its central character. Let $w \in W$, and let φ',γ' be the homomorphisms analogous to φ,γ relative to the basis $w(B)$. The largest weight of V relative to $w(B)$ is $w(\lambda)$. Let $z \in Z(\mathfrak{g})$. Then, from 7.4.4,

$$(\varphi(z))(\lambda) = \chi(z) = (\varphi'(z))(w\lambda),$$

hence, from (a),

$$\begin{aligned}((\gamma \circ \varphi)(z))(w\lambda + w\delta) &= ((\gamma \circ \varphi)(z))(\lambda + \delta) = (\varphi(z))(\lambda)\\ &= (\varphi'(z))(w\lambda) = ((\gamma' \circ \varphi')(z))(w\lambda + w\delta).\end{aligned}$$

Thus the polynomial functions $(\gamma \circ \varphi)(z)$ and $(\gamma' \circ \varphi')(z)$ coincide on $w(P_{++}) + w\delta$, and hence are equal.

7.4.6. Wit the notation of 7.4.5, the mapping $(\gamma \circ \varphi)|Z(\mathfrak{g})$ is termed the *Harish-Chandra isomorphism of* $Z(\mathfrak{g})$ *onto* $S(\mathfrak{h})^W$. Let ω be this isomorphism. If $\lambda \in \mathfrak{h}^*$, then χ_λ can be obtained, from 7.4.4, by forming the composite of ω and the homomorphism $p \mapsto p(\lambda)$ of $S(\mathfrak{h})$ into k.

7.4.7. PROPOSITION (notation as in 7.0). *Let* $\lambda,\lambda' \in \mathfrak{h}^*$. *Then the following conditions are equivalent:*

(i) $\chi_\lambda = \chi_{\lambda'}$;
(ii) $\lambda' \in W\lambda$.

We have (ii) $\Rightarrow$ (i) from 7.4.6. Let us assume that $\lambda' \notin W\lambda$. There exists a polynomial function p on $\mathfrak{h}^*$ such that p is equal to 1 on $W\lambda$ and to 0 on $W\lambda'$. By replacing p by $(\operatorname{Card} W)^{-1} \sum_{w \in W} wp$, we can assume that

$p \in S(\mathfrak{h})^W$. Let $z \in Z(\mathfrak{g})$ such that p is the image of z under the Harish-Chandry isomorphism. Then $\chi_\lambda(z) = 1$ and $\chi_{\lambda'}(z) = 0$, hence $\chi_\lambda \neq \chi_{\lambda'}$.

7.4.8. Proposition (notation as in 7.0). *Assume that k is algebraically closed. Let χ be a homomorphism of $Z(\mathfrak{g})$ into k. There exists $\lambda \in \mathfrak{h}^*$ such that $\chi = \chi_\lambda$.*

Since W is finite, every element of $S(\mathfrak{h})$ is integral on $S(\mathfrak{h})^W$ (AC V, p. 33). Hence every homomorphism of $S(\mathfrak{h})^W$ into k can be extended to a homomorphism of $S(\mathfrak{h})$ into k (AC V, p. 39), that is, to a mapping $p \mapsto p(\lambda)$, where $\lambda \in \mathfrak{h}^*$. The proposition then follows from 7.4.6.

7.4.9. Proposition (notation as in 7.0). *Let w_0 be the element of W which transforms B into $-B$. Let $\lambda \in \mathfrak{h}^*$ and $z \in Z(\mathfrak{g})$. Then $\chi_{\lambda+\delta}(z) = \chi_{-w_0\lambda+\delta}(z^\mathsf{T})$.*

Let ω be the Harish-Chandra isomorphism of $Z(\mathfrak{g})$ onto $S(\mathfrak{h})^W$. We assume that $\lambda \in P_{++}$, and let V be a finite-dimensional simple $\mathfrak{g}$-module with largest weight λ. Then the dual $\mathfrak{g}$-module V^* has largest weight $-w_0\lambda$ (7.2.8). Taking 2.2.19 into account, we have

$$\chi_{\lambda+\delta}(z) = \chi_{-w_0\lambda+\delta}(z^\mathsf{T}). \tag{1}$$

Now the mappings

$$\lambda \mapsto \chi_{\lambda+\delta}(z) = \omega(z)(\lambda + \delta), \qquad \lambda \mapsto \chi_{-w_0\lambda+\delta}(z^\mathsf{T}) = \omega(z^\mathsf{T})(-w_0\lambda + \delta)$$

are polynomial functions. Equality (1) hence remains valid for all $\lambda \in \mathfrak{h}^*$.

7.4.10. Let $\mathfrak{c}$ be a commutative Lie algebra. Then $\mathfrak{g} \times \mathfrak{c}$ is reductive, and $\mathfrak{h} \times \mathfrak{c}$ is a Cartan subalgebra of $\mathfrak{g} \times \mathfrak{c}$. The commutant C of $\mathfrak{h} \times \mathfrak{c}$ in $U(\mathfrak{g} \times \mathfrak{c})$ is $U(\mathfrak{g})_0 \oplus U(\mathfrak{c})$. Let $\varphi: U(\mathfrak{g})_0 \to U(\mathfrak{h})$ be the Harish-Chandra homomorphism of $\mathfrak{g}$ relative to B. Then

$$\varphi \otimes 1: U(\mathfrak{g})_0 \otimes U(\mathfrak{c}) \to U(\mathfrak{h}) \otimes U(\mathfrak{c}) = U(\mathfrak{h} \times \mathfrak{c})$$

is termed the *Harish-Chandra homomorphism of C onto $U(\mathfrak{h} \times \mathfrak{c})$* relative to B. Its properties may be deduced immediately from those of φ.

References: [63], [126].

7.5. Characters

7.5.1. Let $\mathbf{Z}^{\mathfrak{h}^*}$ be the additive group of mappings of $\mathfrak{h}^*$ into $\mathbf{Z}$. For all $f \in \mathbf{Z}^{\mathfrak{h}^*}$, we term the set of the $\lambda \in \mathfrak{h}^*$ such that $f(\lambda) \neq 0$ the *support of f*, and denote it by Supp f. For all $\lambda \in \mathfrak{h}^*$, we denote by e^λ the element of $\mathbf{Z}^{\mathfrak{h}^*}$ with support $\{\lambda\}$ whose value is 1 at λ. The elements of $\mathbf{Z}^{\mathfrak{h}^*}$ of finite support constitute the additive subgroup of $\mathbf{Z}^{\mathfrak{h}^*}$ classically denoted by $\mathbf{Z}[\mathfrak{h}^*]$. We agree to denote an element f of $\mathbf{Z}^{\mathfrak{h}^*}$ also by $\sum_{\lambda\in\mathfrak{h}^*} f(\lambda)e^{\lambda}$; if $f \in \mathbf{Z}[\mathfrak{h}^*]$, this convention is none other than the usual algebraic notation.

We denote by $\mathbf{Z}\langle\mathfrak{h}^*\rangle$ the additive subgroup of $\mathbf{Z}^{\mathfrak{h}^*}$ consisting of those f which satisfy the following condition: Supp f is contained in a finite union of sets of the form $\nu - Q_+$, where $\nu \in \mathfrak{h}^*$. We have $\mathbf{Z}[\mathfrak{h}^*] \subset \mathbf{Z}\langle\mathfrak{h}^*\rangle \subset \mathbf{Z}^{\mathfrak{h}^*}$. We endow $\mathbf{Z}\langle\mathfrak{h}^*\rangle$ with the following multiplication

$$\Big(\sum_{\lambda\in\mathfrak{h}^*} c_\lambda e^{\lambda}\Big)\Big(\sum_{\mu\in\mathfrak{h}^*} c'_\mu e^{\mu}\Big) = \sum_{\nu\in\mathfrak{h}^*}\Big(\sum_{\lambda+\mu=\nu} c_\lambda c'_\mu\Big) e^{\nu} \quad (c_\lambda c'_\mu \in \mathbf{Z})$$

(the sums $\sum_{\lambda+\mu=\nu} c_\lambda c'_\mu$ are in fact finite sums because of the condition on the supports). Thus $\mathbf{Z}\langle\mathfrak{h}^*\rangle$ becomes a ring. We have $e^{\lambda}e^{\mu} = e^{\lambda+\mu}$, so that $\mathbf{Z}[\mathfrak{h}^*]$ is a subring of $\mathbf{Z}\langle\mathfrak{h}^*\rangle$.

7.5.2. Let V be a $\mathfrak{g}$-module. V is said to *have a character* if $V = \oplus_{\mu\in\mathfrak{h}^*} V_\mu$ and if $\dim V_\mu < +\infty$ for all $\mu \in \mathfrak{h}^*$. In this case, the element $\mu \mapsto \dim V_\mu$ of $\mathbf{Z}^{\mathfrak{h}^*}$ is termed the *character of* V and is denoted by $\mathrm{ch}(V)$. This character should not be confused with the central character.

7.5.3. Let V be a $\mathfrak{g}$-module, and V' a sub-$\mathfrak{g}$-module. If V has a character, then V' and V/V' have characters, and we have

$$\mathrm{ch}\, V = \mathrm{ch}\, V' + \mathrm{ch}\,(V/V').$$

7.5.4. Let V_1 and V_2 be $\mathfrak{g}$-modules with characters such that $\mathrm{ch}\, V_1 \in \mathbf{Z}\langle\mathfrak{h}^*\rangle$ and $\mathrm{ch}\, V_2 \in \mathbf{Z}\langle\mathfrak{h}^*\rangle$, Then the $\mathfrak{g}$-module $V_1 \otimes V_2$ has a character, and we have

$$\mathrm{ch}\,(V_2 \otimes V{:}) = (\mathrm{ch}\, V_1)(\,\mathrm{h}\, V_2).$$

7.5.5. The group W operates in $\mathbf{Z}^{\mathfrak{h}^*}$ by transport of structure. Let $w \in W$. Then $w(e^{\lambda}) = e^{w\lambda}$ for all $\lambda \in \mathfrak{h}^*$, and $w(\mathbf{Z}[\mathfrak{h}^*]) = \mathbf{Z}[\mathfrak{h}^*]$, but in general $w(\mathbf{Z}\langle\mathfrak{h}^*\rangle) \neq \mathbf{Z}\langle\mathfrak{h}^*\rangle$.

7.5.6. Lemma (notation as in 7.0). *We set*

$$d = \sum_{w\in W} \varepsilon(w)e^{w\delta} \in \mathbf{Z}[\mathfrak{h}^*],$$

$$K = \sum_{\gamma\in Q_+} \mathfrak{P}(\gamma)e^{-\gamma} \in \mathbf{Z}\langle\mathfrak{h}^*\rangle.$$

In the ring $\mathbf{Z}\langle\mathfrak{h}^*\rangle$, *we have* $Ke^{-\delta}d = 1$; *in particular, d is invertible.*

From 11.1.15, we have

$$e^{-\delta}d = \prod_{\alpha\in R_+}(1 - e^{-\alpha}).$$

On the other hand, the definition of K implies that

$$K = \prod_{\alpha \in R_+} (1 + e^{-\alpha} + e^{-2\alpha} + \ldots),$$

whence the lemma follows.

7.5.7. Proposition (notation as in 7.0). *Let $d = \sum_{w \in W} \varepsilon(w) e^{w\delta}$. Let $\lambda \in \mathfrak{h}^*$. The $\mathfrak{g}$-module $M(\lambda)$ has a character belonging to $\mathbf{Z}\langle \mathfrak{h}^* \rangle$, and we have*

$$\operatorname{ch} M(\lambda) = d^{-1} e^{\lambda}$$

From 7.1.6 (ii), $M(\lambda)$ has a character belonging to $\mathbf{Z}\langle \mathfrak{h}^* \rangle$, and $\operatorname{ch} M(\lambda) = e^{\lambda - \delta} \sum_{\gamma \in Q_+} \mathfrak{P}(\gamma) e^{-\gamma}$. The proposition then follows from 7.5.6.

7.5.8. Lemma (notation as in 7.0). *Let $\lambda_0 \in \mathfrak{h}^*$. Let M be a $\mathfrak{g}$-module such that:*

(a) *M has the central character χ_{λ_0};*

(b) *M has a character which belongs to $\mathbf{Z}\langle \mathfrak{h}^* \rangle$.*

Let D_M be the set of the $\lambda \in W\lambda_0$ such that $\lambda - \delta + Q_+$ intersects the support of $\operatorname{ch}(M)$. *Then* $\operatorname{ch}(M)$ *is a* $\mathbf{Z}$*-linear combination of the* $\operatorname{ch} M(\lambda)$ *for $\lambda \in D_M$.*

Let us assume that $M \neq 0$. Let $\mu - \delta$ be a maximal element (for the ordering $\leqq$) of $\operatorname{Supp} \operatorname{ch} M$, and let us set $\dim M_{\mu-\delta} = m$. There exists a $\mathfrak{g}$-homomorphism φ of $(M(\mu))^m$ into M which maps $(M(\mu)_{\mu-\delta})^m$ bijectively onto $M_{\mu-\delta}$ (7.1.8 (i)). The central character of $M(\mu)$ is thus χ_{λ_0}, whence $\mu \in W\lambda_0$ (7.4.7), and $\mu \in D_M$, whence $D_M \neq \emptyset$.

If $D_M = \emptyset$, M is thus null, and the lemma is obvious. We reason by induction on $\operatorname{Card} D_M$, assuming that $M \neq 0$. Let μ, m and φ be as above, and let L and N be the kernel and the co-kernel of φ; hence we have an exact sequence of $\mathfrak{g}$-homomorphisms

$$0 \to L \to (M(\mu))^m \to M \to N \to 0.$$

From 7.5.3 and 7.5.7, L and N have characters belonging to $\mathbf{Z}\langle \mathfrak{h}^* \rangle$, and

$$\operatorname{ch}(M) = -\operatorname{ch}(L) + m \operatorname{ch}(M(\mu)) + \operatorname{ch}(N).$$

On the other hand, L and N have the central character χ_{λ_0}. Let us define D_L and D_N in a similar way to D_M. Then $D_N \subset D_M$. On the other hand, $(\mu - \delta + Q_+) \cap \operatorname{Supp} \operatorname{ch} M = \{\mu - \delta\}$, and $\mu - \delta \notin \operatorname{Supp} \operatorname{ch} N$, hence $\mu \notin D_N$, and $\operatorname{Card} D_N < \operatorname{Card} D_M$. If $\lambda \in D_L$, $\lambda - \delta + Q_+$ intersects $\operatorname{Supp} \operatorname{ch} L$ and *a fortiori* $\operatorname{Supp} \operatorname{ch} M(\mu)$, and hence $\mu - \delta \in \lambda - \delta + Q_+$; it follows from this that $D_L \subset D_M$. Since $L \cap (M(\mu)_{\mu-\delta})^m = 0$, we have

$\mu \notin D_L$, and hence Card $D_L <$ Card D_M. We can then apply the induction hypothesis to L and M.

7.5.9. THEOREM (notation as in 7.0). *Let V be a finite-dimensional simple $\mathfrak{g}$-module, and λ its largest weight. Then*

$$\Big(\sum_{w\in W} \varepsilon(w)e^{w\delta}\Big) \operatorname{ch} V = \sum_{w\in W} \varepsilon(w)e^{w(\lambda+\delta)}.$$

The module V is isomorphic to $L(\lambda+\delta)$, and hence has the central character $\chi_{\lambda+\delta}$. Let $d = \sum_{w\in W} \varepsilon(w)e^{w\delta}$. From 7.5.7 and 7.5.8, $d \operatorname{ch} V$ is a $\mathbf{Z}$-linear combination of the $e^{w(\lambda+\delta)}$ for $w \in W$. For all $w \in W$, we have $w(d) = \varepsilon(w)d$, $w(\operatorname{ch} V) = \operatorname{ch} V$ (7.2.1), $w\lambda \leq \lambda$ and $w\delta < \delta$ (11.1.7, 11.1.12, 11.1.13). Hence there exists $a \in \mathbf{Z}$ such that

$$d \operatorname{ch} V = a \sum_{w\in W} \varepsilon(w)e^{w(\lambda+\delta)}.$$

Finally, for every weight μ of V which is distinct from λ, we have $\mu < \lambda$, and hence the coefficient of $e^{\lambda+\delta}$ in $d \operatorname{ch} V$ is 1. Consequently, $a = 1$.

7.5.10. COROLLARY (notation as in 7.0). *If $\mu \in \mathfrak{h}^*$, the multiplicity of μ as a weight of V is*

$$\sum_{w\in W} \varepsilon(w)\, \mathfrak{P}\,(w(\lambda+\delta) - (\mu+\delta)).$$

With the notation of 7.5.6, we have

$$\operatorname{ch} V = (Ke^{-\delta})(d \operatorname{ch} V) = \Big(\sum_{\gamma\in Q_+} \mathfrak{P}(\gamma)e^{-\delta-\gamma}\Big)\Big(\sum_{w\in W} \varepsilon(w)e^{w(\lambda+\delta)}\Big),$$

and the required multiplicity is hence

$$\sum_{\substack{\gamma\in Q_+, w\in W,\\ -\delta-\gamma+w(\lambda+\delta)=\mu}} \mathfrak{P}(\gamma)\varepsilon(w) = \sum_{w\in W} \varepsilon(w)\, \mathfrak{P}\,(w(\lambda+\delta) - \mu - \delta).$$

References: [10], [16], [71], [76], [118].

7.6. Submodules of $M(\lambda)$

7.6.1. PROPOSITION (notation as in 7.0). *Let $\lambda \in \mathfrak{h}^*$. Then:*

(i) *$M(\lambda)$ has a Jordan–Hölder series.*

(ii) *Every simple subquotient of $M(\lambda)$ is isomorphic to $L(\mu)$ for some $\mu \in W\lambda \cap (\lambda - Q_+)$.*

(a) Let N and N' be sub-$\mathfrak{g}$-modules of $M(\lambda)$ such that $N' \subset N$ and N/N't is simple. Since $M(\lambda) = \oplus_{\mu\in\mathfrak{h}^*} M(\lambda_\mu)$, we have

$$N/N' = \bigoplus_{\mu\in\mathfrak{h}^*} (N/N')_\mu.$$

Every weight of N/N' is a weight of $M(\lambda)$, and hence belongs to $\lambda - \delta - Q_+$, Consequently, there exists a weight $\mu - \delta$ of N/N' such that, for all $\alpha \in B$. $\mu - \delta + \alpha$ is not a weight of N/N'. If v is a non-zero element of $(N/N')_{\mu-\delta}$, we have $\mathfrak{n}_+ v = 0$, and hence N/N' is isomorphic to $L(\mu)$ (7.1.13). We have $\mu \in \lambda - Q_+$. The central characters of N/N' and $M(\lambda)$ are equal, whence $\chi_\mu = \chi_\lambda$ and $\mu \in W\lambda$ (7.4.7). This proves (ii).

(b) Since $U(\mathfrak{g})$ is Noetherian and $M(\lambda)$ is singly generated, $M(\lambda)$ is a Noetherian $U(\mathfrak{g})$-module. Every non-null sub-$\mathfrak{g}$-module N of $M(\lambda)$ hence contains a sub-$\mathfrak{g}$-module N' such that N/N' is simple. If $M(\lambda)$ had no Jordan—Hölder series, there would exist an infinite decreasing sequence $(N_0, N_1, \ldots)$ of sub-$\mathfrak{g}$-modules of $M(\lambda)$ such that every N_i/N_{i+1} was simple. From (a), infinitely many of these quotients would be isomorphic to each other. Then one of the weights of $M(\lambda)$ would have infinite multiplicity, contrary to 7.1.6. This proves (i).

7.6.2. Proposition (notation as in 7.0). *Let $\lambda, \mu \in \mathfrak{h}^*$. If $M(\mu)$ is isomorphic to a sub-$\mathfrak{g}$-module of $M(\lambda)$, then $\mu \in W\lambda$ and $\mu \leqq \lambda$.*

From 7.1.6, we have $\mu \leqq \lambda$. On the other hand, $\chi_\mu = \chi_\mu$, and hence $\mu \in W\lambda$ (7.4.7).

7.6.3. Proposition (notation as in 7.0). *Let $\lambda \in \mathfrak{h}^*$.*

(i) *There exists in $M(\lambda)$ a smallest non-null sub-$\mathfrak{g}$-module V.*

(ii) *The module V is isomorphic to $M(\mu)$ for some $\mu \in \mathfrak{h}^*$.*

As an $\mathfrak{n}_-$-module, $M(\lambda)$ is isomorphic to $U(\mathfrak{n}_-)$ endowed with the left regular representation (7.1.5). Two left ideals $\neq 0$ in $U(\mathfrak{n}_-)$ have intersection $\neq 0$ (3.6.13). *A fortiori*, two $\mathfrak{g}$-submodules $\neq 0$ of $M(\lambda)$ have intersection $\neq 0$. From 7.6.1 (i), $M(\lambda)$ contains a minimal submodule $V \neq 0$. Then V is the smallest non-null submodule of $M(\lambda)$.

From 7.6.1 (ii), V is isomorphic to $L(\mu)$ for some $\mu \in \mathfrak{h}^*$. For every $u \in U(\mathfrak{n}_-) - \{0\}$, $u_V = u_{M(\lambda)}|V$ is injective. From 7.1.8 (iii), V is isomorphic to $M(\mu)$.

7.6.4. If $\lambda \in \mathfrak{h}^*$, then $M(\lambda)$ may contain infinitely many submodules (7.8.11), and the structure of these submodules is obscure.

7.6.5. From 7.6.3, however, the study of the sub-$\mathfrak{g}$-modules of $M(\lambda)$ which are isomorphic to some $M(\mu)$ is an important first step; now, in this respect, we shall obtain a complete result (7.6.23).

7.6.6. THEOREM (notation as in 7.0). *Let* $\lambda, \mu \in \mathfrak{h}^*$. *The vector space* $\mathrm{Hom}_{\mathfrak{g}}(M(\mu), M(\lambda))$ *is null or one-dimensional over* k. *Every non-zero element of* $\mathrm{Hom}_{\mathfrak{g}}(M(\mu), M(\lambda))$ *is injective.*

(a) Let v and v' be the canonical generators of $M(\lambda)$ and $M(\mu)$ respectively, and let φ be a $\mathfrak{g}$-homomorphism of $M(\mu)$ into $M(\lambda)$. There exists $u \in U(\mathfrak{n}_-)$ such that $\varphi(v') = uv$. If φ is not injective, there exists $u' \in U(\mathfrak{n}_-)$ such that $u'v' \neq 0$ and $0 = \varphi(u'v') = u'\varphi(v') = (u'u)v$; then $u' \neq 0$ and $u'u = 0$, whence $u = 0$, and then

$$\varphi(M(\mu)) = \varphi(U(\mathfrak{n}_-)v') = U(\mathfrak{n}_-)uv = 0.$$

Thus every non-zero element of $\mathrm{Hom}_{\mathfrak{g}}(M(\mu), M(\lambda))$ is injective.

(b) Let $\varphi_1, \varphi_2 \in \mathrm{Hom}_{\mathfrak{g}}(M(\mu), M(\lambda))$ be such that $\varphi_1(M(\mu)) = \varphi_2(M(\mu))$. From the above, there exists a $\mathfrak{g}$-automorphism α of $M(\mu)$ such that $\varphi_2 = \varphi_1 \circ \alpha$. From 7.1.8 (iv), α is scalar, and hence φ_1 and φ_2 are lineraly dependent.

(c) Let us assume that $M(\mu)$ is simple. Let $\varphi_1, \varphi_2 \in \mathrm{Hom}_{\mathfrak{g}}(M(\mu), M(\lambda)) - \{0\}$. From 7.6.3, we have $\varphi_1(M(\mu)) = \varphi_2(M(\mu))$, hence φ_1 and φ_2 are linearly dependent.

(d) We now pass to the general case. From 7.6.3, there exist $\nu \in \mathfrak{h}^*$ and $\psi \in \mathrm{Hom}_{\mathfrak{g}}(M(\nu), M(\mu))$ such that $M(\nu)$ is simple and ψ is injective. We have $\dim \mathrm{Hom}_{\mathfrak{g}}(M(\nu), M(\lambda)) \leq 1$ from (c). The mapping $f: \varphi \mapsto \varphi|\circ \psi$ of $\mathrm{Hom}_{\mathfrak{g}}(M(\mu), M(\lambda))$ into $\mathrm{Hom}\,(M(\nu), M(\lambda))$ is linear; if $f(\varphi) = 0$, then $\varphi|\psi(M(\nu)) = 0$, hence φ is not injective and consequently zero according to (a). This proves that f is injective, whence $\dim \mathrm{Hom}_{\mathfrak{g}}(M(\mu), M(\lambda)) \leq 1$.

7.6.7. Let $\lambda, \mu \in \mathfrak{h}^*$. From 7.6.6, only two cases are possible: either the only $\mathfrak{g}$-homomorphism of $M(\mu)$ into $M(\lambda)$ is 0, or else $M(\mu)$ *is embedded* in $M(\lambda)$, uniquely up to a scalar, and we then write $M(\mu) \subset M(\lambda)$ by abuse of notation. From 7.6.2, λ being fixed, the second case only occurs for a finite set of values of μ. In 7.6.23 we shall give a necessary and sufficient condition for the second case to arise. This requires a fairly difficult proof; nevertheless, an important special case can be rapidly obtained:

7.6.8. PROPOSITION (notation as in 7.0). *Let* $\lambda \in P_{++}$, $w \in W$, *and let* $w = s_{\alpha_n} s_{\alpha_{n-1}} \cdots s_{\alpha_1}$ *be a reduced decomposition of* w (*where* $\alpha_1, \ldots, \alpha_n \in B$). *Let*

$$\lambda_0 = \lambda,\ \lambda_1 = s_{\alpha_1}\lambda_0,\ \lambda_2 = s_{\alpha_2}\lambda_1,\ \ldots,\ \lambda_n = s_{\alpha_n}\lambda_{n-1}.$$

Then:

(i) $\lambda_0 \geqq \lambda_1 \geqq \cdots \geqq \lambda_n$, *and* $\lambda_i(H_{\alpha_{i+1}}) \in \mathbf{N}$ *for* $i = 0,1, \ldots, n-1$.

(ii) $M(\lambda_0) \supset M(\lambda_1) \supset \cdots \supset M(\lambda_n)$. *In particular,* $M(w\lambda) \subset M(\lambda)$.

Let $i \in \{0,1, \ldots, n-1\}$. We set $w' = s_{\alpha_i} \cdots s_{\alpha_1}$. Since the decomposition $s_{\alpha_{i+1}} s_{\alpha_i} \cdots s_{\alpha_1}$ is reduced, we have $w'^{-1}\alpha_{i+1} \in R_+$ (11.1.8). Hence

$$\lambda_i(H_{\alpha_{i+1}}) = (w'\lambda)(H_{\alpha_{i+1}}) = \lambda(H_{w'^{-1}\alpha i+1}) \in \mathbf{N}$$

and in particular $\lambda_{i+1} = \lambda_i - \lambda_i(H_{\alpha_{i-1}})\alpha_{i+1} \leqq \lambda_i$, whence (i). Assertion (ii) follows from 7.1.15.

7.6.9. Lemma. *Let* $\mathfrak{a}$ *be a nilpotent Lie algebra,* $x \in \mathfrak{a}$, $n \in \mathbf{N}$, *and* $p \in \mathbf{N}$. *There exists* $l \in \mathbf{N}$ *such that* $x^l \alpha^n \subset U(\mathfrak{a})x^p$.

Let L_x (or R_x) $\in$ End $U(\mathfrak{a})$ be the left (resp. right) multiplication by x. Then L_x, R_x and ad $x = L_x - R_x$ commute. If $u \in U(\mathfrak{a})$, these exists q such that $(\text{ad } x)^q u = 0$. Then

$$\begin{aligned} x^l u = L_x^l u &= (R_x + \text{ad } x)^l u \\ &= \sum_{i=0}^{l} \tbinom{l}{i} R_x^{l-i} (\text{ad } x)^i\, u \\ &= \sum_{i=0}^{q} \tbinom{l}{i} ((\text{ad } x)^i u) x^{l-i} \in U(\mathfrak{a}) x^{l-q}, \end{aligned}$$

whence the lemma follows.

7.6.10. Lemma (notation as in 7.0). *Let* $\lambda, \mu \in \mathfrak{h}^*$ *and* $\alpha \in B$ *such that* $M(s_\alpha \mu) \subset M(\mu) \subset M(\lambda)$. *Assume that* $p = \lambda(H_\alpha) \in \mathbf{Z}$.

(i) *If* $p \leqq 0$, *then* $M(\lambda) \subset M(s_\alpha \lambda)$.

(ii) *if* $p > 0$, *then* $M(s_\alpha \mu) \subset M(s_\alpha \lambda) \subset M(\lambda)$.

If $p \leqq 0$, then

$$(s_\alpha \lambda)(H_\alpha) = \lambda(H_\alpha) - \lambda(H_\alpha)\alpha(H_\alpha) = -p \in \mathbf{N},$$

hence $M(\lambda) \subset M(s_\alpha \lambda)$ (7.1.15). Let us assume that $p > 0$, whence $M(s_\alpha \lambda) \subset M(\lambda)$. Let v and v' be the canonical generators of $M(\lambda)$ and $M(\mu)$ respectively. Let $X_\alpha \in \mathfrak{g}^\alpha$ and $X_{-\alpha} \in \mathfrak{g}^{-\alpha}$ be such that $[X_\alpha, X_{-\alpha}] = H_\alpha$. Since $M(s_{\alpha\mu}) \subset M(\mu)$, we have $m = \mu(H_\alpha) \in \mathbf{N}$ (7.6.2), and $M(s_\alpha \mu)$ is the sub-$\mathfrak{g}$-module of $M(\mu)$ generated by $X_{-\alpha}^m v'$ (7.1.15 and 7.6.6). Similarly, $M(s_\alpha \lambda)$ is the sub-$\mathfrak{g}$-module of $M(\lambda)$ generated by $X_{-\alpha}^p v$. Since $M(\mu) \subset M(\lambda)$, there exists $u \in U(\mathfrak{n}_-)$ such that we can identify v' with uv and $M(\mu)$ with the sub-$\mathfrak{g}$-module of $M(\lambda)$ generated by uv. From 7.6.9, there exists an integer l such that $X^1_{-\alpha}u \in U(\mathfrak{n}_-)X^p_{-\alpha}$, whence

$$X^l_{-\alpha}v' = X^l_{-\alpha}uv \in U(\mathfrak{n}_-)X^p_{-\alpha}w \subset M(s_\alpha\lambda).$$

Obviously, we may assume that $l > m$. From 7.1.14, we have

$$\begin{aligned}[X_\alpha, X^l_{-\alpha}]v' &= lX^{l-1}_{-\alpha}(H_\alpha - l + 1)v' \\ &= l(\mu(H_\alpha) - \delta(H_\alpha) - l + 1)X^{l-1}_{-\alpha}v' = l(m-l)X^{l-1}_{-\alpha}v',\end{aligned}$$

i.e.

$$l(m-l)X^{l-1}_{-\alpha}v' = X_\alpha X^l_{-\alpha}v' - X^l_{-\alpha}X_\alpha v' = X_\alpha X^l_{-\alpha}v' \in M(s_\alpha\lambda),$$

whence $X^{1-1}_{-\alpha}v' \in M(s_\alpha\lambda)$. In a stepwise fashion we deduce that $X^m_{-\alpha}v' \in M(s_\alpha\lambda)$, whence $M(s_\alpha\mu) \subset M(s_\alpha\lambda)$.

7.6.11. LEMMA (notation as in 7.0). *Let $\lambda \in P, \alpha \in R_+, \mu = s_\alpha\lambda$, and $m = \lambda(H_\alpha)$. Let us assume that $m \in \mathbf{N}$. Then $M(\mu) \subset M(\lambda)$.*

We may assume that $m > 0$. There exist $w \in W$ and $\mu' \in P_{++}$ such that $\mu = w\mu'$. Let $\lambda' \in \mathfrak{h}^*$ such that $\lambda = w\,\lambda'$. Let $w = s_{\alpha_n}s_{\alpha_{n-1}} \cdots s_{\alpha_1}$ be a reduced decomposition of w. Let

$$\lambda_0 = \lambda', \qquad \lambda_1 = s_{\alpha_1}\lambda_0, \qquad \lambda_2 = s_{\alpha_2}\lambda_1, \qquad \ldots, \qquad \lambda_n = s_{\alpha_n}\lambda_{n-1} = \lambda,$$

$$\mu_0 = \mu', \qquad \mu_1 = s_{\alpha_1}\mu_0, \qquad \mu_2 = s_{\alpha_2}\mu_1, \qquad \ldots, \qquad \mu_n = s_{\alpha_n}\mu_{n-1} = \mu.$$

Then $\lambda_0 \in W\mu_0$ and $\mu_0 \in P_{++}$, hence $\mu_0 - \lambda_0 \in Q_+$, and on the other hand $\mu_n - \lambda_n = -m\alpha$. Since the same element of W transforms λ and μ into λ_i and μ_i, respectively, μ_i is transformed from λ_i by a reflexion s_{γ_i} ($\gamma_i \in R_+$), hence $\mu_i - \lambda_i \in Q_+$ or $\lambda_i - \mu_i \in Q_+$. Hence there exists a least integer i such that $\mu_i - \lambda_i \in Q_+$ and $\mu_{i+1} - \lambda_{i+1} \in -Q_+$. Now

$$\mu_i - \lambda_i = s_{\alpha_{i+1}}(\mu_{i+1} - \lambda_{i+1}).$$

Since $\mu_{i+1} - \lambda_{i+1}$ is proportional to γ_{i+1}, it can be seen that $s_{\alpha_{i+1}}\gamma_{i+1} \in R_-$, whence $\gamma_{i+1} = \alpha_{i+1}$. The relations $\lambda_{i+1} - \mu_{i+1} \in Q_+$ and $\mu_{i+1} = s_{\alpha_{i+1}}\lambda_{i+1}$ imply that $M(\mu_{i+1}) \subset M(\lambda_{i+1})$ (7.1.15). On the other hand,

$$M(\mu_{i+2}) = M(s_{\alpha_{i+2}}\mu_{i+1}) \subset M(\mu_{i+1}).$$

(7.6.8). Hence $M(\mu_{i+2}) \subset M(s_{\alpha_{i+2}}\lambda_{i+1}) = M(\lambda_{i+2})$ (7.6.10). Stepwise, we arrive at $M(\mu) \subset M(\lambda)$.

7.6.12. LEMMA (notation as in 7.0). *Let $\mu \in \mathfrak{h}^*$. The set of the $\lambda \in \mathfrak{h}^*$ such that $M(\lambda - \mu) \subset M(\lambda)$ is an algebraic subset of $\mathfrak{h}^*$.*

Let $\alpha_1, \dots, \alpha_l$ be the (pairwise distinct) elements of B. For $i = 1, \dots, l$, let us set $H_{\alpha_i} = H_i$, and let $X_i \in \mathfrak{g}^{\alpha_i}$ and $Y_i \in \mathfrak{g}^{-\alpha_i}$ be such that $[X_i, Y_i] = H_i$. Then $[X_i Y_j] = 0$ for $i \neq j$, and the Lie algebra $\mathfrak{n}_-$ is generated by $Y_1, \dots, Y_l$. Hence

$$\begin{aligned}[X_i, U(\mathfrak{n}_-)] &\subset U(\mathfrak{n}_-)\mathfrak{h}U(\mathfrak{n}_-)\\ &= U(\mathfrak{n}_-) \oplus U(\mathfrak{n}_-)H_1 \oplus \cdots \oplus U(\mathfrak{n}_-)H_l.\end{aligned}$$

Consequently, for all $u \in U(\mathfrak{n}_-)$ there exist unique elements $u_{i,0}, u_{i,1} \dots, u_{i,l}$ of $U(\mathfrak{n}_-)$, which are linearly dependent on u, such that

$$[X_i, u] = u_{i,0} + u_{i,1}H_1 + \cdots + u_{i,l}H_l.$$

For all $\lambda \in \mathfrak{h}^*$, we set

$$f_i^\lambda(u) = u_{i,0} + (\lambda - \delta)(H_1)u_{i,1} + \cdots + (\lambda - \delta)(H_l)u_{i,l},$$

so that $u \mapsto (f_1^\lambda(u), \dots, f_l^\lambda(u))$ is a linear mapping g^λ of $U(\mathfrak{n}_-)$ into $(U(\mathfrak{n}_-))^l$.

Let X be the set of the $u \in U(\mathfrak{n}_-)$ such that $[H, u] = -\mu(H)u$ for all $H \in \mathfrak{h}$. If $\lambda \in \mathfrak{h}^*$, we have, if we denote the canonical generator of $M(\lambda)$ by v_λ:

$$\begin{aligned}
&M(\lambda - \mu) \subset M(\lambda)\\
&\Leftrightarrow \exists e \in M(\lambda)_{\lambda-\delta-\mu} \text{ such that } e \neq 0 \text{ and } \mathfrak{n}_+ e = 0\\
&\Leftrightarrow \exists u \in X - \{0\} \text{ such that } \mathfrak{n}_+ u v_\lambda = 0\\
&\Leftrightarrow \exists u \in X - \{0\} \text{ such that } X_1 u v_\lambda = \cdots = X_l u v_\lambda = 0\\
&\Leftrightarrow \exists u \in X - \{0\} \text{ such that } [X_1, u] v_\lambda = \cdots = [X_l, u] v_\lambda = 0\\
&\Leftrightarrow \exists u \in X - \{0\} \text{ such that } f_1^\lambda(u) v_\lambda = \cdots = f_l^\lambda(u) v_\lambda = 0\\
&\Leftrightarrow \exists u \in X - \{0\} \text{ such that } f_1^\lambda(u) = \cdots = f_l^\lambda(u) = 0\\
&\Leftrightarrow \operatorname{rank} g^\lambda | X < \dim X.
\end{aligned}$$

This condition can be expressed by writing that certain matrices, whose elements are linear functions of λ, have determinant zero, whence the lemma follows.

7.6.13. The following lemma is an essential part of theorem 7.6.23 below.

LEMMA (notation as in 7.0). *Let $\lambda \in \mathfrak{h}^*$, $\alpha \in R_+$, and $m = \lambda(H_\alpha)$. Let us assume that $m \in \mathbf{N}$. Then $M(s_\alpha \lambda) \subset M(\lambda)$.*

Let A be the set of the $\nu \in \mathfrak{h}^*$ such that $M(\nu - m\alpha) \subset M(\nu)$; it is an algebraic subset of $\mathfrak{h}^*$ (7.6.12). Let H be the affine hyperplane of $\mathfrak{h}^*$ consisting of the $\nu \in \mathfrak{h}^*$ such that $\nu(H_\alpha) = m$. Then $P \cap H \subset A$. We identify $\mathfrak{h}^*$ with k^l by taking the set of the fundamental weights as a basis for $\mathfrak{h}^*$. Then P can be identified with $\mathbf{Z}^l$, and H is the set of the $(\xi_1, \ldots, \xi_l) \in k^l$ such that $n_1\xi_1 + \cdots + n_l\xi_l = m$, where $n_1, \ldots, n_l$ are certain integers; consequently, $P \cap H$ is dense in H, hence $H \subset A$, whence $\lambda \in A$ and

$$M(s_\alpha\lambda) = M(\lambda - m\alpha) \subset M(\lambda).$$

7.6.14. Lemma (notation as in 7.0). *Let F be a finite-dimensional $\mathfrak{g}$-module and $(f_1, \ldots, f_s)$ a basis for F such that, for all i, we have $f_i \in F_{\mu_i}$. We assume that the numbering is such that $\mu_j > \mu_i \Rightarrow j < i$ (it is easy to see that such a numbering exists). Let $\lambda \in \mathfrak{h}^*$.*

(i) *The $\mathfrak{g}$-module $M(\lambda) \otimes F$ has a composition series*

$$M(\lambda) \otimes F = M_s \supset M_{s-1} \supset \cdots \supset M_0 = 0$$

such that, for all $i \in \{1, 2, \ldots, s\}$, M_i/M_{i-1} is isomorphic to $M(\lambda + \mu_i)$.

(ii) *The $\mathfrak{g}$-module $M(\lambda) \otimes F$ has a Jordan–Hölder series.*

Let v be the canonical generator of $M(\lambda)$, let $a_i = v \otimes f_i$, let

$$M_i = U(\mathfrak{g})a_1 + \cdots + U(\mathfrak{g})a_i,$$

and let b_i be the canonical image of a_i in M_i/M_{i-1}. Then $v \otimes F \subset M_s$, whence $(U_n(\mathfrak{g})v) \otimes F \subset M_s$ by induction on n, and hence $M_s = M(\lambda) \otimes F$. On the other hand, b_i generates the $\mathfrak{g}$-module M_i/M_{i-1}, and $b_i \in (M_i/M_{i-1})_{\lambda+\mu_i-\delta}$. If $\alpha \in R_+$ and $X_\alpha \in \mathfrak{g}^\alpha$, then

$$X_\alpha a_i = (X_\alpha v) \otimes f_i + v \otimes (X_\alpha f_i) = v \otimes (X_\alpha f_i) \in v \otimes F_{\mu_{i+\alpha}}$$

et $F_{\mu_{i+\alpha}} \subset kf_1 + \cdots + kf_{i-1}$; hence $X_\alpha a_i \in M_{i-1}$, $\mathfrak{n}_+ b_i = 0$ and

$$M_i = U(\mathfrak{n}_-)a_1 + \cdots + U(\mathfrak{n}_-)a_i.$$

Let us show that M_i is a free $U(\mathfrak{n}_-)$-module with basis $(a_1, \ldots, a_i)$. Let $u_1, \ldots, u_i$ be elements of $U(\mathfrak{n}_-)$ which are not all zero, and p the largest of the filtrations of the non-zero u_j; then

$$\begin{aligned} u_1a_1 + \cdots + u_ia_i = (u_1v) \otimes f_1 + \cdots + (u_iv) \otimes f_i \\ + (u_1'v) \otimes f_1 + \cdots + (u_s'v) \otimes f_s \end{aligned}$$

where $u_1', \ldots, u_s'$ have filtration $<p$; if $j \in \{1, \ldots, i\}$ is such that u_j has filtration p, then $u_j + u_j = 0$, hence $(u_j + u_j')v \neq 0$, whence

$$u_1a_1 + \cdots + u_ia_i \neq 0.$$

This proves that M_i/M_{i-1} is a free $U(\mathfrak{n}_-)$-module and consequently is isomorphic to $M(\lambda + \mu_i)$ as a $\mathfrak{g}$-module (7.1.8). Then (i) has been proved; (ii) follows from (i) and 7.6.1.

7.6.15. Notation. For the conclusion of the proof of 7.6.23, we shall require a fairly copious terminology.

The topology used in $\mathfrak{h}_{\mathbf{R}}^*$ will be the usual topology and not the Zariski topology.

If F is a finite-dimensional $\mathfrak{g}$-module, we shall denote the set of its weights by $\Pi(F)$.

If $\mu \in P$, there exists one and only one element μ' of P_{++} in $W\mu$. We shall denote a finite-dimensional simple $\mathfrak{g}$-module of largest weight μ' by F_μ.

A triple (λ,μ,F) such that

(i) $\lambda \in \mathfrak{h}^*$,

(ii) F is a finite-dimensional $\mathfrak{g}$-module,

(iii) $\mu \in \Pi(F)$,

(iv) there exists no pair (w,ν) such that $w \in W$, $\nu \in \Pi(F)$, $w\lambda < \lambda$, $w(\lambda + \nu) \geqq \lambda + \mu$ and $\lambda + \mu \in W(\lambda + \nu)$

will be termed an *admissible triple*.

If $\lambda,\lambda' \in \mathfrak{h}^*$, we shall say that a sequence $(\gamma_1, \ldots, \gamma_n)$ of elements of R_+ *links* λ' *with* λ if

$$\lambda' \geqq s_{\gamma_1}\lambda' \geqq s_{\gamma_2}s_{\gamma_1}\lambda' \geqq \cdots \geqq s_{\gamma_n}s_{\gamma_{n-1}} \cdots s_{\gamma_1}\lambda' = \lambda.$$

We shall denote the Weyl chamber corresponding to B by C_+.

In $\mathfrak{h}_Q^*$, the scalar product $\langle .,. \rangle$ defines a norm, which will be denoted by $||\cdot||$, and a distance, which will be denoted by $d(.,.)$. Then $\langle \delta,\alpha \rangle > 0$ for all $\alpha \in R_{++}$, and we set

$$c_1 = \sup_{\alpha \in B} \frac{||\delta|| \cdot ||\alpha||}{\langle \delta,\alpha \rangle} + 1.$$

We shall denote the fundamental weights by $\tilde{\omega}_1, \ldots, \tilde{\omega}_l$; let

$$c_2 = \sup_{1 \leq i \leq l} ||\tilde{\omega}_i||, \qquad c_3 = \inf_{\substack{\xi \in Q_+ \\ \xi \neq 0}} \xi||.$$

If $\alpha \in R$, we shall denote the orthogonal sobspace of α in $\mathfrak{h}_Q^*$ by Ξ_α. Let $\Xi = \bigcap_{\alpha \in R} \Xi_\alpha$.

We shall choose a decomposition $k = \mathbf{Q} \oplus k'$ of k considered as a vector $\mathbf{Q}$-space. Then $\mathfrak{h}^* = \mathfrak{h}_Q^* \oplus k'\mathfrak{h}_Q^*$. For all $\lambda \in \mathfrak{h}^*$. we shall denote the components of λ corresponding to this decomposition of $\mathfrak{h}^*$ by $\Re\lambda$ and $\Im\lambda$.

An element λ of $\mathfrak{h}^*$ will be termed *far from the walls* if $d(\Re\lambda,\Xi) < 2c_1c_2$. If C is a chamber, we shall denote the set of the $\lambda \in \mathfrak{h}^*$ which are far from the walls such that $\Re\lambda \in C$ by C^0.

If $\lambda \in \mathfrak{h}^*$, we shall denote the following assertion by A(λ):

$$\lambda' \in \mathfrak{h}^* \text{ and } L(\lambda) \in \mathcal{JH}(M(\lambda')) \Rightarrow$$

there exists a sequence of elements of R_+ linking λ' with λ.

Our aim is to prove A(λ) for all $\lambda \in \mathfrak{h}^*$. If $\Re\lambda \in \overline{C}_+$, then A($\lambda$) is true. Indeed, let $\lambda' \in \mathfrak{h}^*$ with $L(\lambda) \in \mathcal{JH}(M(\lambda'))$. Then $\lambda \leq \lambda'$ and there exists $w \in W$ such that $\lambda = w\lambda'$. Hence

$$\Im\lambda = \Im\lambda', \qquad \Re\lambda = w\Re\lambda', \qquad \Im\lambda = w\Im\lambda', \qquad \Re\lambda \leqq \Re\lambda'.$$

Since $\Re\lambda \in \overline{C}_+$, we have $\Re\lambda = \Re\lambda'$, whence $\lambda = \lambda'$.

7.6.16. LEMMA (notation as in 7.6.15). *Let (λ,μ,F) be an admissible triple. Then $L(\lambda + \mu) \in \mathcal{JH}(L(\lambda) \otimes F)$.*

(a) Let K be the largest sub-$\mathfrak{g}$-module of $M(\lambda)$ which is distinct from $M(\lambda)$. From 7.6.1, we have

$$\mathcal{JH}(K) \subset \{L(\nu) \mid \nu \in W\lambda, \nu < \lambda\}$$

$$\subset \bigcup_{\substack{\nu \in W\lambda, \\ \nu < \lambda}} \mathcal{JH}(M(\nu)). \tag{1}$$

If $(T_1, \ldots, T_u)$ is a Jordan–Hölder series of K, then $(T_1 \otimes F, \ldots, T_u \otimes F)$ is a composition series of $K \otimes F$, which can be refined to a Jordan–Hölder series of $K \otimes F$ (from 7.6.14 (ii)). Hence, if $N \in \mathcal{JH}(K \otimes F)$, there exists $S \in \mathcal{JH}(K)$ such that $N \in \mathcal{JH}(S \otimes F)$. From (1), there exists $\nu \in W\lambda$ such that $\nu < \lambda$ and $N \in \mathcal{JH}(M(\nu) \otimes F)$.

Taking 7.6.14 (i) into account, we see that

$$\mathcal{JH}(K \otimes F) \subset \bigcup_{\substack{\nu \in W\lambda, \nu < \lambda, \\ \xi \in \Pi(F)}} \mathcal{JH}\,(M(\nu + \xi)).$$

(b) Let us assume that $L(\lambda + \mu) \in \mathcal{JH}(K \otimes F)$. From (a), there exist $w \in W$ and $\xi \in \Pi(F)$ such that $w\lambda < \lambda$ and

$$L(\lambda + \mu) \in \mathcal{JH}(M(w\lambda + \xi)). \tag{2}$$

We set $\nu = w^{-1}\xi \in \Pi(F)$. From (2), we have

$$\lambda + \mu \leqq w\lambda + \xi = w(\lambda + \nu), \qquad \lambda + \mu \in W(w\lambda + \xi) = W(\lambda + \nu),$$

which contradicts the assumption that (λ,μ,F) is admissible.

(c) Hence $L(\lambda+\mu) \notin \mathscr{J}\mathscr{H}(K \otimes F)$. Now

$$L(\lambda+\mu) \in \mathscr{J}\mathscr{H}(M(\lambda) \otimes F)$$

from 7.6.14, and

$$\mathscr{J}\mathscr{H}(M(\lambda) \otimes F) = \mathscr{J}\mathscr{H}(L(\lambda) \otimes F) \cup \mathscr{J}\mathscr{H}(K \otimes F).$$

Whence the lemma follows.

7.6.17. Lemma (notation as in 7.6.15). *Let (λ,μ,F) be an admissible triple, and $\lambda' \in \mathfrak{h}^*$. We assume that $L(\lambda) \in \mathscr{J}\mathscr{H}(M(\lambda'))$. Then there exists $\nu \in \Pi(F)$ such that $L(\lambda+\mu) \in \mathscr{J}\mathscr{H}(M(\lambda'+\nu))$.*

We have

$$\begin{aligned} L(\lambda+\mu) &\in \mathscr{J}\mathscr{H}(L(\lambda) \otimes F) && \text{from 7.6.16} \\ &\subset \mathscr{J}\mathscr{H}(M(\lambda') \otimes F) && \text{by assumption} \\ &\subset \bigcup_{\nu \in \Pi(F)} \mathscr{J}\mathscr{H}(M(\lambda'+\nu)) && \text{from 7.6.14.} \end{aligned}$$

7.6.18. Lemma (notation as in 7.6.15). *Let C, C' be Weyl chambers, and $\lambda,\lambda',\mu,\mu' \in \mathfrak{h}^*$. We assume that $\mu \in \mathfrak{h}_{\mathbf{Q}}^*$, $\mu' \in P$, $\lambda' \in W\lambda$, $\Re\lambda \in \overline{C}$, $(\Re\lambda)+\mu \in C$, $\Re\lambda' \in \overline{C'}$, $(\Re\lambda')+\mu' \in C'$. Let $(\gamma_1, \ldots, \gamma_n)$ be a sequence of elements of R_+ linking $\lambda'+\mu'$ with $\lambda+\mu$. Then $(\gamma_1, \ldots, \gamma_n)$ links λ' with λ.*

We set

$$\lambda_0 = \lambda',\ \lambda_1 = s_{\gamma_1}\lambda_0,\ \lambda_2 = s_{\gamma_2}\lambda_1,\ \ldots,\ \lambda_n = s_{\gamma_n}\lambda_{n-1}$$
$$\nu_0 = \lambda'+\mu',\ \nu_1 = s_{\gamma_1}\nu_0,\ \nu_2 = s_{\gamma_2}\nu_1,\ \ldots,\ \nu_n = s_{\gamma_n}\nu_{n-1} = \lambda+\mu.$$

Since $\Re\nu_0 \in C'$, we have $\nu_0 > \nu_1 > \cdots > \nu_n$, so that $\nu_i(H_{\gamma_{i+1}})$ is a positive integer for $i = 0, \ldots, n-1$. Since $s_{\gamma_i}s_{\gamma_{i-1}} \cdots s_{\gamma_i}$ transforms λ' into λ_i and $\lambda'+\mu'$ into ν_i, $\Re\lambda_i$ and $\Re\nu_i$ are in the closure of one and the same chamber, and $\lambda_i - \nu_i \in P$, hence $\lambda_i(H_{\gamma_{i+1}})$ is non-negative. Consequently,

$$\lambda_0 \geqq \lambda_1 \geqq \cdots \geqq \lambda_n.$$

Let $w = s_{\gamma_n}s_{\gamma_{n-1}} \cdots s_{\gamma_1}$. Since $w(\Re\lambda'+\mu') = \Re\lambda+\mu$, we have $w(C') = C$, hence $\Re\lambda_n = w(\Re\lambda') \in \overline{C}$. On the other hand, $\lambda \in W\lambda_n$ and $\Re\lambda \in \overline{C}$ from the assumptions, hence $w(\Re\lambda') = \Re\lambda$. Then $w\mu' = \mu$. Since

$$w(\lambda'+\mu') = \lambda+\mu,$$

we finally have $\lambda_n = w\lambda' = \lambda$.

7.6.19. LEMMA (notation as in 7.6.15). *Let* $\lambda,\lambda' \in \mathfrak{h}^*$, $\mu,\mu' \in \mathfrak{h}_Q^*$ *be such that* $\lambda > \lambda'$ *and* $\lambda + \mu \leqq \lambda' + \mu'$. *Then*

$$\|\lambda - \lambda'\| < c_1(\|\mu\| + \|\mu'\|).$$

If $\xi \in Q_+ - \{0\}$, we can write $\xi = \sum_{\alpha \in B} n_\alpha \alpha$, where the n_α are non-negative integers which are not all zero. Since $\|\delta\| \cdot \|\alpha\| < c_1\langle\delta,\alpha\rangle$ for all $\alpha \in B$, we have

$$\|\delta\| \cdot \|\xi\| \leqq \|\delta\| \sum_{\alpha \in B} n_\alpha \|\alpha\| < c_1\langle\delta,\xi\rangle.$$

Now $\lambda - \lambda' \in Q_+ - \{0\}$, and $\mu' - \mu \geqq \lambda - \lambda'$. Hence

$$\|\delta\| \cdot \|\lambda - \lambda'\| < c_1\langle\delta,\lambda - \lambda'\rangle$$
$$\leqq c_1\langle\delta,\mu' - \mu\rangle \leqq c_1 \|\delta\| (\|\mu\| + \|\mu'\|).$$

7.6.20. LEMMA (notation as in 7.6.15). *Let* $\lambda \in \mathfrak{h}^*$, $\mu \in P$, *and let* C *be a chamber. We make the following assumptions:*

(a) $\Re\lambda \in C$;
(b) $2c_1 \|\mu\| < d(\Re\lambda,\Xi)$ (*whence* $\Re\lambda + \mu \in C$);
(c) $\mathrm{A}(\lambda + \mu)$ *is true. Then* $\mathrm{A}(\lambda)$ *is true.*

(a) We show that the triple (λ,μ,F_μ) is admissible. Let $w \in W$ and $\nu \in \Pi(F_\mu)$ be such that $w\lambda < \lambda$ and $w(\lambda + \nu) \geqq \lambda + \mu$. From 7.6.19 and 7.2.2 (iii), we have

$$\|\lambda - w\lambda\| < c_1(\|\mu\| + \|w\nu\|) \leqq 2c_1 \|\mu\| < d(\Re\lambda,\Xi)$$

hence $w(\Re\lambda) \in C$ and $w(\Re\lambda) = \Re\lambda$, contrary to $w\lambda < \lambda$.

(b) Let $\lambda' \in \mathfrak{h}^*$ be such that $L(\lambda) \in \mathcal{JH}(M(\lambda'))$. From (a) and 7.6.17, there exists $\mu' \in \Pi(F_\mu)$ such that $L(\lambda + \mu) \in \mathcal{JH}(M(\lambda' + \mu'))$. Since $\mathrm{A}(\lambda+\mu)$ is true, there exists a sequence $(\gamma_1, \ldots, \gamma_n)$ linking $\lambda' + \mu'$ with $\lambda + \mu$. On the other hand, since $L(\lambda) \in \mathcal{JH}(M(\lambda'))$, we have $\lambda' \in W\lambda$, whence

$$2c_1 \|\mu'\| \leqq 2c_1 \|\mu\| < d(\Re\lambda,\Xi) = d(\Re\lambda',\Xi).$$

Consequently, if C' is the chamber containing $\Re\lambda'$, then $\Re\lambda' + \mu' \in C'$. Then, from 7.6.18, $(\gamma_1, \ldots, \gamma_n)$ links λ' with λ.

7.6.21. LEMMA (notation as in 7.6.15). *Let* $\lambda \in \mathfrak{h}^*$ *and* $\mu \in P$, *let* C *be a chamber,* γ *an element of* R_+ *orthogonal to a wall of* C *and such that* $\langle C,\gamma\rangle < 0$.

We make the following assumptions:

(a) $\Re\lambda \in C$;

(b) $\Re\lambda + \mu \in s_\gamma C$;

(c) $2c_1\,||\mu|| < d(\Re\lambda,\Xi_\beta)$ *for* $\beta \in R_+ - \{\gamma\}$;

(d) $\mathrm{A}(\lambda + \mu)$ *is true.*

Then $\mathrm{A}(\lambda)$ *is true.*

(a) We show that the triple (λ,μ,F) is admissible. Let $w \in W$ and $\nu \in \Pi(F)$ be such that $w\lambda < \lambda$ and $w(\lambda + \nu) \geqq \lambda + \mu$. From 7.6.19 and 7.2.2 (iii), we have

$$||\lambda - w\lambda|| < c_1(||\mu|| + ||w\nu||) \leqq 2c_1\,||\mu|| < d\Re\lambda,\Xi_\beta)$$

for all $\beta \in R_+ - \{\gamma\}$. Hence $w(\Re\lambda) \in C$ or $w(\Re\lambda) \in s_\gamma C$. If $w(\Re\lambda) \in C$, then, $w(\Re\lambda) = \Re\lambda$, contrary to $w\lambda < \lambda$. If $w(\Re\lambda) \in s_\gamma C$, then $w = s_\gamma$. Since $w\lambda < \lambda$, we deduce that $\lambda(H_\gamma) > 0$, which contradicts $\langle\Re\lambda,\gamma\rangle < 0$.

(b) Let $\lambda' \in \mathfrak{h}^*$ be such that $L(\lambda) \in \mathscr{J}\mathscr{H}(M(\lambda'))$. From (a) and 7.6.17, there exists $\mu' \in \Pi(F_\mu)$ such that $L(\lambda + \mu) \in \mathscr{J}\mathscr{H}(M(\lambda' + \mu'))$. Since $\mathrm{A}(\lambda + \mu)$ is true, there exists a sequence $(\gamma_1, \ldots, \gamma_n)$ linking $\lambda' + \mu'$ with $\lambda + \mu$. We set

$$\lambda_0 = \lambda',\ \lambda_1 = s_{\gamma_1}\lambda_0,\ \lambda_2 = s_{\gamma_2}\lambda_1,\ \ldots,\ \lambda_n = s_{\gamma_n}\lambda_{n-1},$$

$$\nu_0 = \lambda' + \mu',\ \nu_1 = s_{\gamma_1}\nu_0,\ \nu_2 = s_{\gamma_2}\nu_1,\ \ldots,\ \nu_n = s_{\gamma_n}\nu_{n-1} = \lambda + \mu.$$

Then $\nu_0,\nu_1, \ldots, \nu_n$ are congruent modulo Q. Since $\lambda_0 - \nu_0 \in P$, the elements $\lambda_0,\lambda_1, \ldots, \lambda_n$ are congruent modulo Q. Since

$$L(\lambda) \in \mathscr{J}\mathscr{H}(M(\lambda')),$$

λ and λ' are congruent modulo Q, and there exists $w \in W$ such that $\lambda = \lambda'$. In particular, the mapping $\mathfrak{I}$ takes the same value at all points $\lambda_0, \ldots, \lambda_n$, $\nu_0, \ldots, \nu_n$, and this value is fixed for $s_{\gamma_1}, \ldots, a_{\gamma_n}$ We have

$$2c_1\,||w\mu'|| \leqq 2c_1\,||\mu|| < d(\Re\lambda,\Xi_\beta) \quad \text{for } \beta \in R_+ - \{\gamma\}.$$

Consequently, $\Re\lambda + w\mu' = \Re w(\lambda' + \mu')$ belongs to C or $s_\gamma C$.

(c) We assume that $\Re w(\lambda' + \mu') \in C$. Then $\Re\lambda' + \mu' \in w^{-1}C$. Since $\Re\lambda' \in w^{-1}C$, we see that $\Re\lambda_n$ and $\Re\nu_n = \Re\lambda + \mu$ belong to the same chamber, namely $s_\gamma C$. Now $\Re(s_\gamma\lambda) \in s_\gamma C$ and $\lambda_n \in W(s_\gamma\lambda)$, hence $\lambda_n = s_\gamma\lambda$. Since λ and λ_n are congruent modulo Q, we have $s_\gamma\lambda = \lambda + p_\gamma$ with $p \in \mathbf{Z}$ (11.1.11). Now $\langle\Re\lambda,\gamma\rangle < 0$, hence $p \leqq 0$, hence $s_\gamma\lambda \geqq \lambda$, and $(\gamma_1, \ldots, \gamma_n,\gamma)$ links λ' with λ (7.5.18).

(d) We assume that $\Re w(\lambda' + \mu') \in s_\gamma C$. Then $\Re w(\lambda' + \mu')$ and $\Re(\lambda + \mu)$ are in the same chamber, hence $w = s_{\gamma_n}s_{\gamma_{n-1}} \cdots s_{\gamma_1}$ and consequently $\lambda_n = \lambda$. Since $\Re\lambda$ belongs to a chamber, what we have seen in (b) proves that

$\lambda_i > \lambda_{i+1}$ or $\lambda_i < \lambda_{i+1}$ for all i. If $\lambda_i > \lambda_{i+1}$ for all i, the proof is concluded. Otherwise, let i_0 be the least integer such that $\lambda_{i_0-1} < \lambda_{i_0}$. We show that the sequence $(\gamma_1, \ldots, \gamma_{i_0-1}, \gamma_{i_0+1}, \ldots, \gamma_n, \gamma)$ links λ' with λ.

First of all, the sequence $(\gamma_1, \ldots, \gamma_{i_0-1})$ links λ' with λ_{i_0+1}.

We show that $\Re\lambda_{i_0-1}$ and $\Re\nu_{i_0}$ are in one and the same chamber. We have $||\lambda_{i_0} - \nu_{i_0}|| \leqq ||\mu||$. From 7.6.19, $||\lambda_{i_0} - \lambda_{i_0-1}|| \leqq 2c_1 ||\mu||$. Now the closed ball with centre $\Re\lambda_{i_0}$ and radius $2c_1 ||\mu||$ intersects at most two chambers. Moreover (setting $s_{\gamma i} = s_i$ for all i), $\Re\lambda_{i_0}$ and $\Re\lambda_{i_0-1} = s_{i_0}\Re\lambda_{i_0}$ belong to distinct chambers; similarly for $\Re\lambda_{i_0}$ and $\mathscr{R}\nu_{i_0}$, which are transforms of $\mathscr{R}\lambda$ and $\mathscr{R}\lambda + \mu$ under one and the same element of W. This proves our assertion.

We deduce from the foregoing that s_{i_0} transforms the chamber containing $\Re\nu_{i_0}$ into the chamber containing $\Re\lambda_{i_0}$. We set $w' = s_n \cdots s_{i_0+1}$. Then $w's_{i_0}w'^{-1}$ transforms the chamber containing $\Re\nu_n = \Re\lambda + \mu$ into the chamber containing $\Re\lambda_n = \Re\lambda$, hence is equal to s_γ. Consequently, $s_\gamma(\Im\lambda) = \Im\lambda$.

The sequence $(\gamma_{i_0+1}, \ldots, \gamma_n)$ links ν_{i_0} with ν_n. Now $\Re\lambda_{i_0-1}$ and $\Re\nu_{i_0}$ are in one and the same chamber, $\Re(s_\gamma\lambda)$ and $\Re\nu_n$ are in one and the same chamber, $\nu_{i_0} - \lambda_{i_0-1} \in P$ and $\nu_n - s_\gamma\lambda \in \mathfrak{h}^*_{\mathbf{Q}}$ from the foregoing. Then $(\gamma_{i_0+1}, \ldots, \gamma_n)$ links λ_{i_0-1} with $s_\gamma\lambda$ (7.6.18). Consequently, $\lambda - s_\gamma\lambda \in \mathbf{Q}$, $s_\gamma\lambda \in \lambda + \mathbf{Z}\gamma$, and, as $\langle\Re\lambda,\gamma\rangle < 0$, we see that $s_\gamma\lambda > \lambda$.

7.6.22. LEMMA (notation as in 7.6.15). *Let λ be an element of $\mathfrak{h}^*$ far from the walls. Then* A(λ) *is true.*

Let C be the chamber such that $\Re\lambda \in C$. Let n be the least non-negative integer such that there exist chambers $C_0 = C, C_1, \ldots, C_n = C_+$ with the following properties:

for $i = 0,1, \ldots, n - 1$, there exists an element γ_i of R_+, orthogonal to a common wall of C_i and C_{i+1}, such that $\langle C_i,\gamma_i\rangle < 0$ and $C_{i+1} = s_{\gamma_i}C_i$

(11.1.8). For $n = 0$, we have seen that A(λ) is true (7.6.15). We reason by induction on n.

Since C_1^0 contains balls of arbitrarily large radius, there exists $\mu \in P$ such that $\Re\lambda + \mu \in C_1^0$.

Let $\tilde{\omega}'_1, \tilde{\omega}'_2, \ldots, \tilde{\omega}'_1$ be the elements which generate the edges of $C = C_0$ and which are transformed under W from fundamental weights. We can assume that the edges of C_1 are generated by $\tilde{\omega}'_1, \tilde{\omega}'_2, \ldots, \tilde{\omega}'_{l-1}, s_{\gamma_0}\tilde{\omega}'_l$. For p a sufficiently large integer, we have

$$2c_1 ||\mu|| < d(\Re\lambda + p(\tilde{\omega}'_1 + \cdots + \tilde{\omega}'_{l-1}), \Xi_\alpha) \quad \text{for } \alpha \in R_+ - \{\gamma_0\}.$$

On the other hand, for any integers $p_1, \ldots, p_{l-1} \geqq 0$, we have

$$\lambda + p_1\tilde{\omega}'_1 + \cdots + p_{l-1}\tilde{\omega}'_{l-1} \in C_0^0, \qquad \lambda + \mu + p_1\tilde{\omega}'_1 + \cdots + p_{l-1}\tilde{\omega}'_{l-1} \in C_1^0.$$

For every $i \in \{1, \ldots, l\}$ and all $\nu \in \mathfrak{h}^*$ far from the walls, we have

$$2c_1 \|\tilde{\omega}_i\| \leqq 2c_1c_2 < d(\Re\nu, \Xi).$$

Lemma 7.6.20, applied stepwise, gives us

$$\mathrm{A}(\lambda) \Leftrightarrow \mathrm{A}(\lambda + p(\tilde{\omega}'_1 + \cdots + \tilde{\omega}'_{l-1})).$$

Lemma 7.6.21 gives us

$$\mathrm{A}(\lambda + \mu + p(\tilde{\omega}'_1 + \cdots + \tilde{\omega}'_{l-1})) \Rightarrow \mathrm{A}(\lambda + p(\tilde{\omega}'_1 + \cdots + \tilde{\omega}'_{l-1})).$$

Finally $\mathrm{A}(\lambda + \mu + p(\tilde{\omega}'_1 + \cdots + \tilde{\omega}'_{l-1}))$ is true by the induction hypothesis.

7.6.23. THEOREM (notation as in 7.0). *Let $\lambda, \lambda' \in \mathfrak{h}^*$. The following conditions are equivalent:*

(i) $M(\lambda) \subset M(\lambda')$;

(ii) $L(\lambda) \in \mathscr{J}\mathscr{H}(M(\lambda'))$;

(iii) *there exist $\gamma_1, \ldots, \gamma_n \in R_+$ such that*

$$\lambda' \geqq s_{\gamma_1}\lambda' \geqq s_{\gamma_2}s_{\gamma_1}\lambda' \geqq \cdots \geqq s_{\gamma_n} \cdots s_{\gamma_2}s_{\gamma_1}\lambda' = \lambda.$$

(iii) $\Rightarrow$ (i). This follows from 7.6.13.

(i) $\Rightarrow$ (ii). Obvious.

It remains to prove that, with the notation of 7.6.15, $\mathrm{A}(\lambda)$ is true.

(a) Let C be a chamber such that $\Re\lambda \in \overline{C}$. There exists $\zeta \in P \cap \overline{C}$ such that $\Re\lambda + \zeta \in C^0$. Then

$$\Re\lambda + \zeta + (P \cap \overline{C}) \subset C^0.$$

On the other hand, $\bigcup_{p=1,2,\ldots}(1/p)(\zeta + (P \cap \overline{C}))$ is dense in $\overline{C}$. Hence there exists a $\mu \in P \cap \overline{C}$ and an integer $n > 0$ such that $\Re\lambda + \mu \in C^0$ and $\|\Re\lambda - (1/n)\mu\| < c_3/2c_1$.

(b) Let $\nu \in \Pi(F_\mu^*)$ be such that $\lambda + \mu \in W(\lambda + \nu)$. We show that there exists an element of W which transforms λ into λ and μ into ν.

Let $w_1, w_2 \in W$ be such that $w_1C = C_+$ and $w_2(\lambda + \nu) = w_1(\lambda + \mu)$. Since $\Re w_1\lambda \in \overline{C}_+$, we have

(1) $$\langle \delta, \Re w_1\lambda \rangle \geqq \langle \delta, \Re w_2\lambda \rangle.$$

Since $w_1\mu \in C_+$, $w_1\mu$ is the dominant weight of F_μ, hence $w_1\mu \geqq w_2\nu$ and consequently

$$\langle \delta, w_1 \mu \rangle \geqq \langle \delta, w_2 \nu \rangle. \tag{2}$$

Now

$$\langle \delta, \mathfrak{R} w_1 \lambda + w_1 \mu \rangle = \langle \delta, \mathfrak{R} w_2 \lambda + w_2 \nu \rangle.$$

Hence we have equality in (1) and (2), which requires that $\mathfrak{R} w_1 \lambda = \mathfrak{R} w_2 \lambda$ and $w_1 \mu = w_2 \nu$. Moreover, $\mathfrak{I} w_1 \lambda = \mathfrak{I} w_2 \lambda$, hence $w_1 \lambda = w_2 \lambda$, which proves our assertion.

(c) We show that the triple (λ, μ, F) is admissible. Let $w \in W$ and $\nu \in \Pi(F_\mu)$ be such that $w\lambda < \lambda$, $w(\lambda + \nu) \geqq \lambda + \mu$ and $\lambda + \mu \in W(\lambda + \nu)$. Then

$$w \mathfrak{R}(n+1)\lambda < \mathfrak{R}(n+1)\lambda, \qquad w(\mathfrak{R}\lambda + \nu) \geqq \mathfrak{R}\lambda + \mu,$$

hence, from 7.6.19, where we replace $\lambda, \lambda', \mu, \mu'$ by $\mathfrak{R}(n+1)\lambda$, $w\mathfrak{R}(n+1)\lambda$, $-\mathfrak{R}n\lambda + \mu$, $-w\mathfrak{R}n\lambda + w\nu$ respectively, we have

$$(n+1)\,\|\mathfrak{R}(w\lambda - \lambda)\| < c_1(\|n\mathfrak{R}\lambda - \mu\| + \|n\mathfrak{R}\lambda - \nu\|).$$

But $w\lambda - \lambda \in Q - \{0\}$; taking (b) into account, this gives

$$(n+1)\, c_3 < c_1 n \frac{c_3}{2c_1} + c_1 n \frac{c_3}{2c_1},$$

whence we have a contradiction, which proves our assertion.

(d) We assume that $L(\lambda) \in \mathscr{J}\mathscr{H}(M(\lambda'))$. From (c) and 7.6.17, there exists $\nu \in \Pi(F_\mu)$ such that $L(\lambda + \mu) \in \mathscr{J}\mathscr{H}(M(\lambda' + \nu))$. From 7.6.22, there exists a sequence $(\gamma_1, \ldots, \gamma_p)$ linking $\lambda' + \nu$ with $\lambda + \mu$. Let $w \in W$ be such that $w\lambda' = \lambda$. Then $\lambda + \mu \in W(\lambda' + \nu) = W(\lambda + w\nu)$. On the other hand, $\mathfrak{R}\lambda + \mu \in C$ and $\mathfrak{R}\lambda \in \bar{C}$. From (b), there exists a chamber C' such that $\mathfrak{R}\lambda + w\nu \in C'$ and $\mathfrak{R}\lambda \in \bar{C}'$, whence $\mathfrak{R}\lambda' + \nu \in w^{-1}C'$ and $\mathfrak{R}\lambda' \in w^{-1}\bar{C}'$. From 7.6.18, $(\gamma_1, \ldots, \gamma_p)$ links λ' with λ.

7.6.24. THEOREM (notation as in 7.0). *Let $\lambda \in \mathfrak{h}^*$. The following conditions are equivalent:*

(i) *the $\mathfrak{g}$-module $M(\lambda)$ is simple;*

(ii) *for all $\alpha \in R_+$, we have $\lambda(H_\alpha) \notin \{1,2,3, \ldots\}$.*

Not (ii) $\Rightarrow$ not (i). Let $\alpha \in R_+$ be such that $\lambda(H_\alpha) \in \{1,2,3,\ldots\}$. Then $M(s_\alpha \lambda) \subset M(\lambda)$ (7.6.13), and $s_\alpha \lambda \neq \lambda$, hence $M(\lambda)$ is not simple.

Not (i) $\Rightarrow$ not (ii). We assume that $M(\lambda)$ is not simple. There exists $\mu \in \mathfrak{h}^*$ such that $\mu \neq \lambda$ and $M(\mu) \subset M(\lambda)$ (7.6.3). There hence exists $\alpha \in R_+$ such that $\lambda > s_\alpha \lambda$ (7.6.23), and then $\lambda(H_\alpha)$ is a positive integer.

References: [10], [11], [126].

7.7. Submodules of $M(\lambda)$ and the ordering relation on the Weyl group

7.7.1. If $\lambda \in P_{++}$, we have seen (7.6.8) that $M(w\lambda) \subset M(\lambda)$ for all $w \in W$. We shall specify the inclusion relations between the $M(w\lambda)$ for λ fixed and w variable. If λ is not in a wall, these inclusion relations, as we shall see, depend only on the values of w and not on those of λ.

7.7.2. LEMMA (notation as in 7.0). *Let $\lambda \in \delta + P_{++}$, $w \in W$ and $\gamma \in R_+$. Then:*
(i) $s_\gamma w\lambda = w\lambda - n\gamma$ *with* $n \in \mathbf{Z} - \{0\}$.
(ii) $n > 0 \Leftrightarrow l(s_\gamma w) > l(w)$.

We have $w\lambda \in P$, whence $s_\gamma w\lambda = w\lambda - n\gamma$ with $n \in \mathbf{Z}$. Since λ is not in a wall, we have $n \neq 0$. We assume that $n > 0$ and show that $l(s_\gamma w) > l(w)$. [If $n < 0$, it can be shown, by interchanging w and $s_\gamma w$, that $l(w) > l(s_\gamma w)$.] For all $v \in W$, we set $R_v = R_+ \cap v(R_-)$. Then we have:
(a) $\gamma \notin R_w$. Indeed, $\langle \lambda, w^{-1}\gamma \rangle = \langle w\lambda, \gamma \rangle > 0$, hence $w^{-1}\gamma \in R_+$.
(b) $\gamma \in R_{s_\gamma w}$. Indeed, $(s_\gamma w)^{-1}\gamma = -w^{-1}\gamma \in R_-$ from (a).
(c) Let S and S' be the sets of the $\zeta \in R_w$ such that $s_\gamma\zeta \in R_+$ and $s_\gamma\zeta \in R_-$, respectively. Then $s_\gamma(S) \subset R_{s_\gamma w}$. Indeed, if $\zeta \in S$, then

$$(s_\lambda w)^{-1}(s_\gamma \zeta) = w^{-1}\zeta \in R_-.$$

We show that $S' \subset R_{s_\gamma w}$. Let $\zeta \in S'$. Then $\zeta \in R_+$, $w^{-1}\zeta \in R_-$ and $s_\gamma\zeta \in R_-$. In the equality $\zeta - s_\gamma\zeta = p_\gamma$ we therefore have $p > 0$. Then

$$(s_\gamma w)^{-1}\zeta = w^{-1}(\zeta - p\gamma) = w^{-1}\zeta - pw^{-1}\gamma \in R_- + pR_-$$

(from (a)), whence $(s_\gamma w)^{-1}\zeta \in R_-$.
(d) Thus R_w is the disjoint union of S and S', and $R_{s_\gamma w}$ contains $s_\gamma(S)$ and S', which are disjoint because $s_\gamma(S') \subset R_-$ and $s_\gamma(s_\gamma(S)) \subset R_+$. Now $\gamma \in R_{s_\gamma w}$ from (b), $\gamma \notin S'$ from (a), and $\gamma \notin s_\gamma(S)$ because $\gamma \in R_+$. Hence Card $(R_{s_\gamma w}) >$ Card(R_w), and consequently $l(s_\gamma w) > l(w)$ (11.1.9).

7.7.3. Let $w, w' \in W$ and $\gamma \in R_+$. We write $w \xleftarrow{\gamma} w'$ if $w = s_\gamma w'$ and $l(w) = l(w') + 1$. We write $w \leftarrow w'$ if there exists some (necessarily unique) $\gamma \in R_+$ such that $w \xleftarrow{\gamma} w'$. We write $w \leqq w'$ if $w = w'$ or if there exist elements $w_1 = w, w_2, \ldots, w_n = w'$ such that $w_1 \leftarrow w_2 \leftarrow \cdots \leftarrow w_n$. We thus obtain an ordering relation on W.

7.7.4. LEMMA (notation as in 7.0). *Let $w_1, w_2 \in W$, $\gamma \in R_+$* and *$\alpha \in B$, with $\alpha \neq \gamma$. The following conditions are equivalent:*
(i) $s_\alpha w_1 \xleftarrow{\alpha} w_1$ *and* $s_\alpha w_1 \xleftarrow{\gamma} w_2$,
(ii) $w_2 \xleftarrow{\alpha} s_\alpha w_2$ *aand* $w_1 \xleftarrow{s_\alpha \lambda} s_\alpha w_2$.

Thus

$$\begin{array}{ccc} w_1 & \xleftarrow{s_\alpha\gamma} & s_\alpha w_2 \\ {\scriptstyle\alpha}\downarrow & & \downarrow{\scriptstyle\alpha} \\ s_\alpha w_1 & \xleftarrow{\gamma} & w_2 \end{array}$$

Let us assume that condition (i) is satisfied. We set $\gamma' = s_\alpha\gamma$, so that $\gamma' \in R_+$ (11.1.7). Since $s_{\gamma'} = s_\alpha s_\gamma s_\alpha$, we have $s_{\gamma'} w_1 = s_\alpha w_2$. Let $\lambda \in \delta + P_{++}$. Since $l(w_2) < l(s_\alpha w_1)$, we have $s_\alpha w_1 \lambda = w_2 \lambda - n\gamma$ with $n > 0$ (7.7.2), whence $w_1\lambda = s_\alpha w_2 \lambda - n\gamma'$, and hence $l(s_\alpha w_2) < l(w_1)$ (7.7.2). Since $l(s_\alpha w_2) = l(w_\alpha) \pm 1$, we obtain

$$l(s_\alpha w_2) = l(w_2) + 1 = l(w_1) + 1 .$$

Thus condition (ii) is satisfied. The converse can be shown in an analogous way.

7.7.5. Lemma (notation as in 7.0). *Let $w \in W$ and $\gamma \in R_+$ be such that $l(w) > l(s_\gamma w)$. Then $w < s_\gamma w$.*

We assume that the lemma has been proved for those elements of W whose length is $< l(w)$. If $l(s_\gamma w) = l(w) - 1$, then $w < s_\gamma w$ by definition. Since $l(w)$ and $l(s_\gamma w)$ have different parities, we shall henceforth assume that $l(s_\gamma w) \leqq l(w) - 3$. There exists $\alpha \in B$ such that $w \leftarrow s_\alpha w$. We have $\alpha \neq \gamma$. We set $\gamma' = s_\alpha \gamma \in R_+$. Then $s_{\gamma'} s_\alpha w = s_\alpha s_\gamma w$, and

$$l(s_\alpha s_\gamma w) \leqq l(s_\gamma w) + 1 \leqq l(w) - 2 \leqq l(s_\alpha w) - 1 .$$

From the induction hypothesis, we can write

$$s_\alpha w \xleftarrow{\gamma_1} w_1 \xleftarrow{\gamma_2} w_2 \leftarrow \cdots \leftarrow w_n \xleftarrow{\gamma_{n+1}} s_\alpha s_\gamma w .$$

Setting $s_\alpha w = w_0$, $w = w_{-1}$ and $\alpha = \gamma_0$, we thus have

$$w_{-1} \xleftarrow{\gamma_0} w_0 \xleftarrow{\gamma_1} w_1 \xleftarrow{\gamma_2} w_2 \leftarrow \cdots \leftarrow w_n \xleftarrow{\gamma_{n+1}} s_\alpha s_\gamma w .$$

If $s_\alpha s_\gamma w \leftarrow s_\gamma w$, the proof is complete. Henceforth we assume that $s_\gamma w \leftarrow s_\alpha s_\gamma w$. Let i be the largest non-negative integer such that $\gamma_i = \alpha$ (we recall that $\gamma_0 = \alpha$). If $i = n + 1$, then $s_\gamma w = w_n$, and the proof is complete. Let us assume that $i < n + 1$. From 7.7.4, we have

$$s_\alpha w_n \leftarrow w_n, \qquad s_\alpha w_n \xleftarrow{s_\alpha \gamma_{n+1}} s_\gamma w .$$

Proceeding in a stepwise fashion, we obtain

$$\begin{array}{ccccccccccccc}
w_{-1} & \xleftarrow{\gamma_0} & w_0 & \leftarrow & \cdots & \xleftarrow{\gamma_i} & w_i & \xleftarrow{\gamma_{i+1}} & w_{i+1} & \leftarrow \cdots \leftarrow & w_n & \xleftarrow{\gamma_{n+1}} & s_\alpha s_\gamma w \\
 & & & & & & \downarrow{\scriptstyle \alpha} & & \downarrow{\scriptstyle \alpha} & & \downarrow{\scriptstyle \alpha} & & \downarrow{\scriptstyle \alpha} \\
 & & & & & & s_\alpha w_i & \xleftarrow{s_\alpha \gamma_{i+1}} & s_\alpha w_{i+1} & \leftarrow \cdots \leftarrow & s_\alpha w_n & \xleftarrow{s_\alpha \gamma_{n+1}} & s_\gamma w .
\end{array}$$

But since $\gamma_i = \alpha$, we have finally

$$w = w_{-1} \leftarrow w_0 \leftarrow \cdots \leftarrow w_{i-1} \leftarrow s_\alpha w_{i+1} \leftarrow s_\alpha w_{i+2} \leftarrow \cdots \leftarrow s_\gamma w .$$

7.7.6. Lemma (notation as in 7.0). *Let $w_1, w_2 \in W$. The number of elements w of W such that $w_1 \leftarrow w \leftarrow w_2$ is 0 or 2.*

We can assume that $l(w_2) = l(w_1) - 2$. We reason by induction on $l(w_1)$. Since $l(w_1) \geqq 2$, we can fix $\alpha \in B$ such that $w_1 \leftarrow s_\alpha w_1$. We then distinguish two cases:

(1) Assume that $w_2 \leftarrow s_\alpha w_2$. Let W' and W'' be the sets of the $w \in W$ such that $w_1 \leftarrow w \leftarrow w_2$ and $s_\alpha w_1 \leftarrow w \leftarrow w s_\alpha \leftarrow w_2$ respectively.

$$\begin{array}{ccccc}
w_1 & \xleftarrow{\gamma_1} & w & \xleftarrow{\gamma_2} & w_2 \\
\uparrow{\scriptstyle \alpha} & & \uparrow{\scriptstyle \alpha} & & \uparrow{\scriptstyle \alpha} \\
s_\alpha w_1 & \xleftarrow{s_\alpha \gamma_1} & s_\alpha w & \xleftarrow{s_\alpha \gamma_2} & s_\alpha w_2
\end{array}$$

We define $\varphi : W' \to W''$ in the following way. If $w \in W'$ and $w \neq s_\alpha w_1$, we set $\varphi(w) = s_\alpha w$. (If $w = s_{\alpha_1} w_1$ and $w_2 = s_{\gamma_2} w$, we have, from 7.7.4, the above diagram. Note that $\alpha \neq \gamma_2$ since $w_2 \leftarrow s_\alpha w_2$; hence we have $\varphi(w) \in W''$ in fact.) If $w \in W'$ and $w = s_\alpha w_1$, we set $\varphi(w) = w_2$. (Since $w \leftarrow w_2 \leftarrow s_\alpha w_2$, we indeed have $\varphi(w) \in W''$.) Clearly, φ is an injective mapping of W' into W''. We show that φ is surjective. If $w_2 \in W''$, then $w_1 \leftarrow s_\alpha w_1 \leftarrow w_2$, hence $s_\alpha w_1 \in W'$ and $w_2 = \varphi(s_\alpha w_1) \in \varphi(W')$. Let us consider an element of W'' distinct from w_2 and write it in the form $s_\alpha w$. Then

$$s_\alpha w_1 \xleftarrow{\gamma_1'} s_\alpha w \xleftarrow{\gamma_2'} s_\alpha w_2 ,$$

and $\gamma_2' \neq \alpha$ since $s_\alpha w \neq w_2$. From 7.7.4, we have the above diagram (note that $\alpha \neq \gamma_1'$ since $w_1 \leftarrow s_\alpha w_1$), with $\gamma_1' = s_\alpha \gamma_1$ and $\gamma_2' = s_\alpha \gamma_2$. Since $\gamma_1' \neq \alpha$, we have $w \neq s_\alpha w_1$, hence also $w_2 \in \varphi(W')$. Thus φ is a bijection of W' onto W''. Now $l(s_\alpha w_1) < l(w_1)$, so that we can apply the induction hypothesis to Card W''.

(2) We assume that $s_\alpha w_2 \leftarrow w_2$. If $w_1 \xleftarrow{\gamma_1} w \xleftarrow{\gamma_2} w_2$ with $\gamma_1 \neq \alpha$ (or $\gamma_2 \neq \alpha$), then $w \leftarrow s_\alpha w$ (or $s_\alpha w \leftarrow w$) from 7.7.4. Hence

$w_1 \xleftarrow{\gamma_1} w \xleftarrow{\gamma_2} w_2$ implies that $\gamma_1 = \alpha$ or $\gamma_2 = \alpha$. Let us assume that $w_1 \xleftarrow{\gamma_1} w \xleftarrow{\gamma_2} w_2$. Since $w_1 \leftarrow s_\alpha w_1$, we have $s_\alpha w_1 \leftarrow s_\alpha w = w_2$ (from 7.7.4), hence $w_1 \xleftarrow{\alpha} s_\alpha w_1 \xleftarrow{\gamma_2} w_2$. Similarly, $w_1 \xleftarrow{\alpha} w \xleftarrow{\gamma_2} w_2$ implies that $w_1 \xleftarrow{\gamma_1} s_\alpha w_2 \xleftarrow{\alpha} w_2$. If there exist elements w such that $w_1 \leftarrow w \leftarrow w_2$, then there exist two, namely $s_\alpha w_1$ and $s_\alpha w_2$.

7.7.7. THEOREM (notation as in 7.0). *Let $\lambda \in \delta + P_{++}$.*

(i) *For $w, w' \in W$ to be such that $M(w\lambda) \subset M(w'\lambda)$, it is necessary and sufficient that $w \leqq w'$.*

(ii) *If $M(w\lambda) \subset M(w'\lambda)$, there exist $w_1, \ldots w_n \in W$ such that $l(w_{i+\gamma}) = l(w_i) - 1$ for $i = 1, \ldots, n-1$, $w_n = w'$, $w_1 = w$, and*

$$M(w\lambda) = M(w_1\lambda) \subset M(w_2\lambda) \subset \cdots \subset M(w_n\lambda) = M(w'\lambda).$$

(iii) *If $l(w) = l(w') + 2$, the number of $w'' \in W$ such that*

$$M(w\lambda) \subset M(w''\lambda) \subset M(w'\lambda), \qquad M(w\lambda) \neq M(w''\lambda) \neq M(w'\lambda)$$

is 0 *or* 2.

We assume that $M(w\lambda) \subset M(w'\lambda)$. There exist $\gamma_1, \ldots, \gamma_{n-1} \in R_+$ such that

$$w'\lambda \geqq s_{\gamma_1} w'\lambda \geqq \cdots \geqq s_{\gamma_{n-1}} s_{\gamma_{n-2}} \cdots s_{\gamma_1} w'\lambda = w\lambda.$$

Then $l(w') < l(s_{\gamma_1} w') < \cdots < l(w)$ (7.7.2), hence $w' \geqq w$ (7.7.5).

Let us now assume that $w \leqq w'$. There exist $\gamma_1, \ldots, \gamma_{n-1} \in R_+$ such that

$$w = w_1 \xleftarrow{\gamma_1} w_2 \xleftarrow{\gamma_2} \cdots \to w_{n-1} \xleftarrow{\gamma_{n-1}} w_n = w'.$$

Then $w_n\lambda \geqq w_{n-1} \geqq \cdots \geqq w_1\lambda$ (7.7.2), hence

$$M(w_1\lambda) \subset M(w_2\lambda) \subset \cdots \subset M(w_n\lambda)$$

(7.6.23). This proves both (i) and (ii). Assertion (iii) then follows from 7.7.6.

Reference: [11].

7.8. Supplementary remarks

7.8.1. There us much to say concerning finite dimensional simple $\mathfrak{g}$-modules. We have merely given the results which are naturally connected with Verma modules, and consequently with enveloping algebras.

The essence of the results in sections 7.1, 7.2, 7.3 is classical. Nonetheless, the methods used here had scarcely appeared before Harish-Chandra; cf.

[63], [126]. Theorem 7.3.5 is due to Chevalley, and theorem 7.4.5 to Harish-Chandra [63]. The methods of sections 7.3 and 7.4 are due to various authors (Harish-Chandra, Kostant, Verma, Varadarajan, etc.). Theorem 7.5.9 is Weyl's celebrated formula for characters; it is here proved by a method due to Bernstein—Gelfand—Gelfand [10]. Results 7.6.2, 7.6.3, 7.6.6, 7.6.8 are due to Verma [126], and theorem 7.6.23 to Bernstein—Gelfand—Gelfand ([10], [11]); the implication (iii) $\Rightarrow$ (i) was also proved by Verma [126] using a very different method (Verma had also conjectured that (i) $\Rightarrow$ (iii)). Theorem 7.7.7 is due to Bernstein—Gelfand—Gelfand [11].

7.8.2. (notation as in 7.2.7) For every $\alpha \in R_+$, let $X_{-\alpha} \in \mathfrak{g}^{-\alpha} - \{0\}$ and $m_\alpha = \lambda(H_\alpha) + 1$. If $\lambda \in P_{++}$, then

$$I' = I + \sum_{\alpha \in R_+} U(\mathfrak{g}) X_{-\alpha}^{m_\alpha}.$$

7.8.3. We adopt the notation of 6.6.9 (a) and (b). Let β be the canonical mapping of $S(\mathfrak{g})$ into $U(\mathfrak{g})$.

(a) Assume that k is algebraically closed, and use the notation of 7.0. Let $\lambda \in \mathfrak{h}^*$. Then for all $u \in Y(\mathfrak{g})$ we have $\chi_\lambda(\beta u) = (D(q^{-1})u)(\lambda')$ (where λ' is the linear extension of λ to $\mathfrak{g}$ which is zero on $\mathfrak{n}_+ + \mathfrak{n}_-$).

(b) Assume that $\mathfrak{g}$ is semi-simple. The mapping $u \mapsto \beta(D(q)u)$ of $Y(\mathfrak{g})$ into $U(\mathfrak{g})$ is an isomorphism of the algebra $Y(\mathfrak{g})$ onto the algebra $Z(\mathfrak{g})$ [43].

7.8.4 (notation as in 7.0). (a) Let γ_0 be the class of the one-dimensional trivial simple $\mathfrak{g}$-module. Then $U(\mathfrak{g})_{\gamma_0} = Z(\mathfrak{g})$ and

$$\sum_{\gamma \in \mathfrak{g}^\wedge, \gamma \neq \gamma_0} U(\mathfrak{g})_\gamma = [U(\mathfrak{g}), U(\mathfrak{g})].$$

Let $u \mapsto u^\natural$ be the projection of $U(\mathfrak{g})$ onto $Z(\mathfrak{g})$ defined by the decomposition $U(\mathfrak{g}) = Z(\mathfrak{g}) \oplus [U(\mathfrak{g}), U(\mathfrak{g})]$. If $u, v \in U(\mathfrak{g})$, then $(uv)^\natural = (vu)^\natural$, and $uv - vu$ cannot be central without being zero. If $u \in U(\mathfrak{g})$ and $z \in Z(\mathfrak{g})$, then $(uz)^\natural = u^\natural z$ [58].

(b) $U(\mathfrak{g}) = U(\mathfrak{h}) \oplus [U(\mathfrak{g}), U(\mathfrak{g})]$.

(c) Let $\lambda \in P_{++}$, let E be a simple $\mathfrak{g}$-module with largest weight λ, and let φ be the Harish-Chandra homomorphism. For all $u \in U(\mathfrak{g})$, we have

$$\frac{1}{\dim E} \operatorname{tr}(u_E) = (\varphi(u^\natural))(\lambda).$$

[63].

7.8.5 (notation as in 7.0). Let $\langle .,. \rangle$ be an invariant non-degenerate symmetric bilinear form on $\mathfrak{g}$. Let $(e_1, \ldots, e_n)$ and $(e_1', \ldots, e_n')$ be two bases for

$\mathfrak{g}$ such that $\langle e_i, e'_j\rangle = \delta_{ij}$. The element $c = \sum_{i=1}^{n} e_i e'_i$ of $U(\mathfrak{g})$ belongs to $Z(\mathfrak{g})$ and depends only on $\langle .,. \rangle$. We also denote by $\langle .,. \rangle$ the inverse form on $\mathfrak{h}^*$ of the restriction of $\langle .,. \rangle$ to $\mathfrak{h}$. If $\lambda \in \mathfrak{h}^*$, then $\chi_\lambda(c) = \langle \lambda, \lambda - 2\,\delta \rangle$. (If $X_\alpha \in \mathfrak{g}^\alpha - \{0\}$ for all $\alpha \in R$, and $(H_1, \ldots, H_l)$ and $(H'_1, \ldots, H'_l)$ are dual bases for $\mathfrak{h}$, then

$$c = \sum_{\alpha \in R} \frac{1}{\langle X_\alpha, X_{-\alpha}\rangle} X_\alpha X_{-\alpha} + H_1 H'_1 + \cdots + H_l H'_l.$$

We can then calculate the image of c under the Harish-Chandra homomorphism.) ([71], p. 247).

7.8.6. Assume that $\mathfrak{g}$ is semi-simple and $k = \mathbf{C}$. Let $\mathfrak{h}$ be a Cartan subalgebra of $\mathfrak{g}$. Every choice of a Weyl chamber C in $\mathfrak{h}^*_{\mathbf{R}}$ defines a Harish-Chandra homomorphism φ_C of $U(\mathfrak{g})_0$ (with the notation of 7.4.1) into $U(\mathfrak{h})$. We identify the elements of $U(\mathfrak{h}) = S(\mathfrak{h})$ with polynomial functions on $\mathfrak{h}^*_{\mathbf{R}}$. For every Weyl chamber C, let f_C be a complex-valued polynomial function on $\mathfrak{h}^*_{\mathbf{R}}$. The following conditions are equivalent:

(i) there exists $u \in U(\mathfrak{g})_0$ such that $\varphi_C(u) = f_C$ for every chamber C;

(ii) for every chamber C and for every reflexions s with respect to a wall L of C, we have $f_C | L = f_{s(C)} | L$.

[40]

7.8.7. (notation as in 7.0) Let C be the commutant of $\mathfrak{h}$, or of $U(\mathfrak{h})$, in $U(\mathfrak{g})$.

(a) Let $\alpha_1, \ldots, \alpha_m$ be the elements of R arranged in a certain order, let $X_{\alpha_i} \in \mathfrak{g}^{\alpha_i} - \{0\}$, let be the set of non-null sequences $(p_1, \ldots, p_m) \in \mathbf{N}^m$ such that $p_1\alpha_1 + \cdots + p_m\alpha_m = 0$, and let $\mathcal{M}$ be the set (which is finite from 2.6.2) of minimal elements of ζ, $\mathbf{N}^m$ being ordered by the product ordering. Then $\mathfrak{h}$ and the $X_{\alpha_1}^{p_1} \cdots X_{\alpha_m}^{p_m}$, where $(p_1, \ldots, p_m) \in \mathcal{M}$, generate the algebra C.

(b) For every non-zero element c of C, there exists a finite-dimensional simple representation ϱ of C such that $\varrho(c) \neq 0$. (Use 2.5.7. On the other hand, if π is a finite-dimensional simple representation of $\mathfrak{g}$ in V, the V_λ are stable under $\pi(C)$, and the subrepresentations of C defined by the V_λ are simple, as may be seen by using the graduation 7.4.1 of $U(\mathfrak{g})$.)

(c) If $\mathfrak{g}$ is simple and of type A_1, then C is commutative. If $\mathfrak{g}$ is simple, but not of type A_1, then C has arbitrarily large finite-dimensional simple representations. (From 7.5.10, $\mathfrak{g}$ has finite-dimensional simple representations whose weights have arbitrarily large multiplicity. We then use the reasoning of (c).)

(d) Let $\lambda \in \mathfrak{h}^*$. Then $U(\mathfrak{g})_\lambda$ is a left (or right) module of finite type over C.

(e) Let V be a simple $\mathfrak{g}$-module such that $V = \oplus_{\lambda\in\mathfrak{h}^*} V_\lambda$. If one of the V_λ is finite-dimensional, then all the V_λ are finite-dimensional.

(f) Let W be a simple-C module. There exists one, and up to isomorphism only one simple $\mathfrak{g}$-module V such that there exists a sub-C-module of V isomorphic to W. (Identify W with C/J, where J is a maximal left ideal of C. Since $U(\mathfrak{g}) \oplus_{\lambda\in\mathfrak{h}^*} U(\mathfrak{g})_\lambda$ and the $U(\mathfrak{g})$ are right C-modules, we have $U(\mathfrak{g})J \neq U(\mathfrak{g})$. Let I be a maximal left ideal of $U(\mathfrak{g})$ such that $I \supset U(\mathfrak{g})J$. Take $V = U(\mathfrak{g})/I$. As in 7.1.11, one sees that $V' = U(\mathfrak{g})/U(\mathfrak{g})J$ has a largest sub-module distinct from V'.) If k is algebraically closed, there exists $\lambda \in \mathfrak{h}^*$ such that $W = V_\lambda$, so that $V = \oplus_{\mu\in\mathfrak{h}^*} V_\mu$. (Use 2.6.4. We could also use 9.1.12.) ([17], [82], [83], [84], [85], [86]).

(g) Choose B as in 7.0. With the notation of (f), V is of the type considered in 7.1.8 if and only if the annihilator of W contains the ideal $U(\mathfrak{g})\mathfrak{n}_+ \cap C$ of C. [14]

7.8.8 (notation as in 7.0). (a) Let $\lambda \in P$, and let $s = (\beta, \ldots, \beta_p)$ be a sequence of elements of B. For $j = 1, \ldots, p$, let $n_j = (s_{\beta_{j+1}} \cdots s_{\beta_p}\lambda)(H_{\beta_j}) \in \mathbf{Z}$. Then $X^{n_1}_{-\beta_1} \cdots X^{n}p_{\beta_\varrho}$ is an element of $K(\mathfrak{n}_-)$ which we denote by $u(s,\lambda)$. (Choose an $X_\alpha \in \mathfrak{g}^\alpha - \{0\}$ for all $\alpha \in R$.)

(b) Let $\lambda \in P$, let $w \in W$, and let $w = s_{\beta_1} \cdots s_{\beta_\varrho} = s_{\gamma_1} \cdots s_{\gamma_\varrho}$ be two reduced decompositions of w. Then $u((\beta_1, \ldots, \beta_p),\lambda) = u((\gamma_1, \ldots, \gamma_p),\lambda)$. We denote this element of $K(\mathfrak{n}_-)$ by $u(w,\lambda)$. If $\lambda \in P_{++}$, then $u(w,\lambda) \in U(\mathfrak{n}_-)$.

(c) We have $u(s_\alpha,\lambda) = X^{1(H_\alpha)}_{-\alpha}$ for $\alpha \in B$, and $u(ww',\lambda) = u(w,w'\lambda)u(w',\lambda)$ for any $w,w' \in W$, $\lambda \in P$.

(d) For $w \in W$, $\lambda \in P$, $u \in K(\mathfrak{n}_-)$, set $\theta(w)(\lambda,u) = (w\lambda,u(w,\lambda)u)$. Then θ is a homomorphism of W into the group of permutations of $P \times K(\mathfrak{n}_-)$.

(e) Let $\lambda \in P$ and $w \in W$. If $M(w\lambda) \subset M(\lambda)$, then $u(w,\lambda) \in U(\mathfrak{n}_-)$. If $u(w,\lambda) \in U(\mathfrak{n}_-)$ and v is the canonical generator of $M(\lambda)$, then $u(w,\lambda)\ v$ generates a sub-module of $M(\lambda)$ which is isomorphic to $M(w\lambda)$.

(f) From the above one can deduce the following identities, which are valid in any ring A. Let $x,y \in A$ be such that $[x,y]$ commutes with y, let and $m,n \in N$. If $x,y]$ commutes with x, then

$$x^m y^{m+n} x^n = y^n x^{m+n} y^m.$$

If $[x,[x,y]]$ commutes with x, then

$$x^m y^{m+n} x^{m+2n} y^n = y^n x^{m+2n} y^{m+n} x^m.$$

If $[x,[x,[x,y]]]$ commutes with x, then

$$x^m y^{m+n} x^{2m+3n} y^{m+2n} x^{m+3n} y^n = y^n x^{m+3n} y^{m+2n} x^{2m+3n} y^{m+n} x^m.$$

[126]

7.8.9 (notation as in 7.0). Let $\lambda \in \mathfrak{h}^*$, $w \in W$, and let $w = s_{\alpha_n} s_{\alpha_{n-1}} \cdots s_{\alpha_1}$ be a reduced decomposition of w (where $\alpha_1, \ldots, \alpha_n \in B$, α_1 distinct from $\alpha_2, \ldots, \alpha_n$). Then

$$M(w\lambda) \subset M(\lambda) \Leftrightarrow M(w\lambda) \subset M(s_{\alpha_1}\lambda).$$

[126]

7.8.10. (notation as in 7.0). Let $\lambda, \lambda' \in \mathfrak{h}^*$. We consider the equivalent conditions (i), (ii), (iii) of 7.6.23 and the condition

(iv) $\lambda \in W\lambda'$ and $\lambda \leqq \lambda'$.

Then:

(a) If $\mathfrak{g}$ is simple of type A_2, then (i) $\Leftrightarrow$ (iv).

(b) If $\mathfrak{g}$ is simple and of rank 2, then conditions (i) and (iv) are not equivalent; however, they are if $\lambda \in P$.

(c) If $\mathfrak{g}$ is simple and of rank 3, then conditions (i) and (iv) are not equivalent, even if $\lambda \in P$. [126]

7.8.11. A Verma module may contain infinitely many sub-modules. [25]

7.8.12. Take for $\mathfrak{g}$ the Lie algebra $\mathfrak{sl}(3, C)$, for $\mathfrak{h}$ the set of diagonal elements of $\mathfrak{g}$, and for $\mathfrak{n}_+$ the set of strictly upper triangular elements of $\mathfrak{g}$. Let $X_\alpha = E_{12}, X_\gamma = E_{23}, X_\beta = E_{23}, X_\alpha = E_{21}, X_{-\gamma} = E_{32}$, and $X_{-\beta} = E_{31}$, which determines a basis (α,γ) for $R(\mathfrak{g},\mathfrak{h})$. Then $\beta = \alpha + \gamma$.

Consider the Verma module $Z_0 = M(\varrho)$ and identify it which $U(\mathfrak{n}_-)$. Then

$$Z_{-\alpha} = M(s_\alpha \beta) = M(\gamma) = U(\mathfrak{n}_-)X_{-\alpha},$$

$$Z_{-\gamma} = M(s_\gamma \beta) = M(\alpha) = U(\mathfrak{n}_-)X_{-\gamma},$$

$$Z_{-2\gamma-\alpha} = M(s_\gamma s_\alpha \beta) = M(-\gamma) = U(\mathfrak{n}_-)X^2_{-\gamma}X_{-\alpha},$$

$$Z_{-2\alpha-\gamma} = M(s_\alpha s_\gamma \beta) = M(-\alpha) = U(\mathfrak{n}_-)X^2_{-\alpha}X_{-\gamma},$$

$$Z_{-2\alpha-2\gamma} = M(s_\alpha s_\gamma s_\alpha \beta) = M(s_\gamma s_\alpha s_\gamma \beta) = M(-\beta)$$
$$= U(\mathfrak{n}_-)X_{-\gamma}X^2_{-\alpha}X_{-\gamma} = U(\mathfrak{n}_-)X_{-\alpha}X^2_{-\gamma}X_{-\alpha}.$$

Let v be the canonical generator of $M(\beta)$. Although $\beta(H_\beta) = 2$, $M(s_\beta \beta) = M(-\beta)$ is not generated by $X^2_{-\beta}v$ (cf. 7.1.15). We have

$$Z_{-\alpha} \cap Z_{-\gamma} = Z_{-2\alpha-\gamma} + Z_{-2\gamma-\alpha}, \qquad Z_{-2\alpha-\gamma} \cap Z_{-2\gamma-\alpha} = Z_{-2\alpha-2\gamma},$$

and $Z_{-\alpha} + Z_{-\gamma}$ is the largest submodule of Z_0 distinct from Z_0 (it has codimension 1). The annihilator I of $Z_0/Z_{-\gamma}$ is generated by $9X_\alpha X_{-\alpha} - (H_\gamma -$

$- H_\alpha + 3)(H_\gamma + 2H_\alpha)$. The annihilator I' of $Z_0/Z_{-\alpha}$ is generated by $9X_\gamma X_{-\gamma} - (H_\alpha - H_\gamma + 3)(H_\alpha + 2H_\gamma)$. We have

$$IZ_0 = Z_{-\gamma}, \qquad I'Z_0 = Z_{-\alpha}, \qquad (I \cap I')Z_0 = Z_{-\alpha} \cap Z_{-\gamma},$$

$$II'Z_0 = Z_{-2\alpha-\gamma}, \qquad I'IZ_0 = Z_{-2\gamma-\alpha}, \qquad (I \cap I')^2 Z_0 = Z_{-2\gamma-2\alpha},$$

$$I^2 = I, \qquad I'^2 = I'.$$

7.8.13. (notation as in 7.8.12) Let J be the annihilator of $M(2\beta)/M(2\alpha)$, and let

$$u = 9X_\gamma X_{-\gamma} - (H_\alpha - H_\gamma)(H_\beta + H_\gamma - 3),$$

$$u' = 9X_\gamma X_{-\gamma} - (H_\alpha - H_\gamma + 3)(H_\beta + H_\gamma - 6).$$

Then $u \notin J$ and $u' \notin J$, but $uu' \in J$.

7.8.14 (notation as in 7.0). (a) To each pair (w_1,w_2) of elements of W such that $w_1 \leftarrow w_2$ we can associate $\sigma(w_1,w_2) = \pm 1$ such that the following property is satisfied:

if $w_1,w_2,w_3,w_4 \in W$ are such that $w_1 \leftarrow w_2 \leftarrow w_4$ and $w_1 \leftarrow w_3 \leftarrow w_4$, where $w_2 \neq w_3$, then

$$\sigma(w_1,w_2)\,\sigma(w_2,w_4) = -\sigma(w_1,w_3)\,\sigma(w_3,w_4).$$

We fix for (b) such a function σ.

(b) Let $\lambda \in P_{++}$, let V be a simple $\mathfrak{g}$-module with largest weight λ, and let $s = \dim \mathfrak{n}_-$. Then $s = \sup_{w \in W} l(w)$. For $i = 0,1, \ldots, s$, let W_i be the set of elements of W which have length i, and let $C_i = \oplus_{w \in W_i} M(w(\lambda + \delta))$. In particular, $C_0 = M(\lambda + \delta)$, so that there exists a $\mathfrak{g}$-homomorphism ε of C_0 onto V. For every $w \in W$, we fix an embedding of $M(w(\lambda + \delta))$ in $M(\lambda + \delta)$. For $i = 1, \ldots, s$, we define a matrix $(d^i_{w_1,w_2})$, where $w_1 \in W_i$ and $w_2 \in W_{i-1}$, by setting $d^i_{w_1,w_2} = \sigma(w_1,w_2)$ if $w_1 \leftarrow w_2$ and $d^i_{w_1,w_2} = 0$ otherwise. A $\mathfrak{g}$-homomorphism d_i of C_i into C_{i-1} corresponds to this matrix (if we observe that $M(w_1(\lambda + \delta)) \subset M(w_2(\lambda + \delta))$ if $w_1 \leftarrow w_2$). Then the sequence

$$0 \longrightarrow C_s \xrightarrow{d_s} C_{s-1} \longrightarrow \cdots \xrightarrow{d_2} C_1 \xrightarrow{d_1} C_0 \xrightarrow{\varepsilon} V \longrightarrow 0$$

is exact. [11]

7.8.15 (notation as in 7.0). We say that a $\mathfrak{g}$-module M is *regular* if it satisfies the following conditions:

(i) M is of finite type as a $U(\mathfrak{g})$-module;

(ii) $M = \oplus_{\lambda \in \mathfrak{h}^*} M_\lambda$;

(iii) for every $m \in M$, $U(\mathfrak{n}_+)\, m$ is finite-dimensional.

Then:

(a) If M is regular, every subquotient of M is regular.

(b) If M is regular, M has a character which belongs to $\mathbf{Z}\langle\mathfrak{h}^*\rangle$.

(c) If M is regular, M has a Jordan–Hölder series.

(d) The $M(\lambda)$ are regular.

(e) The regular simple modules are the $L(\lambda)$.

(f) For every $\mathfrak{g}$-module N, let $\Theta(N)$ be the set of homomorphisms θ of $Z(\mathfrak{g})$ into k such that there exists $n \in \mathbf{N} - \{0\}$ which satisfies $zn = \theta(z)n$ for all $z \in Z(\mathfrak{g})$. If M is regular, then $\Theta(M)$ is finite.

(g) Let us assume that M is regular. For every homomorphism θ of $Z(\mathfrak{g})$ into k and all $n \in \mathbf{N}$, let $M_\theta^{(n)}$ be the set of the $m \in M$ such that $(\operatorname{Ker}\theta)^n m = 0$. Then the sequence $M_\theta^{(1)} \subset M_\theta^{(2)} \ldots$ is stationary; let M_θ be its union. We have $\Theta(M_\theta) = \{\theta\}$, and $M = \oplus_{\theta \in \Theta(M)} M_\theta$ [11].

7.8.16. Let $\mathfrak{g} = \mathfrak{sl}(2,\mathbf{C})$, let e,f,h, be as in 1.8, let $Q = 4ef + h^2 - 2h \in Z(\mathfrak{g})$, and let $\mathfrak{h} = \mathbf{C}h$. A representation ϱ of $\mathfrak{g}$ is said to be $\mathfrak{h}$-diagonalizable if $\varrho(h)$ is diagonalizable. Let $q \in \mathbf{C}$ and $v \in \mathbf{C}/\mathbf{Z}$.

(a) The free complex vector space with basis $\mathbf{C}$ is denoted by S; for all $w \in \mathbf{C}$, let e_w be the corresponding element of S, so that $(e_w)_{w \in \mathbf{C}}$ is a basis for S. Let S_v be the vector subspace of S generated by the e_w for $w \in v$. Define the endomorphisms E,F,H of S_v by

$$He_w = 2we_w, \qquad Ee_w = \sqrt{q - w^2 - w}\, e_{w+1}, \qquad Fe_w = \sqrt{q - w^2 + w}\, e_{w-1}$$

(it is agreed that, for $z \in \mathbf{C}$, $\sqrt{z}$ is the square root of z whose argument belongs to $[0,\pi[$). There exists one and only one representation $\varrho = \varrho_{v,q}$ of $\mathfrak{g}$ in S_v such that $\varrho(h) = H, \varrho(e) = E$ and $\varrho(f) = F$. We designate the corresponding $\mathfrak{g}$-module by $S_{v,q}$. We have $\varrho(Q) = 4q$.

(b) If $q \neq u^2 + u$ for all $u \in v$, then $S_{v,q}$ is simple.

(c) Assume that $2v \neq 0$, and that q is of the form $u^2 + u$, where $u \in v$ (which defines u uniquely). Let $S_{v,q}^-$ and $S_{v,q}^+$ be the vector subspaces of $S_{v,q}$ generated by the e_w for $w \leqq u$ and $w > u$, respectively. Then $S_{v,q}^-$ and $S_{v,q}^+$ are simple sub-$\mathfrak{g}$-modules of $S_{v,q}$.

(d) Assume that $2v = 0$, and that q is of the form $u^2 + u$, where $u \in v$, $u \geqq 0$ (which defines u uniquely). Let $S_{v,q}^-$, $S_{v,q}^0$ and $S_{v,q}^+$ be the vector subspaces of $S_{v,q}$ generated by the e_w for $w < -u$, $-u \leqq w \leqq u$ and $w > u$, respectively. Then $S_{v,q}^-$, $S_{v,q}^0$ and $S_{v,q}^+$ are simple sub-$\mathfrak{g}$-modules of $S_{v,q}$.

(e) Assume that $v = -\frac{1}{2} + \mathbf{Z}$ and $q = -\frac{1}{4}$. Let $S^-_{v,-1/4}$ and $S^+_{v,-1/4}$ be the vector subspaces of $S_{v,-1/4}$ generated by the e_w for $w \leqq -\frac{1}{2}$ and $w > -\frac{1}{2}$, respectively. Then $S^-_{v,-1/4}$ and $S^+_{v,-1/4}$ are simple sub-$\mathfrak{g}$-modules of $S_{v,-1/4}$.

(f) For q fixed and v variable, the simple $\mathfrak{g}$-modules of (b), (c), (d) and (e) are, up to isomorphism, all the $\mathfrak{h}$-diagonalizable simple $\mathfrak{g}$-modules in which Q defines the homothety with ratio $4q$.

(g) Let M be an $\mathfrak{h}$-diagonalizable $\mathfrak{g}$-module in which Q defines the homothety with ratio $4q$. For $v \in \mathbf{C}/\mathbf{Z}$, let M_v be the vector subspace of M generated by the eigenvectors of h_M such that the corresponding eigenvalue belongs to $2v$. It is a sub-$\mathfrak{g}$-module of M, and $M = \oplus_{v \in \mathbf{C}/\mathbf{Z}} M_v$. If $q \neq u^2 + u$ for all $u \in v$, then M_v is the direct sum of the sub-$\mathfrak{g}$-modules isomorphic to $S_{v,q}$.

Assume that $2v \neq 0$, and that q is of the form $u^2 + u$, where $u \in v$, i.e. $v = -\frac{1}{2} + \mathbf{Z}$ and $q = -\frac{1}{4}$. Then M_v is the direct sum of sub-$\mathfrak{g}$-modules isomorphic to $S^-_{v,q}$, $S^+_{v,q}$, $S^-_{v,q} \to S^+_{v,q}$ and $S^-_{v,q} \leftarrow S^+_{v,q}$, the latter two modules (which are indecomposable) being defined in the following way: the underlying vector space is $S_{v,q}$; the formulae which define H,E,F are as in (a) with the following exception:

$$S^-_{v,q} \to S^+_{v,q} : Fe_{u+1} = e_u \qquad \text{instead of } Fe_{u+1} = 0,$$

$$S^-_{v,q} \leftarrow S^+_{v,q} : Ee_u = e_{u+1} \qquad \text{instead of } Ee_u = 0.$$

For all the foregoing, cf. Gabriel, lectures at the Séminaire Godement, Paris, 1959—60, cyclostyled notes, and [95]. If $2v = 0$ and q is of the form $u^2 + u$, where $u \in v$ and $u \geqq 0$, it is also true that M_v is the direct sum of indecomposable $\mathfrak{g}$-modules, and Gabriel has also compiled the list of these indecomposable $\mathfrak{g}$-modules up to isomorphism.

(h) For the non-$\mathfrak{h}$-diagonalizable representations of $\mathfrak{sl}(2,\mathbf{C})$, cf. notably [2], [3]. In [3] a simple representation ϱ of $\mathfrak{sl}(2,C)$ is constructed having the following property:

for every $x \in \mathfrak{g} - \{0\}$, $\varrho(x)$ has no eigenvalue.

7.8.17. Let $\mathfrak{h}$, $\mathfrak{n}_1$, $\mathfrak{n}_2$ be Lie subalgebras of $\mathfrak{g}$ such that

$$\mathfrak{g} = \mathfrak{h} \oplus \mathfrak{n}_1 \oplus \mathfrak{n}_2, \qquad [\mathfrak{h},\mathfrak{n}_1] \subset \mathfrak{n}_1, \qquad [\mathfrak{h},\mathfrak{n}_2] \subset \mathfrak{n}_2.$$

Let $\mathfrak{b} = \mathfrak{h} \oplus \mathfrak{n}_1$.

(a) Let W be a $\mathfrak{b}$-module such that $\mathfrak{n}_1 W = 0$ (and hence can be identified with an $\mathfrak{h}$-module), let $X = \text{coind}(W,\mathfrak{b} \uparrow \mathfrak{g})$, let γ be the projection of $U(\mathfrak{g})$ onto $U(\mathfrak{b})$ defined by the decomposition $U(\mathfrak{g}) = U(\mathfrak{b}) \oplus U(\mathfrak{g})\mathfrak{n}_2$. For every

$w \in W$, let $\omega(w)$ be the element $u \mapsto \gamma(u)w$ of X. Then ω is an injection of W into X, and $\omega(W) = X^{\mathfrak{n}_2}$. Let $W^\natural$ be the sub-$\mathfrak{g}$-module of X generated by $\omega(W)$.

(b) Henceforth we assume that W is simple and that, for every finite-dimensional representation ϱ of $\mathfrak{g}$, $\varrho|\mathfrak{n}_1$ and $\varrho|\mathfrak{n}_2$ are strictly triangularizable. If dim $W^\natural = +\infty$, then X does not contain a finite-dimensional simple sub-$\mathfrak{g}$-module. If dim $W^\natural < +\infty$, then $W^\natural$ is the only finite-dimensional simple sub-$\mathfrak{g}$-module of X.

(c) If V is a finite-dimensional simple $\mathfrak{g}$-module, then V is isomorphic to $(V^{\mathfrak{n}_2})^\natural$.

(d) Assume that $\mathfrak{h}$ is a splitting Cartan subalgebra of $\mathfrak{g}$ (supposed to be semi-simple), and that $\mathfrak{g} = \mathfrak{h} \oplus \mathfrak{n}_1 \oplus \mathfrak{n}_2$ is a corresponding triangular decomposition. If dim $W = 1$, then $W^\natural$ is a simple $\mathfrak{g}$-module, can be canonically embedded in the dual of a module $L(\lambda)$, and is isomorphic to a module $L(\lambda')$ (for opposite bases of the system of roots). ([129], [130])

7.8.18. (notation as in 7.0). Let W be the $\mathfrak{g}$-module coinduced by the one-dimensional null representation of $\mathfrak{n}_+$; the vector space W is the set of the $f \in U(\mathfrak{g})^*$ such that $f(nu) = 0$ for all $n \in \mathfrak{n}_+$ and all $u \in U(\mathfrak{g})$. Let Φ be as in 2.7.12. Then $\Psi = W \cap \Phi$ is a subalgebra of $U(\mathfrak{g})^*$. For all $\delta \in \hat{\mathfrak{g}}$, we have $\mathrm{mtp}(\delta,\Psi) = 1$, and $\Psi = \oplus_{\delta\in\hat{\mathfrak{g}}}\Psi_\delta$. If $\delta,\delta',\delta'' \in \hat{\mathfrak{g}}$ have $\lambda,\lambda',\lambda+\lambda'$ as largest weights, then $\Psi_\delta\Psi_{\delta'} = \Psi_{\delta''}$. The algebra Ψ is of finite type. [24], [54], [130]

7.8.19 (notation as in 7.0). Consider $U(\mathfrak{h})$ as a $Z(\mathfrak{g})$-algebra by virtue of the Harish-Chandra isomorphism. Let A be the algebra $U(\mathfrak{g}) \oplus_{Z(\mathfrak{g})} U(\mathfrak{h})$. The Weyl group W operates naturally on A in such a way that $U(\mathfrak{g})$ is the set of W-invariant elements of A. Let $l = \dim \mathfrak{h}$, let $n = \dim \mathfrak{n}_+$, and let P_l be the algebra of polynomials over k in l unknowns. The algebra A has a field of fractions isomorphic to the field of fractions of $A_n \oplus P_l$. ([24], [54]). Cf. problem 3.

7.8.20. We assume that k is algebraically closed and $\mathfrak{g}$ is semi-simple. Let $\mathfrak{h}$ be a Cartan subalgebra of $\mathfrak{g}$, $R = R(\mathfrak{g},\mathfrak{h})$, A the group of automorphisms of $\mathfrak{h}$ under which R is stable, C the commutant of $\mathfrak{h}$ in $U(\mathfrak{g})$, $\mathscr{A}$ the adjoint group of $\mathfrak{g}$, $\mathscr{H}$ the irreducible subgroup of $\mathscr{A}$ with Lie algebra $\mathrm{ad}_{\mathfrak{g}}\mathfrak{h}$.

(a) For all $\gamma \in A$, there exists $\gamma' \in \mathrm{Aut}(\mathfrak{g})$ such that $\gamma'|\mathfrak{h} = \gamma$ ([71], p. 127). If $\gamma'_1 \in \mathrm{Aut}(\mathfrak{g})$ is such that $\gamma'_1|\mathfrak{h} = \gamma$, then $\gamma'_1 \in \gamma'\mathscr{H}$ (BO, p. 342—343) Hence γ' and γ'_1 operate in the same way in C (cf. 2.4.16). It follows that A operates naturally in C. The action of A in C extends its action in $\mathfrak{h}$, and commutes with the Harish-Chandra isomorphism.

(b) Let $W = W(\mathfrak{g},\mathfrak{h})$. From (a), W operates naturally in C. Its action on $Z(\mathfrak{g})$ is trivial.

(c) The automorphism -1 of $\mathfrak{h}$ defines an automorphism ζ of C. For every $c \in C$, we set $c^0 = \zeta(c^\mathsf{T}) = \zeta(c)^\mathsf{T}$. We have $c^0 = c$ for all $c \in U(\mathfrak{h})$. If φ is the Harish-Chandra homomorphism defined by a basis of R, then $\varphi(c^0) = \varphi(c)$ for all $c \in C$. Consequently, $z^0 = z$ for all $z \in Z(\mathfrak{g})$. [100]

7.8.21 (notation as in 7.0). Let $\alpha \in B$, $X_\alpha \in \mathfrak{g}^\alpha$ and $X_{-\alpha} \in \mathfrak{g}^{-\alpha}$ be such that $[X_\alpha, X_{-\alpha}] = H_\alpha$, let ω be the automorphism of $S(\mathfrak{h})$ defined by the mapping $\mu \mapsto s_\alpha\mu + s_\alpha\delta - \delta$ of $\mathfrak{h}^*$ into $\mathfrak{h}^*$. Let $u \in U(\mathfrak{g})_0$ and $j \in \mathbf{N}$ be such that $[X_{-\alpha}[X_\alpha,u]] = j(j+1)u$. Then the image of u under the Harish-Chandra homomorphism has the form $H_\alpha(H_\alpha - 1) \cdots (H_\alpha - j + 1)h$, where $h \in S(\mathfrak{h})$ is invariant under ω. [100]

7.8.22. (notation as in 7.0, and k algebraically closed). Let V be a simple $\mathfrak{g}$-module. If $V^{\mathfrak{n}_+} \neq 0$, then V is isomorphic to a module $L(\lambda)$. (The space $V' = V^{\mathfrak{n}_+}$ is stable under $\mathfrak{h}$. For all $x \in \mathfrak{h}$, there exists a polynomial p of degree $\geqq 1$ such that $p(x)$ is the image of an element of $Z(\mathfrak{g})$ under the Harish-Chandra homomorphism (7.4.5). Consequently, there exist $\lambda_1, \ldots, \lambda_n \in k$ such that $(x - \lambda_1) \cdots (x - \lambda_n)$ operates in a scalar fashion in V'. We deduce that there exists $\lambda \in \mathfrak{h}^*$ such that $V'_\lambda \neq 0$, and apply 7.1.13.)

7.8.23 (notation as in 7.0). Let P be the projection of $U(\mathfrak{g})$ onto $U(\mathfrak{h})$ defined by the decomposition

$$U(\mathfrak{g}) = U(\mathfrak{h}) \oplus \mathfrak{n}_- U(\mathfrak{g}) \oplus U(\mathfrak{g})\mathfrak{n}_+.$$

Let $\mu, \xi \in \mathfrak{h}^*$. If $M(\mu)$ is simple, the bilinear form $(v,u) \mapsto P(vu)(\mu)$ on $U(\mathfrak{n}_+)_\xi \times U(\mathfrak{n}_-)_{-\xi}$ is non-degenerate. [19]

7.8.24 (notation as in 7.0). Let $\lambda \in \mathfrak{h}^*$. Since $M(\lambda)$ is a $\mathfrak{g}$-module, $\mathrm{End}_k M(\lambda)$ is automatically a $\mathfrak{g}$-module. Let A be the set of the $e \in \mathrm{End}_k M(\lambda)$ such that $\dim U(\mathfrak{g}) \cdot e < +\infty$. Let ϱ_λ be the representation of $\mathfrak{g}$ in $M(\lambda)$. Then $A = \varrho_\lambda(U(\mathfrak{g}))$. [24]

7.8.25. For k algebraically closed, every primitive ideal of $U(\mathfrak{g})$ is the annihilator of some $L(\lambda)$. [203]

CHAPTER 8

THE ENVELOPING ALGEBRA OF A SEMI-SIMPLE LIE ALGEBRA

8.1. The cone of nilpotent elements

8.1.1. LEMMA ($\mathfrak{g}$ semi-simple). *Let r be the rank of $\mathfrak{g}$, $\mathfrak{h}$ a splitting Cartan subalgebra of $\mathfrak{g}$, $R = R(\mathfrak{g},\mathfrak{h})$, B a basis for R, and h the element of $\mathfrak{h}$ such that $\alpha(h) = 2$ for all $\alpha \in B$. Let us set $h = \sum_{\alpha\in B} a_\alpha H_\alpha$. For every $\alpha \in B$, let b_α and c_α be scalars such that $b_\alpha c_\alpha = a_\alpha$, let $X_\alpha \in \mathfrak{g}^\alpha$ and $X_{-\alpha} \in \mathfrak{g}^{-\alpha}$ be such that $[X_\alpha, X_{-\alpha}] = H_\alpha$, and let*

$$x = \sum_{\alpha\in B} b_\alpha X_\alpha, \qquad y = \sum_{\alpha\in B} c_\alpha X_{-\alpha}, \qquad \mathfrak{t} = kx + kh + ky.$$

Then:

(i) $[h,x] = 2x, [h,y] = -2y, [x,y] = h$; *$x$ and y are regular.*

(ii) *Let us consider $\mathfrak{g}$ as a $\mathfrak{t}$ module by virtue of the adjoint representation. Let $\mathfrak{g} = \mathfrak{a}_1 \oplus \cdots \oplus \mathfrak{a}_n$ be a decomposition of $\mathfrak{g}$ into simple $\mathfrak{t}$-modules of dimension $\lambda_1 + 1, \ldots, \lambda_n + 1$, with $\lambda_1 \leqq \cdots \leqq \lambda_n$. Then $n = r$.*

(iii) *Let $f_1, \ldots, f_r$ be algebraically independent homogeneous generators of the algebra $S(\mathfrak{g}^*)^{\mathfrak{g}}$ (7.3.5 and 11.1.14), of degree $\nu_1, \cdots, \nu_r$, with $\nu_1 \leqq \cdots \leqq \nu_r$. Then $\nu_j = 1 + \frac{1}{2}\lambda_j$ for $j = 1, \ldots, r$.*

(iv) *The differentials of $f_1, \ldots, f_r$ are linearly independent at every point of $x + \mathfrak{g}^y$.*

We have

$$[h,x] = \sum_{\alpha\in B} b_\alpha \alpha(h) X_\alpha = 2x,$$

$$[h,y] = \sum_{\alpha\in B} c_\alpha(-\alpha)(h) X_{-\alpha} = -2y,$$

$$[x,y] = \sum_{\alpha,\beta\in B} b_\alpha c_\beta [X_\alpha, X_{-\beta}]$$

$$= \sum_{\alpha\in B} b_\alpha c_\alpha [X_\alpha, X_{-\alpha}] = \sum_{\alpha\in B} a_\alpha H_\alpha = h.$$

All eigenvalues of ad h are even. Hence, from 1.8.5,

$$n = \dim \mathfrak{g}^h = \dim \mathfrak{g}^x = \dim \mathfrak{g}^y.$$

Now $\mathfrak{g}^h = \mathfrak{h}$, which proves at the same time that $n = r$ and that x,y are regular.

From 1.8.5, we have $\mathfrak{g} = [x,\mathfrak{g}] \oplus \mathfrak{g}^y$, and $\mathfrak{g}^y$ has a basis $(y_1, \ldots, y_r)$ such that $[h,y_i] = -\lambda_i y_i$ for $i = 1, \ldots, r$. Let $\mathscr{A}$ be the adjoint group of $\mathfrak{g}$. For $a \in \mathscr{A}$ and $\xi_1, \ldots, \xi_r \in k$, we set

$$\psi(a,\xi_1, \ldots, \xi_r) = a(x + \xi_1 y_1 + \cdots + \xi_r y_r),$$

whence we have a mapping ψ of $\mathscr{A} \times k^r$ into $\mathfrak{g}$. Let $T_{\xi_1,\ldots,\xi_r}$ be the linear mapping tangent to ψ at $(1,\xi_1, \ldots, \xi_r)$. For all $z \in \mathrm{ad}_{\mathfrak{g}}\mathfrak{g}$ and all $(\eta_1, \ldots, \eta_r) \in k^r$, we have

$$(1) \qquad T_{\xi_1,\ldots,\xi_r}(z,\eta_1, \ldots, \eta_r) = z(x + \xi_1 y_1 + \cdots + \xi_r y_r) + \eta_1 y_1 + \cdots + \eta_r y_r$$

and hence the image of $T_{0,\ldots,0}$ is $[\mathfrak{g},x] + ky_1 + \cdots + ky_r = \mathfrak{g}$. It follows that $\psi(\mathscr{A} \times k^r) = \mathscr{A}(x + \mathfrak{g}^y)$ is dense in $\mathfrak{g}$. Consequently, the mapping $f \mapsto f|x + \mathfrak{g}^y$ of $S(\mathfrak{g}^*)^{\mathfrak{g}}$ into the algebra of polynomial functions on $x + \mathfrak{g}^y$ is injective.

Let f be a homogeneous element of $S(\mathfrak{g}^*)^{\mathfrak{g}}$ of degree ν. From the Euler identity, if $u \in \mathfrak{g}$ is considered as a vector tangent to $\mathfrak{g}$ at u, then

$$(2) \qquad \langle u,f \rangle = \nu f(u).$$

We take $u = x + \xi_1 y_1 + \cdots + \xi_r y_r$ and observe that

$$u = \tfrac{1}{2}(2x - \xi_1\lambda_1 y_1 - \cdots - \xi_r\lambda_r y_r) + (1 + \tfrac{1}{2}\lambda_1)\,\xi_1 y_1 + \cdots + (1 + \tfrac{1}{2}\lambda_r)\xi_r y_r$$
$$= T_{\xi_1,\ldots,\xi_r}(\tfrac{1}{2}\mathrm{ad}\, h, (1 + \tfrac{1}{2}\lambda_1)\,\xi_1, \ldots, (1 + \tfrac{1}{2}\lambda_r)\xi_r)$$

from (1). Hence

$$(3) \qquad \langle u,f \rangle = \langle (\tfrac{1}{2}\,\mathrm{ad}\, h, (1 + \tfrac{1}{2}\lambda_1)\,\xi_1, \ldots, (1 + \tfrac{1}{2}\lambda_r)\,\xi_r), f \circ \psi \rangle.$$

For $\zeta_1, \ldots, \zeta_r \in k$, we set

$$g(\zeta_1, \ldots, \zeta_r) = f(x + \zeta_1 y_1 + \cdots + \zeta_r y_r).$$

The function $f \circ \psi$ is defined on $\mathscr{A} \times k^r$, is independent of the first variable because f is $\mathfrak{g}$-invariant, and

$$(f \circ \psi)\,(1, (\zeta_1, \ldots, \zeta_r)) = g(\zeta_1, \ldots, \zeta_r).$$

Taking (2) and (3) into account, we obtain

$$\nu f(u) = (1 + \tfrac{1}{2}\lambda_1)\,\xi_1 \frac{\partial g}{\partial \xi_1}(\xi_1, \ldots, \xi_r) + \cdots + (1 + \tfrac{1}{2}\lambda_r)\,\xi_r \frac{\partial g}{\partial \xi_r}(\xi_1, \ldots, \xi_r)$$

or

$$\text{(4)} \qquad \nu g(\zeta_1, \ldots, \zeta_r) = \sum_{i=1}^{r} (1 + \tfrac{1}{2}\lambda_i)\,\zeta_i \frac{\partial g}{\partial \xi_i}(\zeta_1, \ldots, \zeta_r).$$

In particular, let $g_1, \ldots, g_r$ be the polynomial functions corresponding to $f_1, \ldots, f_r$. From (4), the polynomial g_j is a linear combination of the monomials $\zeta_1^{m_1} \cdots \zeta_r^{m_r}$ such that

$$\text{(5)} \qquad \sum_{i=1}^{r} (1 + \tfrac{1}{2}\lambda_i)\, m_i = \nu_j.$$

Let us assume that $\nu_{j_0} < 1 + \frac{1}{2}\lambda_{j_0}$ for some j_0. Then, for $j \leqq j_0$, we have $\nu_j < 1 + \frac{1}{2}\lambda_{j_0}$; from (5), g_j only depends on $\zeta_1, \ldots, \zeta_{j_0-1}$. Hence $g_1, \ldots, g_{j_0}$ are algebraically dependent. This contradicts the fact that the mapping $f \mapsto f|x + \mathfrak{g}^y$ ($f \in S(\mathfrak{g}^*)^{\mathfrak{g}}$) is injective. Hence

$$\text{(6)} \qquad 1 + \tfrac{1}{2}\lambda_j \leqq \nu_j \quad \text{for } j = 1, \ldots, r.$$

From this we deduce that

$$\begin{aligned} \nu_1 + \cdots + \nu_r &\geqq r + \tfrac{1}{2}(\lambda_1 + \cdots + \lambda_r) = \tfrac{1}{2} r + \tfrac{1}{2}((\lambda_1 + 1) + \cdots + (\lambda_r + 1)) \\ &= \tfrac{1}{2}(r + \dim \mathfrak{g}) \\ &= \nu_1 + \cdots + \nu_r. \end{aligned}$$

(7.3.8). Hence the inequalities (6) are all equalities.

Let $\mu_1 < \mu_2 < \cdots < \mu_p$ be the distinct elements of the set $\{\nu_1, \ldots, \nu_r\}$. For $s = 1, \ldots, p$, let C_s be the set of the j such that $\nu_j = \mu_s$. The equality (5) can now be written in the form

$$\nu_1 m_1 + \cdots + \nu_r m_r = \nu_j.$$

This proves that, for $j \in C_s$, we have

$$g_j(\zeta_1, \ldots, \zeta_r) = \sum_{j' \in C_s} \alpha_{jj'} \zeta_{j'} + g_j^*,$$

where the $\alpha_{jj'}$ are scalars and g_j^* only depends on the ζ_i such that $i \in \bigcup_{t \leqq s-1} C_t$. Let A_s be the matrix $(\alpha_{jj'})_{j,j' \in C_s}$. Then the Jacobian of $g_1, \ldots, g_r$ with respect to $\zeta_1, \ldots, \zeta_r$ is $\prod_{s=1}^{p} \det(A_s)$, and in particular is a constant. Since $g_1, \ldots, g_r$ are algebraically independent, this constant is non-zero. This proves (iv).

8.1.2. LEMMA (k algebraically closed). *Let $\mathfrak{a}$ be a Lie subalgebra of $\mathfrak{g}$, W a vector subspace of $\mathfrak{g}$ such that $\mathfrak{g} = \mathfrak{a} \oplus W$, and $[\mathfrak{a}, W] \subset W$. We assume that* $\mathrm{ad}_{\mathfrak{g}}\mathfrak{g}$ *and* $\mathrm{ad}_{\mathfrak{g}}\mathfrak{a}$ *are the Lie algebras of irreducible algebraic groups $\mathscr{G}$ and $\mathscr{A}$, respectively. Let Ω be a $\mathscr{G}$-orbit. Then the irreducible components of $\Omega \cap \mathfrak{a}$ are $\mathscr{A}$-orbits.*

Let Z be an irreducible component of $\Omega \cap \mathfrak{a}$, and let $x \in Z$. The space T tangent to Ω at x is $[\mathfrak{g},x] = [\mathfrak{a},x] + [W,x]$, whence $T \cap \mathfrak{a} = [\mathfrak{a},x]$. Let T' and T'' be the spaces tangent to Z and $\mathscr{A}x$ at x, respectively. Then

$$T' \subset T \cap \mathfrak{a} = [\mathfrak{a},x] = T'' \subset T',$$

hence

$$T'' = T'.$$

Consequently, $\mathscr{A}x$ is open in Z. Since two intersecting orbits are equal, Z is an $\mathscr{A}$-orbit.

8.1.3. THEOREM (k algebraically closed, $\mathfrak{g}$ semi-simple). *Let P (or N) be the set of the regular (or nilpotent) elements of $\mathfrak{g}$, $\mathscr{A}$ the adjoint group of $\mathfrak{g}$, J the set of invariant polynomial functions on $\mathfrak{g}$, J_+ the set of elements of J which have no constant term, r the rank of $\mathfrak{g}$, and $n = \dim \mathfrak{g}$. Then:*

(i) *The set of zeros of the ideal $J_+S(\mathfrak{g}^*)$ of $S(\mathfrak{g}^*)$ is N.*

(ii) *The algebraic cone N is irreducible and $(n-r)$-dimensional. The set of elements of $S(\mathfrak{g}^*)$ which are zero on N is $J_+S(\mathfrak{g}^*)$; it is a prime ideal.*

(iii) *The group $\mathscr{A}$ has only a finite number of orbits in N.*

(iv) *The set $N \cap P$ is open and dense in N, and consists of a single $\mathscr{A}$-orbit.*

Let $x \in \mathfrak{g}$. Then

$$\begin{aligned} x \in N &\Leftrightarrow \pi(x) \text{ nilpotent for all } \pi \in \mathfrak{g}^\wedge \\ &\Leftrightarrow \mathrm{tr}(\pi(x)^m) = 0 \text{ for all } \pi \in \mathfrak{g}^\wedge \text{ and all } m > 0 \\ &\Leftrightarrow f(x) = 0 \text{ for all } f \in J_+ \quad (7.3.5\ \text{(ii)}), \end{aligned}$$

which proves (i). From this it follows that N is an algebraic cone. From 1.10.21, this cone is irreducible. Then, taking 8.1.1 (iv) and 11.2.3 into account, $J_+S(\mathfrak{g}^*)$ is a prime ideal of $S(\mathfrak{g}^*)$; hence it is exactly the set of elements of $S(\mathfrak{g}^*)$ which are zero on N.

We may assume that $\mathfrak{g}$ is embedded in $\mathfrak{g}' = \mathfrak{gl}(V)$, where V is a finite-dimensional vector space. There exists a supplement W of $\mathfrak{g}$ in $\mathfrak{g}'$ such that $[\mathfrak{g}, W] \subset W$ (1.6.3). The irreducible algebraic group with Lie algebra $\mathrm{ad}_{\mathfrak{g}'}\mathfrak{g}'$

is the set of mappings $x \mapsto gxg^{-1}$, where $g \in \mathrm{GL}(V)$. From the theory of Jordan reductions, the set of nilpotent elements of $\mathfrak{g}'$ only contains a finite number of $\mathrm{GL}(V)$-orbits. Assertion (iii) follows from this and 8.1.2.

From 8.1.1 (i), $N \cap P$ is non-empty. From 1.11.5, $N \cap P$ is open in N, and hence dense in N since N is irreducible. The number of $\mathscr{A}$-orbits $\Omega_1, \ldots, \Omega_l$ contained in $N \cap P$ is finite from (iii), and they all have dimension $n - r$. Hence $\dim N = n - r$ and $\Omega_1, \ldots, \Omega_l$ are dense in N. Since $\Omega_1, \ldots, \Omega_l$ are open in their closure, these orbits pairwise intersect each other, whence $l = 1$.

References: [78], [112], [122].

8.2. The enveloping algebra as a module over its centre

8.2.1. LEMMA (k algebraically closed, $\mathfrak{g}$ semi-simple). *Let $J = S(\mathfrak{g}^*)^{\mathfrak{g}}$, let J_+ be the set of elements of J without constant term, let $h_1, \ldots, h_n$ be elements of $S(\mathfrak{g}^*)$ which are linearly independent modulo $J_+S(\mathfrak{g}^*)$, and let $\mathscr{A}$ be the adjoint group of $\mathfrak{g}$. Then there exists a non-empty open subset Ω of $\mathfrak{g}$ which has the following properties:*

(1) *if $x \in \Omega$, then the functions $h_1|\mathscr{A}x, \ldots, h_n|\mathscr{A}x$ are linearly independent;*

(2) *Ω contains a regular nilpotent element.*

Let N be the set of nilpotent elements of $\mathfrak{g}$, N' the set of regular elements of N, θ the mapping $(a,x) \mapsto a(x)$ of $\mathscr{A} \times \mathfrak{g}$ into $\mathfrak{g}$. Let $l_i = h_i \circ \theta$, which is a regular function on $\mathscr{A} \times \mathfrak{g}$. There exist:

(1) regular functions $\varphi_1, \ldots, \varphi_p$ on $\mathscr{A}$ which are linearly independent;

(2) polynomial functions $\psi_1, \ldots, \psi_q$ on $\mathfrak{g}$ which are linearly independent;

(3) scalars α_{rs}^i ($r = 1, \ldots, p$; $s = 1, \ldots, q$; $i = 1, \ldots, n$) such that

$$l_i(a,x) = \sum_{r=1}^{p} \sum_{s=1}^{q} \alpha_{rs}^i \varphi_r(a) \psi_s(x) \qquad (a \in \mathscr{A}, x \in \mathfrak{g}, i = 1, \ldots, n).$$

Let $x \in \mathfrak{g}$. Then

$$h_1 \,|\, \mathscr{A}x, \ldots, h_n \,|\, \mathscr{A}x \text{ are linearly independent}$$

$$\Leftrightarrow l_1 \,|\, \mathscr{A} \times \{x\}, \ldots, l_n \,|\, \mathscr{A} \times \{x\} \text{ are linearly independent}$$

$$\Leftrightarrow \text{the matrix } \left(\sum_{s=1}^{q} \alpha_{rs}^i \psi_s(x) \right)_{1 \leq i \leq n, 1 \leq r \leq p} \text{ has rank } n.$$

Let $D_1(x), \ldots, D_t(x)$ be the determinants of the submatrices with n rows and n columns of the above matrix. These are polynomial functions of x.

The above conditions are moreover equivalent to

one of the scalars $D_1(x), \ldots, D_l(x)$ is non-zero.

Let $\beta_1, \ldots, \beta_n \in k$ be such that

$$(\beta_1 h_1 + \cdots + \beta_n h_n) \mid N' = 0.$$

Then

$$(\beta_1 h_1 + \cdots + \beta_n h_n) \mid N = 0$$

(8.1.3 (iv)), hence

$$\beta_1 h_1 + \cdots + \beta_n h_n \in J_+ S(\mathfrak{g}^*)$$

(8.1.3 (ii)), hence $\beta_1 = \cdots = \beta_n = 0$. Thus $h_1|N', \ldots, h_n|N'$ are linearly independent. Now N' is an $\mathscr{A}$-orbit (8.1.3 (iv)). Hence there exist an $x_0 \in N'$ and an integer u such that $D_u(x_0) \neq 0$. Let Ω be the set of the $x \in \mathfrak{g}$ such that $D_u(x) \neq 0$. Then Ω has the properties of the lemma.

8.2.2. LEMMA ($\mathfrak{g}$ semi-simple). *Let J and J_+ be as in* 8.2.1. *Let H be a complement of $J_+ S(\mathfrak{g}^*)$ in $S(\mathfrak{g}^*)$ such that $H = \sum_{n \geq 0} H \cap S^n(\mathfrak{g}^*)$. The linear mapping f of $H \otimes J$ into $S(\mathfrak{g}^*)$ such that $f(h \otimes j) = hj$ for $h \in H$ and $j \in J$ is a vector space isomorphism.*

We have $S^0(\mathfrak{g}^*) \subset H \subset HJ$. Let us assume that

$$S^0(\mathfrak{g}^*) + \cdots + S^k(\mathfrak{g}^*) \subset HJ.$$

has been proved. Clearly,

$$S^{k+1}(\mathfrak{g}^*) \subset H + J_+(S^0(\mathfrak{g}^*) + \cdots + S^k(\mathfrak{g}^*)),$$

whence $S^{k+1}(\mathfrak{g}^*) \subset HJ$. Thus $HJ = S(\mathfrak{g}^*)$. Let $h_1, \ldots, h_n$ be elements of H which are linearly independent (over k). We shall prove that they are linearly independent over J, which will complete the proof. We can assume that k is algebraically closed. Let $\mathscr{A}$ and Ω be as in 8.2.1. Let $j_1, \ldots, j_n \in J$ be such that

$$h_1 j_1 + \cdots + h_n j_n = 0.$$

Let $x \in \Omega$. Then $h_1|\mathscr{A}x, \ldots, h_n|\mathscr{A}x$ are linearly independent. Now $j_1, \ldots, j_n$ take the constant values $j_1(x), \ldots, j_n(x)$ on $\mathscr{A}x$. Hence

$$(h_1 j_1(x) + \cdots + h_n j_n(x)) \mid \mathscr{A}x = 0,$$

whence

$$j_1(x) = \cdots = j_n(x) = 0.$$

Thus $j_1, \ldots, j_n$ are zero on Ω, whence $j_1 = \cdots = j_n = 0$.

8.2.3. PROPOSITION ($\mathfrak{g}$ semi-simple). $S(\mathfrak{g})$, *when considered as a module over* $Y(\mathfrak{g})$, *is free.*

With the notation of 8.2.2, a basis for H over k is a basis for $S(\mathfrak{g}^*)$ over J. It is then sufficient to use the Killing isomorphism of $S(\mathfrak{g}^*)$ onto $S(\mathfrak{g})$.

8.2.4. THEOREM ($\mathfrak{g}$ semi-simple). $U(\mathfrak{g})$, *when considered as a module over* $Z(\mathfrak{g})$, *is free. More precisely, let* $Y_+(\mathfrak{g})$ *be the set of elements of* $Y(\mathfrak{g})$ *without constant term,* H *a graded complement of* $Y_+(\mathfrak{g})S(\mathfrak{g})$ *in* $S(\mathfrak{g})$, β *the canonical mapping of* $S(\mathfrak{g})$ *into* $U(\mathfrak{g})$, $K = \beta(H)$, *and* g *the linear mapping of* $K \otimes Z(\mathfrak{g})$ *into* $U(\mathfrak{g})$ *such that* $g(l \otimes z) = lz$ *for* $l \in K$, $z \in Z(\mathfrak{g})$. *Then* g *is a vector space isomorphism.*

The filtration of $U(\mathfrak{g})$ induces filtrations on $Z(\mathfrak{g})$ and K, and we have canonical isomorphisms $\mathrm{gr}(Z(\mathfrak{g})) \to Y(\mathfrak{g})$ and $\mathrm{gr}(K) \to H$. The mapping g is compatible with these filtrations, hence defines a mapping $\mathrm{gr}(g) : \mathrm{gr}(K) \otimes \mathrm{gr}(Z(\mathfrak{g})) \to \mathrm{gr}\, U(\mathfrak{g})$. The latter can be identified with the mapping f of $H \otimes Y(\mathfrak{g})$ into $S(\mathfrak{g})$ such that $f(h \otimes y) = hy$ for $h \in H$ and $y \in Y(\mathfrak{g})$. From 8.2.2, $\mathrm{gr}(g)$ is bijective, hence g is bijective.

8.2.5. With the notation of 8.2.4, we see at the same time that, for all $n \geqq 0$, $U_n(\mathfrak{g})$ is the image under g of

$$\sum_{p+q=n} (K \cap U_p(\mathfrak{g})) \otimes (Z(\mathfrak{g}) \cap U_q(\mathfrak{g})).$$

Reference: [78].

8.3. The adjoint representation in the enveloping algebra

8.3.1. LEMMA ($\mathfrak{g}$ semi-simple). *Let* $\mathfrak{g} = \mathfrak{h} \oplus \mathfrak{n}_+ \oplus \mathfrak{n}_-$ *be a triangular decomposition of* $\mathfrak{g}$, *and let* $x \in \mathfrak{h}$, $y \in \mathfrak{n}_+$.

(i) *If* π *is a finite-dimensional representation of* $\mathfrak{g}$, *then* $\pi(x)$ *and* $\pi(x+y)$ *have the same characteristic polynomial.*

(ii) *If* $f \in S(\mathfrak{g}^*)^{\mathfrak{g}}$, *then* $f(x) = f(x+y)$.

Taking 7.1.2 (i) and 7.2.1 (i) into account, the proof of (i) is practically the same as that of 1.10.17. Assertion (ii) follows from (i) and 7.3.5 (ii).

8.3.2. LEMMA (k algebraically closed, $\mathfrak{g}$ semi-simple). *Let* x *be a generic element of* $\mathfrak{g}$, *and* $\mathscr{A}$ *the adjoint group of* $\mathfrak{g}$. *Then* $\mathscr{A}x$ *is the set of the* $y \in \mathfrak{g}$ *such that* $f(x) = f(y)$ *for all* $f \in S(\mathfrak{g}^*)^{\mathfrak{g}}$.

If $a \in \mathscr{A}$ and $f \in S(\mathfrak{g}^*)^{\mathfrak{g}}$, then $f(ax) = f(x)$. Let $y \in \mathfrak{g}$ be such that $f(x) = f(y)$ for all $f \in S(\mathfrak{g}^*)^{\mathfrak{g}}$. Then ad x and ad y have the same characteristic polynomial, hence y is generic. We prove that $y \in \mathscr{A}x$. We can assume that x and y are in one and the same Cartan subalgebra (1.9.11). Let $W = W(\mathfrak{g},\mathfrak{h})$. If $g \in S(\mathfrak{h}^*)^W$, then $g(x) = g(y)$ (7.3.5 (i)). If $Wx \cap Wy = \emptyset$, there exists a function in $S(\mathfrak{h}^*)^W$ which takes the value 1 on Wx and the value 0 on Wy, which is contradictory. Hence $y \in Wx \subset \mathscr{A}x$ (1.10.19).

8.3.3. Lemma (k algebraically closed, $\mathfrak{g}$ semi-simple). *Let N be the set of nilpotent elements of $\mathfrak{g}$, $\mathscr{A}$ the adjoint group of $\mathfrak{g}$, x a generic element of $\mathfrak{g}$, and $k^* = k - 0$. The closure of $\mathscr{A}(k^*x)$ contains N.*

Let $\mathfrak{h}$ be the centralizer of x in $\mathfrak{g}$; it is a Cartan subalgebra of $\mathfrak{g}$. Let $\mathfrak{g} = \mathfrak{h} \oplus \mathfrak{n}_+ \oplus \mathfrak{n}_-$ be a triangular decomposition of $\mathfrak{g}$. There exists in $\mathfrak{n}_+$ a regular element y (8.1.1). For all $\lambda \in k^*$ we have $\lambda x + y \in \mathscr{A}(\lambda x)$ (8.3.1 and 8.3.2). Hence $\mathscr{A}(k^*x)$ contains $k^*x + y$; consequently, its closure contains y and hence N (8.1.3).

8.3.4. Lemma (k algebraically closed, $\mathfrak{g}$ semi-simple). *Let H be as in* 8.2.2. *Let x be a generic element of $\mathfrak{g}$, and $\mathscr{A}$ the adjoint group of $\mathfrak{g}$.*

(i) *The orbit $\mathscr{A}x$ is closed in $\mathfrak{g}$. Let A be the algebra of regular functions on $\mathscr{A}x$, so that $A = S(\mathfrak{g}^*)|\mathscr{A}x$.*

(ii) *The mapping $h \mapsto h|\mathscr{A}x$ is an isomorphism of the vector space H onto A.*

Assertion (i) follows from 8.3.2. If $f \in S(\mathfrak{g}^*)$, there exist $f_1, \ldots, f_n \in H$ and $j_1, \ldots, j_n \in S(\mathfrak{g}^*)^{\mathfrak{g}}$ such that

$$f = j_1 f_1 + \cdots + j_n f_n$$

(8.2.2). Since the $j_i|\mathscr{A}x$ are constants, $f|\mathscr{A}x$ is a linear combination of $f_1|\mathscr{A}x, \ldots, f_n|\mathscr{A}x$, so that the mapping considered in (ii) is surjective. Let $h_1, \ldots, h_p$ be homogeneous and linearly independent elements of H. We prove that $h_1|\mathscr{A}x, \ldots, h_p|\mathscr{A}x$ are linearly independent, which will complete the proof. Let Ω be as in 8.2.1. From 8.3.3, there exist $a \in \mathscr{A}$ and $\lambda \in k^*$ such that $a(\lambda x) \in \Omega$. Hence $h_1|\lambda\mathscr{A}x, \ldots, h_p|\lambda\mathscr{A}x$ are linearly independent. Since $h_1, \ldots, h_p$ are homogeneous, this implies our assertion.

8.3.5. Lemma (k algebraically closed, $\mathfrak{g}$ semi-simple). *Let $\mathscr{A}$ be the adjoint group of $\mathfrak{g}$, $\mathscr{B}$ an algebraic subgroup of $\mathscr{A}$, $R(\mathscr{A})$ the set of regular functions on $\mathscr{A}$ with values in k, $R(\mathscr{A},\mathscr{B})$ the set of elements of $R(\mathscr{A})$ which are right invariant under $\mathscr{B}$. We consider $R(\mathscr{A})$ as an $\mathscr{A}$-module by setting*

$(ar)(a') = r(a^{-1}a')$ *for* $a,a' \in \mathscr{A}$ *and* $r \in R(\mathscr{A})$. *Let* Ξ *be the set of classes of finite-dimensional simple rational representations of* $\mathscr{A}$. *For every* $\xi \in \Xi$ *and every* $\mathscr{A}$*-module* M, *let* M_ξ *be the isotypic component of type* ξ *of* M. *Then:*

(i) $R(\mathscr{A}) = \bigoplus_{\xi \in \Xi} R(\mathscr{A})_\xi$.

(ii) $R(\mathscr{A},\mathscr{B})$ *is a sub-*$\mathscr{A}$*-module of* $R(\mathscr{A})$.

(iii) *Let* V *be a finite-dimensional vector space,* ξ *a simple rational representation of* $\mathscr{A}$ *in* V, V^* *the dual* $\mathscr{A}$*-module, and* $V^{*\mathscr{B}}$ *the set of* $\mathscr{B}$*-invariant elements of* V^*. *Then* $\dim R(\mathscr{A},\mathscr{B}) = (\dim V)(\dim V^{*\mathscr{B}})$.

For all $r \in R(\mathscr{A})$, the set $\mathscr{A} \cdot r$ has finite rank. Now every finite-dimensional rational representation of $\mathscr{A}$ is semi-simple (from the corresponding property of $\mathfrak{g}$), whence (i). Assertion (ii) is obvious. Let V, ξ and $V^{*\mathscr{B}}$ be as in (iii), and ξ^* the representation in V^* contragradient to ξ.

For $v \in V$, $v' \in V^*$, let $\theta(v,v')$ be the function $a \mapsto \langle \xi(a^{-1})v,v' \rangle$ on $\mathscr{A}$. It is clear that $\theta(v,v') \in R(\mathscr{A})$. If $a,a' \in \mathscr{A}$, then

$$\theta(\xi(a)v,v')\,(a') = \langle \xi(a'^{-1})\xi(a)v,v' \rangle = \langle \xi(a'^{-1}a)v,v' \rangle$$

$$= \theta(v,v')(a^{-1}a') = (a \cdot \theta(v,v'))\,(a')$$

hence $\theta(\xi(a)v,v') = a \cdot \theta(v,v')$. This proves that $v \mapsto \theta(v,v')$ is an $\mathscr{A}$-homomorphism $h_{v'}$ of V into $R(\mathscr{A})$; its image is hence contained in $R(\mathscr{A})_\xi$. If $v \neq 0$ and $v' \neq 0$, then $\theta(v,v') \neq 0$ because V is simple. Hence, for $v' \neq 0$, $h_{v'}$ is an injective $\mathscr{A}$-homomorphism of V into $R(\mathscr{A})_\xi$, and the linear mapping $v' \to h_{v'}$ is injective. Consequently, if $(v_1. \ldots, v_n)$ is a basis for V and $(v'_1, \ldots, v'_n)$ is a basis for V^*, the $\theta(v_i,v'_j)$ are linearly independent elements of $R(\mathscr{A})_\xi$.

Let us assume that $v'_1, \ldots, v'_n$ have been chosen in such a way that $(v'_1, \ldots, v'_p)$ is a basis for $V^{*\mathscr{B}}$. For $i = 1, \ldots, n$ and $j = 1, \ldots, p$, we have $\theta(v_i,v'_j) \in R(\mathscr{A},\mathscr{B})_\xi$. Let l be an $\mathscr{A}$-homomorphism of V in $R(\mathscr{A},\mathscr{B})_\xi$. The mapping $v \mapsto l(v)(e)$ of V into k is an element v' of V^*. For $a \in \mathscr{A}$ and $v \in V$, we have

$$l(v)\,(a) = (a^{-1} \cdot l(v))\,(e) = l(\xi(a^{-1})v)\,(e)$$

$$= \langle \xi(a^{-1})v,v' \rangle = \theta(v,v')(a)$$

and hence $l(v) = \theta(v,v')$. On the other hand, for $b \in \mathscr{B}$ and $v \in V$, we have

$$\langle v,\xi^*(b)v' \rangle = \langle \xi(b^{-1})v,v' \rangle = l(\xi(b^{-1})v)\,(e) = (b^{-1} \cdot l(v))\,(e)$$

$$= l(v)(b) = l(v)\,(e) = \langle v,v' \rangle$$

and hence $v' \in V^{*\mathscr{B}}$. This proves that $R(\mathscr{A},\mathscr{B})_\xi = \sum_{v'\in V^{*\mathscr{B}}} h_{v'}(V)$. Hence the $\theta(v_i,v'_j)$, for $i = 1, \ldots, n$ and $j = 1, \ldots, p$, constitute a basis for $R(\mathscr{A},\mathscr{B})_\xi$.

8.3.6. Proposition ($\mathfrak{g}$ semi-simple) *Let $Y_+(\mathfrak{g})$ be the set of elements of $Y(\mathfrak{g})$ without constant term, and H a complement of $Y_+(\mathfrak{g}) \cdot S(\mathfrak{g})$ in $S(\mathfrak{g})$ which is stable under the adjoint representation and is such that $H = \sum_{n\geq 0} H \cap S^n(\mathfrak{g})$. Then:*

(i) $H = \oplus_{\xi\in\mathfrak{g}^\wedge} H_\xi$.

(ii) *Let $\mathfrak{h}$ be a splitting Cartan subalgebra of $\mathfrak{g}$. For all $\xi \in \mathfrak{g}^\wedge$, let l_ξ be the multiplicity of the weight 0 in ξ. Then* $\operatorname{mtp}(\xi,H) = l_\xi$.

Assertion (i) is obvious. To prove (ii), we may assume that k is algebraically closed. We can identify $S(\mathfrak{g})$ with $S(\mathfrak{g}^*)$ by means of the Killing isomorphism. Let x be a generic element of $\mathfrak{h}$. Let $\mathscr{A}$ and A be as in 8.3.4.

Let $\mathscr{H}$ be the stabilizer of x in $\mathscr{A}$. This is an algebraic subgroup of $\mathscr{A}$ which has $\operatorname{ad}_\mathfrak{g}\mathfrak{h}$ as its Lie algebra. Let $\mathscr{K}$ be the smallest algebraic subgroup of $\mathscr{A}$ whose Lie algebra $\mathfrak{k}$ contains $\operatorname{ad}_\mathfrak{g} x$; it is connected, and $\mathfrak{k} \subset \operatorname{ad}_\mathfrak{g}\mathfrak{h}$, hence $\mathscr{K}$ is a torus. Let $h \in \mathscr{H}$. The centralizer of h in $\mathscr{A}$ is an algebraic subgroup $\mathscr{B}$ of $\mathscr{A}$ whose Lie algebra contains $\operatorname{ad}_\mathfrak{g} x$; consequently, $\mathscr{B} \supset \mathscr{K}$. Thus $\mathscr{H}$ centralizes $\mathscr{K}$. Clearly, then, $\mathscr{H}$ is the centralizer of $\mathscr{K}$ in $\mathscr{A}$ and hence is connected (BO, p. 271). In short, $\mathscr{H}$ is the connected subgroup of $\mathscr{A}$ with Lie algebra $\operatorname{ad}_\mathfrak{g}\mathfrak{h}$.

The mapping $\theta : a \mapsto ax$ of $\mathscr{A}$ into $\mathfrak{g}$ defines a bijection θ' of $\mathscr{A}|\mathscr{H}$ onto $\mathscr{A}x$. This bijection is regular and $\mathscr{A}x$ does not have a singular point, hence θ' is biregular. Consequently, the mapping $f \mapsto f \circ \theta$ is an isomorphism of the vector space A onto the vector space $R(\mathscr{A},\mathscr{H})$ of regular functions on $\mathscr{A}$ which are right invariant under $\mathscr{H}$.

Let us consider $S(\mathfrak{g}) = S(\mathfrak{g}^*)$ (or $S(\mathfrak{g})|\mathscr{A}x$) as an $\mathscr{A}$-module by setting $(af)(y) = f(a^{-1}y)$ for $a \in \mathscr{A}$, $f \in S(\mathfrak{g})$, $y \in \mathfrak{g}$ (or $y \in \mathscr{A}x$). Let us consider H as an $\mathscr{A}$-submodule of $S(\mathfrak{g})$. Let us consider $R(\mathscr{A},\mathscr{H})$ as an $\mathscr{A}$-module by setting $(ar)(b) = r(a^{-1}b)$ for $a,b \in \mathscr{A}$, $r \in R(\mathscr{A},\mathscr{H})$. Then the isomorphisms $H \to A \to R(\mathscr{A},\mathscr{H})$ (defined in lemma 8.3.4 and above) are obviously $\mathscr{A}$-module isomorphisms. Let $\xi \in \mathfrak{g}^\wedge$, and let V be the space of ξ. Since $\mathscr{H}$ is connected with Lie algebra $\operatorname{ad}_\mathfrak{g}\mathfrak{h}$, the set of $\mathscr{H}$-invariant elements of V^* is equal to the set of the elements with weight 0 in V^*; it thus has dimension l_ξ (7.2.8). Given this, (ii) follows from 8.3.5 (iii).

8.3.7. Let V be a $\mathfrak{g}$-module. If $h \in \operatorname{Hom}_\mathfrak{g}(V,S(\mathfrak{g}))$ and $y \in Y(\mathfrak{g})$, then the mapping $v \mapsto yh(v)$ also belongs to $\operatorname{Hom}_\mathfrak{g}(V,S(\mathfrak{g}))$; let us denote it by yh. Thus $\operatorname{Hom}_\mathfrak{g}(V,S(\mathfrak{g}))$ *becomes a* $Y(\mathfrak{g})$*-module.*

Let $n \in \mathbf{N}$. Then h is said to be *homogeneous of degree n* if $h(V) \subset S^n(\mathfrak{g})$.

8.3.8. THEOREM ($\mathfrak{g}$ semi-simple). *Let V be a finite-dimensional vector space, ξ a simple representation of $\mathfrak{g}$ in V, $\mathfrak{h}$ a split Cartan subalgebra of $\mathfrak{g}$, and l_ξ the multiplicity of the weight 0 in ξ. Then the $Y(\mathfrak{g})$-module* $\mathrm{Hom}_{\mathfrak{g}}(V,S(\mathfrak{g}))$ *has a basis formed of l_ξ homogeneous elements.*

Let us choose H as in 8.3.6. Let $(y_\lambda)_{\lambda\in\Lambda}$ be a basis for the vector space $Y(\mathfrak{g})$. From 8.2.2. and 8.3.6, we have

$$S(\mathfrak{g}) = \bigoplus_{\lambda\in\Lambda} y_\lambda H = \bigoplus_{\lambda\in\Lambda,\xi\in\mathfrak{g}^\wedge} y_\lambda H_\xi,$$

and hence

$$S(\mathfrak{g})_\xi = \bigoplus_{\lambda\in\Lambda} y_\lambda H_\xi .$$

We set $l_\xi = l$. From 8.3.6, we have $H_\xi = H_\xi^1 \oplus \cdots \oplus H_\xi^l$, where the H_ξ^i are homogeneous sub-$\mathfrak{g}$-modules of $S(\mathfrak{g})$. Then

$$\begin{aligned}(1)\qquad \mathrm{Hom}_{\mathfrak{g}}(V,S(\mathfrak{g})) &= \mathrm{Hom}_{\mathfrak{g}}(V,S(\mathfrak{g})_\xi)\\ &= \mathrm{Hom}_{\mathfrak{g}}\left(V, \bigoplus_{\lambda\in\Lambda} \bigoplus_{i-1}^{l} y_\lambda H_\xi^i\right).\end{aligned}$$

For $i = 1, \ldots, l$, let h_i be a $\mathfrak{g}$-homomorphism of V onto H^i. From (1), the vector space $\mathrm{Hom}_{\mathfrak{g}}(V,S(\mathfrak{g}))$ has the basis $(y_\lambda h_i)_{\lambda\in\Lambda,1<i<l}$, and hence (taking 8.2.2 into account) the $Y(\mathfrak{g})$-module $\mathrm{Hom}_{\mathfrak{g}}(V,S(\mathfrak{g}))$ has the basis $(h_i)_{1<i<l}$.

8.3.9. PROPOSITION ($\mathfrak{g}$ semi-simple). *We retain the notation of* 8.3.6. *Let β be the canonical mapping of $S(\mathfrak{g})$ into $U(\mathfrak{g})$, and let $K = \beta(H)$. Then:*

(i) $K = \bigoplus_{\xi\in\mathfrak{g}^\wedge} K_\xi$.

(ii) *Let $\mathfrak{h}$ be a splitting Cartan subalgebra of $\mathfrak{g}$. For all $\xi \in \mathfrak{g}^\wedge$, we have* $\mathrm{mtp}\,(\xi,K) = l_\xi$.

This follows from 8.3.6 since β is a $\mathfrak{g}$-module isomorphism.

8.3.10. Let V be a $\mathfrak{g}$-module, $h \in \mathrm{Hom}_{\mathfrak{g}}(V,U(\mathfrak{g}))$, and $z \in Z(\mathfrak{g})$. Then the mapping $v \mapsto zh(v)$ belongs to $\mathrm{Hom}_{\mathfrak{g}}(V,U(\mathfrak{g}))$; we shall denote it by zh. Thus $\mathrm{Hom}_{\mathfrak{g}}(V,U(\mathfrak{g}))$ *becomes a $Z(\mathfrak{g})$-module.*

Let $n \in \mathbf{N}$. Then h is said to be *homogeneous of degree n* if $h(V) \subset U^n(\mathfrak{g})$, in other words, if h is the composition of the canonical mapping of $S(\mathfrak{g})$ into $U(\mathfrak{g})$ and a homogeneous $\mathfrak{g}$-homomorphism of degree n of V into $S(\mathfrak{g})$.

8.3.11. THEOREM ($\mathfrak{g}$-semi-simple). *Let V be a finite-dimensional vector space, ξ a simple representation of $\mathfrak{g}$ in V, $\mathfrak{h}$ a splitting Cartan subalgebra of $\mathfrak{g}$, and*

l_ξ the multiplicity of the weight 0 *in* ξ. *Then the* $Z(\mathfrak{g})$*-module* $\mathrm{Hom}_{\mathfrak{g}}(V,U(\mathfrak{g}))$ *has a basis consisting of* l_ξ *homogeneous elements.*

This follows from 8.3.8.

8.3.12. We resume the notation of 8.3.8 and set $l_\xi = l$. For all $n \in \mathbf{N}$, let $Y^n(\mathfrak{g}) = Y(\mathfrak{g}) \cap S^n(\mathfrak{g})$. Let $(h_1, \ldots, h_l)$ be a basis for $\mathrm{Hom}_{\mathfrak{g}}(V,S(\mathfrak{g}))$ over $Y(\mathfrak{g})$ consisting of homogeneous elements; let p_i be the degree of h_i. Let us set $d_n = \dim Y^n(\mathfrak{g})$, so that the Poincaré series of the graded vector space $Y(\mathfrak{g})$ is $\sum_{n \geqq 0} d_n T^n$ (with $d_0 = 1$). Then the Poincaré series of the graded vector space $\mathrm{Hom}_{\mathfrak{g}}(V,S(\mathfrak{g}))$ is

$$(T^{p_1} + \cdots + T^{p_l}) \sum_{n \geqq 0} d_n T^n.$$

This series is independent of the choice of the basis $(h_1, \ldots, h_l)$, hence *the family of integers* $(p_1, \ldots, p_l)$ *is independent, up to the ordering, of the choice of this basis* (cf. 8.5.2).

Similarly, if $(h_1', \ldots, h_l')$ is a basis for $\mathrm{Hom}_{\mathfrak{g}}(V,U(\mathfrak{g}))$ over $Z(\mathfrak{g})$ consisting of homogeneous elements, the degrees of $h_1', \ldots, h_l'$ are independent of the choice of the basis (and, moreover, are equal to the p_i given above).

Reference: [78].

8.4. The annihilators of Verma modules

8.4.1. LEMMA ($\mathfrak{g}$ semi-simple). *Let* $Y_+(\mathfrak{g})$ *and* H *be as in* 8.3.6. *Let* $\mathfrak{g} = \mathfrak{h} \oplus \mathfrak{n}_+ \oplus \mathfrak{n}_-$ *be a triangular decomposition of* $\mathfrak{g}$, *and let* $\mathfrak{b}_+ = \mathfrak{h} \oplus \mathfrak{n}_+$. *We can consider* H *as a* $\mathfrak{g}$*-module, and hence as an* $\mathfrak{n}_-$*-module. Then:*

(i) $S(\mathfrak{g}) = S(\mathfrak{g})\mathfrak{b}_+ \oplus S(\mathfrak{n}_-)$.

(ii) *Let* P *be the projection of* $S(\mathfrak{g})$ *onto* $S(\mathfrak{n}_-)$ *defined by* (i). *The restriction of* P *to* $H^{\mathfrak{n}_-}$ *is injective.*

Assertion (i) is obvious. To prove (ii), we can assume that k is algebraically closed. Let $\mathscr{A}$ be the adjoint group of $\mathfrak{g}$, $\mathscr{N}$ the irreducible algebraic subgroup of $\mathscr{A}$ with Lie algebra $\mathrm{ad}_{\mathfrak{g}}\mathfrak{n}_-$, and N the set of nilpotent elements of $\mathfrak{g}$. Let $f \in H^{\mathfrak{n}_-} \cap S(\mathfrak{g})\mathfrak{b}_+$. Then f, which can be identified with a polynomial function on $\mathfrak{g}$ by virtue of the Killing isomorphism, is $\mathscr{N}$-invariant and is zero on the orthogonal subspace of $\mathfrak{b}_+$, i.e., $\mathfrak{n}_+$; hence f is zero on $\mathscr{N}\mathfrak{n}_+$ and consequently on N (1.10.21), whence $f \in Y_+(\mathfrak{g})S(\mathfrak{g})$ (8.1.3). Since $f \in H$, we have $f = 0$.

8.4.2. LEMMA ($\mathfrak{g}$ semi-simple). *We use the notation of* 8.4.1. *Let* β *be the canonical mapping of* $S(\mathfrak{g})$ *into* $U(\mathfrak{g})$, $K = \beta(H)$, $\lambda \in \mathfrak{h}^*$, *and* J *the left ideal of* $U(\mathfrak{g})$ *generated by* $\mathfrak{n}_+$ *and the* $h - \lambda(h)$, *where* $h \in \mathfrak{h}$. *Then:*

(i) $U(\mathfrak{g}) = J \oplus U(\mathfrak{n}_-)$.

(ii) *Let* Q *be the projection of* $U(\mathfrak{g})$ *onto* $U(\mathfrak{n}_-)$ *defined by* (i). *If* $u \in K^{\mathfrak{n}_-}$, $p \in \mathbf{N}$ *and* $u \notin U_p(\mathfrak{g})$, *then* $Q(u) \notin U_p(\mathfrak{n}_-)$.

Assertion (i) follows from e.g. 5.1.9 and 7.1.5.

Let $u \in K^{\mathfrak{n}_-}$ be such that $u \notin U_p(\mathfrak{g})$. Then $u = \beta(x)$, where x is an element of $H^{\mathfrak{n}_-}$ whose degree m is $>p$. Let x_m be the homogeneous component of degree m of x. Then $x_m \in H^{\mathfrak{n}_-}$ and $x_m \neq 0$, hence $P(x_m) \neq 0$ (8.4.1), and consequently $P(x)$ is of degree m. Thus $\beta(P(x)) \notin U_{m-1}(\mathfrak{n}_-)$. On the other hand, $x - P(x)$ is an element of $S(\mathfrak{g})\mathfrak{b}_+$ of degree $\leqq m$, and hence is a linear combination of elements of the form $x_1 x_2 \cdots x_q b$, where $q < m$, $x_1, \ldots, x_q \in \mathfrak{g}$, $b \in \mathfrak{h}$ or $b \in \mathfrak{n}_+$; it follows that $\beta(x - P(x)) \in J + U_{m-1}(\mathfrak{g})$, Then

$$Q(u) - \beta(P(x)) = Q(u) - u + \beta(x - P(x)) \in (J + J + U_{m-1}(\mathfrak{g})) \cap U(\mathfrak{n}_-)$$

Finally, it is easily seen that $Q(U_{m-1}(\mathfrak{g})) \subset U_{m-1}(\mathfrak{n}_-)$. Hence

$$Q(u) - \beta(P(x)) \in U_{m-1}(\mathfrak{n}_-)$$

and consequently $Q(u) \notin U_{m-1}(\mathfrak{n}_-)$.

8.4.3. THEOREM ($\mathfrak{g}$ semi-simple). *Let* $\mathfrak{h}$ *be a splitting Cartan subalgebra of* $\mathfrak{g}$, B *a basis for* $R(\mathfrak{g},\mathfrak{h})$, *and* $\lambda \in \mathfrak{h}^*$, *whence we have Verma module* $M(\lambda)$ *with central character* χ_λ. *Then the annihilator of* $M(\lambda)$ *is the ideal* $U(\mathfrak{g}) \operatorname{Ker} \chi_\lambda$.

We shall retain the notation of 8.4.1 and 8.4.2. Let $H = \oplus_{i \in I} H_i$ be a decomposition of H into finite-dimensional simple sub-$\mathfrak{g}$-modules such that the elements of H_i are all homogeneous with a certain degree n_i. Let $K_i = \beta(H_i)$. We choose in K_i a non-zero and $\mathfrak{n}_-$-invariant element t_i (such an element is well-defined up to a scalar). Let $(z_\gamma)_{\gamma \in \Gamma}$ be a basis for the vector space $Z(\mathfrak{g})$.

Let A be the annihilator of $M(\lambda)$ in $U(\mathfrak{g})$, and $A' = U(\mathfrak{g}) \operatorname{Ker} \chi_\lambda$. Clearly, we have $A' \subset A$. Let us assume that $A' \neq A$. The vector subspaces A and A' of $U(\mathfrak{g})$ are $\mathfrak{g}$-stable. There therefore exists a finite-dimensional simple sub-$\mathfrak{g}$-module C of A such that $C \cap A' = 0$. Let u be a non-zero element of $C^{\mathfrak{n}_-}$. There exist $l_\gamma \in K$ such that $u = \sum_{\gamma \in \Gamma} l_\gamma z_\gamma$ (8.2.4). For all $n \in \mathfrak{n}_-$, we have $0 = [n,u] = \sum_{\gamma \in \Gamma} [n,l_\gamma] z_\gamma$, and $[n,l_\gamma] \in K$, hence $[n,l_\gamma] = 0$ for all γ (8.2.4). Thus $l_\gamma \in K^{\mathfrak{n}_-}$ for all γ. Consequently, u can be written in the form $\sum_{i \in I} t_i z'_i$, where $z'_i \in Z(\mathfrak{g})$ for all i.

Let m be the canonical generator of $M(\lambda)$. Then

$$0 = um = \sum_{i \in I} t_i z_i' m = \sum_{i \in I} \chi_\lambda(z_i') t_i m,$$

hence $\sum_{i \in I} \chi_\lambda(z_i') t_i$ is both in $K^{\mathfrak{n}_-}$ and in the left ideal of $U(\mathfrak{g})$ generated by $\mathfrak{n}_+$ and the $h - (\lambda - \delta)(h)$, where $h \in \mathfrak{h}$ (δ denotes the semi-sum of the positive roots). From 8.4.2 (ii), we have $\sum_{i \in I} \chi_\lambda(z_i') t_i = 0$, hence $\chi_\lambda(z_i') = 0$ for all i. Consequently, $u \in A'$, which is a contradiction.

8.4.4. Theorem ($\mathfrak{g}$ semi-simple). *Let $\mathfrak{h}$ be a splitting Cartan subalgebra of $\mathfrak{g}$, $W = W(\mathfrak{g},\mathfrak{h})$ and B a basis for $R(\mathfrak{g},\mathfrak{h})$. For every $\lambda \in \mathfrak{h}^*$, let J_λ be the annihilator of the Verma module $M(\lambda)$. Then:*

(i) *The ideals J_λ are primitive and completely prime.*

(ii) *Let $\lambda, \lambda' \in \mathfrak{h}^*$. Then $J_\lambda = J_{\lambda'}$ if and only if $\lambda' \in W\lambda$.*

(iii) *Let D be a dense subset of $\mathfrak{h}^*$. Then $\bigcap_{\lambda \in D} J_\lambda = 0$.*

(iv) *If k is algebraically closed, the J_λ are the minimal primitive ideals of $U(\mathfrak{g})$.*

Assertion (ii) follows from 8.4.3 and 7.4.7.

Let $\lambda \in \mathfrak{h}^*$. There exists $\mu \in W\lambda$ such that $M(\mu)$ is simple (7.6.2 and 7.6.3). Then J_μ is primitive, hence J_λ is primitive from (ii).

Let $J = \bigcap_{\lambda \in D} J_\lambda$, which is a two-sided ideal of $U(\mathfrak{g})$. Then

$$J \cap Z(\mathfrak{g}) = \sum_{\lambda \in D} \operatorname{Ker} \chi_\lambda .$$

Now, if χ_λ' is the homomorphism of $S(\mathfrak{h})$ into k defined by λ, we have $\bigcap_{\lambda \in D} \operatorname{Ker} \chi_\lambda' = 0$. From 7.4.6, we deduce that $\bigcap_{\lambda \in D} \operatorname{Ker} \chi_\lambda = 0$, whence $J \cap Z(\mathfrak{g}) = 0$ and $J = 0$ (4.2.2).

Let us identify with $S(\mathfrak{g})$ the graded algebra associated with $U(\mathfrak{g})$. We fix $\lambda \in \mathfrak{h}^*$. Let $J_n = J_\lambda \cap U_n(\mathfrak{g})$. Then $J^{\mathrm{gr}} = \oplus_{n \geqq 0}(J_n/J_{n-1})$ is the graded ideal of $S(\mathfrak{g})$ associated with J_n. From 2.4.11,

$$Y(\mathfrak{g}) = \bigoplus_{n \geqq 0} (Z(\mathfrak{g}) \cap U_n(\mathfrak{g}))/(Z(\mathfrak{g}) \cap U_{n-1}(\mathfrak{g})).$$

Let $Y_+(\mathfrak{g})$ be the set of elements of $Y(\mathfrak{g})$ without constant term. With the notation of 8.2.4, let us identify the vector spaces $U(\mathfrak{g})$ and $K \otimes Z(\mathfrak{g})$. Let $\chi = \chi_\lambda$ and $Z' = \operatorname{Ker} \chi$. Then $J_\lambda = K \otimes Z'$ (8.4.3), hence J_λ is the kernel of the mapping $1 \otimes \chi$ of $U(\mathfrak{g}) = K \otimes Z(\mathfrak{g})$ into $K \otimes k = K$. Let $u \in J_n$. From 8.2.5, we have

(1) $$u = u_1 z_1 + \cdots + u_q z_q \qquad [u_1, \ldots, u_q \in K, z_1, \ldots, z_q \in Z(\mathfrak{g})]$$

with $u_iz_i \in U_n(\mathfrak{g})$ for all i. Furthermore,

$$0 = (1 \otimes \chi)(u) = u_1\chi(z_1) + \cdots + u_q\chi(z_q)$$

hence

$$u = u_1(z_1 - \chi(z_1)) + \cdots + u_q(z_q - \chi(z_q)).$$

In (1) we can assume that each z_i has filtration >0. Conversely, if $u_1, \ldots, u_q \in K$, $z_1, \ldots, z_q \in Z(\mathfrak{g})$, and $u_1z_1, \ldots, u_qz_q \in U_n(\mathfrak{g})$, then

$$u_1(z_1 - \chi(z_1)) + \cdots + u_q(z_q - \chi(z_q)) \in J_n.$$

Thus J_n/J_{n-1} is the homogeneous component of degree n of $Y_+(\mathfrak{g})S(\mathfrak{g})$, whence $J^{\text{gr}} = Y_+(\mathfrak{g})S(\mathfrak{g})$. From 8.1.3 and by extension of the scalar field, J^{gr} is a prime ideal of $S(\mathfrak{g})$. Hence the graded algebra associated with the filtered algebra $U(\mathfrak{g})/J_\lambda$ is integral (2.3.10). It follows that the algebra $U(\mathfrak{g})/J_\lambda$ is integral.

Let us assume that k is algebraically closed. Let J be a primitive ideal of $U(\mathfrak{g})$. Since the centre of $U(\mathfrak{g})/J$ is one-dimensional (2.6.8), $J \cap U(\mathfrak{g})$ is the kernel of a homomorphism of $Z(\mathfrak{g})$ into k, hence has the form $\operatorname{Ker} \chi_\lambda$ for some $\lambda \in \mathfrak{h}^*$ (7.4.8). Then $J_\lambda = U(\mathfrak{g}) \cdot \operatorname{Ker} \chi_\lambda \subset J$. If J is minimal primitive, then $J = J_\lambda$. If $J = J_\mu$ for some $\mu \in \mathfrak{h}^*$ and J contains a primitive ideal J', then from the foregoing there exists some $\mu' \in \mathfrak{h}^*$ such that $J' \supset J_{\mu'}$; then $\operatorname{Ker} \chi_\mu \supset \operatorname{Ker} \chi_{\mu'}$, whence $\operatorname{Ker} \chi_\mu = \operatorname{Ker} \chi_{\mu'}$ and $J_\mu = J_{\mu'} = J$; hence J is a minimal primitive ideal.

References: [37], [47].

8.5. Supplementary remarks

8.5.1. The results of sections 8.1, 8.2 and 8.3 are due to Kostant [78]; some of the proofs have been simplified by Varadarajan [122]. The content of section 8.4 is essentially due to Duflo [47].

8.5.2. Assume that k is algebraically closed and $\mathfrak{g}$ semi-simple. A Lie subalgebra $\mathfrak{s}$ of $\mathfrak{g}$ isomorphic to $\mathfrak{sl}(2,k)$ is termed *principal* if the non-zero nilpotent elements of $\mathfrak{s}$ are regular in $\mathfrak{g}$. Such Lie subalgebras of $\mathfrak{g}$ exist and are conjugate under the adjoint group of $\mathfrak{g}$ [77]. We choose a principal Lie subalgebra $\mathfrak{s}$ of $\mathfrak{g}$, which we identify with $\mathfrak{sl}(2,k) = ke + kf + kh$ (with the notation of section 1.8). Let ξ be a finite-dimensional simple representation of $\mathfrak{g}$ in V, having the weight 0 with multiplicity $l > 0$. Let W be the set of elements of V which vanish under $\xi(\mathfrak{g}^e)$. The eigenvalues of $\xi(h)|W$ form an increasing sequence $(m_1, m_2, \ldots, m_l)$ of non-negative integers (each eigenvalue being written the same number of times as its multiplicity). Given this, the integers $p_1, \ldots, p_l$ of 8.3.12 are $m_1, \ldots, m_l$. [78]

8.5.3. Assume that k is algebraically closed and $\mathfrak{g}$ semi-simple. Let $f_1, \ldots, f$ be algebraically independent homogeneous generators of $S(\mathfrak{g}^*)^{\mathfrak{g}}$.

(a) An element x of $\mathfrak{g}$ is regular if and only if the differentials of $f_1, \ldots, f_r$ at x are linearly independent. (Cf. 4.9.24 (c)).

(b) Let $\xi = (\xi_1, \ldots, \xi_r) \in k^r$, let J_ξ be the ideal of $S(\mathfrak{g}^*)$ generated by $f_1 - \xi_1, \ldots, f_r - \xi_r$, and let $V_\xi \subset \mathfrak{g}$ be the set of zeros of J_ξ. Then J_ξ is a prime ideal. If $\mathscr{A}$ is the adjoint group of $\mathfrak{g}$, then V_ξ only contains a finite number of $\mathscr{A}$-orbits. Let V'_ξ be the set of simple points of V_ξ. A point of V_ξ is regular if and only if it belongs to V'_ξ, and V'_ξ is an $\mathscr{A}$-orbit. The $\mathscr{A}$-orbit of an element y of $\mathfrak{g}$ is closed if and only if y is semi-simple. Cf. [78], where much information on $\mathscr{A}$-orbits may be found. Cf. also [80].

8.5.4. Assume that k is algebraically closed and $\mathfrak{g}$ semi-simple. Let V be a finite-dimensional simple $\mathfrak{g}$-module, x a regular element of $\mathfrak{g}$, f the image of x under the Killing isomorphism, and H as in 8.3.6. For all $h \in \operatorname{Hom}_{\mathfrak{g}}(V,H)$, let $\omega_x(h)$ be the linear form $v \mapsto h(v)(f)$ on V. Then ω_x is an isomorphism of the vector space $\operatorname{Hom}_{\mathfrak{g}}(V,H)$ onto the set of $\mathfrak{g}^x$-invariant elements of V^*. [78]

8.5.5. Assume that k is algebraically closed and $\mathfrak{g}$ semi-simple.

(a) Every $p \in S(\mathfrak{g}^*)$ defines a differential operator with constant coefficients $D(p)$ on $\mathfrak{g}^*$ (6.6.9). Let J and J_+ be as in 8.2.1, H the set of the $f \in S(\mathfrak{g})$ such that $D(p)f = 0$ for all $p \in J_+$, and $Y_+(\mathfrak{g})$ the set of elements of $Y(\mathfrak{g})$ without constant term. Then H is a $\mathfrak{g}$-stable graded complement of $Y_+(\mathfrak{g})S(\mathfrak{g})$ in $S(\mathfrak{g})$; it is also the vector subspace of $S(\mathfrak{g})$ generated by the powers of the nilpotent elements of $\mathfrak{g}$.

(b) Let β be the canonical mapping of $S(\mathfrak{g})$ into $U(\mathfrak{g})$. Then $E = \beta(H)$ is the vector subspace of $U(\mathfrak{g})$ generated by the powers of the nilpotent elements of $\mathfrak{g}$. It is a sub-$\mathfrak{g}$-module of $U(\mathfrak{g})$. Let E' be a finite-dimensional vector subspace of E. Then there exists a finite-dimensional simple representation π of $\mathfrak{g}$ such that $\pi|E'$ is injective. [78]

8.5.6. Assume that k is algebraically closed and $\mathfrak{g}$ semi-simple. Let $\mathfrak{g} = \mathfrak{h} \oplus \mathfrak{n}_+ \oplus \mathfrak{n}_-$ be a triangular decomposition of $\mathfrak{g}$.

(a) The commutant of $\mathfrak{n}_+$ in $U(\mathfrak{g})$ is an algebra of finite type [61]. The centre of $U(\mathfrak{n}_+)$ is a polynomial algebra (Kostant, unpublished, and [210]).

(b) Let $\lambda \in \mathfrak{h}^*$, let $I_\lambda = U(\mathfrak{g}) \operatorname{Ker} \chi_\lambda$, let B_γ be the algebra $U(\mathfrak{g})/I_\lambda$, let π be the canonical homomorphism of $U(\mathfrak{g})$ onto B_λ, and let $E = \pi(U(\mathfrak{n}_-))$.

The homomorphism $\pi|U(\mathfrak{n}_-)$ of $U(\mathfrak{n}_-)$ onto E is injective. Let E' be the commutant of E in B_λ. Let us canonically identify $M(\lambda)$ with $U(\mathfrak{n}_-)$. The module $M(\lambda)$ can be considered as a B_λ-module, whence we have a mapping $e' \mapsto e' \cdot 1$ of E' into $M(\lambda) = U(\mathfrak{n}_-)$. This mapping is an injective anti-homomorphism of the algebra E' into the algebra $U(\mathfrak{n}_-)$. The commutant of E' in B_λ is E.

(c) We resume the hypotheses and notation of 7.8.7. The centre of C contains $U(\mathfrak{h})$ and $Z(\mathfrak{g})$. These subalgebras are linearly disjoint. [For all $\lambda \in \mathfrak{h}^*$, the canonical homomorphism of $U(\mathfrak{h})$ into $U(\mathfrak{g})/U(\mathfrak{g}) \operatorname{Ker} \chi_\lambda$ is injective, as may be seen by studying the action of $U(\mathfrak{h})$ in $M(\lambda)$.] They generate the centre of C. [160]

8.5.7. Let $\mathfrak{g}$ be a complex semi-simple Lie algebra.

(a) Let J be a prime ideal of $U(\mathfrak{g})$. The following conditions are equivalent:

(i) J is primitive;

(ii) $\dim Z(\mathfrak{g})/J \cap Z(\mathfrak{g}) = 1$;

(iii) the centre of $U(\mathfrak{g})/J$ is C;

(iv) the intersection of the prime ideals of $U(\mathfrak{g})$ strictly containing J is distinct from J.

(b) If I is a prime ideal of $Z(\mathfrak{g})$, then the set of the prime ideals J of $U(\mathfrak{g})$ such that $J \cap Z(\mathfrak{g}) = I$ is finite. [37]

8.5.8. Let $\mathfrak{g}$ be a complex semi-simple Lie algebra, $\mathfrak{g} = \mathfrak{h} \oplus \mathfrak{n}_+ \oplus \mathfrak{n}_-$ a triangular decomposition of $\mathfrak{g}$, $\lambda \in \mathfrak{h}^*$, and $J_\lambda = U(\mathfrak{g}) \operatorname{Ker} \chi_\lambda$.

(a) The set of the primitive ideals of $U(\mathfrak{g})$ containing J_λ is finite from 8.5.7 (b), and possesses a largest element J'_λ. If $\lambda(H_\alpha) \notin \mathbf{Z} - \{0\}$ for every root α, then $J_\lambda = J'_\lambda$. [37]

(b) Assume that $\lambda(H_\alpha) \notin \{-1, -2, -3, \ldots\}$ for every positive root α. Then J'_λ is the annihilator of $L(\lambda)$. [25]

8.5.9. Assume that k is algebraically closed and $\mathfrak{g}$ semi-simple. Let $\mathfrak{h}$ be a Cartan subalgebra of $\mathfrak{g}$, V and W finite-dimensional simple $\mathfrak{g}$-modules, λ the smallest weight of V, μ and μ^* the largest weights of W and W^* respectively, and L and I the annihilators of W_μ and W in $U(\mathfrak{g})$ respectively.

(a) If h is a $\mathfrak{g}$-homomorphism from V into $U(\mathfrak{g})$, the following conditions are equivalent:

(i) $h(V_\lambda) \subset L$;

(ii) $h(V_0) \subset L$;

(iii) $h(V) \subset I$.

(b) Let $(\alpha_1, \ldots, \alpha_l)$ be a basis for $R(\mathfrak{g},\mathfrak{h})$, $H_i = H_{\alpha_i}$, $X_i \in \mathfrak{g}^{\alpha_i} - \{0\}$, and V' the set of the $v \in V_0$ such that $X_i^{\mu^*(H_i)+1} v = 0$ for $i = 1, \ldots, l$. Then

$$\operatorname{mtp}(V, U(\mathfrak{g})/I) = \dim V'.$$

[100]

8.5.10. Assume that $k = \mathbf{C}$ and that $\mathfrak{g}$ is semi-simple. Let $\mathfrak{h}$ be a Cartan subalgebra of $\mathfrak{g}$, $R = R(\mathfrak{g},\mathfrak{h})$, B a basis for R, R_+ the set of positive roots, $\delta = \frac{1}{2} \sum_{\alpha \in R_+} \alpha$, φ the Harish-Chandra homomorphism defined by B, and C the commutant of $\mathfrak{h}$ in $U(\mathfrak{g})$. Let ϱ be a finite-dimensional simple representation of $\mathfrak{g}$ in V, h and h' $\mathfrak{g}$-homomorphisms from V and V^*, respectively, into $U(\mathfrak{g})$, $(v_1, \ldots, v_n)$ a basis for V, and $(v_1^*, \ldots, v_n^*)$ the dual basis for V^*. Assume that $(v_1, \ldots, v_d)$ is a basis for V_0, so that $(v_1^*, \ldots, v_d^*)$ is a basis for V_0^* and $hv_1, \ldots, hv_d, h'v_1^*, \ldots, h'v_d^* \in C$.

(a) The element $z_{h,h'} = (h'v_1^*)(hv_1) + \cdots + (h'v_n^*)(hv_n)$ belongs to $Z(\mathfrak{g})$; it depends solely on h,h', and not on the choice of the basis $(v_1, \ldots, v_n)$. Let w_0 be the element of $W(\mathfrak{g},\mathfrak{h})$ which transforms B into $-B$, and w_0^A the mapping $\lambda \mapsto w_0(\lambda) + w_0(\delta) - \delta$ of $\mathfrak{h}^*$ into $\mathfrak{h}^*$. Recall (7.8.20) that $W(\mathfrak{g},\mathfrak{h})$ operates canonically in C. Then, for all $\lambda \in \mathfrak{h}^*$, we have

$$\varphi(z_{h,h'})(\lambda) = \sum_{i=1}^{d} \varphi(w_0(h'v_i^*))(\lambda)\varphi(hv_i)(w_0^A\lambda).$$

(b) For all $\alpha \in R_+$, let $X_\alpha \in \mathfrak{g}^\alpha$ and $X_{-\alpha} \in \mathfrak{g}^{-\alpha}$, with $[X_\alpha, X_{-\alpha}] = H_\alpha$. For every integer $j \geqq 1$, let $m_j(\alpha)$ be the multiplicity of the eigenvalue $j(j+1)$ for $\varrho(X_{-\alpha}X_\alpha)|V_0$. Let $(h_1, \ldots, h_d)$ be a basis for the $Z(\mathfrak{g})$-module $\operatorname{Hom}\mathfrak{g}(V, U(\mathfrak{g}))$. Let M be the matrix with d rows and d columns, with elements in $U(\mathfrak{h})$, such that the element in the ith row and j^{th} column is $\varphi(h_i v_j)$. Then $\det M$ is a non-zero element of $U(\mathfrak{h})$ which is proportional to

$$\prod_{\alpha \in R_+} \prod_{j \geqq 1} [(H_\alpha + \delta(H_\alpha) - 1)(H_\alpha + \delta(H_\alpha) - 2) \cdots (H_\alpha + \delta(H_\alpha) - j)]^{m_j(\alpha)}.$$

[100]

(c) Assume that $\mathfrak{g}$ is simple. Let $(h_1, \ldots, h_d)$ be a basis for the $Z(\mathfrak{g})$-module $\operatorname{Hom}_{\mathfrak{g}}(\mathfrak{g}, U(\mathfrak{g}))$ consisting of homogeneous elements. We may assume that h_1 is the identity mapping of $\mathfrak{g}$. Identify $\mathfrak{g}$ with $\mathfrak{g}^*$ by means of the Killing isomorphism. Then $(z_{h_1,h_1}, z_{h_1,h_2}, \ldots, z_{h_1,h_d})$ is an algebraically free generating system of the algebra $Z(\mathfrak{g})$. [134]

8.5.11. Let $(\mathfrak{g},\mathfrak{h})$ be a splitting semi-simple Lie algebra, $\lambda \in \mathfrak{h}^*$, $J_\lambda = U(\mathfrak{g}) \cdot \operatorname{Ker} \chi_\lambda$, and $n = \frac{1}{2}(\dim \mathfrak{g} - \dim \mathfrak{h})$. There exists an injective homomorphism f of $U(\mathfrak{g})/J_\lambda$ into A_n such that $\operatorname{Fract} f(U(\mathfrak{g})/J_\lambda) = \operatorname{Fract} A_n$. [24]

CHAPTER 9

HARISH-CHANDRA MODULES

Let $\mathscr{G}$ be a real semi-simple Lie group with finite centre, $\mathscr{K}$ a maximal compact subgroup, $\mathfrak{g}_0$ and $\mathfrak{k}_0$ the Lie algebras of $\mathscr{G}$ and $\mathscr{K}$, $\mathfrak{g}$ and $\mathfrak{k}$ the complexifications of $\mathfrak{g}_0$ and $\mathfrak{k}_0$, and π a completely irreducible representation of $\mathscr{G}$ in a Banach space (for example, an irreducible unitary representation of $\mathscr{G}$ in a Hilbert space). Let V be the space of vectors which are $\mathscr{K}$-finite under π. Then V is a simple $\mathfrak{g}$-module such that $V = \oplus_{\varrho \in \mathfrak{k}^\wedge} V_\varrho$. The study of π can for the most part be reduced to the study of this $\mathfrak{g}$-module. The modules which we shall study in this chapter are generalizations of the $\mathfrak{g}$-modules just considered.

9.0. Notation

In this chapter, $\mathfrak{k}$ will denote a Lie subalgebra of $\mathfrak{g}$ and G and K the enveloping algebras of $\mathfrak{g}$ and $\mathfrak{k}$. An essential role will be played by the commutant $G^\mathfrak{k}$ of $\mathfrak{k}$ in G. If $\varrho \in \mathfrak{k}^\wedge$, we shall denote the kernel of ϱ in K by I^ϱ. If V is a $\mathfrak{g}$-module, we have defined V_ϱ in 1.2.8.

9.1. The case of a Lie subalgebra which is reductive in $\mathfrak{g}$

9.1.1. For $\varrho, \sigma \in \mathfrak{k}$, we set

$$G^{\varrho,\sigma} = \{u \in G \mid I^\varrho u \subset GI^\sigma\}.$$

We write G^ϱ instead of $G^{\varrho,\varrho}$. The $G^{\varrho,\sigma}$ are vector subspaces of G.

9.1.2. Proposition (notation as in 9.0). (i) *We have* $G^{\varrho,\sigma} \supset GI^\sigma$ *and*

$$G^{\varrho,\sigma}/GI^\sigma = (G/GI^\sigma)_\varrho$$

when we consider G/GI^σ *as a* $\mathfrak{g}$*-module for the left regular representation.*

(ii) *If* V *is a* $\mathfrak{g}$*-module, then* $G^{\varrho,\sigma}(V_\sigma) \subset V_\varrho$.

We have $I^{\varrho}GI^{\sigma} \subset GI_{\sigma}$, hence $GI^{\sigma} \subset G^{\varrho,\sigma}$.

Every representation of K/I^{ϱ} is semi-simple and is the sum of representations of class ϱ (AL VIII, pp. 47—48). Hence, if V is a $\mathfrak{g}$-module, then

$$V_{\varrho} = \{v \in V \mid I^{\varrho}v = 0\}.$$

Firstly this implies (ii), because $I^{\varrho}G^{\varrho,\sigma}(V_{\sigma}) \subset GI^{\sigma}(V_{\sigma}) = 0$. On the other hand,

$$(G/GI^{\sigma})_{\varrho} = \{u + GI^{\sigma} \mid u \in G \quad \text{and} \quad I^{\varrho}u \subset GI^{\sigma}\},$$

whence (i).

9.1.3. Proposition (notation as in 9.0). (i) *G^{ϱ} is a subalgebra of G which contains K as a subalgebra and GI^{ϱ} as a two-sided ideal.*

(ii) *If V is a $\mathfrak{g}$-module, then $G^{\varrho}(V_{\varrho}) \subset V_{\varrho}$ and $(GI^{\varrho})(V_{\varrho}) = 0$; we can thus consider V_{ϱ} as a $(G^{\varrho}/GI^{\varrho})$-module.*

We have

$$I^{\varrho}G^{\varrho}G^{\varrho} \subset GI^{\varrho}G^{\varrho} \subset GI^{\varrho}, \qquad I^{\varrho}K \subset I^{\varrho}, \qquad GI^{\varrho}G^{\varrho} \subset GGI^{\varrho} = GI^{\varrho},$$

whence (i). Assertion (ii) follows from 9.1.2.

9.1.4. A $\mathfrak{g}$-module V is said to be a *Harish-Chandra module* relative to $\mathfrak{k}$ (or simply a Harish-Chandra module if there is no doubt about $\mathfrak{k}$) if $V = \sum_{\varrho \in \mathfrak{k}^{\wedge}} V_{\varrho}$, whence $V = \oplus_{\varrho \in \mathfrak{k}^{\wedge}} V_{\varrho}$. The Verma modules are Harish-Chandra modules relative to a Cartan subalgebra. In this chapter, the subalgebras $\mathfrak{k}$ that we have in view are in general very different from Cartan subalgebras.

9.1.5. Proposition (notation as in 9.0). *We assume that $\mathfrak{k}$ is reductive in $\mathfrak{g}$. Let $\varrho \in \mathfrak{k}^{\wedge}$. Then:*

(i) *G/GI^{ϱ} is a Harish-Chandra $\mathfrak{g}$-module under the left regular representation.*

(ii) *$G/GI^{\varrho} = \oplus_{\sigma \in \mathfrak{k}^{\wedge}} G^{\sigma,\varrho}/GI^{\varrho}$.*

From 1.7.9, $W = \sum_{\sigma \in \mathfrak{k}^{\wedge}} (G/GI^{\varrho})_{\sigma}$ is a sub-$\mathfrak{g}$-module of G/GI^{ϱ}. Now the class of 1 in G/GI^{ϱ} belongs to W, and hence $W = G/GI^{\varrho}$. This proves (i), and (ii) then follows from 9.1.2 (i).

9.1.6. Proposition (notation as in 9.0). *We assume that $\mathfrak{k}$ is reductive in $\mathfrak{g}$. Let V be a simple $\mathfrak{g}$-module, and $\varrho \in \mathfrak{k}^{\wedge}$ such that $V_{\varrho} \neq 0$. Then:*

(i) *V is a Harish-Chandra $\mathfrak{g}$-module.*

(ii) *The G^{ϱ}-module V_{ϱ} is simple.*

From 1.7.9, V is a Harish-Chandra $\mathfrak{g}$-module. Let $v \in V_\varrho - \{0\}$. Then $V = Gv = \sum_\sigma G^{\sigma,\varrho}v$ from 9.1.5 (ii), and $G^{\sigma,\varrho}v \subset V_\sigma$ from 9.1.2 (ii), hence $G^\varrho v = V_\varrho$.

9.1.7. PROPOSITION (notation as in 9.0). *We assume that $\mathfrak{k}$ is reductive in $\mathfrak{g}$. Let V be a Harish-Chandra $\mathfrak{g}$-module, $\varrho \in \mathfrak{k}^\wedge$, P the projection of V onto V_ϱ defined by the decomposition $V = \oplus_{\sigma \in \mathfrak{k}^\wedge} V_\sigma$, and W a sub-G^ϱ-module of V_ϱ. We set*

$$W^{\min} = GW, \qquad W^{\max} = \{v \in V \mid (Gv) \cap V_\varrho \subset W\}.$$

Then:

(i) $W^{\max} = \{v \in V | P(Gv) \subset W\}$.

(ii) $W^{\min} \cap V_\varrho = W^{\max} \cap V_\varrho = W$.

(iii) *If X is a sub-$\mathfrak{g}$-module of V such that $X \cap V_\varrho \supset W$ (or $X \cap V_\varrho \subset W$), then $X \supset W^{\min}$ (or $X \subset W^{\max}$).*

For all $v \in V$, we have $Gv = \oplus_{\sigma \in \mathfrak{k}^\wedge} (Gv) \cap V_\sigma$, whence (i). On the other hand, $GW = \sum_{\sigma \in \mathfrak{k}^\wedge} G^{\sigma,\varrho}W$ from 9.1.5 (ii), and $G^{\sigma,\varrho}W \subset V_\sigma$, hence

$$W^{\min} \cap V_\varrho = G^\varrho W = W.$$

Consequently, $W^{\max} \supset W^{\min}$, and it is then clear that $W^{\max} \cap V_\varrho = W$. Assertion (iii) is obvious.

9.1.8. PROPOSITION (notation as in 9.0). *We assume that $\mathfrak{k}$ is reductive in $\mathfrak{g}$. Let $\varrho \in \mathfrak{k}^\wedge$, and let $\mathscr{M}$ and $\mathscr{L}$ be the sets of the maximal left ideals of G and G^ϱ, respectively, containing GI^ϱ.*

(i) *The mapping $M \mapsto M \cap G^\varrho$ is a bijection φ of $\mathscr{M}$ onto $\mathscr{L}$.*

(ii) *If $L \in \mathscr{L}$, then $\varphi^{-1}(L)$ contains every left ideal M of G such that $M \cap G^\varrho \subset L$.*

(iii) *Let π be the canonical mapping of G onto G/GI^ϱ, and P the projection of G/GI^ϱ onto G^ϱ/GI^ϱ defined by* 9.1.5 (ii). *If $L \in \mathscr{L}$, then*

$$\begin{aligned}\varphi^{-1}(L) &= \{u \in G \mid (Gu) \cap G^\varrho \subset L\} \\ &= \{u \in G \mid P(G\pi(u)) \subset \pi(L)\}.\end{aligned}$$

The mapping $M \mapsto M/GI^\varrho$ is a bijection of the set of the left ideals of G containing GI^ϱ onto the set of sub-$\mathfrak{g}$-modules of G/GI^ϱ; if $M \neq G$, then $1 \notin M$ and hence $M \cap G^\varrho \neq G^\varrho$. The mapping $L \mapsto L/GI^\varrho$ is a bijection of the set of the left ideals of G^ϱ containing GI^ϱ onto the set of sub-G^ϱ-modules of $G^\varrho/GI^\varrho = (G/GI^\varrho)_\varrho$. Given this, the assertions of the proposition are easy consequences of 9.1.7 applied to $V = G/GI^\varrho$.

9.1.9. PROPOSITION (notation as in 9.0). *We assume that* $\mathfrak{k}$ *is reductive in* $\mathfrak{g}$. *Let* $\varrho \in \mathfrak{k}^\wedge$. *Let* $\mathscr{S}$ *be the set of the classes of simple* $\mathfrak{g}$*-modules* V *such that* $V_\varrho \neq 0$. *Let* $\mathscr{S}'$ *be the set of classes of simple* (G^ϱ/GI^ϱ)*-modules. Then the mapping* $V \mapsto V_\varrho$ *defines a bijection of* $\mathscr{S}$ *onto* $\mathscr{S}'$.

If $V \in \mathscr{S}$, then $V_\varrho \in \mathscr{S}'$ (9.1.6 (ii)).

Let $V, W \in \mathscr{S}$, let f be a G^ϱ-isomorphism of V_ϱ onto W_ϱ, v a non-zero element of V_ϱ, and L the annihilator of v in G^ϱ; it is also the annihilator of $f(v)$ in G^ϱ, and it is a maximal left ideal of G^ϱ. Let M and M' be the annihilators of v and $f(v)$ respectively in G. Then M and M' are maximal left ideals of G, and $M \cap G^\varrho = L = M' \cap G^\varrho$, hence $M = M'$ (9.1.8), which proves that V and W are G-isomorphic.

Let Z be a simple G^ϱ-module whose annihilator contains GI^ϱ, let $z \in Z - \{0\}$, and let L be the annihilator of z in G^ϱ. There exists a maximal left ideal M of G such that $M \cap G^\varrho = L$ (9.1.8). Let V be the simple $\mathfrak{g}$-module G/M. Then

$$G/GI^\varrho = \bigoplus_{\sigma \in \mathfrak{k}^\wedge} (G/GI^\varrho)_\sigma, \qquad (G/GI^\varrho)_\varrho = G^\varrho/GI^\varrho$$

(9.1.2, 9.1.5); hence

$$G/M = \bigoplus_{\sigma \in \mathfrak{k}^\wedge} (G/M)_\sigma, \qquad (G/M)_\varrho = G^\varrho/M \cap G^\varrho = G^\varrho/L,$$

so that V_ϱ is G^ϱ-isomorphic to Z.

9.1.10. PROPOSITION (notation as in 9.0). *We assume that* $\mathfrak{k}$ *is reductive in* $\mathfrak{g}$. *Let* $\varrho \in \mathfrak{k}^\wedge$. *Then:*

(i) $(GI^\varrho) \cap K = I^\varrho$, *so that* K/I^ϱ *can be identified with a subalgebra of* G^ϱ/GI^ϱ.

(ii) $G^\mathfrak{k} \subset G^\varrho$, *so that* $G^\mathfrak{k}/G^\mathfrak{k} \cap GI^\varrho$ *can be identified with a subalgebra of* G^ϱ/GI^ϱ*; moreover,* $G^\mathfrak{k} \cap GI^\varrho = G^\mathfrak{k} \cap I^\varrho G$.

(iii) *If* ϱ *is absolutely simple, the canonical mapping of*

$$(K/I^\varrho) \otimes (G^\mathfrak{k}/G^\mathfrak{k} \cap GI^\varrho)$$

into G^ϱ/GI^ϱ *is an isomorphism; we have* $G^\varrho = KG^\mathfrak{k} + GI^\varrho$.

Assertion (i) follows from the fact that G is a free right K-module (2.2.7).

Clearly, $G^\mathfrak{k} \subset G^\varrho$. Let V be a $\mathfrak{g}$-module such that $V = \oplus_{\sigma \in \mathfrak{k}^\wedge} V_\sigma$. Let V^* be the dual $\mathfrak{g}$-module. Let us identify $\oplus_{\sigma \in \mathfrak{k}^\wedge} (V_\sigma)^*$ with a sub-$\mathfrak{k}$-module of V^*. Let $u \in G^\mathfrak{k} \cap I^\varrho G$. Then $u^\mathsf{T} \in G^\mathfrak{k} \cap GI^{\varrho^*}$, and $I^\varrho((V_\varrho)^*) = 0$, hence $u^\mathsf{T}((V_\varrho)^*) = 0$ and

$$\langle u(V_\varrho), (V_\varrho)^* \rangle = \langle V_\varrho, u^\mathsf{T}((V_\varrho)^*) \rangle = 0;$$

since $u(V_\varrho) \subset V_\varrho$, we conclude that $u(V_\varrho) = 0$. Applying this to $V = G/GI^\varrho$, we see that $uG^\varrho \subset GI^\varrho$, whence $u \in GI^\varrho$. Thus $G^\mathfrak{k} \cap I^\varrho G \subset GI^\varrho$, and consequently

$$G^\mathfrak{k} \cap GI^\varrho = (G^\mathfrak{k} \cap I^{\varrho *}G)^\top \subset (GI^{\varrho *})^\top = I^\varrho G.$$

This proves (ii).

The adjoint representation of $\mathfrak{k}$ in G is semi-simple, and G^ϱ and GI^ϱ are sub-$\mathfrak{k}$-modules of G; consequently, every element of G^ϱ/GI^ϱ annihilated by the adjoint representation of $\mathfrak{k}$ is the canonical image of an element of G^ϱ which commutes with $\mathfrak{k}$ (1.2.11), i.e., of an element of $G^\mathfrak{k}$. The commutant of K/I^ϱ in G^ϱ/GI^ϱ is thus $G^\mathfrak{k}/G^\mathfrak{k} \cap GI^\varrho$. Finally, if ϱ is absolutely simple, the algebra K/I^ϱ is central, simple and finite-dimensional, and hence G^ϱ/GI^ϱ is the tensor product of K/I^ϱ and its commutant in G^ϱ/GI^ϱ (11.2.5). This proves (iii).

9.1.11. LEMMA (notation as in 9.0). *Let V be a $\mathfrak{g}$-module, Z an absolutely simple $\mathfrak{k}$-module, ϱ the corresponding representation, and H the vector space* $\mathrm{Hom}_\mathfrak{k}(Z,V) = \mathrm{Hom}_\mathfrak{k}(Z,V_\varrho)$. *We consider H as a left $G^\mathfrak{k}$-module by virtue of the action of $G^\mathfrak{k}$ in V (or V_ϱ). There exists one and only one isomorphism of the vector space $Z \otimes H$ onto the vector space V_ϱ which transforms $z \otimes h$ ($z \in Z$, $h \in H$) into $h(z)$. If we identify $Z \otimes H$ with V_ϱ by means of this isomorphism, then, for $z \in Z$, $h \in H$, $u \in K$ and $v \in G^\mathfrak{k}$,*

$$u \cdot (z \otimes h) = uz \otimes h, \qquad v \cdot (z \otimes h) = z \otimes vh.$$

This follows from AL VIII, pp. 13–15.

9.1.12. THEOREM (notation as in 9.0). *We assume that $\mathfrak{k}$ is reductive in $\mathfrak{g}$. Let ϱ be an absolutely simple element of $\mathfrak{k}^\wedge$, and X a $\mathfrak{k}$-module of class ϱ. Let $\mathscr{S}$ be the set of the classes of simple $\mathfrak{g}$-modules V such that $V_\varrho \neq 0$. Let $\mathscr{T}$ be the set of classes of simple $(G^\mathfrak{k}/G^\mathfrak{k} \cap GI^\varrho)$-modules.*

(i) *If V is a simple $\mathfrak{g}$-module such that $V_\varrho \neq 0$, let us consider $F = \mathrm{Hom}^\mathfrak{k}(X,V_\varrho)$ as a left $G_\mathfrak{k}$-module by virtue of the action of $G^\mathfrak{k}$ in V_ϱ. Then F is a simple $G^\mathfrak{k}$-module whose annihilator contains $G^\mathfrak{k} \cap GI^\varrho$.*

(ii) *The mapping $V \mapsto F$ defines a bijection of $\mathscr{S}$ onto $\mathscr{T}$.*

Let V be a $\mathfrak{g}$-module and $F = \mathrm{Hom}_\mathfrak{k}(X,V)$. Let us identify $F \otimes X$ with V_ϱ by virtue of 9.1.11. If F' is a sub-$G^\mathfrak{k}$-module of F, $F' \otimes X$ is a sub-$G^\mathfrak{k}$-module and a sub-K-module of V_ϱ, and hence a sub-G^ϱ-module of V_ϱ (9.1.10 (iii)). If V is simple as a $\mathfrak{g}$-module and $V_\varrho \neq 0$, then V_ϱ is simple as a G^ϱ-module (9.1.6 (ii)), hence $F' \otimes X = V_\varrho$ and $F' = F$. This proves (i).

Let $V, V' \in \mathscr{S}$, $F = \text{Hom}_{\mathfrak{k}}(X,V)$, $F' = \text{Hom}_{\mathfrak{k}}(X,V')$, and let φ be a $G^{\mathfrak{k}}$-isomorphism of F onto F'. Then $\varphi \otimes 1$ is a K-module and a $G^{\mathfrak{k}}$-module isomorphism, and hence a G^{ϱ}-module isomorphism, of $F \otimes X$ onto $F' \otimes X$ (9.1.10 (iii)).

Hence V and V' are isomorphic (9.1.9).

Let F be a simple $G^{\mathfrak{k}}$-module whose annihilator contains $G^{\mathfrak{k}} \cap GI^{\varrho}$. Then, from 9.1.10 (iii), $F \otimes X$ is a simple G^{ϱ}-module whose annihilator contains GI^{ϱ}. There exists a simple $\mathfrak{g}$-module V such that the G^{ϱ}-module V_{ϱ} can be identified with $F \otimes X$ (9.1.9). Then the $G^{\mathfrak{k}}$-module $\text{Hom}_{\mathfrak{k}}(X,V_{\varrho})$ can be identified with the $G^{\mathfrak{k}}$-module F (AL VIII, p. 15).

9.1.13. With the notation of 9.0, let V be a $\mathfrak{g}$-module, $\varrho \in \mathfrak{k}^{\wedge}$, and X a $\mathfrak{k}$-module with class ϱ. We consider $\text{Hom}_{\mathfrak{k}}(X,V)$ as a $G^{\mathfrak{k}}$-module by virtue of the action of $G^{\mathfrak{k}}$ in V. Let σ be the corresponding representation of $G^{\mathfrak{k}}$. Then σ is said to be the *spherical function of type ϱ of V*. If $\mathfrak{k}$ is reductive in $\mathfrak{g}$, ϱ is absolutely simple and V is a simple Harish-Chandra module such that $V_{\varrho} \neq 0$, then the spherical function of type ϱ of V is simple and characterises V up to isomorphism (9.1.12); we shall see in 9.5.1 that it is finite-dimensional.

9.1.14. PROPOSITION (notation as in 9.0). *We assume that $\mathfrak{k}$ is reductive in $\mathfrak{g}$. Let ϱ be an absolutely simple element of $\mathfrak{k}^{\wedge}$ and X a $\mathfrak{k}$-module with class ϱ. Let V, V' be $\mathfrak{g}$-modules, with V simple and $V_{\varrho} \neq 0$. We assume the existence of sub-$G^{\mathfrak{k}}$-modules Y_1, Y_2 of* $\text{Hom}(X,V')$ *such that $Y_1 \subset Y_2$ and* $\text{Hom}_{\mathfrak{k}}(X,V)$ *is $G^{\mathfrak{k}}$-isomorphic to Y_2/Y_1. Then V is isomorphic to a subquotient of V'.*

We have $V'_{\varrho} \neq 0$. By replacing V' by $\sum_{\tau \in \mathfrak{k}^{\wedge}} V'_{\tau}$, we can assume that V' is a Harish-Chandra module. We have

$$V_{\varrho} = \text{Hom}_{\mathfrak{k}}(X,V) \otimes X, \qquad V'_{\varrho} = \text{Hom}_{\mathfrak{k}}(X,V') \otimes X.$$

Let $Z_1 = Y_1 \otimes X$ and $Z_2 = Y_2 \otimes X$. From 9.1.10 (iii), the G^{ϱ}-module V_{ϱ} is isomorphic to the G^{ϱ}-module Z_2/Z_1. Let $Z'_2 = GZ_2$. Applying 9.1.7, let us consider the largest sub-G-module Z'_1 of Z'_2 such that $Z'_1 \cap V'_{\varrho} = Z_1$. Then Z'_2/Z'_1 is a simple Harish-Chandra $\mathfrak{g}$-module and $(Z'_2/Z'_1)_{\varrho} = Z_2/Z_1$. From 9.1.9, V is isomorphic to Z'_2/Z'_1.

9.1.15. We assume that k is algebraically closed. Let $\mathfrak{h}$ be a reductive Lie algebra, and σ a finite-dimensional semi-simple representation of $\mathfrak{h}$ having the adjoint representation as a subrepresentation. The elements ϱ_1, ϱ_2 of $\mathfrak{h}^{\wedge}$ are said to be *congruent modulo σ* when the following condition is satisfied:

if λ_1 and λ_2 are weights of ϱ_1 and ϱ_2 respectively, then $\lambda_1 - \lambda_2$ belongs to an additive group generated by the weights of σ.

Since σ has the adjoint representation as a subrepresentation, this condition is, from 9.1.8 (ii), independent of the choice of λ_1 and λ_2; from 1.9.11, it is independent of the choice of the Cartan subalgebra of $\mathfrak{h}$ which had to be made in order to speak of weights of representations.

The congruence relation modulo σ is an equivalence relation in $\mathfrak{h}^\wedge$. If $\mathfrak{h}$ is semi-simple, there exists in $\mathfrak{h}^\wedge$ only a finite number of congruence classes modulo σ (11.1.11), but this is not the case in general. Each congruence class modulo σ is denumerable.

If $\varrho \in \mathfrak{h}^\wedge$, all weights of $\varrho \otimes \sigma$ are sums of weights of ϱ and weights of σ, hence all simple subrepresentations of $\varrho \otimes \sigma$ are congruent to ϱ modulo σ.

9.1.16. Given this, let V be a Harish-Chandra $\mathfrak{g}$-module. We assume that k is algebraically closed and that $\mathfrak{k}$ is reductive in $\mathfrak{g}$. Let σ be the adjoint representation of $\mathfrak{k}$ in $\mathfrak{g}$. Let $(\mathfrak{k}_i^\wedge)_{i \in I}$ be the family of congruence classes modulo σ in $\mathfrak{k}^\wedge$. For all $i \in I$, $V_i = \oplus_{\varrho \in \mathfrak{k}_i^\wedge} V_\varrho$ *is a sub-*$\mathfrak{g}$*-module of* V. Indeed, let $\varrho \in \mathfrak{k}_i^\wedge$. Let θ be the linear mapping of $\mathfrak{g} \otimes V_\varrho$ into V such that $\theta(x \otimes w) = xw$ for $x \in \mathfrak{g}$ and $w \in V_\varrho$. It is a $\mathfrak{k}$-module homomorphism from 1.7.9, equation (1). All simple subrepresentations of $\mathfrak{g} \otimes V_\varrho$ (considered as a $\mathfrak{k}$-module) are congruent to ϱ modulo σ from 9.1.15, hence $\mathfrak{g} \cdot V_\varrho \subset V_i$, which proves our assertion.

In particular, if V is simple, all the $\varrho \in \mathfrak{k}^\wedge$ such that $V_\varrho \neq 0$ belong to one and the same congruence class modulo σ, termed the *class modulo σ corresponding to V.*

References: [64], [89].

9.2. Canonical mappings defined by a symmetrizing subalgebra

9.2.1. In sections 9.2 to 9.5 we shall assume that k is algebraically closed, that $\mathfrak{g}$ is semi-simple, and that $\mathfrak{k}$ is a symmetrizing Lie subalgebra of $\mathfrak{g}$. We shall then reduce the study of $G^\mathfrak{k}$ (the commutant of $\mathfrak{k}$ in $U(\mathfrak{g}) = G$) to that of algebras which are easier to handle.

Let $\mathfrak{g} = \mathfrak{k} \oplus \mathfrak{p}$ be the symmetric decomposition corresponding to $\mathfrak{k}$. With the notation of 1.13.11, we choose $\mathfrak{h}$ and B, whence we have $\mathfrak{a},\mathfrak{l},\mathfrak{n},\mathfrak{m},\mathfrak{q},R'$, $B' = B \cap R', R_+$ and $R'_+ = R_+ \cap R'$. We set

$$U(\mathfrak{k}) = K, \qquad U(\mathfrak{h}) = H, \qquad U(\mathfrak{a}) = A,$$

$$U(\mathfrak{l}) = L, \qquad U(\mathfrak{n}) = N, \qquad U(\mathfrak{m}) = M.$$

We identify the algebra H with the algebra $L \otimes A$. Every element of $\mathfrak{h}^*$

(or $\mathfrak{a}^*,\mathfrak{l}^*$) defines a homomorphism of H (resp. A,L) into k, with which it will be identified. If $\varrho \in \mathfrak{k}^\wedge$, the kernel of ϱ in K will always be denoted by I^ϱ. For $n \in \mathbf{N}$, we set

$$U_n(\mathfrak{g}) = G_n, \qquad U_n(\mathfrak{a}) = A_n, \qquad U^n(\mathfrak{a}) = S^n(\mathfrak{a}) = A^n.$$

We denote the commutant of $\mathfrak{m}$ in G by $G^\mathfrak{m}$ and $K^\mathfrak{m}$, $G^\mathfrak{h}$, etc., are defined analogously.

9.2.2. From 2.2.10, the vector space G can be canonically identified with the vector space $N \otimes A \otimes K$. On the other hand, $N = k \cdot 1 \oplus \mathfrak{n}N$, hence

$$G = AK \oplus \mathfrak{n}G.$$

This decomposition defines a projection $\tilde{\omega}$ of G onto AK. On the other hand, the vector space AK can be identified with the vector space $A \otimes K$; the latter is naturally equipped with an algebra structure. We say that $\tilde{\omega}$ is the *canonical mapping of G onto $A \otimes K$* (defined by the Iwasawa decomposition $\mathfrak{g} = \mathfrak{k} \oplus \mathfrak{a} \oplus \mathfrak{n}$).

9.2.3. PROPOSITION (notation as in 9.2.1). *Let $\tilde{\omega}$ be the canonical mapping of G onto $A \otimes K$. Then:*

(i) *For $u \in G$ and $v \in G^\mathfrak{k}$, we have $\tilde{\omega}(uv) = \tilde{\omega}(v)\tilde{\omega}(u)$.*

(ii) $\tilde{\omega}(G^\mathfrak{m}) \subset A \otimes K^\mathfrak{m}$.

(iii) *In particular, $\tilde{\omega}|G^\mathfrak{k}$ is an anti-homomorphism of $G^\mathfrak{k}$ into $A \otimes K^\mathfrak{m}$.*

(iv) *For $u \in G$ and $k \in K$, we have $\tilde{\omega}(uk) = \tilde{\omega}(u)(l \otimes k)$.*

(v) *Let $n \in N$. If $u \in G_n$ and $k \in K$, then $\tilde{\omega}(uk) \in A_n \otimes K$.*

Let $u \in G$ and $v \in G^\mathfrak{k}$, and let us write

$$u = \sum_i u_i u_i' + u'', \qquad v = \sum_j v_j v_j' + v''$$

with $u_i,v_j \in A$, $u_i',v_j' \in K$, $u'',v'' \in \mathfrak{n}G$. Then

$$uv = \sum_i u_i v u_i' + u''v = \sum_{i,j} u_i v_j v_j' u_i' + \sum_i u_i v'' u_i' + u''v.$$

Since $[\mathfrak{a},\mathfrak{n}] \subset \mathfrak{n}$ (1.13.11), we have $\mathfrak{a}\mathfrak{n} \subset \mathfrak{n}\mathfrak{a} + \mathfrak{n} \subset \mathfrak{n}A$, and consequently $A\mathfrak{n} \subset \mathfrak{n}A$. Hence $u_i v'' u_i' \in \mathfrak{n}G$. It is clear that $u''v \in \mathfrak{n}G$. We then have

$$\begin{aligned} \tilde{\omega}(uv) &= \sum_{i,j} (u_i v_j) \otimes (v_j' u_i') \\ &= \Big(\sum_j v_j \otimes v_j'\Big)\Big(\sum_i u_i \otimes u_i'\Big) = \tilde{\omega}(v)\tilde{\omega}(u). \end{aligned}$$

If in addition $u \in G^\mathfrak{m}$, then, for all $m \in \mathfrak{m}$,

$$0 = [m,u] = \sum_i u_i[m,u_i'] + [m,u''].$$

Now $[m,u''] \in [\mathfrak{m},\mathfrak{n},G] \subset \mathfrak{n}G$ since $[\mathfrak{m},\mathfrak{n}] \subset \mathfrak{n}$ (1.13.11). Hence $\sum_i u_i \otimes [m,u_i'] = 0$. By taking the u_i to be linearly independent over k, we conclude that $[m,u_i'] = 0$ for all i, whence $\tilde{\omega}(u) \in A \otimes K^{\mathfrak{m}}$.

Assertion (iv) is obvious. To prove (v), we may, taking (iv) into account, assume that $k = 1$. If $u \in G_n$, then $u \in NA_nK$, whence

$$\tilde{\omega}(u) \in \tilde{\omega}(A_nK) = A_n \otimes K.$$

9.2.4. We say that $\tilde{\omega}|G^{\mathfrak{k}}$ is *the canonical antihomomorphism of* $G^{\mathfrak{k}}$ *into* $A \otimes K^{\mathfrak{m}}$ (defined by the Iwasawa decomposition $\mathfrak{g} = \mathfrak{k} \oplus \mathfrak{a} \oplus \mathfrak{n}$).

9.2.5. Let I be a two-sided ideal of K, π the canonical mapping of K onto K/I, and $\tilde{\omega}$ the canonical mapping of G onto $A \otimes K$. We set

$$\tilde{\omega}_I = (1 \otimes \pi) \circ \tilde{\omega} : G \to A \otimes (K/I).$$

To study $G^{\mathfrak{k}} \cap (\operatorname{Ker} \tilde{\omega}_I)$, we introduce the auxiliary mappings ω, ω_I as follows. Let λ be the canonical mapping of $S(\mathfrak{g})$ onto G. The bilinear mapping $(x,y) \mapsto \lambda(x)y$ of $S(\mathfrak{p}) \times K$ into G defines an isomorphism of the vector space $S(\mathfrak{p}) \otimes K$ onto the vector space G (2.4.15); let Φ be the inverse isomorphism. Let $\mathfrak{b}$ be the orthogonal subspace of $\mathfrak{a}$ in $\mathfrak{p}$. Then $\mathfrak{p} = \mathfrak{a} \oplus \mathfrak{b}$ (1.13.7), hence $S(\mathfrak{p}) = A \oplus \mathfrak{b}S(\mathfrak{p})$, and this decomposition defines a projection r of $S(\mathfrak{p})$ onto A. We set

$$\omega = (r \otimes 1) \circ \Phi : G \to A \otimes K,$$

$$\omega_I = (1 \otimes \pi) \circ \omega : G \to A \otimes (K/I).$$

If $u \in G$ and $u = \sum \lambda(u_i)u_i'$, with $u_i \in S(\mathfrak{p})$ and $u_i' \in K$, then

$$\omega(u) = \sum r(u_i) \otimes u_i'.$$

9.2.6. Lemma (notation as in 9.2.1, 9.2.5). *Let* $u \in \lambda(S^n(\mathfrak{p}))K$. *Then* $\omega(u) \in A^n \otimes K$ *and* $\tilde{\omega}(u) - \omega(u) \in A_{n-1} \otimes K$.

Let $(a_1, \ldots, a_l)$ and $(b_1, \ldots, b_q)$ be bases for $\mathfrak{a}$ and $\mathfrak{b}$ respectively. The orthogonal subspace of $\mathfrak{a}$ in $\mathfrak{g}$ is $\mathfrak{b} \oplus \mathfrak{k}$, but also $\mathfrak{n} \oplus \mathfrak{k}$ (1.13.11). If we write each b_i in the form $n_i + k_i$ ($n_i \in \mathfrak{n}$, $k_i \in \mathfrak{k}$), then $(n_1, \ldots, n_q)$ is a basis for $\mathfrak{n}$. We have

$$u = \sum_{m_1+\cdots+m_{l+q}=n} \lambda(a_1^{m_1} \cdots a_l^{m_l} b_1^{m_{l+1}} \cdots b_q^{m_{l+q}})\, k_{m_1,\ldots,m_{l+q}},$$

where the $k_{m_1,\ldots,m_{l+q}}$ belong to K. Then

$$\omega(u) = \sum_{m_1+\cdots+m_l=n} a_1^{m_1} \cdots a_l^{m_l} \otimes k_{m_1,\ldots,m_l,0,\ldots,0} \in A^n \otimes K.$$

On the other hand, $\lambda(a_1^{m_1} \cdots a_l^{m_l} b_1^{m_{l+1}} \cdots b_q^{m_{l+q}})$ is congruent modulo G_{n-1} to

$$\sum_{0 \leqq m_i' \leqq mi+1} \binom{m_{l+1}}{m_1'} \cdots \binom{m_{l+q}}{m_q'} n_1^{m_1'} \cdots n_q^{m_q'} a_1^{m_1} \cdots a_l^{m_l} k_1^{m_{l+1}-m_1'} \cdots k_p^{m_{l+q}-m_q'}.$$

From 9.2.3 (v), we have, modulo $A_{n-1} \otimes K$,

$$\begin{aligned}\tilde{\omega}(u) &\equiv \tilde{\omega}\Big(\sum_{m_1+\cdots+m_{l+q}=n} a_1^{m_1} \cdots a_l^{m_l} k_1^{m_{l+1}} \cdots k_q^{m_{l+q}} k_{m_1,\ldots,m_{l+q}}\Big) \\ &= \sum_{m_1+\cdots+m_{l+q}=n} a_1^{m_1} \cdots a_l^{m_l} \otimes k_1^{m_{l+1}} \cdots k_q^{m_{l+q}} k_{m_1,\ldots,m_{l+q}} \\ &\equiv \sum_{m_1+\cdots+m_l=n} a_1^{m_1} \cdots a_l^{m_l} \otimes k_{m_1,\ldots,m_l,0,\ldots,0} \\ &= \omega(u).\end{aligned}$$

9.2.7. LEMMA (notation as in 9.2.1, 9.2.5). *Let I be a two-sided ideal of K. Then*

$$G^{\mathfrak{k}} \cap \operatorname{Ker} \tilde{\omega}_I = G^{\mathfrak{k}} \cap GI = G^{\mathfrak{k}} \cap \operatorname{Ker} \omega_I.$$

(a) We have $G^{\mathfrak{k}} \cap GI \subset G^{\mathfrak{k}} \cap \operatorname{Ker} \tilde{\omega}_I$. Indeed

$$GI = (AK + \mathfrak{n}G)I \subset AI + \mathfrak{n}G,$$

hence $\tilde{\omega}(GI) \subset A \otimes I$ and $\tilde{\omega}_I(GI) = 0$.

(b) We have $G^{\mathfrak{k}} \cap \operatorname{Ker} \omega_I \subset G^{\mathfrak{k}} \cap GI$. Indeed, let us identify the $\mathfrak{k}$-module G with the $\mathfrak{k}$-module $S(\mathfrak{p}) \otimes K$ under the isomorphism of 2.4.15. An element of $G^{\mathfrak{k}}$ can be identified with a $\mathfrak{k}$-invariant element of $S(\mathfrak{p}) \otimes K$. Let us assume that this element is u in $\operatorname{Ker} \omega_1$; its image in $S(\mathfrak{a}) \otimes (K/I)$ is zero. Let (x_i) be a basis for K/I. We write the image of u in $S(\mathfrak{p}) \otimes (K/I)$ in the form $\sum_i w_i \otimes x_i$, where $w_i \in S(\mathfrak{p})$ for all i. By virtue of the Killing form, we identify $S(\mathfrak{p})$ and $S(\mathfrak{a})$ with the algebras of polynomial functions over $\mathfrak{p}$ and $\mathfrak{a}$ respectively. Then r is precisely the restriction operation. Hence we have $\sum_i w_i(a) x_i = 0$ for all $a \in \mathfrak{a}$.

Let $\mathscr{K}$ be the smallest algebraic group of automorphisms of $\mathfrak{g}$ whose Lie algebra is $\operatorname{ad}_{\mathfrak{g}} \mathfrak{k}$ (cf. 1.13.13). Then $\mathscr{K}$ operates in $\mathfrak{k}$, $\mathfrak{p}$, K, K/I, $S(\mathfrak{p})$ and $S(\mathfrak{p}) \otimes (K/I)$. For all $k \in \mathscr{K}$, we have

$$\sum_i (kw_i) \otimes (kx_i) = \sum_i w_i \otimes x_i,$$

hence

$$\sum_I w_i(k^{-1}a) \cdot (kx_i) = 0 \quad \text{for all } a \in \mathfrak{a}.$$

Now (kx_i) is a basis for K/I, hence $w_i(\mathscr{K}\mathfrak{a}) = 0$ for all i. Since $\mathscr{K}\mathfrak{a}$ is dense in $\mathfrak{p}$ (1.13.13), we have $w_i = 0$ for all i, hence $u \in S(\mathfrak{p}) \otimes I$.

(c) $G^{\mathfrak{k}} \cap \operatorname{Ker} \tilde{\omega}_I \subset G^{\mathfrak{k}} \cap \operatorname{Ker} \omega_I$. Indeed, let $u \in G^{\mathfrak{k}} \cap \operatorname{Ker} \tilde{\omega}_I$. Write $u = u_0 + u_1 + \cdots + u_n$, with $u_i \in \lambda(S^i(\mathfrak{p}))K$. Each subspace $\lambda(S^i(\mathfrak{p}))K$ is stable under the adjoint action of $\mathfrak{k}$, hence $u_i \in G^{\mathfrak{k}}$ for all i. We give an indirect proof, and assume that the $\omega_I(u_i)$ are not all zero; let $\omega_I(u_i) = 0$ for $i = n, n-1, \ldots, p+1$ but $\omega_I(u_p) \neq 0$. From (a) and (b), we have $\tilde{\omega}(u_i) = 0$ for $i = n, n-1, \ldots, p+1$. For $i < p$, $\tilde{\omega}_I(u_i) \in A_{p-1} \otimes (K/I)$ (9.2.3 (v)). From 9.2.6,

$$\tilde{\omega}_I(u) \in \omega_I(u_p) + A_{p-1} \otimes (K/),$$

whence $\tilde{\omega}_I(u) \neq 0$, which is a contradiction.

9.2.8. PROPOSITION (notation as in 9.2.1). *The canonical antihomomorphism of $G^{\mathfrak{k}}$ into $A \otimes K^{\mathfrak{m}}$ is injective.*

Indeed, $G^{\mathfrak{k}} \cap \operatorname{Ker} \tilde{\omega} = 0$ from 9.2.7.

9.2.9. PROPOSITION (notation as in 9.2.1). *Let φ be the Harish-Chandra homomorphism of $G^{\mathfrak{h}}$ in to $H = A \otimes L$ defined by B, ψ the Harish-Chandra homomorphism of $M^{\mathfrak{l}}$ into L defined by B' and $\tilde{\omega}$ the canonical antihomomorphism of $G^{\mathfrak{k}}$ into $A \otimes K$. We define φ' and ψ' by*

$$\varphi'(z) = (\varphi(z^{\mathsf{T}}))^{\mathsf{T}}, \qquad \psi'(z) = (\psi(z^{\mathsf{T}}))^{\mathsf{T}}.$$

Then:

(i) $\tilde{\omega}(Z(\mathfrak{g})) \subset A \otimes Z(\mathfrak{m})$.

(ii) *If $z \in Z(\mathfrak{g})$, then*

$$\varphi(z) = (1 \otimes \psi)((\tilde{\omega}(z^{\mathsf{T}}))^{\mathsf{T}}), \qquad \varphi'(z) = (1 \otimes \psi')(\tilde{\omega}(z)).$$

(The operation ^{T} in $A \otimes K$ is of course defined as the tensor product of the operations ^{T} in A and K.)

For every $\alpha \in R(\mathfrak{g},\mathfrak{h})$, let $X_\alpha \in \mathfrak{g}^\alpha - \{0\}$. Let $\alpha_1, \ldots, \alpha_n$ be the positive roots, numbered in such a way that $\alpha_1, \ldots, \alpha_t$ are the positive roots which are zero on $\mathfrak{a}$. The elements $X_{\alpha_{t+1}}, \ldots, X_{\alpha_n}$ form a basis for $\mathfrak{n}$, and $X_{\pm\alpha_1}, \ldots, X_{\pm\alpha_t}, \in \mathfrak{m}$. Let $(H_1, \ldots, H_l)$ be a basis for $\mathfrak{l}$, and $(H_{l+1}, \ldots, H_s)$ a basis for $\mathfrak{a}$. The elements

$$u((q_i),(m_i),(p_i)) = X^{q_1}_{-\alpha_1} \cdots X^{q_n}_{-\alpha_n} H_1^{m_1} \cdots H_s^{m_s} X^{p_1}_{\alpha_1} \cdots X^{p_n}_{\alpha_n}$$

$$= X^{q_1}_{-\alpha_1} \cdots X^{q_n}_{-\alpha_n} H_1^{m_1} \cdots H_l^{m_l} X^{p_1}_{\alpha_1} \cdots X^{p_t}_{\alpha_t} H_{l+1}^{m_{l+1}} \cdots H_s^{m_s} X^{p_{t+1}}_{\alpha_{t+1}} \cdots X^{p_n}_{\alpha_n}$$

constitute a basis for G. Let

$$z = \sum \lambda_{(q_i),(m_i),(p_i)} u((q_i),(m_i),(p_i))$$

be an element of $Z(\mathfrak{g})$. If $\lambda_{(q_i),(m_i),(p_i)} \neq 0$, then the relation $[\mathfrak{h},z] = 0$ implies that

$$p_1\alpha_1 + \cdots + p_n\alpha_1 = q_1\alpha_{n_1} + \cdots + q_n\alpha_n,$$

hence

$$(p_{t+1}\alpha_{t+1} + \cdots + p_n\alpha_n) \,\big|\, \mathfrak{a} = (q_{t+1}\alpha_{t+1} + \cdots + q_n\alpha_n) \,\big|\, \mathfrak{a}.$$

If $p_{t+1}, \ldots, p_n$ are not all zero, then $u((q_i),(m_i),(p_i)) \in G\mathfrak{n}$. If they are all zero, then $q_{t+1}, \ldots, q_n$ are also zero. Hence $(\tilde{\omega}(z^{\mathsf{T}}))^{\mathsf{T}}$ is the sum, for $p_{t+1} = \cdots = p_n = q_{t+1} = \cdots = q_n = 0$, of the terms

$$\lambda_{(q_i),(m_i),(p_i)} H_{l+1}^{m_{l+1}} \cdots H_s^{m_s} \otimes \cdot X_{-\alpha_1}^{q_1} \cdots X_{-\alpha_t}^{q_t} H_1^{m_1} \cdots X_l^{m_l} H_{\alpha_1}^{p_1} \cdots HX \cdots X_{\alpha_t}^{p_t},$$

and so belongs to $A \otimes M$. On the other hand, $\tilde{\omega}(z) \in A \otimes K^{\mathfrak{m}}$ from 9.2.3 (ii) This proves (i). Furthermore, $\psi(X_{-\alpha_1}^{q_1} \cdots X_{-\alpha_t}^{q_t} H_1^{m_1} \cdots H_l^{m_l} X_{\alpha_3}^{p_1} \cdots X_{\alpha_t}^{p_t})$ is zero if $q_1 + \cdots + q_t + p_1 + \cdots + p_t > 0$, and is equal to $H_1^{m_1} \cdots H_l^{m_l}$ if $q_1 + \cdots + q_t + p_1 + \cdots + p_t = 0$. Hence

$$(1 \otimes \psi)((\tilde{\omega}(z^{\mathsf{T}}))^{\mathsf{T}}) = \sum \lambda_{(0),(m_i),(0)} H_{l+1}^{m_{l+1}} \cdots H_s^{m_s} \otimes H_1^{m_1} \cdots H_l^{m_l} = \varphi(z),$$

whence

$$\varphi'(z) = ((1 \otimes \psi)((\tilde{\omega}(z))^{\mathsf{T}}))^{\mathsf{T}} = (1 \otimes \psi')(\tilde{\omega}(z)).$$

9.2.10. Let $\sigma \in \mathfrak{k}^\wedge$, and let E^σ be the space of σ. Then K/I^σ can be canonically identified with $\operatorname{End}(E^\sigma)$. If $\tilde{\omega}$ is the canonical antihomomorphism of $G^{\mathfrak{k}}$ into $A \otimes K$, we denote the antihomomorphism $(1 \otimes \sigma) \circ \tilde{\omega}$ of $G^{\mathfrak{k}}$ into $A \otimes (K/I^\sigma) = A \otimes \operatorname{End}(E^\sigma)$ by $\tilde{\omega}_\sigma$. *Its kernel is*

$$G^{\mathfrak{k}} \cap GI^\sigma = G^{\mathfrak{k}} \cap I^\sigma G$$

(9.1.10 (ii) and 9.2.7). Its image is contained in $A \otimes (K^{\mathfrak{m}}/K^{\mathfrak{m}} \cap I^\sigma)$ (9.2.3 (ii)).

9.2.11. Moreover, let $\varrho \in \mathfrak{m}^\wedge$, let E^ϱ be the space of ϱ, and let $H_{\varrho,\sigma} = \operatorname{Hom}_{\mathfrak{m}}(E^\varrho,E^\sigma)$. From 9.1.12, $H_{\varrho,\sigma}$ is null or is a simple $K^{\mathfrak{m}}$-module of dimension $\operatorname{mtp}(\varrho,\sigma)$ (we recall that $\mathfrak{m}$ is reductive in $\mathfrak{g}$ from 1.13.7). Let $J^{\varrho,\sigma}$ be the kernel in $K^{\mathfrak{m}}$ of the corresponding representation of $K^{\mathfrak{m}}$. The algebra $K^{\mathfrak{m}}/J^{\varrho,\sigma}$ can be canonically identified with $\operatorname{End}(H_{\varrho,\sigma})$.

9.2.12. As in 9.2.10, we fix only σ and E^σ. From 9.1.11, the vector space E^σ can be canonically identified with $\bigoplus_{\varrho\in\mathfrak{m}^\wedge} (H_{\varrho,\sigma} \otimes E^\varrho)$ (this sum only comprises a finite number of non-zero terms). Under the identification $K/I^\sigma = \operatorname{End}(E^\sigma)$, $M/M \cap I^\sigma$ is identified with $\prod_{\varrho\in\mathfrak{m}^\wedge} (1 \otimes \operatorname{End}(E^\varrho))$, and, from 1.2.11, $K^{\mathfrak{m}}/K^{\mathfrak{m}} \cap I^{\sigma}$ is identified with the commutant of $M/M \cap I^{\sigma}$, that is, with $\prod_{\varrho \in \mathfrak{m}^{\wedge}} (\mathrm{End}(H_{\varrho,\sigma}) \otimes 1)$. Thus

$$K^{\mathfrak{m}} \cap I^{\sigma} = \bigcap_{\varrho \in \mathfrak{m}^{\wedge}} J^{\varrho,\sigma}, \qquad K^{\mathfrak{m}}/K^{\mathfrak{m}} \cap I^{\sigma} = \prod_{\varrho \in \mathfrak{m}^{\wedge}} (K^{\mathfrak{m}}/J^{\varrho,\sigma}) = \prod_{\varrho \in \mathfrak{m}^{\wedge}} \mathrm{End}(H_{\varrho,\sigma}).$$

Consequently, $\tilde{\omega}_{\sigma}$ is identified with an anti-homomorphism:

$$\tilde{\omega}_{\sigma} : G^{\mathfrak{k}} \to \prod_{\varrho \in \mathfrak{m}^{\wedge}} A \otimes (\mathrm{End}(H_{\varrho,\sigma}) \otimes 1) = \prod_{\varrho \in \mathfrak{m}^{\wedge}} A \otimes (K^{\mathfrak{m}}/J^{\varrho,\sigma}).$$

We denote by $\tilde{\omega}_{\varrho,\sigma}$ the antihomomorphism of $G^{\mathfrak{k}}$ into $A \otimes \mathrm{End}\, H_{\varrho,\sigma}$, or into $A \otimes (K^{\mathfrak{m}}/J^{\varrho,\sigma})$, deduced from $\tilde{\omega}_{\sigma}$ by composition with the projection onto the factor with index ϱ. We have

$$G^{\mathfrak{k}} \cap GI^{\sigma} = \mathrm{Ker}\, \tilde{\omega}_{\sigma} = \bigcap_{\varrho \in \mathfrak{m}^{\wedge}} \mathrm{Ker}\, \tilde{\omega}_{\varrho,\sigma}.$$

9.2.13. Lemma (notation as in 9.2.1, 9.2.12). *Let φ be the Harish-Chandra homomorphism of $G^{\mathfrak{h}}$ into $H = A \otimes L$ defined by B. Let $\sigma \in \mathfrak{k}^{\wedge}$, $\varrho \in \mathfrak{m}^{\wedge}$, and μ be the smallest weight of ϱ, identified with a homomorphism of L into k. Then for all $z \in Z(\mathfrak{g})$, we have*

$$\tilde{\omega}_{\varrho,\sigma}(z) = (1 \otimes \mu)((\varphi(z^{\mathsf{T}}))^{\mathsf{T}}).$$

Let ψ and ψ' be as in 9.2.9, and $\pi_{\varrho,\sigma}$ the canonical homomorphism of $K^{\mathfrak{m}}$ onto $K^{\mathfrak{m}}/J^{\varrho,\sigma}$; we consider $\pi_{\varrho,\sigma}$ as a representation of $K^{\mathfrak{m}}$ in $H_{\varrho,\sigma}$. The elements of $H_{\varrho,\sigma}$ are $\mathfrak{m}$-homomorphisms; hence for $y \in Z(\mathfrak{m})$, $\pi_{\varrho,\sigma}(y)$ is the scalar defined by $\varrho(y)$, or by $\varrho^*(y^{\mathsf{T}})$, i.e., $(-\mu)(\psi(y^{\mathsf{T}})) = \mu(\psi'(y))$ (7.2.8 and 7.4.4). For $z \in Z(\mathfrak{g})$, we have $\tilde{\omega}(z) \in \mathrm{A} \otimes Z(\mathfrak{m})$ from 9.2.9(i), and then

$$\tilde{\omega}_{\varrho,\sigma}(z) = (1 \otimes \pi_{\varrho,\sigma})\, \tilde{\omega}(z) = (1 \otimes \mu)(1 \otimes \psi')\, \tilde{\omega}(z)$$
$$= (1 \otimes \mu)(\varphi(z^{\mathsf{T}})^{\mathsf{T}}) \qquad \text{from 9.3.9 (ii).}$$

9.2.14. Let us retain the notation of 9.2.1, and let $\tilde{\omega}$ still be the canonical antihomomorphism of $G^{\mathfrak{k}}$ into $A \otimes K$. If $\lambda \in \mathfrak{a}^*$, $\sigma \in \mathfrak{k}^{\wedge}$ and $\varrho \in \mathfrak{m}^{\wedge}$, we set

$$\tilde{\omega}_{\lambda} = (\lambda \otimes 1) \circ \tilde{\omega} : G^{\mathfrak{k}} \to K^{\mathfrak{m}},$$
$$\tilde{\omega}_{\sigma,\lambda} = (\lambda \otimes 1) \circ \tilde{\omega}_{\sigma} : G^{\mathfrak{k}} \to K^{\mathfrak{m}}/K^{\mathfrak{m}} \cap I^{\sigma} = \prod_{\varrho \in \mathfrak{m}^{\wedge}} (K^{\mathfrak{m}}/J^{\varrho,\sigma}),$$
$$\tilde{\omega}_{\varrho,\sigma,\lambda} = (\lambda \otimes 1) \circ \tilde{\omega}_{\varrho,\sigma} : G^{\mathfrak{k}} \to K^{\mathfrak{m}}/J^{\varrho,\sigma} = \mathrm{End}(H^{\varrho,\sigma}).$$

9.2.15. Proposition (notation as in 9.2.1), (i) $G = A \oplus (G\mathfrak{k} + \mathfrak{n}G)$.

(ii) *Let p be the projection of G onto A defined by the preceding decomposition. Then $p|G^{\mathfrak{k}}$ is a homomorphism of the algebra $G^{\mathfrak{k}}$ into the algebra A whose kernel is $G^{\mathfrak{k}} \cap G\mathfrak{k} = G^{\mathfrak{k}} \cap \mathfrak{k}G$.*

We have

$$G = AK \oplus \mathfrak{n}G = A \oplus (AK\mathfrak{k} \oplus \mathfrak{n}G),$$

$$G\mathfrak{k} \subset (AK + \mathfrak{n}G)\mathfrak{k} \subset AK\mathfrak{k} + \mathfrak{n}G,$$

whence (i). Given this, (ii) follows from 9.1.10 (ii), 9.2.3 (iii) and 9.2.7 for $I = K\mathfrak{k}$.

9.2.16. With the notation of 9.2.15, $p|G^{\mathfrak{k}}$ is termed the *canonical homomorphism of $G^{\mathfrak{k}}$ into A.*

References: [87], [103].

9.3. The principal series

9.3.1. We retain the notation of 9.2.1.

Let $\varrho \in (\mathfrak{m} \oplus \mathfrak{a})^{\wedge}$; in particular, $\varrho|\mathfrak{a}$ can be identified with a linear form on $\mathfrak{a}$, and $\varrho|\mathfrak{m} \in \mathfrak{m}^{\wedge}$. Let τ_{ϱ} be the representation of $\mathfrak{q}$ which extends ϱ and is zero on $\mathfrak{n}$. We denote the $\mathfrak{g}$-module corresponding to $\mathrm{ind}(\tau_{\varrho},\mathfrak{g})$ by $M'(\varrho)$. The $\mathfrak{g}$-module corresponding to $\mathrm{coind}(\tau_{\varrho},\mathfrak{g})$, which is a set of linear mappings of G into the space of ϱ, can be canonically identified with the dual of $M'(\varrho^*)$ (5.5.4). In this coinduced $\mathfrak{g}$-module, the sum of the finite-dimensional simple sub-$\mathfrak{k}$-modules is a sub-$\mathfrak{g}$-module (1.7.9), which we denote by $X(\varrho)$.

Let τ be the adjoint representation of $\mathfrak{k}$ in $\mathfrak{g}$. Let Γ be the set of congruence classes modulo τ in $\mathfrak{k}^{\wedge}$. For $\gamma \in \Gamma$, let $X(\varrho,\gamma)$ be the sum of the simple sub-$\mathfrak{k}$-modules of $X(\varrho)$ which belong to γ. Then $X(\varrho)$ is the direct sum of the $X(\varrho,\gamma)$, and each $X(\varrho,\gamma)$ is a sub-$\mathfrak{g}$-module of $X(\varrho)$ (9.1.16). The family of $\mathfrak{g}$-modules $X(\varrho,\gamma)$, for $\varrho \in (\mathfrak{m} \oplus \mathfrak{a})^{\wedge}$ and $\gamma \in \Gamma$, is termed the *algebraic principal series of $\mathfrak{g}$-modules* (relative to $\mathfrak{k},\mathfrak{h},B$).

9.3.2. LEMMA (notation as in 9.2.1). *Let $\varrho \in (\mathfrak{m} \oplus \mathfrak{a})^{\wedge}$, let ω be the largest weight of ϱ (relative to $\mathfrak{h}$ and B'), and let $\delta = \frac{1}{2}\sum_{\alpha\in R_+}\alpha$. Then $M'(\varrho)$ can be identified with a quotient $\mathfrak{g}$-module of the Verma module $M(\omega + \delta)$ (constructed relatively to $\mathfrak{h}$ and B).*

Let $\mathfrak{b} = \mathfrak{h} + \sum_{\alpha\in R_+}\mathfrak{g}^{\alpha}$ be the Borel subalgebra of $\mathfrak{g}$ defined by $\mathfrak{h}$ and B; then $\mathfrak{b} \subset \mathfrak{q}$. Let ω' be the linear form on $\mathfrak{b}$ which extends ω and is zero on $\sum_{\alpha\in R'_+}\mathfrak{g}^{\alpha}$. We set

$$\sigma = \mathrm{ind}(\omega',\mathfrak{q}), \qquad \tau = \mathrm{ind}(\sigma,\mathfrak{g}) = \mathrm{ind}(\omega',\mathfrak{g}).$$

Then ω', and hence σ, is zero on the ideal $\mathfrak{n}$ of $\mathfrak{q}$, and σ can be identified with $\mathrm{ind}(\omega'|\mathfrak{h} + \sum_{\in R_{\alpha}'}\mathfrak{g}^{x},\mathfrak{m} \oplus \mathfrak{a})$ (5.1.2). From 7.2.2 (extended to the case of reductive Lie algebras), the space V_σ of σ contains a sub-$(\mathfrak{m} \oplus \mathfrak{a})$-module V'_σ such that the quotient representation of $\mathfrak{m} \oplus \mathfrak{a}$ (or of $\mathfrak{q}$) in V_σ/V'_σ is equivalent to ϱ. Hence the $\mathfrak{g}$-module corresponding to τ, which is precisely $M(\omega + \delta)$, contains a sub-$\mathfrak{g}$-module M such that the representation of $\mathfrak{g}$ in $M(\omega + \delta)/M$ is equivalent to $\mathrm{ind}(\tau_\varrho,\mathfrak{g})$ (with the notation of 9.3.1).

9.3.3. THEOREM (notation as in 9.2.1). *Let* $\varrho \in (\mathfrak{m} \oplus \mathfrak{a})^\wedge$.

(i) *The* $\mathfrak{g}$-*module* $X(\varrho)$ *is a Harish-Chandra module.*

(ii) *If* $\sigma \in \mathfrak{k}^\wedge$, *the multiplicity of* σ *in* $X(\varrho)$ *is* $\mathrm{mtp}(\varrho|\mathfrak{m},\sigma)$.

(iii) *Let* μ *be the smallest weight of* ϱ *(relative to* $\mathfrak{h}$ *and* B'*), and* $\delta = \frac{1}{2}\sum_{\alpha \in R_+}\alpha$. *Let* w_0 *be the element of* $W(\mathfrak{g},\mathfrak{h})$ *which transforms* B *into* $-B$. *Then* $X(\varrho)$ *has a central character, which is equal to* $\chi_{w_0\mu+\delta}$.

Assertion (i) is obvious, and (ii) follows from 5.5.7 and 5.5.8. With the notation of (iii), the largest weight of ϱ^* is $-\mu$, hence $M'(\varrho^*)$ has the central character $\chi_{-\mu+\delta}$ (9.3.2, 7.1.9). Hence $M'(\varrho^*)^*$ and *a fortiori* $X(\varrho)$ have a central character χ; moreover, for $z \in Z(\mathfrak{g})$ we have, from 7.4.9,

$$\chi(z) = \chi_{-\mu+\delta}(z^\mathsf{T}) = \chi_{w_0\mu+\delta}(z).$$

9.3.4. LEMMA (notation as in 9.2.1). *Let* $\varrho \in (\mathfrak{m} \oplus \mathfrak{a})^\wedge$ *and* $\sigma \in \mathfrak{k}^\wedge$, *let* E^ϱ *and* E^σ *be the corresponding modules, let*

$$H = \mathrm{Hom}_\mathfrak{k}(E^\sigma,X(\varrho)), \qquad H' = \mathrm{Hom}_\mathfrak{m}(E^\varrho,E^\varrho),$$

and let ε *be the restriction to* $X(\varrho)$ *of the canonical projection*

$$\mathrm{Hom}_{U(\mathfrak{q})}(U(\mathfrak{g}),E^\varrho) \to E^\varrho.$$

(i) *For* $h \in H$ *and* $h' \in H'$, *the mapping* $\varepsilon \circ h \circ h'$ *of* E^ϱ *into* E^ϱ *has the form* $\varphi(h,h') \cdot 1$, *where* $\varphi(h,h') \in k$.

(ii) *The bilinear form* φ *enables us to identify each of the vector spaces* H *and* H' *with the dual of the other.*

If $h \in H$ and $h' \in H'$, the mapping $\varepsilon \circ h \circ h'$ is an $\mathfrak{m}$-homomorphism of E^ϱ into E^ϱ; since E^ϱ is a simple $\mathfrak{m}$-module, we deduce (i). Let $h \in H - \{0\}$. Then

$$\begin{aligned}\varepsilon(h(E^\sigma)) &= \varepsilon(NAh(E^\sigma)) && \text{because } \varepsilon \text{ is a } \mathfrak{q}\text{-homomorphism}\\ &= \varepsilon\,(NAKh(E^\sigma)) && \text{because } h \text{ is a } \mathfrak{k}\text{-homomorphism}\\ &= \varepsilon(Gh(E^\sigma)),\end{aligned}$$

and this is non-zero from the uniqueness assertion of 5.5.3 [where we take

$V' = Gh(E^\sigma)$]. Hence $\varepsilon \circ h \neq 0$. Now E^σ is a semi-simple $\mathfrak{m}$-module, and $\varepsilon \circ h$ is an $\mathfrak{m}$-homomorphism, so there exists $h' \in H'$ such that $(\varepsilon \circ h) \circ h' \neq 0$, whence $\varphi(h,h') \neq 0$. Since H and H' have the same dimension (9.3.3 (ii)), this proves (ii).

9.3.5. Given this, the spherical functions of $X(\varrho)$ can be calculated in the following way:

PROPOSITION (notation as in 9.2.1). *Let ϱ, σ, E^ϱ, E^σ, H and H' be as in* 9.3.4. *We identify each of the vector spaces H and H' with the dual of the other. Let us consider H as a left $G^{\mathfrak{k}}$-module and H' as a left $K^{\mathfrak{m}}$-module (by virtue of the action of $G^{\mathfrak{k}}$ in $X(\varrho)$ and of $K^{\mathfrak{m}}$ in E^σ). Let λ be the linear form on $\mathfrak{a}$ which can be identified with $\varrho|\mathfrak{a}$ and $\tilde{\omega}_\lambda$ the corresponding anti-homomorphism of $G^{\mathfrak{k}}$ into $K^{\mathfrak{m}}$* (9.2.14).

(i) *For every $u \in G^{\mathfrak{k}}$, the action of u in H and the action of $\tilde{\omega}_\lambda(u)$ in H' are transpositions of each other.*

(ii) *The annihilator of the $G^{\mathfrak{k}}$-module H is the kernel of $\tilde{\omega}_{\varrho|\mathfrak{m},\sigma,\lambda}$* (9.2.14).

Let $u \in G^{\mathfrak{k}}$. We write $u = \sum u_i u_i' + u''$ with $u_i \in A$, $u_i' \in K$ and $u'' \in \mathfrak{n}G$. Let π be the representation of $\mathfrak{g}$ defined by $X(\varrho)$. For $h \in H$ and $h' \in H'$, we have, with the notation of 9.3.4,

$$\begin{aligned}
\varphi(uh,h') \cdot 1 &= \varepsilon \circ \pi(u) \circ h \circ h' \\
&= \sum_i u_i(\lambda)\varepsilon \circ \pi(u_i') \circ h \circ h' && \text{because } \varepsilon \text{ is a } \mathfrak{q}\text{-homomorphism} \\
&= \sum_i u_i(\lambda)\varepsilon \circ h \circ \sigma(u_i') \circ h' && \text{because } h \text{ is a } \mathfrak{k}\text{-homomorphism} \\
&= \varepsilon \circ h \circ \sigma\Big(\sum_i u_i(\lambda)u_i'\Big) \circ h' \\
&= \varepsilon \circ h \circ \sigma(\tilde{\omega}_\lambda(u)) \circ h' \\
&= \varepsilon \circ h \circ (\tilde{\omega}_\lambda(u)h') \\
&= \varphi(h,\tilde{\omega}_\lambda(u)h') \cdot 1
\end{aligned}$$

whence (i). Assertion (ii) follows from (i) by the definition of $\tilde{\omega}_{\varrho|\mathfrak{m},\sigma,\lambda}$ (cf. 9.2.12, where the module H' is denoted by $H_{\varrho|\mathfrak{m},\sigma}$).

References: [64], [87].

9.4. The subquotient theorem

9.4.1. We retain the notation of 9.2.1, 9.2.12 and 9.2.14. Let $\sigma \in \mathfrak{k}^\wedge$ and $u \in G^{\mathfrak{k}}$. Then $\tilde{\omega}(u) \in A \otimes K^{\mathfrak{m}}$. Let us identify $K^{\mathfrak{m}}/K^{\mathfrak{m}} \cap I^\sigma$ with $\prod_{\varrho \in \mathfrak{m}^\wedge} \mathrm{End}(H_{\varrho,\sigma})$ (cf. 9.2.12). For all $\varrho \in \mathfrak{m}^\wedge$, $\tilde{\omega}_{\varrho,\sigma}(u)$ can be identified with an endomorphism of the free A-module $A \otimes H_{\varrho,\sigma}$ with finite basis; we can consider the characteristic polynomial of this endomorphism

$$\det\,(T - \tilde{\omega}_{\varrho,\sigma}(u)),$$

where T denotes an indeterminate. This polynomial belongs to $A[T]$; it is equal to 1 if $\operatorname{mtp}(\varrho,\sigma) = 0$. Let

$$f(T;\sigma,u) = \prod_{\varrho} \det\,(T - \tilde{\omega}_{\varrho,\sigma}(u)) \in A[T].$$

Then

(1) $f(\tilde{\omega}_{\varrho,\sigma}(u);\sigma,u) = 0$ for all $\varrho \in \mathfrak{m}^{\wedge}$ such that $\operatorname{mtp}(\varrho,\sigma) > 0$.

Let $W = W(\mathfrak{g},\mathfrak{h})$, and let φ and φ' be defined as in 9.2.9. From 7.4.5, there exists a homomorphism $w \mapsto w^{\sim}$ of W onto a group $W^{\sim}$ of automorphisms of H such that $\varphi'|Z(\mathfrak{g})$ is an isomorphism of $Z(\mathfrak{g})$ onto $H^{W^{\sim}}$ (the set of elements of H which are invariant under $W^{\sim}$); for every $w \in W$, we also denote by $w^{\sim}$ the permutation of $\mathfrak{h}^*$ corresponding to the automorphism $w^{\sim}$ of H. Since $A[T] \subset H[T]$, we can define

$$\bar{f}(T;\sigma,u) = \prod_{w \in W} w^{\sim}(f(T;\sigma,u)) \in H^{W^{\sim}}[T].$$

Let q be the degree of $\bar{f}$. There exist $z_0, \ldots, z_{q-1} \in Z(\mathfrak{g})$ such that

(2) $$\bar{f}(T;\sigma,u) = T^q + \varphi'(z_{q-1})T^{q-1} + \cdots + \varphi'(z_1)T + \varphi'(z_0).$$

We set

$$v = u^q + z_{q-1}\,u^{q-1} + \cdots + z_1 u + z_0 \in G^{\mathfrak{k}}.$$

9.4.2. Lemma (notation as in 9.4.1). *We have* $v \in G^{\mathfrak{k}} \cap GI^{\sigma}$.

Let $\varrho \in \mathfrak{m}^{\wedge}$ be such that $\operatorname{mtp}(\varrho,\sigma) > 0$. We set

$$v' = \tilde{\omega}_{\varrho,\sigma}(u)^q + \varphi'(z_{q-1})\tilde{\omega}_{\varrho,\sigma}(u)^{q-1} + \cdots + \varphi'(z_1)\tilde{\omega}_{\varrho,\sigma}(u) + \varphi'(z_0)$$
$$\in H \otimes \operatorname{End} H_{\varrho,\sigma} = A \otimes L \otimes \operatorname{End} H_{\varrho,\sigma}.$$

From (1) and (2) we have

$$v' = \bar{f}(\tilde{\omega}_{\varrho,\sigma}(u);\sigma,u) = 0.$$

On the other hand, let $\mu \in \mathfrak{l}^*$ be the smallest weight of ϱ, and let us identify μ with a homomorphism of L into k. Then

$$1 \otimes \mu \otimes 1 : A \otimes L \otimes \operatorname{End} H_{\varrho,\sigma} \to A \otimes \operatorname{End} H_{\varrho,\sigma}$$

is a homomorphism, and we have

$$\begin{aligned} 0 &= (1 \otimes \mu \otimes 1)v' \\ &= \tilde{\omega}_{\varrho,\sigma}(u)^q + (1 \otimes \mu)(\varphi'(z_{q-1}))\tilde{\omega}_{\varrho,\sigma}(u)^{q-1} + \cdots + (1 \otimes \mu)(\varphi'(z_0)) \\ &= \tilde{\omega}_{\varrho,\sigma}(v) \end{aligned}$$

from 9.2.13. Hence $v \in G^{\mathfrak{k}} \cap GI^{\sigma}$ (9.2.10).

9.4.3. Lemma (notation as in 9.2.1). *Let $\sigma \in \mathfrak{k}^{\wedge}$, let τ be a simple representation of $G^{\mathfrak{k}}$, and let E and Y be the modules corresponding to σ and τ. We assume that* $\operatorname{Ker} \tau \supset G^{\mathfrak{k}} \cap GI^{\sigma}$. *Then:*

(i) *Y is finite-dimensional.*

(ii) *There exists $\pi \in (\mathfrak{m} \oplus \mathfrak{a})^{\wedge}$ such that Y is isomorphic to a subquotient of* $\operatorname{Hom}_{\mathfrak{k}}(E, X(\pi))$.

The algebra $G^{\mathfrak{k}}$ has a filtration such that the associated graded algebra is commutative and of finite type (1.7.10). Hence $\tau(Z(\mathfrak{g})) \subset k \cdot 1$ (2.6.4). Let φ and φ' be as in 9.2.9. From 7.4.8, there exists $\mu \in \mathfrak{h}^*$ such that

$$\tau(z) = \mu(\varphi'(z)) \cdot 1 \quad \text{for all } z \in Z(\mathfrak{g}). \tag{1}$$

With the notation of 9.4.1, let L be the set of the $u \in G^{\mathfrak{k}}$ such that

$$\tilde{\omega}_{\varrho,\sigma,w^{\sim}}(u) = 0 \quad \text{for all } \varrho \in \mathfrak{m}^{\wedge} \text{ and all } w \in W. \tag{2}$$

This is a two-sided ideal of $G^{\mathfrak{k}}$. Let $u \in L$. In $\bar{f}(T;\sigma,u)$, the coefficients of the monomials in $T^{q-1}, T^{q-2}, \ldots$ (with the notation of 9.4.1) are zero if we apply the homomorphism $\mu : H \to k$ to them (because of (2)). Hence $\mu(\varphi'(z_i)) = 0$ for all i. We have

$$\begin{aligned} 0 &= \tau(v) \quad \text{from 9.4.2} \\ &= \tau(u)^q + \mu(\varphi'(z_{q-1}))\tau(u)^{q-1} + \cdots + \mu(\varphi'(z_0)) \quad \text{from (1)} \\ &= \tau(u)^q. \end{aligned}$$

Thus, every element of the two-sided ideal $\tau(L)$ of $\tau(G^{\mathfrak{k}})$ is nilpotent. From 3.1.14, $\tau(L)$ is nilpotent. Hence there exists an integer n such that

$$\operatorname{Ker} \tau \supset \Bigg[\bigcap_{\substack{\varrho \in \mathfrak{m}^{\wedge}, \\ \operatorname{mtp}(\varrho,\sigma) > 0, \\ w \in W}} (\operatorname{Ker} \tilde{\omega}_{\varrho,\sigma,w^{\sim}(\mu)|\mathfrak{a}}) \Bigg]^n .$$

Since $\operatorname{Ker} \tau$ is primitive, there exist $\varrho \in \mathfrak{m}^{\wedge}$ and $w \in W$ such that

$$\operatorname{Ker} \tau \supset \operatorname{Ker} \tilde{\omega}_{\varrho,\sigma,w^{\sim}(\mu)|\mathfrak{a}} .$$

Let π be the element of $(\mathfrak{m} \oplus \mathfrak{a})^\wedge$ which extends ϱ and $w^\sim(\mu)|\mathfrak{a}$. From 9.3.5 (ii), Ker τ contains the annihilator J of the $G^\mathfrak{k}$-module $H = \mathrm{Hom}_\mathfrak{k}(E,X(\pi))$. From 9.3.3 (ii), H is finite-dimensional, hence Ker τ has finite codimension in $G^\mathfrak{k}$, which proves (i). Let

$$H = H_0 \supset H_1 \supset \cdots \supset H_p = 0$$

be a Jordan—Hölder series of the $G^\mathfrak{k}$-module H. Let J_i be the annihilator of the simple $G^\mathfrak{k}$-module H_i/H_{i+1}. Then Ker $\tau \supset J \supset (\cap J_i)^p$, hence there exists i such that Ker $\tau \supset J_i$. Then Y is isomorphic to the $G^\mathfrak{k}$-module H_i/H_{i+1}.

9.4.4. THEOREM (notation as in 9.2.1). *Let V be a simple Harish-Chandra $\mathfrak{g}$-module. There exists $\pi \in (\mathfrak{m} \oplus \mathfrak{a})^\wedge$ such that V is isomorphic to a subquotient of $X(\pi)$.*

Let us choose $\sigma \in \mathfrak{k}^\wedge$ such that $V_\sigma \neq 0$; let E be a $\mathfrak{k}$-module of class σ. Let $Y = \mathrm{Hom}_\mathfrak{k}(E,V)$, which is a simple $G^\mathfrak{k}$-module whose annihilator contains $G^\mathfrak{k} \cap GI^\sigma$ (9.1.12 (i)). From 9.4.3, there exists $\pi \in (\mathfrak{m} \oplus \mathfrak{a})^\wedge$ such that Y is isomorphic to a subquotient of $\mathrm{Hom}_\mathfrak{k}(E,X(\pi))$. Then V is isomorphic to a subquotient of $X(\pi)$ (9.1.14).

References: [65], [87], [103].

9.5. Finiteness theorems

9.5.1. THEOREM (notation as in 9.2.1). *Let V be a simple $\mathfrak{g}$-module, $\sigma \in \mathfrak{k}^\wedge$, and $p = \sup_{\varrho \in \mathfrak{m}^\wedge} \mathrm{mtp}(\varrho,\sigma)$. Then $\mathrm{mtp}(\sigma,V) \leq p$.*

We can assume that $V_\sigma \neq 0$, so that V is a Harish-Chandra module (9.1.6). The theorem then follows from 9.4.4 and 9.3.3 (ii).

9.5.2. LEMMA (notation as in 9.2.1, 9.2.10). *Let $\sigma \in \mathfrak{k}^\wedge$, so that*

$$\tilde{\omega}_\sigma(Z(\mathfrak{g})) \subset \tilde{\omega}_\sigma(G^\mathfrak{k}) \subset A \otimes (K/I^\sigma).$$

Then:

(i) *$A \otimes (K/I^\sigma)$ is a module of finite type over $\tilde{\omega}_\sigma(Z(\mathfrak{g}))$.*
(ii) *$G^\mathfrak{k}/G^\mathfrak{k} \cap GI^\sigma$ is a module of finite type over $Z(\mathfrak{g})/Z(\mathfrak{g}) \cap GI^\sigma$.*

We introduce the notation of 9.2.9. From 7.4.5 and AC V, p. 33, H is a module of finite type over $\varphi'(Z(\mathfrak{g}))$, which is a Noetherian ring. Now

$$H \supset A \otimes \psi'(Z(\mathfrak{m})) \supset \varphi'(Z(\mathfrak{g}))$$

from 9.2.9, hence $A \otimes \psi'(Z(\mathfrak{m}))$ is a module of finite type over

$$\varphi'(Z(\mathfrak{g})) = (1 \otimes \psi')(\tilde{\omega}(Z(\mathfrak{g}))).$$

Since $\psi'|Z(\mathfrak{m})$ is injective (7.4.5), $A \otimes Z(\mathfrak{m})$ is a module of finite type over $\tilde{\omega}(Z(\mathfrak{g}))$.

It is clear that $A \otimes (K/I^\sigma)$ is a module of finite type over A and *a fortiori* over $A \otimes (Z(\mathfrak{m})/Z(\mathfrak{m}) \cap I^\sigma)$. Hence $A \otimes (K/I^\sigma)$ is a module of finite type over $\tilde{\omega}_\sigma(Z(\mathfrak{g}))$. *A fortiori*, $\tilde{\omega}_{I^\sigma}(G^{\mathfrak{k}})$ is a module of finite type over $\tilde{\omega}_\sigma(Z(\mathfrak{g}))$. Since the kernel of $\tilde{\omega}_{I^\sigma}|G^{\mathfrak{k}}$ is $G^{\mathfrak{k}} \cap GI^\sigma$ (9.2.10), we have proved the lemma.

9.5.3. THEOREM (notation as in 9.2.1). *Let $\sigma \in \mathfrak{k}^\wedge$ and let χ be a homomorphism of $Z(\mathfrak{g})$ into k. Up to isomorphism, there exist only a finite number of simple $\mathfrak{g}$-modules V with the following properties:*

(i) $V_\sigma \neq 0$;

(ii) *the central character of V is χ.*

(a) Let $J = (\text{Ker}\,\chi)G$, which is a two-sided ideal of G. From 9.5.2, the algebra $G^{\mathfrak{k}}/G^{\mathfrak{k}} \cap (GI^\sigma + J)$ is finite-dimensional. Hence, up to isomorphism, it only has a finite number of simple modules.

(b) Let X be a simple $\mathfrak{k}$-module of class σ. For every simple $\mathfrak{g}$-module V having properties (i) and (ii), let H_V be the $G^{\mathfrak{k}}$-module $\text{Hom}_{\mathfrak{k}}(X,V)$. From 9.1.12 (i), H_V is simple and its annihilator contains $G^{\mathfrak{k}} \cap (G^{\mathfrak{k}} + J)$. On the other hand, if V' is a simple $\mathfrak{g}$-module having properties (i) and (ii), and H_V and $H_{V'}$ are isomorphic as $G^{\mathfrak{k}}$-modules, then V and V' are isomorphic as $\mathfrak{g}$-modules (9.1.12 (ii)). The theorem then follows from (a).

9.5.4. We retain the notation of 9.2.1. Let σ_0 be the one-dimensional null representation of $\mathfrak{k}$. If V is a $\mathfrak{g}$-module, then V_{σ_0} is the set $V^{\mathfrak{k}}$ of $\mathfrak{k}$-invariant elements of V. We say that V is *spherical* if $\dim V^{\mathfrak{k}} = 1$. If V is simple and $V^{\mathfrak{k}} \neq 0$, then V is spherical (9.5.1).

Let $\varrho \in (\mathfrak{m} \oplus \mathfrak{a})^\wedge$. If $\varrho|\mathfrak{m} \neq 0$, then $X(\varrho)^{\mathfrak{k}} = 0$; if $\varrho|\mathfrak{m} = 0$, then $X(\varrho)$ is spherical. This follows from 9.3.3 (ii).

9.5.5. PROPOSITION (notation as in 9.2.1). *Let V be a spherical $\mathfrak{g}$-module, and p the canonical homomorphism of $G^{\mathfrak{k}}$ into A. There exists $\lambda \in \mathfrak{a}^*$ such that, for all $u \in G^{\mathfrak{k}}$, $u_V|V^{\mathfrak{k}}$ is the homothety with ratio $(p(u))(\lambda)$.*

Let $v \in V^{\mathfrak{k}} - \{0\}$. There exists a homomorphism ζ of $G^{\mathfrak{k}}$ into k such that $u \cdot v = \zeta(u)v$ for all $u \in G^{\mathfrak{k}}$ (9.1.3). We have $\text{Ker}\,\zeta \supset G^{\mathfrak{k}} \cap G\mathfrak{k}$. From 9.2.15, there exists a homomorphism ζ' of $p(G^{\mathfrak{k}})$ into k such that $\zeta = \zeta' \circ p$. Now A is a $p(G^{\mathfrak{k}})$-module of finite type (9.5.1 (i)), hence ζ' can be extended to a homomorphism of A into k (AC V, p. 39). Such a homomorphism has the form $a \mapsto a(\lambda)$, where λ is a fixed element of $\mathfrak{a}^*$, whence we have proved the proposition.

9.5.6. PROPOSITION (notation as in 9.2.1). *Let $\lambda \in \mathfrak{a}^*$. There exists one, and up to isomorphism only one, spherical simple $\mathfrak{g}$-module V such that for every $u \in G^{\mathfrak{k}}$, $u_V | V^{\mathfrak{k}}$ is the homothety with ratio $(p(u))(\lambda)$ (where p denotes the canonical homomorphism of $G^{\mathfrak{k}}$ into A).*

This follows from 9.1.12 and 9.2.15.

References: [64], [87].

9.6. Spherical modules in the diagonal case

9.6.1. Let us assume that k is algebraically closed. Let $\mathfrak{v}$ be a semi-simple Lie algebra, $\mathfrak{w}$ a Cartan subalgebra of $\mathfrak{v}$, $\mathfrak{g} = \mathfrak{v} \times \mathfrak{v}$, and let us define θ, $\mathfrak{k}$, $\mathfrak{p}$, $\mathfrak{a}$, $\mathfrak{m}$, $\mathfrak{h}$, S, R, C, B, S_+, S_-, R_+, R_-, $\mathfrak{x}$, $\mathfrak{x}_-$, $\mathfrak{n}$ and $\mathfrak{n}_-$ as in 1.13.14. We set

$$\varepsilon = \frac{1}{2} \sum_{\beta \in S_+} \beta \qquad \delta = \frac{1}{2} \sum_{\alpha \in S_+} \alpha = (\varepsilon, -\varepsilon).$$

We denote the isomorphism $x \mapsto (x,x)$ of $\mathfrak{v}$ onto $\mathfrak{k}$ by i; the elements of $\mathfrak{v}^\wedge$ are the $\sigma \circ i$, where $\sigma \in \mathfrak{k}^\wedge$.

An element of $(\mathfrak{m} + \mathfrak{a})^\wedge$ is a linear form on $\mathfrak{h}$, i.e. a pair (λ_1,λ_2), where $\lambda_1,\lambda_2 \in \mathfrak{w}^*$. For $\lambda_1\lambda_2 \in \mathfrak{w}^*$, we can thus consider the $\mathfrak{g}$-modules

$$M(\lambda_1,\lambda_2), \qquad M'(\lambda_1,\lambda_2) = M(\lambda_1 + \varepsilon, \lambda_2 - \varepsilon),$$

and $X(\lambda_1,\lambda_2)$, which is the sum of the finite-dimensional sub-$\mathfrak{k}$-modules of $M'(-\lambda_1,-\lambda_2)^*$.

9.6.2. PROPOSITION (notation as in 9.6.1). *Let $\lambda_1,\lambda_2 \in \mathfrak{w}^*$, and $\sigma \in \mathfrak{k}^\wedge$. Let us define Γ as in 9.3.1. Then:*

(i) $\mathrm{mtp}(\sigma, X(\lambda_1,\lambda_2)) = \mathrm{mtp}(\lambda_1 + \lambda_2, \sigma \circ i)$.

(ii) *If $\lambda_1 + \lambda_2 \notin P(S)$, then $X(\lambda_1,\lambda_2) = 0$.*

(iii) *If $\lambda_1 + \lambda_2 \in P(S)$, then $X(\lambda_1,\lambda_2)$ is non-zero, and there exists $\gamma \in \Gamma$ such that $X(\lambda_1,\lambda_2) = X(\lambda_1,\lambda_2,\gamma)$.*

(iv) *$X(\lambda_1,\lambda_2)$ is spherical if and only if $\lambda_2 = -\lambda_1$.*

The restriction of (λ_1,λ_2) to $\mathfrak{m}$ is the linear form $(x,x) \mapsto (\lambda_1 + \lambda_2)(x)$, and hence (i) follows from 9.3.3 (ii). Assertion (i) implies that

$$X(\lambda_1,\lambda_2) \neq 0 \Leftrightarrow \lambda_1 + \lambda_2 \in P(S).$$

Let us assume that $\lambda_1 + \lambda_2 \in P(R)$. From (i), two finite-dimensional simple sub-$\mathfrak{k}$-modules of $X(\lambda_1,\lambda_2)$ are congruent modulo the adjoint representation of $\mathfrak{k}$ in $\mathfrak{g}$, and hence $X(\lambda_1,\lambda_2)$ is equal to $X(\lambda_1,\lambda_2,\gamma)$ for some $\gamma \in \Gamma$. Finally, (iv) follows from 9.5.4.

9.6.3. LEMMA (notation as in 9.6.1). *Let $\lambda \in \mathfrak{w}^*$, let m be the canonical generator of $M(-\lambda + \varepsilon)$, I the annihilator of m in $U(\mathfrak{v})$, and φ_λ the linear* *form on* $U(\mathfrak{v})$ *which is the composite of the mappings*

$$U(\mathfrak{v}) \to U(\mathfrak{v})/I \to U(\mathfrak{x}_-) \to k.$$

Here the first arrow is the canonical mapping, the second is the isomorphism defined in 7.1.5, *and the third is the mapping which relates every element of* $U(\mathfrak{x}_-)$ *to its constant term.) Then*

$$\begin{aligned}\operatorname{Ker}\varphi_\lambda &= \mathfrak{x}_-U(\mathfrak{v}) + U(\mathfrak{v})\mathfrak{x} + \sum_{h\in\mathfrak{w}} U(\mathfrak{v})(h+\lambda)(h)) \\ &= \mathfrak{x}_-U(\mathfrak{v}) + U(\mathfrak{v})\mathfrak{x} + \sum_{h\in\mathfrak{w}} (h+\lambda(h))U(\mathfrak{v}).\end{aligned}$$

We know that $I = U(\mathfrak{v})\mathfrak{x} + \sum_{h\in\mathfrak{w}} U(\mathfrak{v})(h+\lambda(h))$ (7.2.7 (i)) and that $U(\mathfrak{v}) = I \oplus U(\mathfrak{x}_-)$ (7.1.5), hence

$$\operatorname{Ker}\varphi_\lambda = I \oplus \mathfrak{x}_-U(\mathfrak{x}_-) = U(\mathfrak{v})\mathfrak{x} + \sum_{h\in\mathfrak{w}} U(\mathfrak{v})(h+\lambda(h)) + \mathfrak{x}_-U(\mathfrak{x}_-).$$

On the other hand,

$$\mathfrak{x}_-U(\mathfrak{v}) = \mathfrak{x}_-(U(\mathfrak{x}_-) + I) \subset \mathfrak{x}_-U(\mathfrak{x}_-) + I \subset \operatorname{Ker}\varphi_\lambda,$$

whence we have the first equality of the lemma. Lastly, if $h \in \mathfrak{w}$, then

$$\begin{aligned}(h+\lambda(h))U(\mathfrak{v}) &= (h+\lambda(h))(\mathfrak{x}_-U(\mathfrak{v}) + U(\mathfrak{v})\mathfrak{x} + U(\mathfrak{w})) \\ &\subset \mathfrak{x}_-U(\mathfrak{v}) + U(\mathfrak{v})\mathfrak{x} + U(\mathfrak{w})(h+\lambda(h)),\end{aligned}$$

hence

$$\mathfrak{x}_-U(\mathfrak{v}) + U(\mathfrak{v})\mathfrak{x} + (h+\lambda(h))U(\mathfrak{v}) \subset \mathfrak{x}_-U(\mathfrak{v}) + U(\mathfrak{v})\mathfrak{x} + U(\mathfrak{v})(h+\lambda(h)),$$

and we prove the opposite inclusion in the same way, whence we have the second equality of the lemma.

9.6.4. LEMMA (notation as in 9.6.1, 9.6.3). *Let* f_λ *be the linear form on* $U(\mathfrak{g}) = U(\mathfrak{v}) \otimes U(\mathfrak{v})$ *such that* $f_\lambda(a\otimes b) = \varphi_\lambda(ba^{\mathsf{T}})$ *for* $a,b \in U(\mathfrak{v})$. *Then* $f_\lambda \in X(\lambda, -\lambda)^{\mathfrak{k}}$.

Let $x \in \mathfrak{x}$, $y \in \mathfrak{x}_-$, $h \in \mathfrak{w}$, $z \in \mathfrak{v}$, $a \in U(\mathfrak{v})$ and $b \in U(\mathfrak{v})$. Then

$$\begin{aligned}f_\lambda((x\otimes 1)(a\otimes b)) &= \varphi_\lambda(-ba^{\mathsf{T}}x) = 0,\\ f_\lambda((1\otimes y)(a\otimes b)) &= \varphi_\lambda(yba^{\mathsf{T}}) = 0,\\ f_\lambda((h\otimes 1)(a\otimes b)) &= \varphi_\lambda(-ba^{\mathsf{T}}h) = \lambda(h)\varphi_\lambda(ba^{\mathsf{T}}) = \lambda(h)f_\lambda(a\otimes b),\\ f_\lambda((1\otimes h)(a\otimes b)) &= \varphi_\lambda(hba^{\mathsf{T}}) = -\lambda(h)\varphi_\lambda(ba^{\mathsf{T}}) = -\lambda(h)f_\lambda(a\otimes b),\end{aligned}$$

and hence $f_\lambda \in X(\lambda, -\lambda)$. On the other hand,

$$f_\lambda((a\otimes b)(z\otimes 1 + 1\otimes z)) = \varphi_\lambda(-bza^{\mathsf{T}}) + \varphi_\lambda(bza^{\mathsf{T}}) = 0,$$

whence the lemma follows.

9.6.5. Lemma (notation as in 9.6.1, 9.6.3, 9.6.4). (i) *For* $x,y \in \mathfrak{v}$ *and* $u \in U(\mathfrak{v})$, *we set* $(x,y)u = yu - ux$. *This defines a* $\mathfrak{g}$*-module structure on* $U(\mathfrak{v})$.

(ii) *For every* $u \in U(\mathfrak{v})$, *let* $\Phi_\lambda(u)$ *be the linear form on* $U(\mathfrak{g}) = U(\mathfrak{v}) \otimes U(\mathfrak{v})$ *such that* $\Phi_\lambda(u)(a \otimes b) = \varphi_\lambda(bua^\mathsf{T})$ *for* $a,b \in U(\mathfrak{v})$. *Then* Φ_λ *is a* $\mathfrak{g}$*-homomorphism of* $U(\mathfrak{v})$ *onto* $X(\lambda,-\lambda)$ *such that* $\Phi_\lambda(1) = f_\lambda$.

(iii) $\Phi_\lambda(U(\mathfrak{v})) = U(\mathfrak{g})f_\lambda$, *and* Ker Φ_λ *is the annihilator of* $L(-\lambda+\varepsilon)$.

Assertion (i) is obvious. Let $x,y \in \mathfrak{v}$ and $u,a,b \in U(\mathfrak{v})$. Then

$$\begin{aligned}\Phi_\lambda((x,y)u)(a \otimes b) &= \Phi_\lambda(yu - ux)(a \otimes b)\\ &= \varphi_\lambda(byua^\mathsf{T}) - \varphi_\lambda(buxa^\mathsf{T})\\ &= \Phi_\lambda(u)(a \otimes by) + \Phi_\lambda(u)(ax \otimes b)\\ &= \Phi_\lambda(u)((a \otimes b)(x \otimes 1 + 1 \otimes y))\\ &= ((x,y)\Phi_\lambda(u))(a \otimes b)\end{aligned}$$

and hence Φ_λ is a $\mathfrak{g}$-homomorphism. Clearly, $\Phi_\lambda(1) = f_\lambda$, and hence

$$\Phi_\lambda(U(\mathfrak{v})) = \Phi_\lambda(U(\mathfrak{g}) \cdot 1) = U(\mathfrak{g})f_\lambda \subset X(\lambda,-\lambda) \quad \text{from 9.6.4.}$$

Let K be the largest sub-$\mathfrak{v}$-module of $M(-\lambda+\varepsilon)$ which is distinct from $M(-\lambda+\varepsilon)$. Let A be the annihilator of $L(-\lambda+\varepsilon) = M(-\lambda+\varepsilon)/K$. If we identify $M(-\lambda+\varepsilon)$ with $U(\mathfrak{x}_-)$ canonically, then $K \subset \mathfrak{x}_-U(\mathfrak{x}_-)$ from 7.1.11 (i). Let $u \in A$, and $a,b \in U(\mathfrak{v})$. Then $bua^\mathsf{T} \in A$, and hence bua^T transforms the canonical generator of $M(-\lambda+\varepsilon)$ into an element of K and hence of $\mathfrak{x}_-U(\mathfrak{x}_-)$. From 9.6.3 we deduce that $\varphi_\lambda(bua^\mathsf{T}) = 0$, whence $\Phi_\lambda(u)=0$ and $A \subset \operatorname{Ker}\Phi_\lambda$.

Finally, Ker Φ_λ is a sub-$\mathfrak{g}$-module of $U(\mathfrak{v})$, i.e., a two-sided ideal of $U(\mathfrak{v})$. Hence $(\operatorname{Ker}\Phi_\lambda) \cdot M(-\lambda+\varepsilon)$ is a sub-$\mathfrak{v}$-module of $M(-\lambda+\varepsilon)$. If $u \in \operatorname{Ker}\Phi_\lambda$ and $a \in U(\mathfrak{v})$, then

$$\varphi_\lambda(ua) = \Phi_\lambda(u)(a^\mathsf{T} \otimes 1) = 0;$$

hence $(\operatorname{Ker}\Phi_\lambda)\, M(-\lambda+\varepsilon) \neq M(-\lambda+\varepsilon)$; consequently, $(\operatorname{Ker}\Phi_\lambda)\, M(-\lambda+\varepsilon) \subset K$, whence $\operatorname{Ker}\Phi_\lambda \subset A$.

9.6.6. Lemma (notation as in 9.6.1). *Let* $\lambda \in \mathfrak{w}^*$, *and let* J *and* J' *be the annihilators of* $M(-\lambda+\varepsilon)$ *and* $L(-\lambda+\varepsilon)$, *respectively. The following conditions are equivalent:*

(i) $J = J'$;

(ii) $X(\lambda,-\lambda)^\mathfrak{k}$ *generates the* $\mathfrak{g}$*-module* $X(\lambda,-\lambda)$.

If these conditions are satisfied, $X(\lambda,-\lambda)$ *is isomorphic to the* $\mathfrak{g}$*-module* $U(\mathfrak{v})/J$ *(the* $\mathfrak{g}$*-module structure of* $U(\mathfrak{v})$ *being defined as in* 9.6.5 (i)).

From 9.6.5, whose notation we are using, we have a diagram of $\mathfrak{g}$-homomorphisms

(1) $$U(\mathfrak{v})/J \xrightarrow{\psi_1} U(\mathfrak{v})/J' \xrightarrow{\psi_2} U(\mathfrak{g})f_\lambda \xrightarrow{\psi_3} X(\lambda,-\lambda),$$

where ψ_1 is surjective, ψ_2 is bijective, and ψ_3 is injective. Let $\sigma \in \mathfrak{k}^\wedge$. From 9.6.2 (i), $\mathrm{mtp}(\sigma,X(\lambda,-\lambda))$ is the multiplicity of the weight 0 in σ. On the other hand, the $\mathfrak{g}$-module structure on $U(\mathfrak{v})$ defines, by restriction to $\mathfrak{k}$, a representation of $\mathfrak{k}$ in $U(\mathfrak{v})$ which can be identified with the adjoint representation of $\mathfrak{v}$ in $U(\mathfrak{v})$. From 8.4.3 and 8.3.9, $\mathrm{mtp}(\sigma,U(\mathfrak{v})/J)$ is the multiplicity of the weight 0 in σ. Hence

(2) $$\mathrm{mtp}(\sigma,U(\mathfrak{v})/J) = \mathrm{mtp}(\sigma,X(\lambda,-\lambda)).$$

Given this, we have

$$\begin{aligned} J = J' &\Leftrightarrow \psi_1 \text{ injective} \Leftrightarrow \psi_3 \circ \psi_2 \circ \psi_1 \text{ injective} \\ &\Leftrightarrow \psi_3 \circ \psi_2 \circ \psi_1 \text{ bijective} && \text{(from (2))} \\ &\Leftrightarrow \psi_3 \circ \psi_2 \circ \psi_1 \text{ surjective} && \text{(from (2))} \\ &\Leftrightarrow \psi_3 \text{ surjective} \\ &\Leftrightarrow U(\mathfrak{g})f_\lambda = X(\lambda,-\lambda). \end{aligned}$$

The last assertion then follows at once.

9.6.7. Proposition (notation as in 9.6.1). *Let* $\lambda \in \mathfrak{w}^*$. *If*

$$(\lambda - \varepsilon)(H_\alpha) \notin \{-1,-2,\ldots,\} \quad \textit{for all } \alpha \in S_+,$$

then $X(\lambda,-\lambda)^{\mathfrak{k}}$ *generates the* $\mathfrak{g}$-*module* $X(\lambda,-\lambda)$.

This follows from 7.6.24 and 9.6.6.

9.6.8. Lemma (notation as in 9.6.1). *Let* ϱ_0 *be the one-dimensional trivial representation of* $\mathfrak{k}$. *Then* $\sum_{\varrho\in\mathfrak{k}^\wedge,\varrho\neq\varrho_0} X(2\varepsilon,-2\varepsilon)_\varrho$ *is a sub-*$\mathfrak{g}$*-module of* $X(2\varepsilon,-2\varepsilon)$.

For all $\lambda \in \mathfrak{w}^*$, let us denote the annihilator of $M(\lambda)$ by J_λ. If $\lambda = 2\varepsilon$, the conditions of 9.6.6 are satisfied from 7.6.24. Hence the $\mathfrak{g}$-modules $X(2\varepsilon,-2\varepsilon)$ and $U(\mathfrak{v})/J_{-\varepsilon}$ are isomorphic. On the other hand, $J_\varepsilon = J_{-\varepsilon}$ from 8.4.4, and $M(\varepsilon)$ has a one-dimensional $\mathfrak{v}$-module as a quotient (7.2.6). Hence $J_{-\varepsilon}$ is contained in a two-sided ideal of codimension 1 of $U(\mathfrak{v})$. Consequently, $X(2\varepsilon,-2\varepsilon)$ has a sub-$\mathfrak{g}$-module W of codimension 1. Then W is a sub-$\mathfrak{k}$-module of $X(2\varepsilon,-2\varepsilon)$, and $X(2\varepsilon,-2\varepsilon)/W$ is a trivial $\mathfrak{g}$-module, hence $X(2\varepsilon,-2\varepsilon)$ is a complement of W in $X(2\varepsilon,-2\varepsilon)$, and

$$W = \sum_{\in\varrho\mathfrak{k}^\wedge,\varrho\neq\varrho_0} X(2\varepsilon,-2\varepsilon).$$

9.6.9. LEMMA (notation as in 9.6.1). *Let* $\lambda \in \mathfrak{w}^*$. *There exists a* $\mathfrak{g}$-*invariant bilinear form* γ *on* $\mathfrak{X}(\lambda,-\lambda)\times X(-\lambda+2\varepsilon,\lambda-2\varepsilon)$ *such that:*

(a) *for every* $\varrho \in \mathfrak{k}^\wedge$, *the restriction of* γ *to* $X(\lambda,-\lambda)_\varrho \times X(-\lambda+2\varepsilon,\lambda-2\varepsilon)_{\varrho^*}$ *is non-degenerate;*

(b) *if* $\varrho,\varrho' \in \mathfrak{k}^\wedge$ *and* $\varrho' \neq \varrho^*$, *then* $X(\lambda,-\lambda)_\varrho$ *and* $X(-\lambda+2\varepsilon,\lambda-2\varepsilon)_{\varrho'}$ *are orthogonal to each other with respect to* γ.

(a) We recall that we defined a multiplication on $U(\mathfrak{g})^*$ (2.7.4). We have

$$M'(-\lambda,\lambda)^* \cdot M'(\lambda-2\varepsilon,-\lambda+2\varepsilon)^* \subset M'(-2\varepsilon,2\varepsilon)^*.$$

Indeed, if $f \in M'(-\lambda,\lambda)^*$, $g \in M'(\lambda-2\varepsilon,-\lambda+2\varepsilon)^*$, $x \in \mathfrak{m}$, $y \in \mathfrak{a}$ and $z \in \mathfrak{n}$, we have, from 2.7.7,

$$\begin{aligned}{}^tL(y)(fg) &= ({}^tL(y)f)g + f({}^tL(y)g)\\ &= (\lambda,-\lambda)(y)f\cdot g + f(-\lambda+2\varepsilon,\lambda-2\varepsilon)(y)g\\ &= (2\varepsilon,-2\varepsilon)(y)(fg)\end{aligned}$$

and similarly ${}^tL(x)(fg) = {}^tL(z)(fg) = 0$. Again from 2.7.7, we have, for all $r \in \mathfrak{g}$,

$$r(fg) = (rf)g + f(rg)$$

hence

$$X(\lambda,-\lambda)X(-\lambda+2\varepsilon,\lambda-2\varepsilon) \subset X(2\varepsilon,-2\varepsilon).$$

(b) Let Ψ be the sum of the finite-dimensional simple sub-$\mathfrak{k}$-modules of $U(\mathfrak{k})^*$ for the left or right coregular representation (2.7.12). In 2.7.13 we defined the fundamental linear form φ_0 and the fundamental bilinear form β_0 on Ψ and $\Psi\times\Psi$ respectively. Let T be the set of the $g \in U(\mathfrak{k})$ such that $g(\mathfrak{m}U(\mathfrak{k})) = 0$. Let $T' = T \cap \Psi$. Then the mapping $f \mapsto f|U(\mathfrak{k})$, where f takes all values in $M'(-\lambda,\lambda)^*$, is a $\mathfrak{k}$-isomorphism of $M'(-\lambda,\lambda)^*$ onto T endowed with the right coregular representation (5.5.8), and hence defines by restriction a $\mathfrak{k}$-isomorphism of $X(\lambda,-\lambda)$ onto T'. We identify $X(\lambda,-\lambda)$ with T' under this isomorphism. Then the multiplication

$$X(\lambda,-\lambda)\times X(-\lambda+2\varepsilon,\lambda-2\varepsilon) \to X(2\varepsilon,-2\varepsilon)$$

can be identified with the multiplication in T' (2.7.4). The form $\varphi_0|T'$ can be identified with a linear form on $X(2\varepsilon,-2\varepsilon)$ which, from 9.6.8, is $\mathfrak{g}$-invariant.

(c) The bilinear form $\gamma : (f,g) \mapsto \varphi_0(fg)$ on

$$X(\lambda,-\lambda)\times X(-\lambda+2\varepsilon,\lambda-2\varepsilon)$$

is hence $\mathfrak{g}$-invariant. It can be identified with β_0, and hence, from 2.7.15 and 2.7.17, satisfies properties (a) and (b) of the lemma.

9.6.10. PROPOSITION (notation as in 9.6.1). *Let* $\lambda \in \mathfrak{w}^*$. *If*

$$(\lambda - \varepsilon)(H_\alpha) \notin \{1,2,\ldots\} \quad \textit{for all } \alpha \in S_+,$$

then every non-null sub-$\mathfrak{g}$*-module of* $X(\lambda,-\lambda)$ *contains* $X(\lambda,-\lambda)^{\mathfrak{k}}$; *in particular, the sub-*$\mathfrak{g}$*-module of* $X(\lambda,-\lambda)$ *generated by* $X(\lambda,-\lambda)^{\mathfrak{k}}$ *is simple.*

From 9.6.7, $X(-\lambda + 2\varepsilon, \lambda - 2\varepsilon)^{\mathfrak{k}}$ generates the $\mathfrak{g}$-module

$$X(-\lambda + 2\,\varepsilon, \lambda - 2\,\varepsilon).$$

Let W be a sub-$\mathfrak{g}$-module of $X(\lambda,-\lambda)$. Then $W = \oplus_{\varrho \in \mathfrak{k}^\wedge} W \cap X(\lambda,-\lambda)_\varrho$. Let γ be as in 9.6.9, and $W^\perp$ the orthogonal subspace of W in $X(-\lambda + 2\varepsilon, \lambda - 2\varepsilon)$. From 9.6.9, W is the orthogonal subspace of $W^\perp$ in $X(\lambda,-\lambda)$. If $X(\lambda,-\lambda)^{\mathfrak{k}}$ is not contained in W, then $X(-\lambda + 2\varepsilon, \lambda - 2\varepsilon)^{\mathfrak{k}}$ is contained in $W^\perp$, hence $W^\perp = X(-\lambda + 2\varepsilon, \lambda - 2\varepsilon)$, and $W = 0$.

9.6.11. THEOREM (notation as in 9.6.1). *Let* $\lambda \in \mathfrak{w}^*$. *If*

$$(\lambda - \varepsilon)(H_\alpha) \notin \mathbf{Z} - \{0\} \quad \textit{for all } \alpha \in S,$$

then $X(\lambda,-\lambda)$ *is simple.*

This follows from 9.6.7 and 9.6.10.

9.6.12. THEOREM (k algebraically closed, $\mathfrak{g}$ semi-simple). *Let* $\mathfrak{h}$ *be a Cartan subalgebra of* $\mathfrak{g}$, B *a basis for* $R = R(\mathfrak{g},\mathfrak{h})$, $\lambda \in \mathfrak{h}^*$, *whence we have a central character* χ_λ, *and* $J_\lambda = U(\mathfrak{g})\,\mathrm{Ker}\,\chi_\lambda$. *We assume that* $\lambda(H_\alpha) \notin \mathbf{Z} - \{0\}$ *for all* $\alpha \in R$. *Then* J_λ *is the only two-sided ideal* J *of* $U(\mathfrak{g})$ *such that* $J \cap Z(\mathfrak{g}) = \mathrm{Ker}\chi_\lambda$. *In particular,* J_λ *is a maximal two-sided ideal of* $U(\mathfrak{g})$.

Indeed, the $(\mathfrak{g}\times\mathfrak{g})$-module $U(\mathfrak{g})/J_\lambda$ is simple from 8.4.3, 9.6.6 and 9.6.11.

9.6.13. THEOREM (notation as in 9.6.1). *Let* χ *be a homomorphism of* $Z(\mathfrak{v})$ *into* k, *and let us identify* $Z(\mathfrak{g})$ *with* $Z(\mathfrak{v}) \otimes Z(\mathfrak{v})$. *There exists one, and up to isomorphism only one, spherical simple* $\mathfrak{g}$*-module whose central character* χ' *satisfies* $\chi'|Z(\mathfrak{g}) \otimes 1 = \chi$.

We set $G = U(\mathfrak{g})$, and $\mathfrak{v}_1 = \mathfrak{v}\times 0 \subset \mathfrak{v}\times\mathfrak{v} = \mathfrak{g}$. Since $\mathfrak{g} = \mathfrak{v}_1 \oplus \mathfrak{k}$, we have $G = G\mathfrak{k} \oplus U(\mathfrak{v}_1)$. For every $x \in \mathfrak{v}$, the inner derivation of G defined by (x,x) leaves $G\mathfrak{k}$ and $U(\mathfrak{v}_1)$ stable, and in $U(\mathfrak{v}_1)$ induces the inner derivation defined by x. Hence $G^{\mathfrak{k}} = (G^{\mathfrak{k}} \cap G\mathfrak{k}) \oplus Z(\mathfrak{v}_1)$. Let Φ be the set of homomorphisms of $G^{\mathfrak{k}}$ into k which are zero on $G^{\mathfrak{k}} \cap G\mathfrak{k}$. Let Φ' be the set of homomorphisms of $Z(\mathfrak{v}_1)$ into k. Then $\varphi \mapsto \varphi|Z(\mathfrak{v}_1)$ is a bijection of Φ onto Φ'. Given this, the theorem follows from 9.1.12.

References: [37], [79], [100].

9.7. Supplementary remarks

9.7.1. The representations of the principal series were first considered by Gelfand and Naimark for the classical complex semi-simple groups. For the case of general real semi-simple groups, the theorems are, roughly, of the following type: (1) a representation of the principal series is most often simple; (2) two representations of the principal series are in general equivalent if the initial data are conjugate under the Weyl group (intertwining operators); (3) we give conditions for a representation of the principal series to be unitary. In spite of the number and significance of the papers dedicated to these questions, the situation is not yet very clear. Some special cases (e.g. complex groups, groups with symmetric rank 1, spherical representations, unitary representations) are closer to a complete solution. In particular, we may cite [79], [100], [131], [133].

In this chapter, we have only given those theorems which can be presented in an entirely algebraic form. The essence is due to Harish-Chandra [64], [65]. We have followed Lepowsky and McCollum [89] in 9.1, and Lepowsky [87] and Rader [103] in 9.2, 9.3, 9.4 and 9.5. The canonical mapping of $G^{\mathfrak{k}}$ into $A \otimes K^{\mathfrak{m}}$ was introduced by Lepowsky and Rader. Some important special cases have been considered previously; for example, the canonical homomorphism of $G^{\mathfrak{k}}$ into A was introduced by Harish-Chandra. Theorems 9.6.11 and 9.6.13 are due to Parthasarathy, Ranga Rao and Varadarajan [100]; the method of proof of 9.6.11 given here was indicated to me orally by Duflo.

9.7.2. We adopt the notation of 9.0, with $\mathfrak{k}$ reductive in $\mathfrak{g}$.

(a) $G^{\varrho,\sigma}$ is the set of the $u \in G$ such that $u(V_\sigma) \subset V_\varrho$ for every $\mathfrak{g}$-module V.

(b) $G^{\varrho,\sigma}G^{\sigma,\tau} \subset G^{\varrho,\tau}$ for all $\varrho,\sigma,\tau \in \mathfrak{k}^\wedge$.

(c) G^ϱ is the largest subalgebra of G containing GI^ϱ as a two-sided ideal.

(d) Let V be a $\mathfrak{g}$-module, and $S \subset V_\varrho$. Then $(GS) \cap V_\sigma = G^{\sigma,\varrho}$. [89]

9.7.3. (notation as in 9.2.1). Let $\mathfrak{k}' = [\mathfrak{k},\mathfrak{k}]$, and let $\mathfrak{k}'^\vee$ be the set of the $\xi \in \mathfrak{k}'^\wedge$ having the following property: there exists $\pi \in \mathfrak{g}^\wedge$ such that $\pi|\mathfrak{k}'$ has a subrepresentation equivalent to ξ. Let L be the left ideal of $U(\mathfrak{k}')$ such that the left regular representation of $\mathfrak{k}'$ in $U(\mathfrak{k}')/L$ belongs to $\mathfrak{k}'^\vee$. Then $U(\mathfrak{g})L$ is the intersection of the maximal left ideals of finite codimension of $U(\mathfrak{g})$ which contain $U(\mathfrak{g})L$. [65]

9.7.4. (notation as in 7.0). Assume that $k = \mathbf{C}$ and that $\mathfrak{g}$ is simple. Let $\lambda \in \mathfrak{h}^*$ and let $\mathfrak{k}$ be a symmetrizing subalgebra of $\mathfrak{g}$ containing $\mathfrak{h}$. Let $\mathfrak{c}$ be the centre of $\mathfrak{k}$. Then $\dim \mathfrak{c} = 0$ or 1.

(a) If $\dim \mathfrak{c} = 0$, then $L(\lambda)$ is only a Harish-Chandra module relative to $\mathfrak{k}$ if $L(\lambda)$ is finite-dimensional, i.e., if $\lambda - \delta \in P_{++}$.

(b) If $\dim \mathfrak{c} = 1$, the following conditions are equivalent:

(i) $L(\lambda)$ is a Harish-Chandra module relative to $\mathfrak{k}$;

(ii) $(\lambda - \delta)(H_\alpha) \in \mathbf{N}$ for every positive root α such that $\mathfrak{g}^\alpha \subset \mathfrak{k}$. [66]

9.7.5 (notation as in 9.2.1). Assume that $k = \mathbf{C}$. Let θ be the automorphism of $\mathfrak{g}$ corresponding to $\mathfrak{k}$, W the Weyl group of $(\mathfrak{g},\theta,\mathfrak{a})$ (1.14.14), $\delta = \frac{1}{2} \sum_{\alpha \in R_+} \alpha$, and $\delta' = \delta|\mathfrak{a}$. If $w \in W$, let w_* be the affine automorphism $\lambda \mapsto w(\lambda - \delta') + \delta'$ of the dual of $\mathfrak{a}$, whence there is an automorphism of A, which is also denoted by w_*. Let A^{W_*} be the set of the elements of A which are invariant under w_* for all $w \in W$.

(a) Let $\sigma \in \mathfrak{k}^\wedge$, let τ be the linear form $u \mapsto \operatorname{tr} \sigma(u)$ on K, $\tilde{\omega}$ the canonical antihomomorphism of $G^{\mathfrak{k}}$ into $A \otimes K$, and $\tilde{\omega}^\sigma$ the linear mapping $(1 \otimes \tau) \circ \tilde{\omega}$ of $G^{\mathfrak{k}}$ into $A \otimes k = A$. Then $\tilde{\omega}^\sigma(G^{\mathfrak{k}}) = A^{W_*}$. If β is the canonical mapping of $S(\mathfrak{g})$ into $U(\mathfrak{g})$, then $\tilde{\omega}^\sigma|\beta(S(\mathfrak{p})^{\mathfrak{k}})$ is a bijection of $\beta(S(\mathfrak{p})^{\mathfrak{k}})$ onto A^{W^*}.

(b) In particular, the canonical homomorphism of $G^{\mathfrak{k}}$ into A has A^{W_*} as its image. Then 9.5.6 defines a canonical bijective correspondence between classes of spherical simple $\mathfrak{g}$-modules and W_*-orbits in $\mathfrak{a}^*$. ([87], [88])

9.7.6 (notation as in 9.2.1). Let V be a spherical simple $\mathfrak{g}$-module. Then $V^{\mathfrak{n}}$ has dimension 0 or 1, and $\mathfrak{m}V^{\mathfrak{n}} = 0$. [79]

9.7.7 (notation as in 9.2.1). Assume that $k = \mathbf{C}$. Let L be a left ideal of K. Assume that K/L is finite-dimensional and that the left regular representation of $\mathfrak{k}$ in K/L is semi-simple.

(a) The $\mathfrak{g}$-module $V = G/GL$ is a Harish-Chandra module.

(b) Let $\sigma \in \mathfrak{k}^\wedge$. Then V_σ is a module of finite type over $Z(\mathfrak{g})$. [64]

(c) Let $\varrho,\sigma \in \mathfrak{k}^\wedge$. Then $G^{\varrho,\sigma}/GI^\sigma$ is a module of finite type over $Z(\mathfrak{g})$. (This is a special case of (b), as Lepowsky pointed out to me.)

9.7.8 (notation as in 9.2.1). Assume that $k = \mathbf{C}$. Let V be a Harish-Chandra $\mathfrak{g}$-module having a central character. Consider the following conditions:

(i) for all $\xi \in \mathfrak{k}^\wedge$, $\operatorname{mtp}(\xi, V) < +\infty$;

(ii) there exists a constant C such that, for all $\xi \in \mathfrak{k}^\wedge$, we have $\operatorname{mtp}(\xi, V) \leq C \dim \xi$;

(iii) V is a $U(\mathfrak{g})$-module of finite type;

(iv) V has a Jordan-Hölder series.

(a) From 9.5.1 and 9.7.7, we have (iv) $\Rightarrow$ (ii) $\Rightarrow$ (i), and (iv) $\Rightarrow$ (iii) $\Rightarrow$ (i). From 7.8.16, (ii) does not imply (iii) and (i) does not imply (ii).

(b) Let $\mathfrak{c}$ be the centre of $\mathfrak{k}$. The weights of the adjoint representation of $\mathfrak{c}$ in $\mathfrak{g}$ generate a subgroup Θ in $\mathfrak{c}^*$; this subgroup is free and of rank dim $\mathfrak{c}$. For every integer $n > 0$, let $\mathfrak{k}_n^\wedge$ be the set of the $\xi \in \mathfrak{k}^\wedge$ such that the linear form with which $\xi|\mathfrak{c}$ can be identified belongs to $(1/n)\Theta$. If there exists an n such that $V = \oplus_{\xi \in \mathfrak{k}_n^\wedge} V_\xi$ (a property which is automatically satisfied if $\mathfrak{k}$ is semi-simple), then conditions (i) to (iv) are equivalent. Cf. problem 34. ([87], [90])

9.7.9 (notation as in 9.2.1). Assume that $k = \mathbf{C}$. Let χ be a homomorphism of $Z(\mathfrak{g})$ into $\mathbf{C}$, and n a positive integer. Up to isomorphism, there only exist a finite number of simple Harish-Chandra $\mathfrak{g}$-modules V such that:

(1) the central character of V is χ;

(2) with the notation of 9.7.8 (b), $V = \oplus_{\xi \in \mathfrak{k}_n^\wedge} V_\xi$.

Cf. problem 34. ([87], [90])

9.7.10. We now abandon the notation of 9.0. Let G be a simply connected semi-simple real Lie group, and $G = KAN$ an Iwasawa decomposition of G. Let $\mathfrak{g}_\mathbf{R}$, $\mathfrak{k}_\mathbf{R}$, $\mathfrak{a}_\mathbf{R}$, $\mathfrak{n}_\mathbf{R}$ be the Lie algebras of G, K, A, N, and $\mathfrak{g}$, $\mathfrak{k}$, $\mathfrak{a}$, $\mathfrak{n}$ the complexifications of $\mathfrak{g}_\mathbf{R}$, $\mathfrak{k}_\mathbf{R}$, $\mathfrak{a}_\mathbf{R}$, $\mathfrak{n}_\mathbf{R}$. Let $\mathfrak{m}_\mathbf{R}$ be the commutant of $\mathfrak{a}_\mathbf{R}$ in $\mathfrak{k}_\mathbf{R}$, and $\mathfrak{m}$ the complexification of $\mathfrak{m}_\mathbf{R}$. Let M_0 be the connected Lie subgroup of G with Lie algebra $\mathfrak{m}_\mathbf{R}$, let Z be the centre of G, which is contained in K, let $M_1 = M_0 Z$, and let M_2 be the centralizer of $\mathfrak{a}$ in K; then $M_0 \subset M_1 \subset M_2$, and these three groups have $\mathfrak{m}_\mathbf{R}$ as their Lie algebra. For $i = 0,1,2$, let $Q_i = M_i AN$, which is a Lie subgroup of G with Lie algebra $\mathfrak{m}_\mathbf{R} \oplus \mathfrak{a}_\mathbf{R} \oplus \mathfrak{n}_\mathbf{R}$, A finite-dimensional irreducible representation ϱ of Q_i is trivial over N and hence can be identified with a pair (λ,σ), where $\lambda \in \mathfrak{a}^*$ and $\sigma \in M_i^\wedge$. Let π be the representation of G induced by ϱ.

We shall leave the case where $i = 0$ aside. (It would then seem reasonable to limit ourselves to the case where ϱ is unitary, taking π in the Mackey sense. Even in this case, the relations with the algebraic coinduced representation are not satisfactory.)

If $i = 1$, then G/Q_1 is compact, and we can define π (which is in general not unitary) in $L^2(K/M_1)$. The corresponding representation of $U(\mathfrak{g})$ in the space of K-finite vectors gives us a Harish-Chandra module which is one of the modules $X(\varrho,\gamma)$ of 9.3.1. Since we often have $M_1 \neq M_2$, the representation π is often reducible, and consequently the module $X(\varrho,\gamma)$ is also reducible.

If $i = 2$, then π belongs to what is classically termed the principal series. Let $X = X^{\lambda,\sigma}$ be the corresponding Harish-Chandra module. Here are certain results concerning the $X^{\lambda,\sigma}$ (which unfortunately do not apply to the modules $X(\varrho,\gamma)$ in the text except in certain cases). We shall restrict ourselves to the case where $(\operatorname{Ker}\sigma) \cap Z$ has finite index in Z.

(i) There exists a subset A of $\mathfrak{a}^*$, whose complement is the denumerable union of algebraic manifolds distinct from $\mathfrak{a}^*$, such that, if $\lambda \in A$, then $X^{\lambda,\sigma}$ is simple. (Kostant, Wallach)

(ii) Let us assume that σ is trivial and one-dimensional. Let us define δ' as in 9.7.5. Let D be the set of the linear forms on $\mathfrak{a}$ which are real-valued on $\mathfrak{a}_{\mathbf{R}}$ and whose scalar product with the elements of R'_+ (with the notation of 9.2.1) is non-negative. If $\lambda - \delta' \in D + i\mathfrak{a}^*_{\mathbf{R}}$, then $X^{\lambda,\sigma}$ is spherical and $(X^{\lambda,\sigma})^{\mathfrak{k}}$ generates $X^{\lambda,\sigma}$ as a $U(\mathfrak{g})$-module. [79]

(iii) There exists a non-degenerate bilinear form which is canonically defined on $X^{\lambda,\sigma} \times X^{2\delta'-\lambda,\sigma^*}$, is $\mathfrak{g}$-invariant, and is such that every $\mathfrak{k}$-stable vector subspace of $X^{\lambda,\sigma}$ is equal to its bi-orthogonal subspace. [79]

(iv) We adopt the notation of 9.7.5. If M' is the normalizer of $\mathfrak{a}$ in K, then W can be identified with M'/M_2, whence we have an action of W in $M_2^{\wedge}$. Then, if $w \in W$, the Jordan–Hölder series of $X^{\lambda,\sigma}$ and of $X^{w_*\lambda,w\sigma}$ have the same quotients with the same multiplicities. [87]

(v) If $\xi \in \mathfrak{g}^{\wedge}$, then there exist λ and σ such that $\operatorname{mtp}(\xi,X^{\lambda,\sigma}) > 0$; we then have $\operatorname{mtp}(\xi,X^{\lambda,\sigma}) = 1$. [90]

(vi) Let $G_{\mathbf{C}}$ be the simply connected complex Lie group with Lie algebra $\mathfrak{g}$. Then there is a canonical homomorphism r of G into $G_{\mathbf{C}}$. If $\operatorname{Ker}\sigma \supset \operatorname{Ker} r$, then $X^{\lambda,\sigma}$ is generated as a $U(\mathfrak{g})$-module by a single element. [90]

9.7.11. For the Harish-Chandra modules over $\mathfrak{sl}(2,\mathbf{C})$ (with $\mathfrak{k} = \mathbf{C}h$), cf. 7.8.16. For the Harish-Chandra modules over $\mathfrak{sl}(2,\mathbf{C}) \times \mathfrak{sl}(2,\mathbf{C})$ (with $\mathfrak{k}$ equal to the diagonal subalgebra), cf. [57]. Cf. also [186].

CHAPTER 10

PRIMITIVE IDEALS (THE GENERAL CASE)

10.1. Some canonical homomorphisms

The statements in this section are actually lemmas, although some of them are graced with the name "proposition" to alleviate the monotony.

10.1.1. NOTATION. Throughout the present section, we assume that k is algebraically closed. We fix a nilpotent Lie algebra $\mathfrak{n}$, and $f \in \mathfrak{n}^*$. We set $A = U(\mathfrak{n})/I(f)$, and we denote the canonical homomorphism of $U(\mathfrak{n})$ onto A by $u \mapsto \bar{u}$. If π is a representation of $U(\mathfrak{n})$ with kernel $I(f)$, we denote the representation of A deduced from π by passage to the quotient by $\bar{\pi}$.

From 10.1.3 on, we also fix a Lie algebra $\mathfrak{s}$ and a homomorphism j of $\mathfrak{s}$ into the Lie algebra of derivations of $\mathfrak{n}$. We denote the semi-direct product of $\mathfrak{s}$ and $\mathfrak{n}$ defined by j by $\mathfrak{b}$. We assume that, for all $s \in \mathfrak{s}$, we have $f(j(s)\mathfrak{n}) = 0$.

10.1.2. LEMMA (notation as in 10.1.1). *Let D be a derivation of $\mathfrak{n}$ such that $f(D\mathfrak{n}) = 0$.*

(i) *Let us also denote by D the derivation of $U(\mathfrak{n})$ which extends D. Then $D(I(f)) \subset I(f)$, so that D defines a derivation D' of A by passage to the quotient.*

(ii) *Let $\mathfrak{h}$ be an element of $P(f)$ such that $D(\mathfrak{h}) \subset \mathfrak{h}$ (recall that such an element exists from* 1.12.10). *Let $\sigma_{\mathfrak{h}} = \mathrm{ind}(f|\mathfrak{h},\mathfrak{n})$ and $\lambda = \mathrm{tr}_{n/\mathfrak{h}} D$. There exists one and only one $a \in A$ such that*

$$\bar{\sigma}_{\mathfrak{h}}(a)(u \otimes 1) = (D + \tfrac{1}{2}\lambda)\, u \otimes 1 \quad \textit{for all } u \in U(\mathfrak{n}).$$

In particular,

$$\bar{\sigma}_{\mathfrak{h}}(a - \tfrac{1}{2}\lambda)(1 \otimes 1) = 0.$$

(iii) *The element a of* (ii) *only depends on D and not on the choice of $\mathfrak{h}$. The inner derivation of A defined by a is D'.*

Let $\mathfrak{g} = kx \oplus \mathfrak{n}$ be the semi-direct product of k and $\mathfrak{n}$ defined by D. Let $g \in \mathfrak{g}^*$ such that $g|\mathfrak{n} = f$ and $g(x) = 0$. It is sufficient to apply 6.2.6 to $\mathfrak{g}$.

10.1.3. In the situation of 10.1.2, a is said to be *the element of A defined by D*. With the notation of 10.1.1, we can thus consider, for all $s \in \mathfrak{s}$, the element of A defined by $j(s)$; let us denote it by $\theta(s)$.

10.1.4. PROPOSITION (notation as in 10.1.1, 10.1.3). *The mapping θ is a Lie algebra homomorphism of $\mathfrak{s}$ into A.*

This is obvious if $\mathfrak{n} = 0$. We reason by induction on $\dim \mathfrak{n}$. Let $\mathfrak{z}$ be the centre of $\mathfrak{n}$, and $\mathfrak{z}_0 = \mathfrak{z} \cap \operatorname{Ker} f$, which is stable under $j(\mathfrak{s})$.

(a) If $\mathfrak{z}_0 \neq 0$, it is sufficient to apply the induction hypothesis to $\mathfrak{n}/\mathfrak{z}_0$. Henceforth we shall assume that $\mathfrak{z}_0 = 0$. Then $\dim \mathfrak{z} = 1$ and $f(\mathfrak{z}) \neq 0$.

(b) Let us assume that there exists a commutative ideal $\mathfrak{a}$ of $\mathfrak{n}$, which is distinct from 0 and $\mathfrak{z}$, and stable under $j(\mathfrak{s})$. Let $\mathfrak{n}' = \mathfrak{a}^f$ and $f' = f|\mathfrak{n}'$. Then $[\mathfrak{n},\mathfrak{a}] \neq 0$, hence, from 1.3.17, $[\mathfrak{n},\mathfrak{a}] \supset \mathfrak{z}$, and hence $f([\mathfrak{n},\mathfrak{a}]) \neq 0$ and $\mathfrak{n}' \neq \mathfrak{n}$. Moreover, $\mathfrak{n}'$ is stable under $j(\mathfrak{s})$. Let $A' = U(\mathfrak{n}')/I(f')$. For all $s \in \mathfrak{s}, j'(s) = j(s)|\mathfrak{n}'$ defines an element $\theta'(s)$ of A'. From the induction hypothesis, θ' is a Lie algebra homomorphism of $\mathfrak{s}$ into A'.

We note that, if $\mathfrak{p} \in P(f')$, then $\mathfrak{p} \in P(f)$. Indeed, $\mathfrak{a} \subset \mathfrak{n}'^{f'}$, and hence $\mathfrak{a} \subset \mathfrak{p}$; consequently, every element of $\mathfrak{p}^f$ belongs to $\mathfrak{a}^f = \mathfrak{n}'$, and then to $\mathfrak{p}$, whence our assertion follows.

Let us fix a representation σ' of $U(\mathfrak{n}')$ in a space W', with kernel $I(f')$; it defines a representation $\bar{\sigma}'$ of A'. Let $\sigma = \operatorname{ind}(\sigma',\mathfrak{n})$. Then $\operatorname{Ker} \sigma = I(f)$. Indeed, if $\mathfrak{p} \in P(f')$, then $\sigma_0' = \operatorname{ind}(f|\mathfrak{p},\mathfrak{n}')$ has $I(f')$ as its kernel, and hence $\sigma_0 = \operatorname{ind}(f|\mathfrak{p},\mathfrak{n})$ has the same kernel as σ; now the kernel of σ_0 is $I(f)$ since $\mathfrak{p} \in P(f)$.

Let us fix an element s of $\mathfrak{s}$ up to (1). Let $\mathfrak{g} = kx \oplus \mathfrak{n}$ be the semi-direct product of k and $\mathfrak{n}$ defined by $j(s)$. Then the Lie subalgebra $\mathfrak{g}' = kx \oplus \mathfrak{n}'$ of $\mathfrak{g}$ is the semi-direct product of k and $\mathfrak{n}'$ defined by $j'(\mathfrak{s})$. Let g be the linear form on $\mathfrak{g}$ such that $g|\mathfrak{n} = f$ and $g(x) = 0$. Let $g' = g|\mathfrak{g}'$.

$$\begin{array}{ccc} \tau' & & \tau \\ \mathfrak{g}' & \subset & \mathfrak{g} \\ \cup & & \cup \\ \mathfrak{n}' & \subset & \mathfrak{n} \\ \sigma' & & \sigma \end{array}$$

From 6.2.6 (ii), there exists one and only one representation τ' of $\mathfrak{g}'$ in W' extending σ' and with kernel $I(g')$. We have $\tau'(x) = \bar{\sigma}'(\theta'(s))$.

Let $\tau = \operatorname{ind}^{\sim}(\tau',\mathfrak{g})$. Let $\mathfrak{p} \in P(f')$, which is stable under $j(s)$, and let $\mathfrak{h} = kx \oplus \mathfrak{p}$. Then $\mathfrak{h} \in P(g')$ and $\mathfrak{h} \in P(g)$ (6.2.6 (vi)). Let $\tau_0' = \operatorname{ind}^{\sim}(g'|\mathfrak{h},\mathfrak{g}')$. Then

$$\operatorname{Ker} \tau_0' = I(g') = \operatorname{Ker} \tau'.$$

Hence

$$\operatorname{Ker} \tau = \operatorname{Ker} \operatorname{ind}^{\sim}(\tau_0',\mathfrak{g}) = \operatorname{Ker} \operatorname{ind}^{\sim}(g' \,|\, \mathfrak{h},\mathfrak{g}) = \operatorname{Ker} \operatorname{ind}^{\sim}(g \,|\, \mathfrak{h},\mathfrak{g}).$$

From 6.2.6 (vi), and since τ extends σ, we have

$$\tau(x) = \bar{\sigma}(\theta(s)).$$

On the other hand, if $u \in U(\mathfrak{n})$ and $w \in W'$, and we moreover denote the derivation of $U(\mathfrak{n})$ which extends $j(s)$ by $j(s)$, then

$$\begin{aligned}\tau(x)(u \otimes w) &= xu \otimes w = ux \otimes w + j(s)u \otimes w \\ &= u \otimes \tau'(x)w + \tfrac{1}{2}(\operatorname{tr}_{\mathfrak{n}/\mathfrak{n}'} j(s))u \otimes w + j(s)u \otimes w,\end{aligned}$$

whence

$$\bar{\sigma}(\theta(s;)(u \otimes w) = u \otimes \bar{\sigma}'(\theta'(s))w + \tfrac{1}{2}(\operatorname{tr}_{\mathfrak{n}/\mathfrak{n}'} j(s))u \otimes w + j(s)u \otimes w,$$

whence

(1) $$\bar{\sigma}(\theta(s)) = 1 \otimes \bar{\sigma}'(\theta'(s)) + j(s) \otimes 1 + \tfrac{1}{2}\operatorname{tr}_{\mathfrak{n}/\mathfrak{n}'} j(s).$$

We deduce from (1) that $\bar{\sigma} \circ \theta$, and hence θ, are Lie algebra homomorphisms.

(c) Let us assume that the only commutative ideals of $\mathfrak{n}$ which are stable under $j(\mathfrak{s})$ are 0 and $\mathfrak{z}$. Then $\mathfrak{n}$ is a Heisenberg algebra (4.6.2). Let $\mathfrak{m} = \operatorname{Ker} f$. Then $\mathfrak{n} = \mathfrak{z} \oplus \mathfrak{m}$. The restriction B of B_f to $\mathfrak{m}$ is a non-degenerate alternating bilinear form. Let $z \in \mathfrak{z}$ be such that $f(z) = 1$. Let $(x_1, \ldots, x_n, y_1, \ldots, y_n)$ be a basis for $\mathfrak{m}$ such that

$$[x_i,x_j] = [y_i,y_j] = [x_i,y_j] = 0 \quad \text{for } i \neq j,$$

$$[x_i,y_i] = z.$$

Every derivation $\delta \in j(\mathfrak{s})$ is zero on $\mathfrak{z}$, leaves $\mathfrak{m}$ stable, and satisfies $B(\delta x,y) + B(x,\delta y) = 0$ for all $x,y \in \mathfrak{m}$. Henceforth we may assume that $\mathfrak{s}$ is the set of all derivations of $\mathfrak{n}$ satisfying these conditions, and that j is the identity mapping.

We have $z - l \in I(f)$, hence $I(f)$ is the two-sided ideal of $U(\mathfrak{n})$ generated by $z - 1$, and we can identify A with the Weyl algebra A_n constructed over the generators $x_1, \ldots, x_n, y_1, \ldots, y_n$. Let T be the vector subspace of A generated by the x_ix_j, the y_iy_j, and the $x_iy_j + y_jx_i$. From 4.6.9, T is a Lie subalgebra of A, and, for all $s \in \mathfrak{s}$, there exists one and only one $\psi(s) \in T$ such that s and $\mathrm{ad}_A \psi(s)$ coincide on $\mathfrak{m}$; furthermore, ψ is an isomorphism of the Lie algebra $\mathfrak{s}$ onto the Lie algebra T. For all $s \in \mathfrak{s}$ $\mathrm{ad}_A \psi(s)$ and $\mathrm{ad}_A \theta(s)$ therefore coincide on $\mathfrak{m}$, and consequently $\psi(s) - \theta(s) \in k$. We shall now show that $\psi(s) = \theta(s)$, which will complete the proof. Henceforth, we fix $s \in \mathfrak{s}$.

There exists $\mathfrak{h} \in P(f)$ which is stable under s. Then $\mathfrak{h} \cap \mathfrak{m}$ is a maximal totally isotropic vector subspace in $\mathfrak{m}$ endowed with B. We can impose on the basis $(x_1, \ldots, y_n)$ chosen above the condition that

$$\mathfrak{h} \cap \mathfrak{m} = kx_1 + \cdots + kx_n.$$

Let $\mathfrak{s}'$ be the Lie subalgebra of $\mathfrak{s}$ consisting of those elements of $\mathfrak{s}$ under which $\mathfrak{h}$ is stable. From 10.1.2 (ii) and (iii), the restriction of θ to $\mathfrak{s}'$ is a Lie algebra homomorphism and in particular is linear. Taking 4.6.9 (v) into account, it is sufficient to prove that $\psi(s) = \theta(s)$ when $\psi(s) = x_i x_j$ and when $\psi(s) = x_i y_j + y_j x_i$. With the notation of 10.1.2, we shall now establish that

$$\bar{\sigma}_{\mathfrak{h}}(\theta(s))(1 \otimes 1) = \bar{\sigma}_{\mathfrak{h}}(\psi(s))(1 \otimes 1),$$

which will suffice since

$$\psi(s) - \theta(s) \in k.$$

We note that

$$\bar{\sigma}_{\mathfrak{h}}(\theta(s))(1 \otimes 1) = \tfrac{1}{2}\, \mathrm{tr}_{\mathfrak{n}/\mathfrak{h}}(s)(1 \otimes 1).$$

If $\psi(s) = x_i x_j$, then

$$[x_i x_j, y_k] = \delta_{ik} x_j + \delta_{jk} x_i, \qquad [x_i x_j, \delta_{ik} x_j + \delta_{jk} x_i] = 0;$$

hence $\mathrm{tr}_{\mathfrak{n}/\mathfrak{h}}(s) = 0$. On the other hand,

$$\bar{\sigma}_{\mathfrak{h}}(\psi(s))(1 \otimes 1) = x_i x_j \otimes 1 = x_i \otimes x_j \cdot 1 = 0.$$

If $\psi(s) = x_i y_j + y_j x_i$, then

$$[x_i y_j + y_j x_i, y_k] = 2\delta_{ik} y_j$$

hence $\mathrm{tr}_{\mathfrak{n}/\mathfrak{h}}(s) = 2\delta_{ij}$. On the other hand,

$$\bar{\sigma}_{\mathfrak{h}}(\psi(s))(1 \otimes 1) = (x_i y_j + y_j x_i) \otimes 1 = (2 y_j x_i + \delta_{ij}) \otimes 1 = \delta_{ij}(1 \otimes 1).$$

10.1.5. Lemma (notation as in 10.1.1, 10.1.3). *There exists one and only one homomorphism r of $U(\mathfrak{b})$ into $U(\mathfrak{s}) \otimes A$ such that $r(y) = 1 \otimes \bar{y}$ for $y \in \mathfrak{n}$,* *and* $r(x) = x \otimes 1 + 1 \otimes \theta(x)$ *for* $x \in \mathfrak{s}$. *This homomorphism is surjective. Its kernel is the two-sided ideal* $U(\mathfrak{b})I(f) = I(f)U(\mathfrak{b})$ *of* $U(\mathfrak{b})$ *generated by* $I(f)$ *(recall that* $[\mathfrak{b}, I(f)] \subset I(f)$*).*

The uniqueness of r is obvious. If $x, x' \in \mathfrak{s}$ and $y, y' \in \mathfrak{n}$, then

$$[1 \otimes \bar{y}, 1 \otimes \bar{y}'] = 1 \otimes [y, y']^-,$$

$$[x \otimes 1 + 1 \otimes \theta(x), x' \otimes 1 + 1 \otimes \theta(x')] = [x, x'] \otimes 1 + 1 \otimes \theta([x, x']),$$

$$[x \otimes 1 + 1 \otimes \theta(x), 1 \otimes \bar{y}] = [1 \otimes \theta(x), 1 \otimes \bar{y}]$$
$$= 1 \otimes [\theta(x), \bar{y}] = 1 \otimes [x, y]^-,$$

whence we have proved the existence of r. Clearly, r is surjective and $U(\mathfrak{b})I(f) \subset \operatorname{Ker} r$. We prove the reverse inclusion. If $x_1, \ldots, x_m \in \mathfrak{s}$ and $u \in U(\mathfrak{n})$, then

$$r(x_1 \cdots x_m u) = \sum_{0 \leq k_1, \ldots, k_m \leq 1} x_1^{k_1} \cdots x_m^{k_m} \otimes \theta(x_1)^{1-k_1} \cdots \theta(x_m)^{1-k_m} \bar{u}. \tag{1}$$

Indeed, (1) is obvious for $m = 0$ and the passage from $m - 1$ to m follows from the equation $r(x_1 \cdots x_m u) = r(x_1)r(x_2 \cdots x_m u)$. Let us now assume that $(x_1, \ldots, x_m)$ is a basis for $\mathfrak{s}$. Every element v of $U(\mathfrak{b})$ can be uniquely written in the form $\sum_{\nu \in \mathbf{N}^m} x^\nu v_\nu$, where $x^{(\nu_1, \ldots, \nu_m)} = x_1^{\nu_1} \cdots x_m^{\nu_m}$ and $v_\nu \in U(\mathfrak{n})$ for all ν. For $s = 0, 1, \ldots$, let $U_s = \sum_{|\nu| \leq s} x_\nu U(\mathfrak{n})$. We assume that $r(v) = 0$, and prove that $v_\nu \in I(f)$ for all ν. This is obvious if $v \in U_0 = U(\mathfrak{n})$. Let us assume that our assertion has been established for $v \in U_{s-1}$, and consider the case where $v \in U_s$. From (1), we have

$$0 = r(v) = \sum_{|\nu| = s} x^\nu \otimes \bar{v}_\nu + \sum_{|\nu| < s} x^\nu \otimes w_\nu$$

for certain $w_\nu \in A$. For $|\nu| = s$, we therefore have $\bar{v}_\nu = 0$, $v_\nu \in I(f)$. Then $v' = v - \sum_{|\nu| = s} x^\nu v_\nu \in \operatorname{Ker} r$, and $v' \in U_{s-1}$ so that it is sufficient to apply the induction hypothesis.

10.1.6. With the notation of 10.1.3 and 10.1.5, θ (or r) is said to be the *canonical homomorphism of* $\mathfrak{s}$ *into* A *(or of* $U(\mathfrak{b})$ *onto* $U(\mathfrak{s}) \otimes A$*) defined by* j.

10.1.7. We retain the notation of 10.1.1, 10.1.3 and 10.1.5. Let σ be a representation of $\mathfrak{n}$ in a space V, with kernel $I(f)$. We consider the representation $i \otimes \bar{\sigma}$ of $U(\mathfrak{s}) \otimes A$, where i denotes the one-dimensional trivial representation of $\mathfrak{s}$. Then $(i \otimes \bar{\sigma}) \circ r$ is a representation of $\mathfrak{b}$ in V, which we denote by $\sigma^\cdot$, such that $\sigma^\cdot | \mathfrak{n} = \sigma$. We term it the *extension of* σ *to* $\mathfrak{b}$ *defined by* j. For $x \in \mathfrak{s}$, we have $\sigma^\cdot(x) = \bar{\sigma}(\theta(x))$.

On the other hand, for every representation τ of $\mathfrak{s}$, we denote by $\tau^\cdot$ the representation of $\mathfrak{b}$ deduced from τ by virtue of the canonical homomorphism of $\mathfrak{b}$ onto $\mathfrak{s}$.

10.1.8. PROPOSITION (notation as in 10.1.1, 10.1.3, 10.1.5). *Let σ be a simple representation of $\mathfrak{n}$ in a space V with kernel $I(f)$.*

(i) *If τ is a representation of $\mathfrak{s}$, then the representation $\tau^{\cdot} \otimes \sigma^{\cdot}$ of $\mathfrak{b}$ is equal to $(\tau \otimes \bar{\sigma}) \circ r$, its kernel is $r^{-1}(\mathrm{Ker}\,\tau \otimes A)$, and its restriction to $\mathfrak{n}$ is a multiple of σ.*

(ii) *Every representation of $\mathfrak{b}$ whose restriction to $\mathfrak{n}$ is a multiple of σ is equivalent to a representation of the form $\tau^{\cdot} \otimes \sigma^{\cdot}$, where τ is a representation of $\mathfrak{s}$.*

(iii) *$\tau^{\cdot} \otimes \sigma^{\cdot}$ is simple if and only if τ is simple.*

(iv) *Let τ and τ' be representations of $\mathfrak{s}$ in the spaces W and W' respectively. The mapping $\varphi \mapsto \varphi \otimes 1$ is a bijection of $\mathrm{Hom}_{\mathfrak{s}}(W,W')$ onto $\mathrm{Hom}_{\mathfrak{s}}(W \otimes V,\, W' \otimes V)$ (where $W \otimes V$, $W' \otimes V$ are endowed with $\tau^{\cdot} \otimes \sigma^{\cdot}$, $\tau'^{\cdot} \otimes \sigma^{\cdot}$).*

We verify immediately that $\tau^{\cdot} \otimes \sigma^{\cdot}$ and $(\tau \otimes \sigma) \circ r$ coincide on $\mathfrak{n}$ and on $\mathfrak{s}$, and so are equal, whence we have proved (i).

Let π be a representation of $\mathfrak{b}$ such that $\pi|\mathfrak{n}$ is a multiple of σ. By replacing π by an equivalent representation, we may assume that the space of π is of the form $W \otimes V$, where W is a vector space, and that $\pi(x) = 1 \otimes \sigma(x)$ for $x \in \mathfrak{n}$. Since the commutant of $\sigma(\mathfrak{n})$ in End V is k, the commutant of $1 \otimes \sigma(\mathfrak{n})$ in $\mathrm{End}(W \otimes V)$ is $(\mathrm{End}\, W) \otimes 1$ (AL VIII, p. 15). Now, if $x \in \mathfrak{s}$ and $y \in \mathfrak{n}$, then

$$[\pi(x),\, 1 \otimes \sigma(y)] = \pi([x,y]) = 1 \otimes \sigma([x,y]),$$

$$[1 \otimes \bar{\sigma}(\theta(x)),\, 1 \otimes \sigma(y)] = 1 \otimes [\bar{\sigma}(\theta(x)),\sigma(y)] = 1 \otimes \sigma([x,y]),$$

hence $\pi(x) - 1 \otimes \bar{\sigma}(\theta(x))$ is of the form $\tau(x) \otimes 1$, where $\tau(x) \in \mathrm{End}\, W$, and it is clear that τ is a representation of $\mathfrak{s}$ in W. The representation $\tau^{\cdot} \otimes \sigma^{\cdot}$ coincides with π on $\mathfrak{n}$, and, if $x \in \mathfrak{s}$, then

$$\begin{aligned}\pi(x) &= \tau(x) \otimes 1 + 1 \otimes \bar{\sigma}(\theta(x)) \\ &= \tau^{\cdot}(x) \otimes 1 + 1 \otimes \sigma^{\cdot}(x) = (\tau^{\cdot} \otimes \sigma^{\cdot})(x).\end{aligned}$$

Hence $\pi = \tau^{\cdot} \otimes \sigma^{\cdot}$, which proves (ii).

Since σ is simple, the vector subspaces of $W \otimes V$ which are stable under $\tau^{\cdot} \otimes \sigma^{\cdot}$ are the $W_1 \otimes V$, where W_1 is a vector subspace of W which is stable under τ (AL VIII, p. 43). This proves (iii).

Let W, W', τ and τ' be as in (iv). Let $\varphi \in \mathrm{Hom}_{\mathfrak{s}}(W,W')$. Clearly,

$\varphi \otimes 1 : W \otimes V \to W' \otimes V$ is an $\mathfrak{n}$-homomorphism; on the other hand, if $s \in \mathfrak{s}$, $v \in V$ and $w \in W$, then

$$(1) \qquad (\varphi \otimes 1)(s(w \otimes v)) = (\varphi \otimes 1)(sw \otimes v + w \otimes \bar{\sigma}(\theta(s))v)$$
$$= \varphi(sw) \otimes v + \varphi(w) \otimes \bar{\sigma}(\theta(s))v,$$

$$(2) \qquad s(\varphi \otimes 1)(w \otimes v) = s\varphi(w) \otimes v + \varphi(w) \otimes \bar{\sigma}(\theta(s))v,$$

hence $\varphi \otimes 1 \in \mathrm{Hom}_{\mathfrak{b}}(W \otimes V, W' \otimes V)$. The mapping $\varphi \mapsto \varphi \otimes 1$ is obviously injective. Since the commutant of $\sigma(\mathfrak{n})$ is k, every $\mathfrak{n}$-homomorphism of $W \otimes V$ into $W' \otimes V$ is of the form $\varphi \otimes 1$, where $\varphi \in \mathrm{Hom}(W,W')$ (AL VIII, p. 15); and, from (1) and (2), $\varphi \otimes 1$ is an $\mathfrak{s}$-homomorphism if and only if φ is an $\mathfrak{s}$-homomorphism. This proves (iv).

Reference: [45].

10.2. Application to induced representations

10.2.1. PROPOSITION (k algebraically closed). *Let $\mathfrak{n}$ be a nilpotent ideal of $\mathfrak{g}$, $g \in \mathfrak{g}^*$, and $\mathfrak{l}$ a Lie subalgebra of $\mathfrak{g}$ subordinate to g and such that $\mathfrak{l} \cap \mathfrak{n} \in P(g|\mathfrak{n})$. Let $\sigma = \mathrm{ind}(g|\mathfrak{l} \cap \mathfrak{n},\mathfrak{n})$. Let $\mathfrak{g}_1 = \mathfrak{n}^g$ and $\tau = \mathrm{ind}^\sim(\mathfrak{g}|\mathfrak{l} \cap \mathfrak{g}_1,\mathfrak{g}_1)$. Let j be the adjoint representation of $\mathfrak{g}_1$ in $\mathfrak{n}$, $\mathfrak{b}$ the semi-direct product of $\mathfrak{g}_1$ and $\mathfrak{n}$ defined by j, $\sigma^\cdot$ and $\tau^\cdot$ the representations of $\mathfrak{b}$ deduced from σ and τ as in* 10.1.7. *Then:*

(i) *Let φ be the homomorphism $(x,y) \mapsto x + y$ of $\mathfrak{b}$ onto $\mathfrak{g}_1 + \mathfrak{n}$ ($x \in \mathfrak{g}_1$, $y \in \mathfrak{n}$). There exists one and only one representation π of $\mathfrak{g}_1 + \mathfrak{n}$ such that $\pi \circ \varphi = \tau^\cdot \otimes \sigma^\cdot$. The restriction of π to $\mathfrak{n}$ is a multiple of σ.*

(ii) $\mathfrak{l} = (\mathfrak{l} \cap \mathfrak{g}_1) + (\mathfrak{l} \cap \mathfrak{n})$ *and* $\varphi^{-1}(\mathfrak{l}) = (\mathfrak{l} \cap \mathfrak{g}_1) \oplus (\mathfrak{l} \cap \mathfrak{n})$.

(iii) *The representations* $\mathrm{ind}^\sim(g|\mathfrak{l},\mathfrak{g})$ *and* $\mathrm{ind}^\sim(\pi,\mathfrak{g})$ *are equivalent.*

We set $f = g|\mathfrak{n}$, $A = U(\mathfrak{n})/I(f)$.

(a) Let $x \in \mathfrak{g}_1 \cap \mathfrak{n}$. Then, with the notation of 10.1.1 and 10.1.3,

$$\tau^\cdot((0,x)) = 0, \qquad \sigma^\cdot((0,x)) = \sigma(x),$$
$$\tau^\cdot((x,0)) = \tau(x), \qquad \sigma^\cdot((x,0)) = \bar{\sigma}(\theta(x)).$$

Now $\theta(x)$ is of the form $x + \lambda$, where $\lambda \in k$, since $\theta(x)$ and $\bar{x}$ define the same inner derivation of A. Let $\mathfrak{h} \in P(f)$ and $\sigma_{\mathfrak{h}} = \mathrm{ind}(f|\mathfrak{h},\mathfrak{n})$. Then $x \in \mathfrak{h}$ and

$$\sigma_{\mathfrak{h}}(x)(1 \otimes 1) = x \otimes 1 = g(x) \cdot 1 \otimes 1;$$

moreover, from 10.1.2 (ii), $\bar{\sigma}_{\mathfrak{h}}(\theta(x)) \cdot 1 \otimes 1 = 0$ hence $0 = g(x) + \lambda$ and $\theta(x) = \bar{x} - g(x)$. Then

$$(\tau^\cdot \otimes \sigma^\cdot)((x,-x)) = \tau(x) \otimes 1 + 1 \otimes (\sigma(x) - g(x)) + 1 \otimes \sigma(-x).$$

Now $\mathfrak{g}_1 \cap \mathfrak{n} = \mathfrak{n}^f \subset \mathfrak{l} \cap \mathfrak{n} \subset \mathfrak{l}$, whence $\mathfrak{g}_1 \cap \mathfrak{n} \subset \mathfrak{l} \cap \mathfrak{g}_1$; and $g([\mathfrak{g}_1,\mathfrak{g}_1 \cap \mathfrak{n}]) = 0$, hence $\tau(x) = g(x) \cdot 1$ (5.1.13). Thus we see that $(\tau^{\cdot} \otimes \sigma^{\cdot})(x,-x)) = 0$, which proves (i).

(b) Since $\mathfrak{g}_1 \cap \mathfrak{n} \subset \mathfrak{l}$, we have

$$\mathfrak{l} \subset \mathfrak{l}^g \subset \mathfrak{g}_1^g + \mathfrak{n}^g = (\mathfrak{n} + \mathfrak{g}^g) + \mathfrak{n}^g = \mathfrak{n} + \mathfrak{g}_1.$$

Let $\mathfrak{l}' = \varphi^{-1}(\mathfrak{l})$. Clearly, $\mathfrak{l} \cap \mathfrak{g}_1 \subset \mathfrak{l}'$ and $\mathfrak{l} \cap \mathfrak{n} \subset \mathfrak{l}'$. If $x \in \mathfrak{g}_1$ and $y \in \mathfrak{n}$ are such that $x + y \in \mathfrak{l}$, then y is orthogonal to $\mathfrak{l} \cap \mathfrak{n}$, and hence $y \in \mathfrak{l} \cap \mathfrak{n}$, so that $x,y \in \mathfrak{l}$; thus,

$$\mathfrak{l}' = (\mathfrak{l} \cap \mathfrak{g}_1) \oplus (\mathfrak{l} \cap \mathfrak{n}),$$

whence

$$\mathfrak{l} = \varphi(\varphi^{-1}(\mathfrak{l})) = (\mathfrak{l} \cap \mathfrak{g}_1) + (\mathfrak{l} \cap \mathfrak{n}).$$

We have now proved (ii).

(c) Let $g' = g \circ \varphi$, let J be the left ideal of $U(\mathfrak{b})$ generated by the

$$h - g'(h) - \tfrac{1}{2} \operatorname{tr}_{\mathfrak{b}/\mathfrak{l}'} \operatorname{ad}_{\mathfrak{b}} h,$$

where h takes all values in $\mathfrak{l}'$, let J_1 be the left ideal of $U(\mathfrak{g}_1)$ generated by the

$$h - g(h) - \tfrac{1}{2} \operatorname{tr}_{\mathfrak{g}_1/\mathfrak{l} \cap \mathfrak{g}_1} \operatorname{ad}_{\mathfrak{g}_1} h,$$

where h takes all values in $\mathfrak{l} \cap \mathfrak{g}_1$, let J_2 be the left ideal of $U(\mathfrak{n})$ generated by the $h - f(h)$, where h takes all values in $\mathfrak{l} \cap \mathfrak{n}$, let

$$J' = J_1 \otimes A + U(\mathfrak{g}_1) \otimes \bar{J_2},$$

and let r be the homomorphism of $U(\mathfrak{b})$ onto $U(\mathfrak{g}_1) \otimes A$ defined by the adjoint representation of $\mathfrak{g}_1$ in $\mathfrak{n}$.

If $h \in \mathfrak{l} \cap \mathfrak{n}$, then

$$r(h - g'(h) - \tfrac{1}{2} \operatorname{tr}_{\mathfrak{b}/\mathfrak{l}'} \operatorname{ad}_{\mathfrak{b}} h) = \bar{h} - f(h) \in \bar{J_2}.$$

Let $h \in \mathfrak{l} \cap \mathfrak{g}_1$. Since $[h,\mathfrak{l} \cap \mathfrak{n}] \subset \mathfrak{l} \cap \mathfrak{n}$, from 10.1.2 (ii) we have

$$\theta(h) - \tfrac{1}{2} \operatorname{tr}_{\mathfrak{n}/\mathfrak{l} \cap \mathfrak{n}} \operatorname{ad}_{\mathfrak{b}} h \in \bar{J_2},$$

and hence

$$\begin{aligned} r(h - g'(h) - \tfrac{1}{2} \operatorname{tr}_{\mathfrak{b}/\mathfrak{l}'} \operatorname{ad}_{\mathfrak{b}} h) &= \\ = h \otimes 1 + 1 \otimes \theta(h) - g'(h) - \tfrac{1}{2} \operatorname{tr}_{\mathfrak{g}_1/\mathfrak{l} \cap \mathfrak{g}_1} \operatorname{ad}_{\mathfrak{b}} h - \tfrac{1}{2} \operatorname{tr}_{\mathfrak{n}/\mathfrak{l} \cap \mathfrak{n}} \operatorname{ad}_{\mathfrak{b}} h \\ = (h - g(h) - \tfrac{1}{2} \operatorname{tr}_{\mathfrak{g}_1/\mathfrak{l} \cap \mathfrak{g}_1} \operatorname{ad}_{\mathfrak{g}_1} h) \otimes 1 + 1 \otimes (\theta(h) - \tfrac{1}{2} \operatorname{tr}_{\mathfrak{n}/\mathfrak{l} \cap \mathfrak{n}} \operatorname{ad}_{\mathfrak{b}} h) \\ \in J_1 \otimes A + U(\mathfrak{g}_1) \otimes \bar{J_2} = J'. \end{aligned}$$

The above proves that $r(J) \subset J'$.

(d) Let $(x_1, \ldots, x_p)$ be a basis for a complement of $\mathfrak{l} \cap \mathfrak{n}$ in $\mathfrak{n}$, and $(y_1, \ldots, y_q)$ a basis for a complement of $\mathfrak{l} \cap \mathfrak{g}_1$ in $\mathfrak{g}_1$. If $\alpha = (\alpha_1, \ldots, \alpha_p) \in \mathbf{N}^p$ and $\beta = (\beta_1, \ldots, \beta_q) \in \mathbf{N}^q$, we set

$$y^\beta x^\alpha = y_1^{\beta_1} \cdots y_q^{\beta_q} x_1^{\alpha_1} \cdots x_p^{\alpha_p}$$

calculated in $U(\mathfrak{b})$. The $y^\beta x^\alpha$ constitute a basis for a complement E of J in $U(\mathfrak{b})$. For all $n \in \mathbf{N}$, let E_n be the vector subspace of E generated by the $y^\beta x^\alpha$ such that $|\beta| \leq n$. Let $\sum_\alpha \lambda_\alpha x^\alpha$ be an element of E_0 (the λ_α are elements of k). If $r(\sum_\alpha \lambda_\alpha x^\alpha) \in J'$, then

$$\sum_\alpha 1 \otimes \lambda_\alpha \bar{x}^\alpha \in J_1 \otimes A + U(\mathfrak{g}_1) \otimes \bar{J}_2,$$

whence $\lambda_\alpha \bar{x}_\alpha \in \bar{J}_2$ and $\lambda_\alpha = 0$ for all α; this proves that $r^{-1}(J') \cap E_0 = 0$. Let us assume that $r^{-1}(J') \cap E_{n-1} = 0$ has been proved. Let $u = \sum_{|\beta| \leq n} y^\beta w_\beta$ be an element of $U(\mathfrak{b})$ such that $r(u) \in J'$ (the w_β are elements of E_0). From formula (1) of 10.1.5, we have

$$r(u) = \sum_{|\beta| = n} y^\beta \otimes \bar{w}_\beta + \sum_{|\beta| < n} y^\beta \otimes v_\beta.$$

where the v_β are elements of A. Then, for $|\beta| = n$, we have $\bar{w}_\beta \in \bar{J}_2$, whence $w_\beta = 0$ from the anove. Hence $u \in E_{n-1}$, and then $u = 0$ from the induction hypothesis. Thus it can be seen that $r^{-1}(J') \cap E = 0$.

(e) The conclusions drawn from (c) and (d) imply that $J = r^{-1}(J')$. Now $\operatorname{ind}^\sim(g'|\mathfrak{l}',\mathfrak{b})$ is equivalent to the regular representation of $U(\mathfrak{b})$ in $U(\mathfrak{b})/J$ (5.1.9). On the other hand, $\tau \otimes \bar{\sigma}$ is equivalent to the regular representation of $U(\mathfrak{g}_1) \otimes A$ in $(U(\mathfrak{g}_1) \otimes A)/J'$. Consequently, $\operatorname{ind}^\sim(g'\mathfrak{l}',\mathfrak{b})$ is equivalent to $(\tau \otimes \sigma^-) \circ r = \tau^\cdot \otimes \sigma^\cdot$ (10.1.8 (i)), i.e. to $\pi \circ \varphi$ from (i). Hence $\operatorname{ind}^\sim(g|\mathfrak{l},\mathfrak{g}_1 + \mathfrak{n})$ is equivalent to π (5.2.4). Assertion (iii) then follows from 5.2.3.

10.2.2. Theorem (k algebraically closed). *Let* $\mathfrak{n}$ *be a nilpotent ideal of* $\mathfrak{g}$, $g \in \mathfrak{g}^*$, $\mathfrak{l}$ *a Lie subalgebra of* $\mathfrak{g}$ *subordinate to* g *such that* $\mathfrak{l} \cap \mathfrak{n} \in P(g|\mathfrak{n})$, *and* $\mathfrak{g}_1 = \mathfrak{n}^g$. *Then:*

(i) $\operatorname{ind}^\sim(g|\mathfrak{l},\mathfrak{g})$ *is simple if and only if* $\operatorname{ind}^\sim(g|\mathfrak{l} \cap \mathfrak{g}_1,\mathfrak{g}_1)$ *is simple.*

(ii) *The representations* $\operatorname{ind}^\sim(g|\mathfrak{l},\mathfrak{g})$ *and* $\operatorname{ind}^\sim(g|\mathfrak{l} \cap \mathfrak{g}_1,\mathfrak{g}_1)$ *have isomorphic commutants.*

(iii) *Let* $g' \in \mathfrak{g}^*$ *be such that* $g'|\mathfrak{n} = g|\mathfrak{n}$, *whence* $\mathfrak{n}^{g'} = \mathfrak{g}_1$, *and let* $\mathfrak{l}'$ *be a Lie subalgebra of* $\mathfrak{g}$ *subordinate to* g' *and such that* $\mathfrak{l}' \cap \mathfrak{n} \in P(g'|\mathfrak{n})$. *If* $\operatorname{ind}^\sim(g|\mathfrak{l} \cap \mathfrak{g}_1,\mathfrak{g}_1)$ *and* $\operatorname{ind}(g'|\mathfrak{l}' \cap \mathfrak{g}_1,\mathfrak{g}_1)$ *have the same kernel, then ind*$^\sim(g|\mathfrak{l},\mathfrak{g})$ *and* $\operatorname{ind}^\sim(g'|\mathfrak{l}',\mathfrak{g})$ *have the same kernel.*

Let us introduce σ, τ, $\mathfrak{b}$, $\sigma^\cdot$, $\tau^\cdot$ and π as in 10.2.1. From 6.2.9, σ is simple.

Then

$$\begin{aligned}\operatorname{ind}^{\sim}(g|\mathfrak{l},\mathfrak{g}) \text{ simple} &\Leftrightarrow \operatorname{ind}^{\sim}(\pi,\mathfrak{g}) \text{ simple} && (10.2.1\text{ (iii)})\\ &\Leftrightarrow \pi \text{ simple} && (10.2.1\text{ (i)}, 6.2.8, 5.3.6)\\ &\Leftrightarrow \tau^{\cdot} \otimes \sigma^{\cdot} \text{ simple} &&\\ &\Leftrightarrow \tau \text{ simple} && (10.1.8\text{ (iii)}).\end{aligned}$$

This proves (i).

The commutant of $\operatorname{ind}^{\sim}(g|\mathfrak{l},\mathfrak{g})$ is isomorphic to that of $\operatorname{ind}^{\sim}(\pi,\mathfrak{g})$ (10.2.1 (iii)), hence to that of π (10.2.1 (i), 6.2.8, 5.3.7), hence to that of $\tau^{\cdot} \otimes \sigma^{\cdot}$, and hence to that of τ (10.1.8 (iv)).

Let τ', $\tau^{\cdot\cdot}$ and π' be the objects analogous to τ, $\tau^{\cdot}$ and π deduced from g' and $\mathfrak{l}'$. If τ and τ' have the same kernel, then $\tau^{\cdot} \otimes \sigma^{\cdot}$ and $\tau^{\cdot\cdot} \otimes \sigma^{\cdot}$ have the same kernel (10.1.8 (i)), hence π and π' have the same kernel, hence $\operatorname{ind}^{\sim}(\pi,\mathfrak{g})$ and $\operatorname{ind}^{\sim}(\pi',\mathfrak{g})$ have the same kernel (5.2.6), and hence $\operatorname{ind}^{\sim}(g|\mathfrak{l},\mathfrak{g})$ and $\operatorname{ind}^{\sim}(g'|\mathfrak{l}',\mathfrak{g})$ have the same kernel (10.2.1 (iii)).

Reference: [45].

10.3. The ideals $I(f)$

10.3.1. LEMMA (k algebraically closed). *Let $\mathfrak{n}$ be a nilpotent ideal of $\mathfrak{g}$, $f \in \mathfrak{g}^*$, $\mathfrak{h} = \mathfrak{n}^f$, and $\mathfrak{l} \in \mathrm{PR}(f)$. Then there exists $\mathfrak{l}_1 \in \mathrm{PR}(f)$ such that:*

(a) $\mathfrak{l}_1 \cap \mathfrak{n} \in P(f|\mathfrak{n})$;

(b) $\mathfrak{l}_1 \cap \mathfrak{h} \in P(f|\mathfrak{h})$;

(c) $\operatorname{ind}^{\sim}(f|\mathfrak{l},\mathfrak{g})$ *and* $\operatorname{ind}^{\sim}(f|\mathfrak{l}_1,\mathfrak{g})$ *have the same kernel.*

Let $\mathfrak{g}' = \mathfrak{l} + \mathfrak{n}$, which is a solvable Lie subalgebra of $\mathfrak{g}$, and $f' = f|\mathfrak{g}'$. Then $\mathfrak{l} \in P(f')$, and hence every element of $P(f')$ belongs to $P(f)$. Now, from 1.12.10, there exists $\mathfrak{l}_1 \in P(f')$ such that $\mathfrak{l}_1 \cap \mathfrak{n} \in P(f|\mathfrak{n})$, and we then have $\mathfrak{l}_1 \cap \mathfrak{h} \in P(f|\mathfrak{h})$ (1.12.6). From 6.1.4, $\operatorname{ind}^{\sim}(f|\mathfrak{l},\mathfrak{g}')$ and $\operatorname{ind}^{\sim}(f|\mathfrak{l}_1,\mathfrak{g}')$ have the same kernel. Hence $\operatorname{ind}^{\sim}(f|\mathfrak{l},\mathfrak{g})$ and $\operatorname{ind}^{\sim}(f|\mathfrak{l}_1,\mathfrak{g})$ have the same kernel (5.2.3, 5.2.6).

10.3.2. LEMMA (k algebraically closed, $\mathfrak{g}$ semi-simple). *Let K be the Killing form of $\mathfrak{g}$, $x \in \mathfrak{g}$, s and n the semi-simple and nilpotent components of x, and g the linear form $y \mapsto K(x,y)$ on $\mathfrak{g}$. Then:*

(i) $\mathfrak{g}^x = \mathfrak{g}^s \cap \mathfrak{g}^n$.

(ii) *The algebra $\mathfrak{g}^s$ is reductive in $\mathfrak{g}$ and has same rank as $\mathfrak{g}$.*

(iii) *If $\mathfrak{b} \in \mathrm{PR}(g)$ (so that x is regular from* 1.12.18), *then $\mathfrak{b} \cap \mathfrak{g}^s$ is a Borel subalgebra of $\mathfrak{g}^s$.*

Since $\operatorname{ad}_{\mathfrak{g}}s$ and $\operatorname{ad}_{\mathfrak{g}}n$ are polynomials without constant term in $\operatorname{ad}_{\mathfrak{g}}x$, (i) is obvious. The algebra $\mathfrak{g}^s$ is reductive in $\mathfrak{g}$ (1.7.7). Let $\mathfrak{h}$ be a Cartan subalgebra of $\mathfrak{g}$ containing s (1.10.6 (iii)). Then $\mathfrak{h}$ is contained in $\mathfrak{g}^s$, is reductive in $\mathfrak{g}^s$ and is maximal and commutative in $\mathfrak{g}^s$, and therefore is a Cartan subalgebra of $\mathfrak{g}^s$ (1.10.6 (iii)), whence we have proved (ii). Let $\mathfrak{q}$ be the orthogonal subspace of $\mathfrak{g}^s$ in $\mathfrak{g}$ with respect to K. From 1.7.7, we have $\mathfrak{q} = [s,\mathfrak{g}]$ and $\mathfrak{g} = \mathfrak{g}^s \oplus \mathfrak{q}$. Let $\mathfrak{b} \in \mathrm{PR}(\mathfrak{g})$. Then $\mathfrak{g}^x \subset \mathfrak{b}$, hence $s \in \mathfrak{b}$. Since $\mathfrak{b}$ is stable under $\mathrm{ad}_{\mathfrak{g}} s$, we have $\mathfrak{b} = (\mathfrak{b} \cap \mathfrak{g}^s) \oplus (\mathfrak{b} \cap \mathfrak{q})$. On the other hand,

$$B_g(\mathfrak{g}^s,\mathfrak{q}) = K(x,[\mathfrak{g}^s,\mathfrak{q}]) = K([x,\mathfrak{g}^s],\mathfrak{q}) \subset K(\mathfrak{g}^s,\mathfrak{q}) = 0.$$

Since $\mathfrak{g}^g \subset \mathfrak{g}^s$, we see that $\mathfrak{g}^s$ is the orthogonal subspace of $\mathfrak{q}$ with respect to B_g. Since $\mathfrak{b}$ is maximal totally isotropic with respect to B_g, $\mathfrak{b} \cap \mathfrak{g}^s$ is maximal totally isotropic with respect to $B_{g|\mathfrak{g}^s}$. Since $\mathfrak{b} \cap \mathfrak{g}^s$ is a solvable Lie subalgebra of $\mathfrak{g}^s$, this proves (iii) (1.12.18).

10.3.3. THEOREM (k algebraically closed). *Let $g \in \mathfrak{g}^*$ be such that* $\mathrm{PR}(g) \neq \emptyset$.

(i) *There exists* $\mathfrak{h} \in \mathrm{PR}(g)$ *such that* $\mathrm{ind}^{\sim}(g|\mathfrak{h},\mathfrak{g})$ *is simple.*

(ii) *If* $\mathfrak{h}_1,\mathfrak{h}_2 \in \mathrm{PR}(g)$, *the representations* $\mathrm{ind}^{\sim}(g|\mathfrak{h}_1,\mathfrak{g})$ *and* $\mathrm{ind}^{\sim}(g|\mathfrak{h}_2,\mathfrak{g})$ *have the same kernel.*

This is obvious if $\dim \mathfrak{g} \leq 1$. We reason by induction on $\dim \mathfrak{g}$. Let $\mathfrak{n}$ be the largest nilpotent ideal of $\mathfrak{g}$, $f = g|\mathfrak{n}$, and $\mathfrak{g}_1 = \mathfrak{n}^g$.

(a) Let us assume that $\mathfrak{g} = \mathfrak{g}_1$ and that $\mathrm{Ker} f \neq 0$. Then $\mathrm{Ker} f$ is an ideal of $\mathfrak{g}$ contained in $\mathfrak{g}^g$. It is sufficient to apply the induction hypothesis to $\mathfrak{g}/\mathrm{Ker} f$ and to the linear form on $\mathfrak{g}/\mathrm{Ker} f$ deduced from g by passage to the quotient.

(b) Let us assume that $\mathfrak{g} = \mathfrak{g}_1$ and that $\mathrm{Ker} f = 0$. If $\mathfrak{n} = 0$, then $\mathfrak{g}$ is semi-simple. If $\mathfrak{n} \neq 0$, then $\mathfrak{n}$ is one-dimensional and f is non-zero, and hence $\mathfrak{n}$ is central in $\mathfrak{g}$. Then the radical of $\mathfrak{g}$ is nilpotent and consequently equal to $\mathfrak{n}$. Finally, $\mathfrak{g}$ is the product of a semi-simple Lie algebra and a one-dimensional Lie algebra. This case may be easily reduced to the semi-simple case.

Let us therefore assume that $\mathfrak{g}$ is semi-simple. Let $\mathfrak{b}_1,\mathfrak{b}_2 \in \mathrm{PR}(g)$. Then $\mathfrak{b}_1$ and $\mathfrak{b}_2$ are Borel subalgebras of $\mathfrak{g}$. There exist a Cartan subalgebra $\mathfrak{k}$ of $\mathfrak{g}$ such that $\mathfrak{k} \subset \mathfrak{b}_1 \cap \mathfrak{b}_2$ (1.10.18), bases B_1,B_2 for $R(\mathfrak{g},\mathfrak{k})$ corresponding to $\mathfrak{b}_1$ and $\mathfrak{b}_2$, a $w \in W(\mathfrak{g},\mathfrak{k})$ such that $w(B_1) = B_2$ (11.1.6), and an element a of the adjoint group of $\mathfrak{g}$ which induces w on $\mathfrak{k}$ (1.10.19). The kernel of $\mathrm{ind}^{\sim}(g|\mathfrak{b}_1,\mathfrak{g})$ is transformed by a into the kernel of $\mathrm{ind}^{\sim}(a \cdot g|\mathfrak{b}_2,\mathfrak{g})$, and so is equal to it (2.4.17). Now w transforms $g|\mathfrak{k}$ into $a \cdot g|\mathfrak{k}$, hence the kernels of $\mathrm{ind}^{\sim}(g|\mathfrak{b}_2,\mathfrak{g})$ and of $\mathrm{ind}^{\sim}(a \cdot g|\mathfrak{b}_2,\mathfrak{g})$ are equal (8.4.4 (ii)). This proves (ii).

Let x be the element of $\mathfrak{g}$ corresponding to g under the Killing isomorphism, and let us introduce the notation of 10.3.2. Let $\mathfrak{b} \in \mathrm{PR}(g)$. Then

$\mathfrak{b} \cap \mathfrak{g}^s$ is a Borel subalgebra of $\mathfrak{g}^s$ (10.3.2 (iii)). Let $\mathfrak{k}$ be a Cartan subalgebra of $\mathfrak{g}^s$ contained in $\mathfrak{b}$. It is a Cartan subalgebra of $\mathfrak{g}$ (10.3.2 (ii)). We have $s \in \mathfrak{k}$. Let $R = R(\mathfrak{g},\mathfrak{k})$ and $R_1 = R(\mathfrak{g}^s,\mathfrak{k})$. Then R_1 is the set of those elements of R which are zero at s. We have $x \in \mathfrak{g}^x \subset \mathfrak{b} \cap \mathfrak{g}^s$, hence every Borel subalgebra of $\mathfrak{g}$ containing $\mathfrak{b} \cap \mathfrak{g}^s$ belongs to PR(g) (1.12.18). The Borel subalgebra $\mathfrak{b} \cap \mathfrak{g}^s$ of $\mathfrak{g}^s$ defines a set R_{1+} of positive roots in R_1. To prove (i), it is then sufficient, taking 7.6.24 into account, to construct a basis for R, defining a set R_+ of positive roots in R, in such a way that:

(1) $R_{1_+} \subset R_+$;

(2) $g(H_\alpha) \notin \{1,2,3, \ldots\}$ for all $\alpha \in R_+$.

We note that

$$g(H_\alpha) = K(x,H_\alpha) = K(s,H_\alpha) + K(n,H_\alpha) = K(s,H_\alpha)$$

since n is orthogonal to every Borel subalgebra which contains it.

Let us consider k as a vector space over $\mathbf{Q}$. There exists a projection of k onto $\mathbf{Q}$ whose kernel contains none of the non-zero scalars $\alpha(s)$ ($\alpha \in R$). This defines a projection of $\mathfrak{k}$ onto $\mathfrak{k}_{\mathbf{Q}}$; let $\mathfrak{h}$ be the image of s under this projection. Then

(1) $\alpha(h) \in \mathbf{Q}$ for all $\alpha \in R$;

(2) $\alpha(h) = \alpha(s)$ if $\alpha \in R$ and $\alpha(s) \in \mathbf{Q}$;

(3) let $\alpha \in R$; then

$$\alpha(h) = 0 \Leftrightarrow \alpha(s) = 0 \Leftrightarrow \alpha \in R_1 .$$

Let $R_+ = R_{1+} \cup \{\alpha \in R | \alpha(h) < 0\}$. It is easily verified that

$$R_+ \cup (-R_+) = R,$$

that $R_+ \cap (-R_+) = \emptyset$, and that the sum of two elements of R_+ belongs to R_+ if it belongs to R. Hence R_+ is the set of positive roots of R relative to a certain basis (11.1.7). Finally, if $\alpha \in R_{1+}$, then $K(s,H_\alpha) = 0$ because $\alpha(s) = 0$; and, if $\alpha \in R$ with $\alpha(h) < 0$, then $K(s,H_\alpha)$ is not a positive integer (otherwise, $\alpha(s)$ is a positive rational number, whence $\alpha(h) = \alpha(s) > 0$).

(c) Let us assume that $\mathfrak{g} \neq \mathfrak{g}_1$. There exists $\mathfrak{l} \in$ PR(g) such that $\mathfrak{l} \cap \mathfrak{n} \in P(f)$ and $\mathfrak{l} \cap \mathfrak{g}_1 \in$ PR($g|\mathfrak{g}_1$) (10.3.1). From the induction hypothesis, there exists $\mathfrak{l}_1 \in$ PR($g|\mathfrak{g}_1$) such that ind$^\sim(g|\mathfrak{l}_1,\mathfrak{g}_1)$ is simple. There exists $\mathfrak{p} \in$ P(f) such that $[\mathfrak{l}_1,\mathfrak{p}] \subset \mathfrak{p}$ (1.12.13), and then $\mathfrak{l}_1 + \mathfrak{p} \in$ PR(g) (1.12.13), $(\mathfrak{l}_1 + \mathfrak{p}) \cap \mathfrak{g}_1 = \mathfrak{l}_1$ and $(\mathfrak{l}_1 + \mathfrak{p}) \cap \mathfrak{n} = \mathfrak{p}$ (1.12.4). From 10.2.2 (i), ind$^\sim(g|\mathfrak{l}_1 + \mathfrak{p},\mathfrak{g})$ is simple. This proves (i).

Let $\mathfrak{h}_1,\mathfrak{h}_2 \in \mathrm{PR}(g)$. We prove (ii). From 10.3.1, we may assume that $\mathfrak{h}_1 \cap \mathfrak{n},\mathfrak{h}_2 \cap \mathfrak{n} \in \mathrm{P}(f)$ and that $\mathfrak{h}_1 \cap \mathfrak{g}_1,\mathfrak{h}_2 \cap \mathfrak{g}_1 \in P(g|\mathfrak{g}_1)$. From the induction hypothesis, $\mathrm{ind}^{\sim}(g|\mathfrak{h}_1 \cap \mathfrak{g}_1,\mathfrak{g}_1)$ and $\mathrm{ind}^{\sim}(g|\mathfrak{h}_2 \cap \mathfrak{g}_1,\mathfrak{g}_1)$ have the same kernel. Then $\mathrm{ind}^{\sim}(g|\mathfrak{h}_1,\mathfrak{g})$ and $\mathrm{ind}^{\sim}(g|\mathfrak{h}_2,\mathfrak{g})$ have the same kernel (10.2.2 (iii)).

10.3.4. We assume that k is algebraically closed. For all $g \in \mathfrak{g}^*$ such that $\mathrm{PR}(g) \neq \emptyset$ (e.g., for every regular element of $\mathfrak{g}^*$), there exists $\mathfrak{h} \in \mathrm{PR}(g)$ such that $\mathrm{ind}^{\sim}(g|\mathfrak{h},\mathfrak{g})$ is simple (10.3.3 (i)); moreover, $\mathrm{Ker\,ind}^{\sim}(g|\mathfrak{h},g)$ depends solely on g and not on the choice of $\mathfrak{h}$ (10.3.3 (ii)). We denote this ideal, which is primitive, by $I(f)$. This notation generalises that of 6.1.5 (for k algebraically closed).

10.3.5. PROPOSITION (k algebraically closed). *Let $\mathfrak{g}_r^*$ be the set of regular elements of $\mathfrak{g}^*$. The mapping $g \mapsto I(g)$ of $\mathfrak{g}_r^*$ into* Prim $U(\mathfrak{g})$ *is continuous (for the Zariski and Jacobson topologies).*

Let $u \in U(\mathfrak{g})$, and let F be the set of those $f \in \mathfrak{g}_r^*$ such that $u \in I(f)$. We must show that F is closed in $\mathfrak{g}_r^*$. Let d' be the index of $\mathfrak{g}$ and $d = \frac{1}{2}(d' + \dim \mathfrak{g})$. Let T_d and $E_{d,u}$ be as in 6.4.1, and let p be the canonical projection of $\mathfrak{g}^* \times \mathrm{Gr}(\mathfrak{g},d)$ onto $\mathfrak{g}^*$. Let $E'_{d,u}$ be the set of the $(f,\mathfrak{h}) \in E_{d,u}$ such that $\mathfrak{h}$ is solvable. Since $\mathrm{Gr}(\mathfrak{g},d)$ is a complete manifold and $E'_{d,u}$ is closed (6.4.1, 1.11.9), $p(E'_{d,u})$ is closed in $\mathfrak{g}^*$. We show that $F = \mathfrak{g}_r^* \cap p(E'_{d,u})$. If $f \in F$, let $\mathfrak{h} \in \mathrm{PR}(f)$; then

$$\dim \mathfrak{h} = d, \qquad u \in \mathrm{Ker\,ind}^{\sim}(f|\mathfrak{h},\mathfrak{g}),$$

hence $(f,\mathfrak{h}) \in E'_{d,u}$ and $f \in \mathfrak{g}_r^* \cap p(E'_{d,u})$. If $g \in \mathfrak{g}_r^* \cap p(E'_{d,u})$, there exists a solvable Lie subalgebra $\mathfrak{h}$ of $\mathfrak{g}$ such that $g([\mathfrak{h},\mathfrak{h}]) = 0$, $\dim \mathfrak{h} = d$ and $u \in \mathrm{Ker\,ind}^{\sim}(g|\mathfrak{h},\mathfrak{g})$. Then $\mathfrak{h} \in \mathrm{PR}(g)$ because $g \in \mathfrak{g}_r^*$, whence $u \in I(g)$ and $g \in F$.

10.3.6. Let $\mathfrak{a}$ be a commutative ideal of $\mathfrak{g}$, $f \in \mathfrak{a}^*$, and N the set of the elements of $S(\mathfrak{a}) = U(\mathfrak{a})$ which are zero at f. Up to 10.3.8, we shall denote by $J_{a,f}$ the left ideal $U(\mathfrak{g})N$ of $U(\mathfrak{g})$, that is, the left ideal generated by the $a - f(a)$, where a takes all values in $\mathfrak{a}$. Let (u_i) be a basis for the right $U(\mathfrak{a})$-module $U(\mathfrak{g})$. Then

$$J_{\mathfrak{a},f} = \Big(\sum_i u_i U(\mathfrak{a})\Big)N = \sum_i u_i N.$$

10.3.7. LEMMA. *Let $\mathfrak{a}$ and $\mathfrak{b}$ be commutative ideals of $\mathfrak{g}$ such that $\mathfrak{b} \supset \mathfrak{a}$. Let $f \in \mathfrak{a}^*$ and $u \in U(\mathfrak{g})$ be such that $u \notin J_{a,f}$ (cf. 10.3.6). There exists $f' \in \mathfrak{b}^*$ such that $f'|\mathfrak{a} = f$ and $u \notin J_{\mathfrak{b},f'}$.*

Let (u_i) be a basis for the right $U(\mathfrak{b})$-module $U(\mathfrak{g})$, $\mathfrak{c}$ a vector subspace complementary to $\mathfrak{a}$ in $\mathfrak{b}$, and (c_j) a basis for $S(\mathfrak{c})$. There exist unique $a_{ij} \in S(\mathfrak{a})$ such that $u = \sum_{i,j} u_i c_j a_{ij}$, whence $u = \sum_i u_i b_i$ on writing $b_i = \sum_j c_j a_{ij}$. There exist i_0 and j_0 such that $a_{i_0 j_0}(f) \neq 0$. Then the element $\sum_j c_j a_{i_0 j}(f)$ of $S(\mathfrak{c})$ is non-zero. Let $f'' \in \mathfrak{c}^*$ be such that $\sum_j c_j(f'') a_{i_0 j}(f) \neq 0$. Let f' be the element of $\mathfrak{b}^*$ which extends f and f''. Then $b_{i_0}(f') \neq 0$, whence $u \notin J_{\mathfrak{b},f}$.

10.3.8. LEMMA (k algebraically closed). *Let $\mathfrak{a}$ be a commutative ideal of $\mathfrak{g}$, $u \in U(\mathfrak{g})$ and $f \in \mathfrak{a}^*$ such that $u \notin J_{\mathfrak{a},f}$ (cf. 10.3.6). Let A be the affine space formed from the $g \in \mathfrak{g}^*$ such that $g|\mathfrak{a} = f$, and P be an open non-empty subset of A such that* $\mathrm{PR}(g) \neq \emptyset$ *for all $g \in P$. There exists $g \in P$ such that $u \notin I(g)$.*

We reason by induction on $\dim \mathfrak{g}$.

(a) We assume that $\mathfrak{g}$ is semi-simple (hence $\mathfrak{a} = 0$). Let $\mathscr{A}$ be the adjoint group of $\mathfrak{g}$. Since $I(af) = I(f)$ for all $f \in \mathfrak{g}$ and all $a \in \mathscr{A}$ (from 2.4.17), we may assume that $\mathscr{A}P = P$. Let $\mathfrak{h}$ be a Cartan subalgebra of $\mathfrak{g}$, B a basis for $R(\mathfrak{g},\mathfrak{h})$, and $\mathfrak{b}_+$ the corresponding Borel subalgebra. We identify $\mathfrak{g}$ with $\mathfrak{g}^*$, and $\mathfrak{h}$ with $\mathfrak{h}^*$ by virtue of the Killing form. Then $P \cap \mathfrak{h}$ is non-empty and open in $\mathfrak{h}$. If $\lambda \in P \cap \mathfrak{h}$, then $\mathfrak{b}_+ \in \mathrm{PR}(\lambda)$, and $I(\lambda)$ is the annihilator of the Verma module $M(\lambda)$. The lemma then follows from 8.4.4 (iii).

(b) We assume that $\mathfrak{g} = \mathfrak{s} \times \mathfrak{a}$ with $\mathfrak{s}$ semi-simple, $\dim \mathfrak{a} = 1$, and $f \neq 0$. Let (s_i) be a basis for $U(\mathfrak{s})$ and write $u = \sum_i s_i a_i$ for some a_i in $S(\mathfrak{a})$. Since $u \notin J_{a,f}$, there exists i_0 such that $a_{i_0}(f) \neq 0$. Then the element $s = \sum_i a_i(f) s_i$ of $U(\mathfrak{s})$ is non-zero. Let $P' \subset \mathfrak{s}^*$ be the set of restrictions to $\mathfrak{s}$ of elements of P. From (a) there exists $g' \in P'$ such that $s \notin I(g')$. Let $\mathfrak{b} \in \mathrm{PR}(g';\mathfrak{s})$ and $\tau' = \mathrm{ind}^\sim(g'|\mathfrak{b},\mathfrak{s})$. Let $g \in P$ be such that $g|\mathfrak{s} = g'$. Then $\mathfrak{l} = \mathfrak{b} + \mathfrak{a}$ belongs to $\mathrm{PR}(g;\mathfrak{g})$. Let $\tau = \mathrm{ind}^\sim(g|\mathfrak{l},\mathfrak{g})$. Then

$$\tau(u) = \sum_i \tau(s_i)\tau(a_i) = \sum_i \tau'(s_i)a_i(f) = \tau'(s) \neq 0,$$

hence $u \notin I(g)$.

(c) We assume that there exists a commutative ideal $\mathfrak{b}$ of $\mathfrak{g}$ and $f' \in \mathfrak{b}^*$ with the following properties:

(i) $\mathfrak{b} \supset \mathfrak{a}$;

(ii) $f'|\mathfrak{a} = f$;

(iii) the set P' of the $g \in P$ such that $g|\mathfrak{b} = f'$ is non-empty;

(iv) $u \in J_{\mathfrak{b},f'}$;

(v) the set $\mathfrak{g}'$ of the $x \in \mathfrak{g}$ such that $f'([x,\mathfrak{b}]) = 0$ is distinct from $\mathfrak{g}$.

The Lie algebra $\mathfrak{g}'$ contains $\mathfrak{b}$. Let $J'_{\mathfrak{b},f'}$ be the left ideal of $U(\mathfrak{g}')$ generated by the $b \in U(\mathfrak{b})$ such that $b(f') = 0$. Let (v_i) be a basis for the right $U(\mathfrak{g}')$-

module $U(\mathfrak{g})$. Then $u = \sum_i v_i u_i$, with $u_i \in U(\mathfrak{g}')$. Since $u \notin J_{\mathfrak{b},f'}$, there exists i_0 such that $u_{i_0} \notin J'_{\mathfrak{b},f'}$. Let $Q = P'|\mathfrak{g}'$. If $g \in P'$, then $\mathrm{PR}(g)$ is non-empty; from 10.3.1 (applied with $\mathfrak{n} = \mathfrak{b}$), we have $\mathrm{PR}(g|\mathfrak{g}') \neq \emptyset$. We can hence apply the induction hypothesis to $\mathfrak{g}'$, $\mathfrak{b}$, u_{i_0}, f' and $Q + \theta_{\mathfrak{g},\mathfrak{g}'}$ (cf. 5.2.1). There consequently exists a $g' \in Q$ such that $u_{i_0} \notin I(g'')$, where $g'' = g' + \theta_{\mathfrak{g},\mathfrak{g}'}$.

Let g be an element of P' which extends g'; we show that $u \notin I(g)$. Let $\mathfrak{l} \in \mathrm{PR}(g';\mathfrak{g}')$. Then $\mathfrak{l} \supset \mathfrak{b}$, hence $\mathfrak{l}$ is maximal totally isotropic with respect to B_g, so that $\mathfrak{l} \in \mathrm{PR}(g;\mathfrak{g})$. Let

$$\tau = \mathrm{ind}^{\sim}(g' \mid \mathfrak{l},\mathfrak{g}'), \qquad \varrho = \mathrm{ind}^{\sim}(g \mid \mathfrak{l},\mathfrak{g}) = \mathrm{ind}^{\sim}(\tau,\mathfrak{g}).$$

Let W be the space of $\tau' = \tau \otimes \theta_{\mathfrak{g},\mathfrak{g}'}$. Then $\mathrm{Ker}\,\tau' = I(g'')$ (5.1.15). There exists $w \in W$ such that $\tau'(u_{i_0})w \neq 0$. Then $\varrho(u)w = \sum_i \varrho(v_i)\tau'(u_i)w \neq 0$ (5.1.6). Thus, $\varrho(u) \neq 0$ and $u \notin I(g)$.

(d) We assume the existence of a commutative ideal $\mathfrak{b}$ of $\mathfrak{g}$ and of $f' \in \mathfrak{b}^*$ with the following properties:

(i), (ii), (iii) and (iv) are as in (c);

(v) $f'([\mathfrak{g},\mathfrak{b}]) = 0$;

(vi) $\mathrm{Ker}\,f' \neq 0$.

Then $\mathrm{Ker}\,f'$ is an ideal of $\mathfrak{g}$, and it is sufficient to apply the induction hypothesis to $\mathfrak{g}/\mathrm{Ker}\,f'$ and $\mathfrak{b}/\mathrm{Ker}\,f'$.

(e) We assume that none of the four preceding cases applies. Let $\mathfrak{z}$ be the centre of $\mathfrak{g}$, and $\mathfrak{n}$ the largest nilpotent ideal of $\mathfrak{g}$.

Let $\mathfrak{b}$ be a commutative ideal of $\mathfrak{g}$ containing $\mathfrak{a}$. From 10.3.7, there exists $f' \in \mathfrak{b}^*$ such that $f'|\mathfrak{a} = f$ and $u \notin J_{\mathfrak{b},f'}$. The set of the $f' \in \mathfrak{b}^*$ with these properties is open in the set of the elements of $\mathfrak{b}^*$ whose restriction to $\mathfrak{a}$ is f. Hence we can choose f' in such a way that it can be extended to an element of P. Since case (c) does not apply here, we have $f'([\mathfrak{g},\mathfrak{b}]) = 0$. Since case (d) does not apply here, we have $\mathrm{Ker}\,f' = 0$, and hence $\mathfrak{b} = 0$, or $\dim \mathfrak{b} = 1$ and $f'(\mathfrak{b}) \neq 0$. Consequently, $\mathfrak{b} \subset \mathfrak{z}$. Since we can take $\mathfrak{b} \neq 0$ (as case (a) does not apply here), we have $\mathfrak{z} \neq 0$.

If we take $\mathfrak{b} = \mathfrak{a} + \mathfrak{z}$, we see that $\mathfrak{a} \subset \mathfrak{z}$, $\dim \mathfrak{z} = 1$ and $f(\mathfrak{z}) \neq 0$. We may assume that $\mathfrak{a} = \mathfrak{z}$.

If $\mathfrak{n} = \mathfrak{z}$, the derived algebra of the radical of $\mathfrak{g}$ is contained in $\mathfrak{z}$, and hence this radical is nilpotent and consequently equal to $\mathfrak{z}$. Case (b) then applies, which has been ruled out. Hence $\mathfrak{n} \neq \mathfrak{z}$. From the above, every commutative characteristic ideal of $\mathfrak{n}$ is null or equal to $\mathfrak{z}$. From 4.6.2, $\mathfrak{n}$ is a Heisenberg algebra.

(f) Henceforth, therefore, we shall assume that $\mathfrak{a} = \mathfrak{z}$, $\dim \mathfrak{z} = 1$, $f(\mathfrak{z}) \neq 0$, and that $\mathfrak{n}$ is a Heisenberg algebra distinct from $\mathfrak{z}$. We may also assume that P is invariant under the algebraic adjoint group $\mathscr{A}$ of $\mathfrak{g}$.

We choose $f' \in \mathfrak{n}^*$ such that $f'|\mathfrak{z} = f$. Let $z \in \mathfrak{z}$ such that $f(z) = 1$. The primitive ideal $I(f')$ of $U(\mathfrak{n})$ associated with f' contains $z - 1$, and the quotient of $U(\mathfrak{n})$ by $U(\mathfrak{n})(z - 1)$ is a Weyl algebra, hence $I(f') = U(\mathfrak{n})(z - 1)$. We have

$$J_{\mathfrak{a},f} = J_{\mathfrak{z},f} = U(\mathfrak{g})(z - 1);$$

hence $u \notin U(\mathfrak{g})I(f')$.

Let $\mathfrak{k} = \operatorname{Ker} f'$, and let $\mathfrak{g}'$ be the set of the $x \in \mathfrak{g}$ such that $[x,\mathfrak{k}] \subset \mathfrak{k}$. If g is any element of $\mathfrak{g}^*$ which extends f', then $\mathfrak{g}'$ is the orthogonal subspace of $\mathfrak{k}$ (or of $\mathfrak{n}$) with respect to B_g, and the restriction of B_g to $\mathfrak{k}$ is non-degenerate, hence $\mathfrak{g} = \mathfrak{g}' \oplus \mathfrak{k}$. The adjoint representation of $\mathfrak{g}'$ in $\mathfrak{n}$ defines a semi-direct product $\mathfrak{d}$ of $\mathfrak{g}'$ and $\mathfrak{n}$; the canonical homomorphism p of $\mathfrak{d}$ into $\mathfrak{g}$ is surjective, and its kernel is $k(z,-z)$. Let $\mathscr{N}$ be the adjoint group of $\mathfrak{n}$. If λ is any element of $\mathfrak{n}^*$ which is non-zero on $\mathfrak{z}$, it is easily verified that $\mathscr{N} \cdot \lambda$ is the set of the elements of $\mathfrak{n}^*$ which have the same restriction to $\mathfrak{z}$ as λ. Let P_0 be the set of the elements of P whose restriction to $\mathfrak{k}$ is zero, and let $P' = P_0|\mathfrak{g}'$. Since $\mathscr{A}P = P$, it can be seen that P' is non-empty and open in the set of the elements of $\mathfrak{g}'^*$ whose restriction to $\mathfrak{z}$ is f. On the other hand, if $g' \in P'$, then $\mathrm{PR}(g') \neq \varnothing$ from 10.3.1.

Let D be the algebra $U(\mathfrak{d})/U(\mathfrak{d})I(f')$, and q the canonical homomorphism of $U(\mathfrak{d})$ onto D. The canonical homomorphism of $U(\mathfrak{d})$ onto

$$U(\mathfrak{g}') \otimes U(\mathfrak{n})/I(f')$$

has $U(\mathfrak{d})I(f')$ as its kernel (10.1.5), whence we have an isomorphism

$$r : D \to U(\mathfrak{g}') \otimes U(\mathfrak{n})/I(f').$$

We choose $u' \in U(\mathfrak{d})$ such that $U(p)u' = u$, and let $v = r(q(u'))$. Let (n_i) be a basis for $U(\mathfrak{n})/I(f')$. We write $v = \sum_i s_i \otimes n_i$, where the s_i belong to $U(\mathfrak{g}')$. Let us assume that $s_i \in U(\mathfrak{g}')(z - 1)$ for all i. Then there exists $v' \in U(\mathfrak{g}') \otimes U(\mathfrak{n})/I(f')$ such that $v = v'((z - 1) \otimes 1)$. It follows from the definition of r that

$$r(q((z,0)-1)) = (z - 1) \otimes 1.$$

We note that $(0,z) - 1 \in I(f')$, whence $q((z,0) - 1) = q((z,-z))$. Hence

$$q(u') = r^{-1}(v) = r^{-1}(v')q((z,-z)).$$

Consequently, $u' \in U(\mathfrak{d})(z,-z) + U(\mathfrak{d})I(f')$, whence $u = U(p)(u') \in U(\mathfrak{g})I(f')$, which is contrary to what we have seen.

There therefore exists i_0 such that $s_{i_0} \notin U(\mathfrak{g}')(z-1)$. We can apply the induction hypothesis to $\mathfrak{g}'$, $\mathfrak{a}$, s_{i_0}, f and P'. Hence there exists $g' \in P'$ such that $s_{i_0} \notin I(g')$. Let g be the element of P which extends g' and f'. We shall show that $u \notin I(g)$. Let $\mathfrak{l}' \in \mathrm{PR}(g')$. There exists $\mathfrak{h} \in \mathrm{P}(f')$ such that

$$[\mathfrak{l}',\mathfrak{h}] \subset \mathfrak{h}, \qquad \mathfrak{l} = \mathfrak{l}' + \mathfrak{h} \in \mathrm{PR}(g)$$

(1.12.13). Let $\sigma = \mathrm{ind}(f'|\mathfrak{h},\mathfrak{n})$ and $\tau = \mathrm{ind}^{\sim}(g'|\mathfrak{l}',\mathfrak{g}')$. From 10.2.1 and 10.1.8 (i), there exists a representation π of $\mathfrak{g}$, which is equivalent to $\mathrm{ind}^{\sim}(g|\mathfrak{l},\mathfrak{g})$, such that

$$\pi(u) = \sum_i \tau(s_i) \otimes \bar{\sigma}(n_i),$$

where σ denotes the representation of $U(\mathfrak{n})/I(f')$ deduced from σ by passage to the quotient. The $\bar{\sigma}(n_i)$ are linearly independent. The kernel of τ is $I(g')$, hence $\tau(s_{i_0}) \neq 0$. Then $\pi(u) \neq 0$ so that $u \notin I(g)$.

10.3.9. THEOREM (k algebraically closed). *Let $\mathfrak{g}_r^*$ be the set of regular elements of $\mathfrak{g}^*$. Let Q be a subset of $\mathfrak{g}_r^*$ which is dense in $\mathfrak{g}^*$. Then $\bigcap_{g\in Q} I(g) = 0$.*

We have $\bigcap_{g\in Q} I(g) = \bigcap_{g\in \mathfrak{g}_r^*} I(g)$ from 10.3.5 and $\bigcap_{g\in \mathfrak{g}_r^*} I(g) = 0$ from 10.3.8 applied with $\mathfrak{a} = 0$.

Reference: [47].

10.4. Application to the centre of the enveloping algebra

10.4.1. We assume that k is algebraically closed. Let $\mathfrak{g}_r^*$ be the set of the regular elements of $\mathfrak{g}^*$. For every $g \in \mathfrak{g}_r^*$, $I(g)$ is a primitive ideal of $U(\mathfrak{g})$, hence $I(\mathfrak{g}) \cap Z(\mathfrak{g})$ is the kernel of a unique homomorphism $\chi_{\mathfrak{g}}$ of $Z(\mathfrak{g})$ into k. For $u \in Z(\mathfrak{g})$ and $g \in \mathfrak{g}_r^*$, we set

$$\tilde{u}(g) = \chi_g(u),$$

so that $\hat{u}$ is a scalar function defined on $\mathfrak{g}_r^*$.

10.4.2. THEOREM (k algebraically closed). *Let $\mathfrak{g}_r^*$ be the set of regular elements of $\mathfrak{g}^*$. Let $u \in Z(\mathfrak{g})$. There exists one and only one $\hat{u} \in S(\mathfrak{g})$ such that $\hat{u}|\mathfrak{g}_r^* = \tilde{u}$ (cf. 10.4.1). We have $\hat{u} \in Y(\mathfrak{g})$.*

Let d' be the index of $\mathfrak{g}$, $d = \frac{1}{2}(d' + \dim \mathfrak{g})$, and T_d the set of the $(g, \mathfrak{h}) \in \mathfrak{g}^* \times \mathrm{Gr}(\mathfrak{g},d)$ such that $g([\mathfrak{h},\mathfrak{h}]) = 0$ and $\mathfrak{h}$ is a solvable Lie subalgebra of $\mathfrak{g}$. Then T_d is closed in $\mathfrak{g}^* \times \mathrm{Gr}(\mathfrak{g},d)$ (1.11.9). Let p be the canonical projection of $\mathfrak{g}^* \times \mathrm{Gr}(\mathfrak{g},d)$ onto $\mathfrak{g}^*$. Then

$$p(T_d) = \mathfrak{g}^*$$

(1.2.17). Consequently, there exists an irreducible component T of T_d such that $p(T) = \mathfrak{g}^*$. Let T' be the set of those $(g,\mathfrak{h}) \in T$ such that $g \in \mathfrak{g}_r^*$. For $t = (g,\mathfrak{h}) \in T$, let $\varrho(t) = \mathrm{ind}^{\sim}(g|\mathfrak{h},\mathfrak{g})$; if $t \in T'$, then $\varrho(t)(u) = \tilde{u}(g) \cdot 1$. From 6.4.1, there thus exists a rational function $\dot{u}$ which is defined everywhere on T such that $\varrho(t)(u) = \dot{u}(t) \cdot 1$ for all $t \in T$. Let $\Gamma \subset \mathfrak{g}^* \times \mathrm{Gr}(\mathfrak{g},d) \times k$ be the graph of $\dot{u}$; it is a closed subset of $\mathfrak{g}^* \times \mathrm{Gr}(\mathfrak{g},d) \times k$, and its projection Δ onto $\mathfrak{g}^* \times k$ is closed. Let Γ' and Δ' be the sets of elements of Γ and Δ, respectively, whose projections onto $\mathfrak{g}^*$ belong to $\mathfrak{g}_r^*$. Then Δ' is the graph of $\tilde{u}$, and is closed in $\mathfrak{g}_r^* \times k$. Let q be the restriction to Δ' of the projection of $\mathfrak{g}_r^* \times k$ onto $\mathfrak{g}_r^*$. Then q is a bijective morphism of the algebraic manifold Δ' onto the algebraic manifold $\mathfrak{g}_r^*$, and hence is an isomorphism (BO, pp. 78–80). Since $\tilde{u}$ can be obtained as the composite of q^{-1} and the projection of $\mathfrak{g}_r^* \times k$ onto k, it can be seen that $\tilde{u}$ is a rational function which is defined everywhere on $\mathfrak{g}_r^*$. Let $\hat{u}$ be the rational function on $\mathfrak{g}^*$, possibly not everywhere defined, such that $\hat{u}|\mathfrak{g}_r^* = \tilde{u}$ (BO, p. 36). Let p_1 be the restriction of p to T. Then the function $\hat{u} \circ p_1$ on T has a rational extension which is everywhere defined, namely $\dot{u}$. Since $p_1(T) = \mathfrak{g}^*$ and $\mathfrak{g}^*$ is a normal manifold, $\hat{u}$ is defined everywhere on $\mathfrak{g}^*$ (BO, p. 79), so that $\hat{u} \in S(\mathfrak{g})$.

The uniqueness assertion of 10.4.2 is obvious. Finally, let $\mathscr{A}$ be the adjoint algebraic group of $\mathfrak{g}$. The function $\tilde{u}$ is invariant under $\mathscr{A}$ (from 2.4.17), hence $\hat{u} \in Y(\mathfrak{g})$.

10.4.3. LEMMA (k algebraically closed). *Let $\mathfrak{g}_r^*$ be the set of regular elements of $\mathfrak{g}^*$. Let V be the set of the $f \in \mathfrak{g}_r^*$ such that there exists $\mathfrak{l} \in \mathrm{PR}(f)$ with $f|\mathfrak{l} \neq 0$. Then V is dense in $\mathfrak{g}^*$ (if $\mathfrak{g} \neq 0$).*

Let U be a non-empty open subset of $\mathfrak{g}_r^*$. We prove that $U \cap V \neq \emptyset$. Let $g \in U$ and $\mathfrak{l} \in \mathrm{PR}(g)$. If $g|\mathfrak{l} \neq 0$, the proof is complete. We assume that $g|\mathfrak{l} = 0$. Since $\mathfrak{g} \neq 0$, we have $\mathfrak{l} \neq 0$, hence $\mathfrak{l} \neq [\mathfrak{l},\mathfrak{l}]$, and there exists $h \in \mathfrak{g}^*$ such that $h|\mathfrak{l} \neq 0$ and $h|[\mathfrak{l},\mathfrak{l}] = 0$. There exists $\lambda \in k - \{0\}$ such that $f = g + \lambda h \in U$. Then $f|[\mathfrak{l},\mathfrak{l}] = 0$; hence $\mathfrak{l} \in \mathrm{PR}(f)$ because $f \in \mathfrak{g}_r^*$. Since $f|\mathfrak{l} \neq 0$, we have $f \in U \cap V$.

10.4.4. LEMMA (k algebraically closed). *Let β be the canonical mapping of $S(\mathfrak{g})$ onto $U(\mathfrak{g})$. Let $u \in Z(\mathfrak{g}) - \{0\}$, and let $\hat{u}$ be as in* 10.4.2. *Then $\hat{u}$ and $\beta^{-1}(u)$ have the same non-null homogeneous component of greatest degree.*

Let $x \in Y(\mathfrak{g})$ be an element of degree d. We define V as in 10.4.3. Let $g \in V$. We shall prove that the function $\lambda \mapsto x(\lambda g) - \beta(x)^{\wedge}(\lambda g)$ on k has degree $<d$, which will establish the lemma. There exists $\mathfrak{l} \in \mathrm{PR}(g)$ such that $g|\mathfrak{l} \neq 0$. We have $\mathfrak{l} + \operatorname{Ker} g = \mathfrak{g}$, hence there exists a basis $(x_1, \ldots, x_n)$ for $\mathfrak{g}$ such that $x_1, \ldots, x_p \in \operatorname{Ker} g$ and $(x_{p+1}, \ldots, x_n)$ is a basis for $\mathfrak{l}$. For $\mu = (\mu_1, \ldots, \mu_n) \in \mathbf{N}^n$, we denote by x^μ the element $x_1^{\mu_1} \cdots x^{\mu_n}$ calculated in $S(\mathfrak{g})$ or $U(\mathfrak{g})$ according to the context. We write $x = \sum_{\mu \in N} \alpha_\mu x^\mu$ $(\alpha_\mu \in k)$. Then

$$x(\lambda g) = \Bigg(\sum_{\substack{|\mu| = d, \\ \mu_1 = \cdots = \mu_p = 0}} \alpha_\mu x^\mu(g) \Bigg) \lambda^d + \cdots, \tag{1}$$

where the omitted terms have degree $<d$ in λ. Let $\varrho_\lambda = \operatorname{ind}^\sim(\lambda g|\mathfrak{l}, \mathfrak{g})$, which operates in $U(\mathfrak{g}) \otimes_{U(\mathfrak{l})} k$; this space has the $\varrho_\lambda(x_1^{\mu_1} \cdots x_p^{\mu_p})(1 \otimes 1)$ as a basis. Now

$$\beta(x) \equiv \sum_{|\mu| = d} \alpha_\mu x^\mu \; [\operatorname{mod} U_{d-1}(\mathfrak{g})],$$

hence $\varrho_\lambda(\beta(x))(1 \otimes 1)$ has as its coordinate corresponding to $1 \otimes 1$ an expression of the form

$$\sum_{\substack{|\mu| = d, \\ \mu_1 = \cdots = \mu_p = 0}} \alpha_\mu x^\mu(\lambda g + \theta_{\mathfrak{g}, \mathfrak{l}}) + \cdots,$$

where the omitted terms have degree $<d$ in λ. For $\lambda \neq 0$, λg is regular, hence we have, with the notation of 10.4.1,

$$\beta(x)^\wedge(\lambda g) = \chi_{\lambda g}(\beta(x)) = \Bigg(\sum_{\substack{|\mu| = d, \\ \mu_1 = \cdots = \mu_p = 0}} \alpha_\mu x^\mu(g) \Bigg) \lambda^d + \cdots, \tag{2}$$

where the omitted terms have degree $<d$ in λ. It is now sufficient to compare (1) and (2).

10.4.5. THEOREM (k algebraically closed). *For all $u \in Z(\mathfrak{g})$, we define $\hat{u} \in Y(\mathfrak{g})$ as in* 10.4.2. *Then the mapping $u \mapsto \hat{u}$ is an isomorphism of the algebra $Z(\mathfrak{g})$ onto the algebra $Y(\mathfrak{g})$.*

Let $\mathfrak{g}_r^*$ be the set of regular elements of $\mathfrak{g}^*$, and let $g \in \mathfrak{g}_r^*$. If $u_1, u_2 \in Z(\mathfrak{g})$, then, with the notation of 10.4.1,

$$(u_1 + u_2)^\wedge(g) = \chi_g(u_1 + u_2) = \chi_g(u_1) + \chi_g(u_2) = \hat{u}_1(g) + \hat{u}_2(g).$$

The polynomial function $(u_1 + u_2)^\wedge - \hat{u}_1 - \hat{u}_2$ is hence zero on $\mathfrak{g}_r^*$ and consequently identically zero. Similarly, we see that $(u_1 u_2)^\wedge = \hat{u}_1 \hat{u}_2$ and that $(\lambda u_1)^\wedge = \lambda \hat{u}_1$ for all $\lambda \in k$. Hence the mapping $\varphi : u \mapsto \hat{u}$ of $Z(\mathfrak{g})$ into $Y(\mathfrak{g})$ is an algebra homomorphism. For all $n \in \mathbf{N}$, let $Z_n(\mathfrak{g}) = Z(\mathfrak{g}) \cap U_n(\mathfrak{g})$, and let $Y_n(\mathfrak{g})$ be the set of elements of $Y(\mathfrak{g})$ of degree $\leq n$. From 10.4.4, we have $\varphi(Z_n(\mathfrak{g})) \subset Y_n(\mathfrak{g})$, and φ and β define the same mapping of gr $Z(\mathfrak{g}) = \oplus_{n\geq 0} Z_n(\mathfrak{g})/Z_{n-1}(\mathfrak{g})$ into gr $Y(\mathfrak{g}) = \oplus_{n\geq 0} Y_n(\mathfrak{g})/Y_{n-1}(\mathfrak{g})$. Now the mapping defined by β is bijective (2.3.6, 2.4.11). Hence φ is bijective (AC III, p. 37).

Reference: [47].

10.5. Supplementary remarks

10.5.1. The material of this chapter has been entirely taken from papers by Duflo ([45], [47]). Some of the methods used in sections 10.1 and 10.2 were inspired by analogous methods elaborated by Auslander and Kostant for the representations of solvable Lie groups [4].

10.5.2. Let Δ be the Lie algebra of derivations of A_n, π the homomorphism $x \mapsto \mathrm{ad}_{A_n}x$ of A_n onto Δ, $\mathfrak{g}$ a Lie algebra, and φ a homomorphism of $\mathfrak{g}$ into Δ.

(a) If $\mathfrak{g}$ is semi-simple, there exists one and only one homomorphism φ' of $\mathfrak{g}$ into A_n such that $\varphi = \pi \circ \varphi'$. (This follows from the fact that $H^1(\mathfrak{g}) = H^2(\mathfrak{g}) = 0$.)

(b) Take $\mathfrak{g} = \mathbf{C}^2$, $n = 1$, $\varphi((1,0)) = \mathrm{ad}_{A_1}p$, and $\varphi((0,1)) = \mathrm{ad}_{A_1}q$. Then there exists no homomorphism φ' of $\mathfrak{g}$ into A_1 such that $\varphi = \pi \circ \varphi'$.

10.5.3. Assume that k is algebraically closed. Let $\mathfrak{n}$ be a nilpotent ideal of $\mathfrak{g}$, $g \in \mathfrak{g}^*$, $\mathfrak{g}_1 = \mathfrak{n}^g$, $f = g|\mathfrak{n}$, and σ a simple representation of $\mathfrak{n}$ with kernel $I(f)$. Then σ can be extended to a representation of $\mathfrak{g}_1 + \mathfrak{n}$ if and only if $f|\mathfrak{n}^f$ can be extended to a one-dimensional representation of $\mathfrak{g}_1$.

10.5.4. Assume that k is algebraically closed and that $\mathfrak{g}$ is semi-simple. Let f be a regular linear form on $\mathfrak{g}$, let $\mathfrak{h} \in \mathrm{PR}(f)$ and let $\varrho = \mathrm{ind}^{\sim}(f|\mathfrak{h},\mathfrak{g})$. The module corresponding to ϱ is a Verma module, and we thus obtain all Verma $\mathfrak{g}$-modules.

10.5.5. Assume that k is algebraically closed. Let f be a regular linear form on $\mathfrak{g}$. The ideal $I(f)$ is completely prime (cf. 5.6.3 (c)). [24]

10.5.6. Assume that k is algebraically closed. Let $K'(\mathfrak{g})$ be the field ot fractions of $S(\mathfrak{g})$. For every $x \in \mathfrak{g}$, let $\varepsilon(x)$ and $\varepsilon'(x)$ be the derivations of $K(\mathfrak{g})$ and $K'(\mathfrak{g})$ respectively, which extends ad x. The set of $\mathfrak{g}$-invarianf elements of $K(\mathfrak{g})$ is $C(\mathfrak{g})$; let $C'(\mathfrak{g})$ be the set of $\mathfrak{g}$-invariant elements of $K'(\mathfrak{g})$. Let $u \in C(\mathfrak{g})$. We denote by V_u the set of the regular elements $f \in \mathfrak{g}^*$ such that u can be written in the form $u_1u_2^{-1}$ $(u_1,u_2 \in U(\mathfrak{g}))$ with $u_2 \notin I(f)$; it is an open non-empty subset of $\mathfrak{g}^*$. With the above notation, if $f \in V_u$, then $(u_1 \bmod I(f))(u_2 \bmod I(f))^{-1}$ is central in the field of fractions of

$U(\mathfrak{g})/I(f)$; it is a scalar $\hat{u}(f)$ which is independent of the choice of u_1 and u_2. The set V_u is stable under the adjoint algebraic group of $\mathfrak{g}$. The function $\hat{u}$, defined on V_u, is rational. Let us also denote by $\hat{u}$ the element of $K'(\mathfrak{g})$ which extends $\hat{u}$. Then $u \mapsto \hat{u}$ is an isomorphism of the field $C(\mathfrak{g})$ onto the field $C'(\mathfrak{g})$. We deduce that $C(\mathfrak{g})$ is an extension of finite type of k. ([108], [111])

10.5.7. Assume that k is algebraically closed. Let $\mathfrak{g}'$ be an ideal of $\mathfrak{g}$, and f a regular element of $\mathfrak{g}^*$. There exists $\mathfrak{h} \in \mathrm{PR}(f)$ such that:

(1) $\mathfrak{h} \cap \mathfrak{g}' \in \mathrm{PR}(f|\mathfrak{g}')$;

(2) $\mathrm{ind}^{\sim}(f|\mathfrak{h},\mathfrak{g})$ is simple;

(3) $\mathrm{ind}^{\sim}(f|\mathfrak{h} \cap \mathfrak{g}',\mathfrak{g})$ is simple. [111]

10.5.8. Assume that k is algebraically closed. Let Γ be the adjoint algebraic group of $\mathfrak{g}$. There exists an open non-empty Γ-stable set W of regular elements of $\mathfrak{g}^*$ such that, if $f_1, f_2 \in W$ and $I(f_1) = I(f_2)$, then $f_2 \in \Gamma f_1$. [111]

CHAPTER 11

APPENDIX

11.1. Root systems

As a reference work, we use N. Bourbaki, *Groupes et algèbres de Lie*, chapters V and VI (which we refer to as GL).

11.1.1. Let V be a vector space of finite dimension l, and let $R \subset V$. Then R is said to be a *reduced system of roots* in V if the following conditions are satisfied:

(i) R is finite, does not contain 0, and generates V;

(ii) for all $\alpha \in R$, there exists an $\alpha^\vee \in V^*$ such that $\langle \alpha, \alpha^\vee \rangle = 2$ and such that R is stable under the reflexion $s_\alpha : v \mapsto v - \langle \alpha^\vee, v \rangle \alpha$ (the element $\alpha^\vee$ of V^* is then unique (GL, p. 142));

(iii) for all $\alpha \in R$, we have $\alpha^\vee(R) \subset \mathbf{Z}$;

(iv) if $\alpha \in R$, the only elements of R which are proportional to α are α and $-\alpha$.

11.1.2. The group W of automorphisms of V generated by the s_α is termed the Weyl group of R. The notation V, l, R, $\alpha \mapsto \alpha^\vee$, $\alpha \mapsto s_\alpha$ and W will remain fixed for the rest of this appendix.

11.1.3. The set of the $\alpha^\vee$, where α takes all values in R, is a reduced system $R^\vee$ of roots in V^*, termed the inverse system of R. We have $\alpha^{\vee\vee} = \alpha$ for all $\alpha \in R$. By transport of structure, W operates in V^* and preserves $R^\vee$ (GL, p. 144).

11.1.4. Let $V_\mathbf{Q}$ and $V_\mathbf{Q}^*$ be the vector sub-$\mathbf{Q}$-spaces of V and V^* generated by R and $R^\vee$ respectively. Then V can be canonically identified with $V_\mathbf{Q} \otimes_\mathbf{Q} k$, V^* can be canonically identified with $V_\mathbf{Q}^* \otimes_\mathbf{Q} k$, and $V_\mathbf{Q}^*$ can be canonically identified with the dual of $V_\mathbf{Q}$ (GL, p. 143). We denote the real vector space $V_\mathbf{Q} \otimes_\mathbf{Q} \mathbf{R}$ by $V_\mathbf{R}$, and in this appendix we shall use the usual topology on it and not the Zariski topology.

11.1.5. Let us consider the $\alpha^\vee$ ($\alpha \in R$) as linear forms on V_R, and let X be the union of their kernels. The connected components of $V_R - X$ are termed *Weyl chambers*. These are open simplicial cones (GL, p. 85). The group W operates in a simply transitive fashion in the set of Weyl chambers (GL, p. 74). Every orbit of W in V_R intersects $\bar{C}$ in one and only one point(GL, p. 75).

11.1.6. Let C be a Weyl chamber. There exists a family $(\alpha_1, \ldots, \alpha_l)$ of elements of R, which is unique up to the ordering, such that C is the set of the $x \in V_R$ which satisfy $\langle \alpha_1^\vee, x\rangle > 0, \ldots, \langle \alpha_l^\vee, x\rangle > 0$. The family $(\alpha_1, \ldots, \alpha_l)$ is a basis for V, for $V_{\mathbf{Q}}$ and for $V_{\mathbf{R}}$; it is termed the basis for R associated with C. The family $(\alpha_1^\vee, \ldots, \alpha_l^\vee)$ is a basis for $R^\vee$. The hyperplanes of V_R which are orthogonal to the $\alpha_i^\vee$ are termed the *walls* of C. The group W operates in a simply transitive fashion on the set of bases for R (GL, p. 153).

11.1.7. Let C and B be as in 11.1.6. Then the ordering relation, compatible with the vector space structure of $V_{\mathbf{R}}$, for which the elements $\geqq 0$ are the linear combinations of elements of B with coefficients $\geqq 0$ is called the *ordering relation defined by* C (or B) in $V_{\mathbf{R}}$. Every root is either positive or negative for C. We denote by R_+ (or R_-) the set of positive (or negative) roots. Every positive root is a linear combination of elements of B with non-negative integer coefficients (GL, p. 156). Let $v \in V_{\mathbf{R}}$; then $v \in C$ if and only if $w(v) < v$ for all $w \in W - \{1\}$ (GL, p. 159). If $\alpha \in B$, then s_α transforms every positive root different from α into a positive root (GL, p. 157). A subset P of R has the form R_+ relative to a suitable chamber if and only if P satisfies the following properties:

(1) $P \cup (-P) = R$;

(2) $P \cap (-P) = \emptyset$;

(3) if the sum of two elements of P belongs to R, then it belongs to P. (GL, p. 161)

11.1.8. Let C and B be as in 11.1.6. The group W is generated by the family $(s_\alpha)_{\alpha \in B}$ (GL, p. 153). If $w \in W$, every decomposition of w into a product $s_1 \cdots s_q$ of reflexions $s_\alpha(\alpha \in B)$ with minimal q is termed a *reduced decomposition of* w; the integer q is termed the length of w (relative to C or B), and is denoted by $l(w)$. Let $s_i = s_{\alpha_i}$, where $\alpha_i \in B$. Then $s_1 s_2 \cdots s_{q-1}\alpha_q \in R_+$ (GL, p. 158). We set

$$t_q = (s_1 \cdots s_{q-1}) s_q (s_1 \cdots s_{q-1})^{-1},$$
$$t_{q-1} = (s_1 \cdots s_{q-2}) s_{q-1} (s_1 \cdots s_{q-2})^{-1},$$
$$t_1 = s_1,$$

so that $w = t_q t_{q-l} \cdots t_1$. Let

$$C_j = (t_j \cdots t_1)(C) = (s_1 \cdots s_j)(C),$$

so that

$$C_0 = C, \quad C_1 = t_1(C), \quad C_2 = t_2(C_1), \ldots,$$

$$w(C) = C_n = t_n(C_{n-1}).$$

Then

$$t_{j+1} = s_\gamma, \quad \text{where} \quad \gamma = (s_1 \cdots s_j)(\alpha_{j+1}),$$

so that γ is orthogonal to a wall of C_j and

$$\langle \gamma, C_j \rangle = \langle \alpha_{j+1}, C \rangle > 0.$$

11.1.9. Let $w \in W$. With the notation of 11.1.6—11.1.8, $l(w)$ is the number of elements α of R_+ such that $w(\alpha) \in R_-$ (GL, p. 158).

11.1.10. Let $(\beta_1, \ldots, \beta_n)$ be a sequence of elements of R^+ such that $\beta_1 + \cdots + \beta_n \in R$. There exists a permutation π of $\{1, \ldots, n\}$ such that, for all $i \in \{1, \ldots, n\}$, we have $\beta_{\pi(1)} + \cdots + \beta_{\pi(i)} \in R$ (GL, p. 159).

11.1.11. We denote the subgroup of V generated by R by $Q(R)$. The elements of $Q(R)$ are termed the *radicial weights* of R. The group $Q(R)$ is free and of rank l, and every basis for R is a basis for $Q(R)$. The set of the $\lambda \in V$ such that $\langle \lambda, \alpha^\vee \rangle \in \mathbf{Z}$ for all $\alpha \in R$ is a free subgroup of V of rank l, and is denoted by $P(R)$; we have $Q(R) \subset P(R) \subset V_\mathbf{Q}$, and $Q(R)$ has finite index in $P(R)$ (GL, p. 166—167); the elements of $P(R)$ are termed the *weights* of R. For all $\alpha \in R$, we have $Q(R) \cap \mathbf{Q}\alpha = \mathbf{Z}\alpha$ (GL, p. 156).

11.1.12. Let B be a basis for R corresponding to a Weyl chamber C. Let $(\tilde{\omega}_\alpha)_{\alpha \in B}$ be the dual basis for $B^\vee$ in V. Then $(\tilde{\omega}_\alpha)_{\alpha \in B}$ is a basis for the group $P(R)$ (GL, p. 167). The $\tilde{\omega}_\alpha$ are termed the fundamental weights relative to B. Let $\lambda \in V$. The following conditions are equivalent:

(i) $\langle \lambda, \alpha^\vee \rangle \in \mathbf{N}$ for all $\alpha \in B$;

(ii) the coordinates of λ with respect to $(\tilde{\omega}_\alpha)_{\alpha \in B}$ belong to $\mathbf{N}$;

(iii) $\lambda \in P(R) \cap \bar{C}$;

(iv) for every $w \in W$, the coordinates of $\lambda - w\lambda$ with respect to B belong to $\mathbf{N}$ (GL, p. 167—168).

If these conditions are satisfied, λ is said to be a *dominant weight* of R.

11.1.13. Let B be a basis for R, $(\tilde{\omega}_\alpha)_{\alpha \in B}$ the family of corresponding fundamental weights, and δ the semi-sum of positive roots. Then $\delta = \sum_{\alpha \in B} \tilde{\omega}_\alpha$, and $\langle \delta, \alpha^\vee \rangle = 1$ for all $\alpha \in B$ (GL, p. 168).

11.1.14. The group W operates in a natural way in the symmetric algebra $S(V)$. Let $S(V)^W$ be the set of the W-invariant elements of $S(V)$. There exist algebraically independent homogeneous elements $f_1, \ldots, f_l$ of $S(V)^W$ which generate the algebra $S(V)^W$ (GL, p. 107). The degrees $\nu_1, \ldots, \nu_l$ of $f_1, \ldots, f_l$ are independent of the choice of $f_1, \ldots, f_l$ (GL, p. 103). We have

$$\nu_1 + \cdots + \nu_l = l + \tfrac{1}{2} \operatorname{Card} R$$

(GL, p. 111).

11.1.15. Let $\mathbf{Z}[V]$ be the algebra over $\mathbf{Z}$ which has the elements e^λ, where $\lambda \in V$ and with multiplication defined by $e^\lambda e^\mu = e^{\lambda+\mu}$, as a basis. Let $\delta = \frac{1}{2} \sum_{\alpha \in R_+} \alpha$. Then in $Z[V]$ we have (GL, p. 185)

$$\prod_{\alpha \in R_+} (e^{\alpha/2} - e^{-\alpha/2}) = e^\delta \prod_{\alpha \in R_+} (1 - e^{-\alpha})$$
$$= e^{-\delta} \prod_{\alpha \in R_+} (e^\alpha - 1) = \sum_{w \in W} \varepsilon(w) e^{w\delta}.$$

[For all $w \in W$, the determinant of w, which is equal to 1 or -1, is denoted by $\varepsilon(w)$.]

11.1.16. Let V' be a vector subspace of V, and let $R' = R \cap V'$. Then R' is a system of roots in the vector subspace which it generates (GL, p. 145).

LEMMA. *Let R'' be the complement of R' in R. Let θ be an automorphism of V such that $\theta^2 = 1$, $\theta(R) = R$, and V' is the set of fixed points of θ. Then there exists a Weyl chamber of R such that, if the corresponding basis and set of positive roots are denoted by B and R_+ respectively (and if we set $R_- = -R_+$, $R''_+ = R'' \cap R_+$, and $R''_- = R'' \cap R_-$), we have the following properties:*

(a) $\theta(R''_+) = R''_-$;

(b) *if* $\alpha \in R''_+$, $\gamma \in R$ *and* $\gamma - \alpha \in V'$, *then* $\gamma \in R''_+$;

(c) $(R''_+ + R''_+) \cap R \subset R''_+$;

(d) $B \cap V'$ *is a basis for* R'.

Indeed, let

$$V'' = \{x \in V \mid \theta x = -x\}, \qquad V'_{\mathbf{Q}} = V' \cap V_{\mathbf{Q}}, \qquad V''_{\mathbf{Q}} = V'' \cap V_{\mathbf{Q}}.$$

Since $\theta(V_{\mathbf{Q}}) = V_{\mathbf{Q}}$, we have $V' = V'_{\mathbf{Q}} \otimes_{\mathbf{Q}} k$ and $V'' = V''_{\mathbf{Q}} \otimes_{\mathbf{Q}} k$. Let $(e_1, \ldots, e_p)$ be a basis for $V''_{\mathbf{Q}}$, and $(e_{p+1}, \ldots, e_n)$ a basis for $V'_{\mathbf{Q}}$; we equip $V_{\mathbf{Q}}$ with the lexicographic ordering defined by the basis $(e_1, \ldots, e_n)$. Let R_+ (or R_-) be the set of positive (negative) elements of R under this ordering. There exists a Weyl chamber of R such that R_+ (or R_-) is the set of positive (or negative) roots relative to this chamber (GL, p. 162). Let $R''_+ = R'' \cap R_+$ and $R''_- = R'' \cap R_-$. If $\alpha \in R''_+$, the first non-zero coordinate of α is positive and of index $\leqq p$; hence the first non-zero coordinate of $\theta\alpha$ is negative and of index $\leqq p$; this proves that $\theta(R''_+) = R''_-$. If, in addition, $\gamma \in R$ and $\gamma - \alpha \in V'$, then the first non-zero coordinate of γ is positive and of index $\leqq p$, hence $\gamma \in R''_+$. It is obvious that $(R''_+ + R''_+) \cap R \subset R''_+$.

Let B be the basis corresponding to R_+. Then every element of $R_+ \cap V'$ is a linear combination of the elements of B with non-negative coefficients, and only the elements of $B \cap V'$ may have a positive coefficient. Hence $B \cap V'$ is a basis for R' (GL, p. 162).

11.2. Miscellaneous results

11.2.1. LEMMA. *Let $X_1, \ldots, X_a, Y_1, \ldots, Y_b$ be indeterminates. Let $(\alpha_{vw})_{1 \leqq v,w \leqq a}$ be an invertible matrix with elements in k. In $Z = k[X_1, \ldots, X_a, Y_1, \ldots, Y_b]$, we consider the elements*

$$P_1 = \alpha_{11}X_1 + \cdots + \alpha_{1a}X_a + Q_1, \quad \ldots, \quad P_a = \alpha_{1a}X_1 + \cdots + \alpha_{aa}X_a + Q_a,$$

where

$$Q_1, \ldots, Q_a \in k[Y_1, \ldots, Y_b].$$

Then $(P_1, \ldots, P_a, Y_1, \ldots, Y_b)$ are algebraically independent and generate Z.

There exists an automorphism ω of Z such that

$$\omega(Y_1) = Y_1. \ \ldots, \ \omega(Y_b) = Y_b,$$

$$\omega(X_1) = \alpha_{11}X_1 + \cdots + \alpha_{1a}X_a, \ \ldots, \ \omega(X_a) = \alpha_{a1}X_1 + \cdots + \alpha_{aa}X_a.$$

We thus return to the case where the matrix (α_{vw}) is the unity matrix. It is clear that $X_1 + Q_1, \ldots, X_a + Q_a, Y_1, \ldots, Y_b$ generate the algebra Z; if they were algebraically dependent, the degree of transcendance of $k(X_1, \ldots, X_a, Y_1, \ldots, Y_b)$ over k would be $<a + b$, which is impossible.

11.2.2. LEMMA. *Let $A = \bigoplus_{n \geqq 0} A^n$ be a graded k-algebra, k' an extension of k, $A'^n = A^n \otimes k'$, and $A' = A \otimes k'$. We assume that A' is a graded polynomial k-algebra* (GL, p. 103). *Then A is a graded polynomial k-algebra, and the family of the characteristic degrees is the same for A and for A'.*

Let $d_1, \ldots, d_m, p_1, \ldots, p_m$ be non-negative integers such that:
(1) $d_1 < \cdots < d_m$;

(2) the family of the characteristic degrees of A' consists of $p_1, \dots, p_m$ integers equal to $d_1, \dots, d_m$ respectively. Let $f_1^i, \dots, f_{p_i}^i$ (where $i = 1, \dots, m$) be homogeneous elements of A' of degree d_i such that $f_1^1, \dots, f_{p_1}^1, \dots, f_1^m, \dots, f_{p_m}^m$ are algebraically independent and generate A'. Let A_i' be the subalgebra of A' generated by $f_1^1, \dots, f_{p_1}^1, \dots, f_1^i, \dots, f_{p_i}^i$. Let $A_i'^n = A_i' \cap A'^n$.

Let us assume that we have proved the existence of homogeneous elements $g_1^1, \dots, g_{p_1}^1, \dots, g_1^i, \dots, g_{p_i}^i$ of A of degrees $d_1, \dots, d_1, \dots, d_i, \dots, d_i$, respectively, which have the following property:

for $u = 1, \dots, i$,

$$g_1^u \equiv \alpha_{11} f_1^u + \cdots + \alpha_{1,p_u} f_{p_u}^u \pmod{A_{u-1}'^{d_u}},$$

$$\vdots$$

$$g_{p_u}^u \equiv \alpha_{u1} f_1^u + \cdots + \alpha_{u,p_u} f_{p_u}^u \pmod{A_{u-1}'^{d_u}},$$

where (α_{vw}) is an invertible matrix with elements in k'.

By repeated application of 11.2.1, we see that $g_1^1, \dots, g_{p_i}^i$ are algebraically independent over k' and generate over k a subalgebra A_i of A such that $A'_i = A_i \otimes k'$.

Then $A'^{d_{i+1}} = A^{d_{i+1}} \otimes k'$ and $A_i'^{d_{i+1}} = A_i^{d_i+1} \otimes k'$, and $f_1^{i+1}, \dots, f_{p_{i+1}}^{i+1}$ generate a vector subspace which is complementary to $A_i'^{d_{i+1}}$ in $A'^{d_{i+1}}$. Hence there exist $g_1^{i+1}, \dots, g_{p_{i+1}}^{i+1} \in A^{d_{i+1}}$ and an invertible matrix (β_{vw}) with elements in k' such that

$$g_1^{i+1} \equiv \beta_{11} f_1^{i+1} + \cdots + \beta_{1,p_{i+1}} f_{p_{i+1}}^{i+1} \pmod{A_i'^{d_{i+1}}},$$

$$\vdots$$

$$g_{p_{i+1}}^{i+1} \equiv \beta_{p_{i+1},1} f_1^{i+1} + \cdots + \beta_{p_{i+1},p_{i+1}} f_{p_{i+1}}^{i+1} \pmod{A_i'^{d_{i+1}}}.$$

Proceeding in a stepwise fashion, we can hence assume that the g_r^u are constructed up to $u = m$, which establishes the lemma since $A_m' = A'$.

11.2.3. Let $f_1, \dots, f_l \in k[X_1, \dots, X_n]$, let I be the ideal of $k[X_1, \dots, X_n]$ generated by $f_1, \dots, f_l$, and let $V \subset k^n$ be the set of zeros of I. We assume that k is algebraically closed, that V is irreducible and that the differentials of $f_1, \dots, f_l$ are linearly independent at at least one point of V. Then I is a prime ideal ([78], p. 345).

11.2.4. Proposition. *Let V be an irreducible affine algebraic manifold, A the algebra of regular functions on V, and $\mathcal{N}$ a unipotent irreducible algebraic group which operates regularly in V. Let $A^{\mathcal{N}}$ be the set of $\mathcal{N}$-invariant elements of A.*

(i) *There exists $t \in A^{\mathcal{N}} - \{0\}$ with the following properties:*

(a) *denoting the set of the $\mathcal{N}$-invariant elements of A_t by $A_t^{\mathcal{N}}$, A_t is an algebra of polynomials over $A_t^{\mathcal{N}}$;*

(b) *denoting the set of elements of V where t is not zero by V_t, $A_t^{\mathcal{N}}$ separates the $\mathcal{N}$-orbits in V_t.*

(ii) *The $\mathcal{N}$-orbits in V are closed.*

(iii) *The algebra of regular functions on an $\mathcal{N}$-orbit is an algebra of polynomials over k.*

(The proof given below of this result of Rosenlicht is taken from [109].)

(i) We can assume that $\mathcal{N} \neq \{1\}$, and we prove (i) by induction on $\dim \mathcal{N}$. Let $\mathfrak{n}$ be the Lie algebra of $\mathcal{N}$, z a non-zero central element of $\mathfrak{n}$, and $\mathcal{Z}$ the algebraic subgroup of $\mathcal{N}$ of the Lie algebra kz. Let $A^{\mathcal{Z}}$ be the set of $\mathcal{Z}$-invariant elements of A. If $A = A^{\mathcal{Z}}$, it is sufficient to apply the induction hypothesis to $\mathcal{N}/\mathcal{Z}$. Hence we shall assume that $A \neq A^{\mathcal{Z}}$. Since A is the union of finite-dimensional $\mathfrak{n}$-stable vector subspaces in which $\mathcal{N}$ operates in a unipotent fashion, there exists $a \in A$ such that $za \neq 0$ and $\mathfrak{n}a \subset A^{\mathcal{Z}}$. We set $t_1 = za$ and $b = at_1^{-1}$. If $n \in \mathfrak{n}$, then

$$nt_1 = [n,z]a + z(na) = 0,$$

hence $t_1 \in A^{\mathcal{N}}$. On the other hand, $zb = 1$. We may assume that $A = A^{\mathcal{Z}}[b]$, with $zb = 1$, by replacing A by A_{t_1} if necessary (4.7.5). Then V can be written as $W \times k$, where W is an irreducible affine algebraic manifold which has $A^{\mathcal{Z}}$ as its algebra of regular functions. The orbits of $\mathcal{Z}$ in V are the sets of the form $\{w\} \times k$, where $w \in W$. The group $\mathcal{N}/\mathcal{Z}$ operates regularly in W by passage to the quotient. From the induction hypothesis, there exists $t \in A^{\mathcal{N}} - \{0\}$ such that $A_t^{\mathcal{Z}} = A_t^{\mathcal{N}}[b_1, \ldots, b_m]$ and $A_t^{\mathcal{N}}$ separates the $\mathcal{N}$-orbits in W_t (the set of the elements of W where t is not zero). Then $A_t = A_t^{\mathcal{N}}[b_1, \ldots, b_m, b]$ and $A_t^{\mathcal{N}}$ separates the $\mathcal{N}$-orbits in $W_t \times k$.

(ii) We prove (ii) by induction on $\dim V$. The set $V - V_t$ is stable under $\mathcal{N}$; its irreducible components are stable under $\mathcal{N}$ and of dimension $< \dim V$, hence the $\mathcal{N}$-orbits contained in $V - V_t$ are closed in $V - V_t$ and consequently in V, from the induction hypothesis. Let $v_0 \in V_t$ and $\alpha = t(v_0) \in k - \{0.\}$ From (i), $\mathcal{N} v_0$ is the set of the $v \in V$ such that $t(v) = \alpha$ and $f(v) = f(v_0)$ for all $f \in A^{\mathcal{N}}$, hence $\mathcal{N} v$ is closed in V.

(iii) Let us assume that V is an $\mathcal{N}$-orbit. Then $A^{\mathcal{N}} = k$, hence A is an algebra of polynomials over k. Taking (ii) into account, this proves (iii).

11.2.5. Let A be an algebra, B a subalgebra, and C the commutant of B in A. If B is simple and central of finite dimension, the canonical homomorphism $B \otimes C \rightarrow A$ is bijective ([71], p. 118).

PROBLEMS

In parts I and II, we assume that k is algebraically closed. The problems in part I deal with more or less general Lie algebras. The problems in part II deal with special classes of Lie algebras.

PART I

1. Does there exist a linear bijection of $S(\mathfrak{g})$ onto $U(\mathfrak{g})$ which transforms the maximal invariant ideals of $S(\mathfrak{g})$ into the primitive ideals of $U(\mathfrak{g})$ (at least for $\mathfrak{g}$ nilpotent)?

2. Is the space Prim $U(\mathfrak{g})$ the union of a sequence of algebraic varieties?

3. Assume that $\mathfrak{g}$ is algebraic. Do there exist integers n and p such that $K(\mathfrak{g})$ is isomorphic to the field of fractions of $A_n(k) \otimes k[X_1, \ldots, X_p]$ (a conjecture of Gelfand–Kirillov [52].) Cf. 4.7.18, 4.9.21 and 7.8.19. The answer is affirmative if $\mathfrak{g}$ is solvable ([15], [72], [93]).

4. Is $C(\mathfrak{g})$ a pure extension of k? Cf. 4.2.4, 4.4.8 and 10.5.5.

5. Let I be a prime ideal of $U(\mathfrak{g})$. Is $C(\mathfrak{g}; I)$ an extension of finite type of k? Cf. 4.4.11.

6. Is a minimal primitive ideal of $U(\mathfrak{g})$ completely prime? Cf. 8.4.4.

7. Does every non-null two-sided ideal of $U(\mathfrak{g})$ contain a non-null eigenvector for the adjoint representation? In particular, if the radical of $\mathfrak{g}$ is nilpotent, does every non-null two-sided ideal of $U(\mathfrak{g})$ have a non-null intersection with $Z(\mathfrak{g})$? Cf. 4.2.2 and 4.4.1.

8. Let I be a prime ideal of $U(\mathfrak{g})$. Can every element of $C(\mathfrak{g}; I)$ be written as ab^{-1} with a, b elements of a certain $(U(\mathfrak{g})/I)_\lambda$? Cf. 4.4.2 and 4.9.7. In particular, if $\mathfrak{g}$ is semi-simple, is it true that $C(\mathfrak{g}; I) = \text{Fract}\, Z(\mathfrak{g}; I)$?

9. We assume that the radical of $\mathfrak{g}$ is nilpotent. Does there exist a dense subset Ω of Prim $U(\mathfrak{g})$ such that, for all $I \in \Omega$, $U(\mathfrak{g})/I$ is simple? Is the algebra $U(\mathfrak{g}) \otimes_{Z(\mathfrak{g})} C(\mathfrak{g})$ simple?

10. Does there exist a dense open subset R of $\mathfrak{g}^*$ with the following property: if $f \in R$, every polarization of $\mathfrak{g}$ at f is solvable? Cf. [200].

11. Let $\mathfrak{k}$ be an ideal of $\mathfrak{g}$, and I a primitive ideal of $U(\mathfrak{g})$. Are the following properties true:

(a) there exists a primitive ideal of $U(\mathfrak{k})$ which is generic for $U(\mathfrak{k}) \cap I$;

(b) two such ideals are conjugate with respect to the algebraic adjoint group of $\mathfrak{g}$;

(c) let L be such an ideal; there exists a simple representation σ of $\mathfrak{k}$ with kernel L and a simple representation ϱ of $\mathfrak{st}(\sigma,\mathfrak{g})$ such that $\varrho|\mathfrak{k}$ is a multiple of σ and $\text{ind}(\varrho,\mathfrak{g})$ is simple with kernel I.

Cf. 4.5.9, 5.4.3, 5.4.4 and 5.6.5. Cf. also [201].

12. Let $\mathfrak{g}'$ be a Lie subalgebra of $\mathfrak{g}$, I' a two-sided ideal of $U(\mathfrak{g}')$, and $I = \text{ind}(I',\mathfrak{g})$ (cf. 5.6.3). We assume that I and I' are prime, so that Fract $(U(\mathfrak{g})/I)$ (or Fract $(U(\mathfrak{g}')/I')$) is an algebra of matrices with n rows and n columns (or n' rows and n' columns). One has $n \leq n'$, and it can happen that $n < n'$ [143], [198]. Give sufficient conditions for $n = n'$.

13. Let G be the algebraic adjoint group of $\mathfrak{g}$. Let $f \in \mathfrak{g}^*$ such that Gf is of maximal dimension. Is $I(f)$ a minimal primitive ideal of $U(\mathfrak{g})$? (This is true for $\mathfrak{g}$ semi-simple).

14. Let q be the element of $S(\mathfrak{g})^*$ defined in 6.6.9. Let $D(q)$ be the corresponding differential operator of infinite order, operating in $S(\mathfrak{g})$. For every regular element f of $\mathfrak{g}^*$, let χ_f be the homomorphism of $Z(\mathfrak{g})$ into k with kernel $I(f) \cap Z(\mathfrak{g})$. Let β be the canonical mapping of $S(\mathfrak{g})$ into $U(\mathfrak{g})$. Is it true that $\chi_f(\beta u) = (D(q^{-1})u)(f)$ for all $u \in Y(\mathfrak{g})$? Cf. 6.6.9, and 10.4.5 [47].)

15. Let $\mathfrak{k}$ be an ideal of $\mathfrak{g}$, and f a regular element of $\mathfrak{k}^*$. What is the stabilizer of $I(f)$ in $\mathfrak{g}$? Cf 6.2.8.

16. Determine all the primitive ideals of $U(\mathfrak{g})$; do this first for $\mathfrak{g}$ semi-simple. If f_1 and f_2 are regular elements of $\mathfrak{g}^*$ such that $I(f_1) = I(f_2)$, are f_1 and f_2 conjugate under the algebraic adjoint group of $\mathfrak{g}$? Cf. 10.5.8.

17. Generalize 10.3.3 (ii) to non-solvable polarizations. (Impossible! Cf. [200].)

18. Let G be an algebraic subgroup of Aut($\mathfrak{g}$) and Ω a G-orbit in Prim $U(\mathfrak{g})$. Is Ω open in its closure? The answer is affirmative if $\mathfrak{g}$ is solvable [15].

19. In $U(\mathfrak{g})$, are (left) primitive ideals identical to right primitive ideals? Cf. 6.6.5, [45], [195].

20. Let I be a primitive ideal of $U(\mathfrak{g})$, Δ the set of derivations of $U(\mathfrak{g})/I$, and Δ' the set of inner derivations of $U(\mathfrak{g})/I$. Is $\dim \Delta/\Delta' < +\infty$? Cf. 4.6.8 and [41].

21. An element u of $U(\mathfrak{g})$ is said to be irreducible if, for every factorization $u = u_1u_2$ with $u_1, u_2 \in U(\mathfrak{g})$, we have $u_1 \in k$ or $u_2 \in k$. Every element of $U(\mathfrak{g})$ has factorizations into a product of irreducible elements. Is the number of these factorizations finite (up to multiplication by scalars)? (Answer: yes; cf. [221]).

22. Does a version of the "Hauptidealsatz" hold in $U(\mathfrak{g})$?

23. Let A be an algebra of finite type, and V a simple A-module. Are the A-endomorphisms of V scalar? Cf. 2.6.5 and 2.8.9.

24. Let A be an algebra of finite type having a ring of fractions K. Are the following conditions equivalent:

(a) the ideal 0 of A is primitive;

(b) the centre of K is k;

(c) the ideal 0 is prime, and the intersection of non-null primitive ideals is non-null. Cf. 4.5.7.

PART II

25. Assume that $\mathfrak{g}$ is nilpotent. Let $\mathscr{A}$ be the adjoint group of $\mathfrak{g}$, P a prime ideal of $U(\mathfrak{g})$, Q the invariant prime ideal of $S(\mathfrak{g})$ associated with P, and β the canonical isomorphism of $C(\mathfrak{g}; P)$ onto $\mathrm{Fract}(S(\mathfrak{g})/Q)^{\mathscr{A}}$. Is it true that $\beta(Z(\mathfrak{g}; P)) = (S(\mathfrak{g})/Q)^{\mathscr{A}}$? Cf. 6.6.11.

26. Assume that $\mathfrak{g}$ is solvable. Let $\mathfrak{h}$ be a Lie subalgebra of $\mathfrak{g}$, and $f \in \mathfrak{g}^*$. Assume that $f([\mathfrak{h},\mathfrak{h}]) = 0$ and that $\mathrm{ind}^{\sim}(f|\mathfrak{h},\mathfrak{g})$ is simple. Is $\mathfrak{h} \in P(f)$? Cf. 6.6.6 and 6.6.7. (Answer: yes; cf. [229]).

27. Assume that $\mathfrak{g}$ is solvable. Let $\mathscr{A}$ be the algebraic adjoint group of $\mathfrak{g}$. Is the canonical bijection of $\mathfrak{g}^*/\mathscr{A}$ onto Prim $U(\mathfrak{g})$, which is continuous, bicontinuous? Cf. 6.6.1.

28. Let $(\mathfrak{g},\mathfrak{h})$ be a split semi-simple Lie algebra. Let $x \in \mathfrak{h}$, let $\mathfrak{p}_1$ and $\mathfrak{p}_2$ be polarizations of $\mathfrak{g}$ at x, I_1 and I_2 the ideals of $U(\mathfrak{g})$ deduced from $(x,\mathfrak{p}_1)$ and $(x,\mathfrak{p}_2)$ by twisted induction. Is $I_1 = I_2$?

29. We adopt the notation of 7.8.7. Let $(p_1, \ldots, p_m) \in \mathscr{M}$. Is the image of $X_{\alpha_1}^{p_1} \cdots X_{\alpha_m}^{p_m}$ under the Harish-Chandra homomorphism proportional to some H_α? ([40], [184]). (Answer: no; Marcovici, unpublished).

30. We adopt the notation of section 7.0. Let $\lambda \in \mathfrak{h}^*$. Let $\mathscr{I}$ be the set of two-sided ideals of $U(\mathfrak{g})$ containing $\operatorname{Ker} \chi_\lambda$. Let $\mathscr{M}$ be the set of submodules of $M(\lambda)$. It would be convenient to study the mappings $I \mapsto IM(\lambda)$ and $M \mapsto \operatorname{ann}(M(\lambda)/M)$ of $\mathscr{I}$ into $\mathscr{M}$ and of $\mathscr{M}$ into $\mathscr{I}$ (where ann denotes the annihilator). The former is not surjective. Is it injective for λ in the chamber corresponding to B? Cf. [25].

31. Assume that $\mathfrak{g}$ is semi-simple. Let $\mathfrak{g} = \mathfrak{k} \oplus \mathfrak{p}$ be a symmetric decomposition of $\mathfrak{g}$ such that $\mathfrak{p}$ contains a Cartan subalgebra of $\mathfrak{g}$. Let $I \in \operatorname{Prim} U(\mathfrak{g})$. Does a Harish-Chandra module M exist (relative to $\mathfrak{k}$) such that I is the annihilator of M?

32. Assume that $\mathfrak{g}$ is semi-simple. Let $\mathfrak{k}$ be a symmetrizing subalgebra of $\mathfrak{g}$. Classify the simple Harish-Chandra $\mathfrak{g}$-modules relative to $\mathfrak{k}$.

33. What is the annihilator in $U(\mathfrak{g})$ of a module of the principal series?

34. Assume that $\mathfrak{g}$ is semi-simple. Let $\mathfrak{k}$ be a semi-simple symmetrising subalgebra of $\mathfrak{g}$.

(a) Let V be a Harish-Chandra $\mathfrak{g}$-module having a central character and such that $\operatorname{mtp}(\xi, V) < +\infty$ for all $\xi \in \mathfrak{k}^\wedge$. Does V posses a Jordan–Hölder series?

(b) Let χ be a homomorphism of $Z(\mathfrak{g})$ into k. Does there exist, up to isomorphism, a finite number of simple Harish-Chandra $\mathfrak{g}$-modules whose central character is χ? (If $\mathfrak{k}$ is not semi-simple, the problems must be stated somewhat differently; cf. 9.7.8 and 9.7.9.)

35. Let $(\mathfrak{g}, \theta)$ be a symmetric semi-simple Lie algebra. We adopt the notation of 9.2.1. The image under the canonical homomorphism of $G^{\mathfrak{k}}$ into A is the subalgebra of the invariants of the Weyl group (operating as in 9.7.5). Cf. [88], [169]. Find a purely algebraic proof of this result. (This is now done; cf. [217]).

36. Prove algebraically, and for any algebraically closed k, the assertions of 8.5.10.

37. We adopt the notation of 9.2.1. Let V be a simple Harish-Chandra $\mathfrak{g}$-module. Does there exist $\pi \in (\mathfrak{m} \oplus \mathfrak{a})^\wedge$ such that V is isomorphic to a sub-$\mathfrak{g}$-module of $X(\pi)$? Cf. 9.4.4.

PART III

In the problems of part III, the essential point is that k is not assumed to be algebraically closed.

38. Let V a simple $\mathfrak{g}$-module. If $\mathfrak{g}$ is solvable, is the ring of $\mathfrak{g}$-endomorphisms of V commutative? Is the same true for for a simple A_n-module? Cf. 3.8.10 (b).

39. Let $\mathfrak{k}$ be an ideal of $\mathfrak{g}, \sigma$ a simple representation of $\mathfrak{k}$, $\mathfrak{h} = \mathfrak{st}(\sigma,\mathfrak{g})$, ϱ a representation of $\mathfrak{h}$ such that $\varrho|\mathfrak{k}$ is a multiple of σ, and $\pi = \operatorname{ind}(\varrho,\mathfrak{g})$. If ϱ is simple, is π simple? (Cf. 5.3.6).

40. Assume that $\mathfrak{g}$ is solvable. Is $C(\mathfrak{g})$ a pure transcendental extension of k? Cf. 4.4.8 and 4.9.11.

BIBLIOGRAPHY

[1] S. ARAKI, On root systems and an infinitesimal classification of irreducible symmetric spaces, *J. Math. Osaka City Univ.* **14** (1967) 1–23.

[2] D. ARNAL and G. PINCZON, Sur certaines représentations de l'algèbre de Lie $\mathfrak{sl}(2)$, *C. R. Acad. Sci. Paris* (A) **272** (1971) 1369–1372.

[3] D. ARNAL and G. PINCZON, On algebraically irreducible representations of the Lie algebra $\mathfrak{sl}(2)$, *J. Math. Phys.* **15** (1974) 350–359.

[4] L. AUSLANDER and B. KOSTANT, Polarization and unitary representations of solvable Lie groups, *Inv. Math.* **14** (1971) 255–354.

[5] F. A. BEREZIN, Some remarks on the associative envelope of a Lie algebra (in Russian), *Funkcion. Anal. Priloz.* **1** (1967) 1–14.

[6] P. BERNAT, Sur le corps des quotients de l'algèbre enveloppante d'une algèbre de Lie, *C. R. Acad. Sci. Paris* **254** (1962) 1712–1714.

[7] P. BERNAT, Sur le corps enveloppant d'une algèbre de Lie résoluble, *C. R. Acad. Sci. Paris* **258** (1964) 2713–2715.

[8] P. BERNAT, Sur le corps enveloppant d'une algèbre de Lie résoluble, *Bull. Soc. Math. France*, Mém. No. 7 (1966).

[9] P. BERNAT, N. CONZE *et al.*, *Représentations des Groupes de Lie Résolubles*, Monographies Soc. Math. France (Dunod, Paris, 1972).

[10] I. N. BERNSTEIN, I. M. GELFAND and S. GELFAND, The structure of representations generated by vectors of largest weight (in Russian), *Funkcion. Anal. Prilož.* **5** (1971) 1–19.

[11] I. N. BERNSTEIN, I. M. GELFAND and S. I. GELFAND, Differential operators on the base affine space and a study of $\mathfrak{g}$-modules, in: I. M. Gelfand, ed., *Publ. of 1971 Summer School in Math.*, Janos Bolyai Math. Soc., Budapest, 21–64.

[12] G. BIRKHOFF, Representability of Lie algebras and Lie groups by matrices, *Ann. Math.* **38** (1937) 526–532.

[13] R. J. BLATTNER, Induced and produced representations of Lie algebras, *Trans. Amer. Math. Soc.* **144** (1969) 457–474.

[14] O. BOREL, Sur les représentations $\mathfrak{h}$-diagonales des algèbres de Lie semi-simples complexes, *C. R. Acad. Sci. Paris* (A) **275** (1972) 1289–1292.

[15] W. BORHO, P. GABRIEL and R. RENTSCHLER, *Primideale in Einhüllenden auflösbarer Lie-Algebren*, Lecture Notes in Math. **357** (Springer, Berlin, 1973).

[16] N. BOURBAKI, Groupes et algèbres de Lie, chap. I (2nd ed.), II-III, IV-V-VI, *Act. Sci. Ind.* **1285**, **1349**, **1337** (Hermann, Paris, 1971, 1972, 1968); and VII-VIII, to appear.

[17] I. Z. Bouwer, Standard representations of simple Lie algebras, *Can. J. Math.* **20** (1968) 344–361.

[18] J. Brezin, Unitary representation theory for solvable Lie groups, *Mem. Amer. Math. Soc.* **79** (1968).

[19] N. N. Chapovalov, Certain bilinear forms over the enveloping algebra of a complex semi-simple Lie algebra (in Russian), *Funktion. Anal. Priložž.* **6** (1972) 65–70.

[20] N. Conze, Idéaux primitifs de l'algèbre enveloppante d'une algèbre de Lie nilpotente et orbites dans l'espace dual, *C. R. Acad. Sci. Paris* (A) **267** (1968) 325–327.

[21] N. Conze, Sur certaines représentations des algèbres de Lie nilpotentes, *C. R. Acad. Sci. Paris* (A) **272** (1971) 460–461.

[22] N. Conze, Action d'un groupe algèbrique dans l'espace des idéaux primitifs d'une algèbre enveloppante, *J. Algebra* **25** (1973) 100–105.

[23] N. Conze, Espace des idéaux primitifs de l'algèbre enveloppante d'une algèbre de Lie nilpotente, *J. Algebra* **34** (1975) 444–450.

[24] N. Conze, Quotients primitifs des algèbres enveloppantes et algèbres d'opérateurs différentiels, *C. R. Acad. Sci Paris* (A) **277** (1973) 1033–1036.

[25] N. Conze and J. Dixmier, Idéaux primitifs dans l'algèbre enveloppante d'une algèbre de Lie semi-simple, *Bull. Sci. Math.* **96** (1972) 339–351.

[26] N. Conze and M. Duflo, Sur l'algèbre enveloppante d'une algèbre de Lie résoluble, *Bull. Sci. Math.* **94** (1970) 201–208.

[27] N. Conze and M. Vergne, Idéaux primitifs des algèbres enveloppantes des algèbres de Lie résolubles, *C. R. Acad. Sci. Paris* (A) **272** (1971) 985–988.

[28] J. Dixmier, Sur les représentations unitaires des groupes de Lie nilpotents, II, *Bull. Soc. Math. France* **85** (1957) 325–388.

[29] J. Dixmier, Sur l'algèbre enveloppante d'une algèbre de Lie nilpotente, *Arch. Math.* **10** (1959) 321–326.

[30] J. Dixmier, Représentations irréductibles des algèbres de Lie nilpotents, *An. Acad. Brasil. Ci.* **35** (1963) 491–519.

[31] J. Dixmier, Représentations irréductibles des algèbres de Lie résolubles. *J. Math. Pures Appl.* **45** (1966) 1–66.

[32] J. Dixmier, Sur le dual d'un groupe de Lie nilpotent, *Bull. Sci. Math.* **90** (1966) 113–118.

[33] J. Dixmier, Sur le centre de l'algèbre enveloppante d'une algèbre de Lie, *C. R. Acad. Sci. Paris* (A) **265** (1967) 408–410.

[34] J. Dixmier, Sur les algèbres de Weyl, II, *Bull. Sci. Math.* **94** (1970) 289–301.

[35] J. Dixmier, Sur les représentations induites des algèbres de Lie, *J. Math. Pures Appl.* **50** (1971) 1–24.

[36] J. Dixmier, Polarisations dans les algèbres de Lie. *Ann. Sci. Ecole Norm. Sup.* **4** (1971) 321–336.

[37] J. Dixmier, Idéaux primitifs dans l'algèbre enveloppante d'une algèbre de Lie semi-simple complexe; (a) *C. R. Acad. Sci. Paris* (A) **271** (1970) 134–136; (b) *Ibid.* (A) **272** (1971) 1628–1630.

[38] J. Dixmier, Idéaux maximaux dans l'algèbre enveloppante d'une algèbre de Lie semi-simple complexe, *C. R. Acad. Sci. Paris* (A) **274** (1972) 228–230.

[39] J. Dixmier, Sur les idéaux génériques dans les algébres enveloppantes, *Bull. Sci. Math.* **96** (1972) 17–26.

[40] J. Dixmier, Sur les homomorphismes d'Harish-Chandra, *Inv. Math.* **17** (1972) 167–176.

[41] J. Dixmier, Quotients simples de l'algèbre enveloppante de $\mathfrak{sl}_2$, *J. Algebra* **24** (1973) 551–564.

[42] M. Duflo, Caractères des algèbres de Lie résolubles, *C. R. Acad. Sci. Paris* (A) **269** (1969) 437–438.

[43] M. Duflo, Caractères des groupes et des algèbres de Lie résolubles, *Ann. Sci. Ecole Norm. Sup.* **3** (1970) 23–74.

[44] M. Duflo, Sur les représentations irréductibles des algèbres de Lie contenant un idéal nilpotent, *C. R. Acad. Sci. Paris* (A) **270** (1970) 504–506.

[45] M. Duflo, Sur les extensions des représentations irréductibles des groupes de Lie nilpotents, *Ann. Sci. Ecole Norm. Sup.* **5** (1972) 71–120.

[46] M. Duflo, Représentations induites d'algèbres de Lie, *C. R. Acad. Sci. Paris* (A) **272** (1971) 1157–1158.

[47] M. Duflo, Construction of primitive ideals in an enveloping algebra, in: I. M. Gelfand, ed., *Publ. of 1971 Summer School in Math.*, Janos Bolyai Math. Soc., Budapest, 77 – 93.

[48] M. Duflo, Certaines algèbres de type fini sont des algèbres de Jacobson, *J. Algebra* **27** (1973) 358–365.

[49] M. Duflo and M. Vergne, Une propriété de la représentation coadjointe d'une algèbre de Lie, *C. R. Acad. Sci. Paris* (A) **268** (1969) 583–585.

[50] P. Gabriel, Des catégories abéliennes, *Bull. Soc. Math. France* **90** (1962) 323–448.

[51] P. Gabriel, Représentations des algèbres de Lie résolubles, Sém. Bourbaki **347** (1968–1969) 1–22, Lecture Notes in Math. (Springer, Berlin).

[52] I. M. Gelfand and A. A. Kirillov, Sur les corps liés aux algèbres enveloppantes des algèbres de Lie, *Publ. Inst. Hautes Etudes Sci.* **31** (1966) 5–19.

[53] I. M. Gelfand and A. A. Kirillov, On the structure of the field of quotients of the enveloping algebra of a semi-simple Lie algebra (in Russian), *Dokl. Akad. Nauk SSSR* **180** (1968) 775–777.

[54] I. M. Gelfand and A. A. Kirillov, The structure of the enveloping field of a semi-simple Lie algebra (in Russian), *Funktion. Anal. Prilož.* **3** (1969) 7–26.

[55] I. M. Gelfand and V. A. Ponomarev, The category of Harish-Chandra modules over the Lie algebra of the Lorentz group (in Russian), *Dokl. Akad. Nauk SSSR* **176** (1967) 243–246.

[56] I. M. Gelfand and V. A. Ponomarev, The classification of the indecomposable infinitesimal representations of the Lorentz group (in Russian), *Dokl. Akad. Nauk SSSR* **176** (1967) 502–505.

[57] I. M. Gelfand and V. A. Ponomarev, Indecomposable representations of the Lorentz group (in Russian), *Uspehi Mat. Nauk* **23** (1968) 3–60; English Transl. in *Russian Math. Surv.*

[58] R. Godement, Mémoire sur la theorie des caractères dans les groupes localement compacts unimodulaires, *J. Math. Pures Appl.* **30** (1951) 1–110.

[59] A. W. Goldie, Lectures given at the Sem. of the Can. Math. Congr. Toronto, 1967, cyclostyled. *The structure of noetherian rings*, Lecture Notes in Math. **246** (Springer, Berlin) 213–321.

[60] R. Goodman, Differential operators of finite order on a Lie group, II, *Indiana Math. J.* **21** (1971) 383–409.

[61] Dz. Hadziev, Some questions in the theory of invariants (in Russian), *Mat. Sb.* **72** (1967) 420–435.

[62] Harish-Chandra, On representations of Lie algebras, *Ann. Math.* **50** (1949) 900–915.

[63] Harish-Chandra, On some applications of the universal enveloping algebra of a semisimple Lie algebra, *Trans. Amer. Math. Soc.* **70** (1951) 28–96.

[64] Harish-Chandra, Representations of a semisimple Lie group on a Banach space, I, *Trans. Amer. Math. Soc.* **75** (1953) 185–243.

[65] Harish-Chandra, Representations of semisimple Lie groups, II, *Trans. Amer. Math. Soc.* **76** (1954) 26–65.

[66] Harish-Chandra, Representations of semisimple Lie groups, IV, *Amer. J. Math.* **77** (1955) 743–777.

[67] D. G. Higman, Induced and produced modules, *Can. J. Math.* **7** (1955) 490–508.

[68] G. Hochschild, Algebraic Lie algebras and representative functions. *Illinois J. Math.* **3** (1959) 499–523.

[69] G. Hochschild, Algebraic Lie algebras and representative functions, Supplements, *Illinois J. Math.* **4** (1960) 609–618.

[70] G. Hochschild, Algebraic groups and Hopf algebras, *Illinois J. Math.* **14** (1970) 52–65.

[71] N. Jacobson, *Lie Algebras* (Interscience, New York, 1962).

[72] A. Joseph, Proof of the Gelfand-Kirillov conjecture for solvable Lie algebras, *Proc. Amer. Math. Soc.* **45** (1974) 1–10.

[73] A. Joseph, A generalization of the Gelfand-Kirillov conjecture, to appear.

[74] H. Kimura, On some infinite dimensional representations of semisimple Lie algebras, *Nagoya Math. J.* **25** (1965) 211–220.

[75] A. A. Kirillov, Unitary representations of nilpotent Lie groups (in Russian), *Uspehi Mat. Nauk* **17** (1962) 57–110.

[76] B. Kostant, A formula for the multiplicity of a weight, *Trans. Amer. Math. Soc.* **93** (1959) 53–73.

[77] B. Kostant, The principal three-dimensional subgroup and the Betti numbers of a complex simple Lie group, *Amer. J. Math.* **81** (1959) 973–1032.

[78] B. Kostant, Lie group representations on polynomial ring, *Amer. J. Math.* **85** (1963) 327–404.

[79] B. Kostant, On the existence and irreducible of certain series of representations, *Bull. Amer. Math. Soc.* **75** (1969) 627–642; and in: I. M. Gelfand, ed., *Publ. of 1971 Summer School in Math.*, Janos Bolyai Math. Soc., Budapest, 231–329.

[80] B. Kostant and S. Rallis, Orbits and representations associated with symmetric spaces, *Amer. J. Math.* **93** (1971) 753–809.

[81] J. L. Koszul, Sur les modules de représentations des algèbres de Lie résolubles, *Amer. J. Math.* **76** (1954) 535–554.

[82] F. W. Lemire, Irreducible representations of a simple Lie algebra admitting a one-dimensional weight space, *Proc. Amer. Math. Soc.* **19** (1968) 1161–1164.

[83] F. W. Lemire, Note on weight spaces of irreducible linear representations, *Can. Math. Bull.* **11** (1968) 399–403.

[84] F. W. Lemire, Weight spaces and irreducible representations of simple Lie algebras, *Proc. Amer. Math. Soc.* **22** (1969) 192–197.

[85] F. W. Lemire, One dimensional representations of the cycle subalgebra of a semisimple Lie algebra, *Can. Math. Bull.* **13** (1970) 463–467.

[86] F. W. Lemire, Existence of weight space decompositions for irreducible representations of simple Lie algebra, *Can. Math. Bull.* **14** (1971) 113–115.

[87] J. Lepowsky, Algebraic results on representations of semisimple Lie groups, *Trans. Amer. Math. Soc.* **176** (1973) 1–44.

[88] J. Lepowsky, On the Harish-Chandra homomorphism, *Trans. Amer. Math. Soc.* **208** (1955) 193–218.

[89] J. Lepowsky, and G. W. McCollum On the determination of irreducible modules by restriction to a subalgebra, *Trans. Amer. Math. Soc.* **176** (1973) 45–57.

[90] J. Lepowsky and N. R. Wallach, Finite and infinite dimensional representations of linear semisimple groups, *Trans. Amer. Math. Soc.* **184** (1973) 223–246.

[91] J. C. McConell, The intersection theorem in a class of non-commutative rings, *Proc. London Math. Soc.* **17** (1967) 487–498.

[92] J. C. McConnell, Localisation in enveloping rings, *J. London Math. Soc.* **43** (1968) 421–428, **3** (1971) 409–410.

[93] J. C. McConnell, Representations of solvable Lie algebras and the Gelfand–Kirillov conjecture, *Proc. London Math. Soc.* **29** (1974) 453–484.

[94] J. C. McConnell, Representations of solvable Lie algebras, II, Twisted group rings, *Ann. Sci. Ecole Norm. Sup.* **8** (1975) 157–178.

[95] W. Miller, On Lie algebras and some special functions of mathematical physics, *Mem. Amer. Math. Soc.* **50** (1964).

[96] Nghiêm Xuan Hai, Sur certains sous-corps commutatifs du corps enveloppant d'une algèbre de Lie résoluble, *Bull. Sci. Math.* **96** (1972), 111–128.

[97] Nghiêm Xuân Hai, Sur certaines représentations d'une algèbre de Lie résoluble complexe, *Bull. Sci. Math.* **97** (1973) 105–128.

[98] Y. Nouazé, Remarques sur «Idéaux premiers de l'algèbre enveloppante d'une algèbre de Lie nilpotente», *Bull. Sci. Math.* **91** (1967) 117–124.

[99] Y. Nouazé, P. Gabriel, Idéaux premiers de l'algèbre enveloppante d'une algèbre de Lie nilpotente, *J. Algebra* **6** (1967) 77–99.

[100] K. P. Parthasarathy, R. Ranga Rao and V. S. Varadarajan, Representations of complex semisimple Lie groups and Lie algebras, *Ann. Math.* **85** (1967) 383–429.

[101] D. Parvizi, Représentations de l'algèbre de Lie $\mathfrak{sl}(2, \mathbf{C})$, *C. R. Acad. Sci. Paris* (A) **276** (1973) 905–908.

[102] D. Quillen, On the endomorphism ring of a simple module over an enveloping algebra, *Proc. Amer. Math. Soc.* **21** (1969) 171–172.

[103] C. Rader, Spherical functions on semisimple Lie groups, to appear.

[104] M. Raïs, Sur les idéaux primitifs des algèbres enveloppantes, *C. R. Acad. Sci. Paris* (A) **272** (1971) 989–991.

[105] R. Rentschler, Sur le centre du quotient de l'algèbre enveloppante d'une algèbre de Lie nilpotente par un idéal premier, *C. R. Acad. Sci. Paris* (A) **268** (1969) 689–692.

[106] R. Rentschler, Propriétés fonctorielles de la bijection canonique entre Spec $U(\mathfrak{g})$ et (Spec $S(\mathfrak{g}))^{\mathfrak{g}}$ pour les algèbres de Lie nilpotentes, *C. R. Acad. Sci. Paris* (A) **271** (1970) 868–871.

[107] R. Rentschler, Sur la topologie des ensembles fermés irréductibles de Prim $U(\mathfrak{g})$ pour les algèbres de Lie nilpotentes, *C. R. Acad. Sci. Paris* (A) **274** (1972) 27–30.

[108] R. Rentschler, Sur le centre du corps enveloppant d'une algèbre de Lie résoluble, *C. R. Acad. Sci. Paris* (A) **276** (1973) 21–24.

[109] R. Rentschler, L'injectivité de l'application de Dixmier pour des algèbres de Lie résolubles, *Inv. Math.* **23** (1974) 49–71.

[110] R. Rentschler, and P. Gabriel, Sur la dimension des anneaux et ensembles ordonnés, *C. R. Acad. Sci. Paris* (A) **265** (1967) 712–715.

[111] R. Rentschler and M. Vergne, Sur le semi-centre du corps enveloppant d'une algèbre de Lie, *Ann. Sci. Ecole Norm. Sup.* **6** (1973) 380–405.

[112] R. W. Richardson, Conjugacy classes in Lie algebras and algebraic groups, *Ann. Math.* **86** (1967) 1–15.

[113] J. E. Roos, Détermination de la dimension homologieque globale des algèbres de Weyl, *C. R. Acad. Sci. Paris* (A) **274** (1972) 23–26.

[114] J. E. Roos, Propriétés homologiques des quotients primitifs des algèbres enveloppantes des algèbres de Lie semi-simples, *C. R. Acad. Sci. Paris* (A) **276** (1973) 351–354.

[115] J. E. Roos, Compléments a l'étude des quotients primitifs des algèbres enveloppantes des algèbres de Lie semi-simples, *C. R. Acad. Sci. Paris* (A) **276** (1973) 447–450.

[116] S. Sakai, On the representations of semisimple Lie groups, *Proc. Jap. Acad.* **30** (1954) 14–18; On infinite dimensional representations of semisimple Lie algebras and some functionals on the universal enveloping algebras, *Ibid.* **30** (1954) 305–312.

[117] H. Samelson, *Notes on Lie Algebras* (Van Nostrand-Reinhold, New York, 1969).

[118] *Séminaire Sophus Lie*, 1re année, 1954–1955, Éc. Norm. Sup., Paris (1955).

[119] L. Solomon, On the Poincaré–Birkhoff–Witt theorem, *J. Combin. Theory* **4** (1968) 363–375.

[120] L. Solomon and D. N. Verma, Sur le corps des quotients de l'algèbre enveloppante d'une algèbre de Lie, *C. R. Acad. Sci. Paris* (A) **264** (1967) 985–986.

[121] D. Tamari, On the embedding of Birkhoff–Witt rings in quotient fields, *Proc. Amer. Math. Soc.* **4** (1953) 197–202.

[122] V. S. Varadarajan, On the ring of invariant polynomials on a semisimple Lie algebra, *Amer. J. Math.* **90** (1968) 308–317.

[123] M. Vergne, La structure de Poisson sur l'algèbre symétrique d'une algèbre de Lie nilpotente, *C. R. Acad. Sci. Paris* (A) **269** (1969) 950–952.

[124] M. Vergne, Construction de sous-algèbres subordonnées a un élément du dual d'une algèbre de Lie résoluble, *C. R. Acad. Sci. Paris* (A) **270** (1970) 173–175, 704–707.

[125] M. Vergne, La structure de Poisson sur l'algèbre symétrique d'une algèbre de Lie nilpotente, *Bull. Soc. Math. France* **100** (1972) 301–335.

[126] D. N. VERMA, Structure of certain induced representations of complex semi-simple Lie algebras, Dissertation, Yale University (1966); *Bull. Amer. Math. Soc.* **74** (1968) 160–166, 628.

[127] D. N. VERMA, Möbius inversion for the Bruhat ordering on a Weyl group, *Ann. Sci. Ecole Norm. Sup.* **4** (1971) 393–398.

[128] M. WAKIMOTO, On the irreducibility of some series of representation, *Hiroshima Math. J.* **2** (1972) 71–98.

[129] N. R. WALLACH, Induced representations of Lie algebras and a theorem of Borel–Weil, *Trans. Amer. Math. Soc.* **136** (1969) 181–187.

[130] N. R. WALLACH, Induced representations of Lie algebras, II, *Proc. Amer. Maht. Soc.* **21** (1969) 161–166.

[131] N. R. WALLACH, Cyclic vectors and irreducibility for principal series representations, I, *Trans. Amer. Math. Soc.* **158** (1971) 107–113; II, *Ibid.* **164** (1972) 389–396.

[132] E. WITT, Treue Darstellung Lieschen Ringe, *J. Reine Angew. Math.* **177** (1937) 152–160.

[133] D. ZELOBENKO, Analysis of irreducibility in the class of elementary representations of a complex semi-simple Lie group (in Russian), *Izv. Akad. Nauk SSSR*, Sér. Mat. **32** (1968) 108–133.

[134] D. ZELOBENKO, Operational calculus over a complex semi-simple Lie group (in Russian), *Izv. Akad. Nauk SSSR*, Sér. Mat. **33** (1969) 931–973.

SUPPLEMENTARY BIBLIOGRAPHY

[135] D. Arnal, and G. Pinczon, Certaines représentations de l'algèbre de Lie $\mathfrak{sl}(2)$, *C. R. Acad. Sci. Paris* (A) **274** (1972) 248–250.

[136] D. Arnal and G. Pinczon, Idéaux à gauche dans les quotients simples de l'algèbre enveloppante de $\mathfrak{sl}(2)$, *Bull. Soc. Math. France* **101** (1973) 381–395.

[137] R. E. Block, Irreducible representations of Lie algebra extensions, *Bull. Amer. Math. Soc.* **80** (1974) 868–872.

[138] W. Borho and H. Kraft, Über die Gelfand–Kirillov Dimension, *Math. Ann.* **220** (1976) 1–24.

[139] W. Borho and R. Rentschler, Oresche Teilmengen in Einhüllenden Algebren, *Math. Ann.* **217** (1975) 201–210.

[140] W. Borho, Primitive vollprime Ideale in der Einhüllenden von $\mathfrak{so}(5, \mathbf{C})$, to appear.

[141] R. W. Carter and G. Lusztig, On the modular representations of the general linear and symmetric groups, *Math. Z.* **136** (1974) 193–242.

[142] W. Casselman and M. S. Orborne, The $\mathfrak{n}$-cohomology of representations with an infinitesimal character, *Compositio Math.* **31** (1975) 219–227.

[143] N. Conze, Algèbres d'opérateurs différentiels et quotients des algèbres enveloppantes, *Bull. Soc. Math. France* **102** (1974) 379–415.

[144] J. Dixmier, Idéaux primitifs complètement premiers dans l'algèbre enveloppante de $\mathfrak{sl}(3, \mathbf{C})$, in: *Lecture Notes in Math.* (Springer, Berlin, **466** 1975) 38–54.

[145] J. Dixmier, Polarisations dans les algèbres de Lie semi-simples complexes, *Bull. Sci. Math.* **99** (1975) 45–63.

[146] J. Dixmier, M. Duflo and M. Vergne, Sur la représentation coadjointe d'une algèbre de Lie, *Compositio Math.* **25** (1974) 309–323.

[147] B. Gruber and A. U. Klimyk, Properties of linear representations with a highest weight for the semi-simple Lie algebras, *J. Math. Phys.* **16** (1975) 1816–1832.

[148] G. Hochschild, Lie algebra cohomology and affine algebraic groups, *Illinois J. Math.* **18** (1974) 170–176.

[149] A. Hole, Invariant differential operators and polynomials of Lie transformation groups, *Math. Scand.* **34** (1974) 109–123.

[150] R. L. Hudson, A new system of Casimir operators for $U(n)$, *J. Math. Phys.* **15** (1974) 1067–1070.

[151] J. C. Jantzen, Zur Charakterformel gewisser Darstellungen halbeinfacher Gruppen und Lie-Algebren, *Math. Z.* **140** (1975) 127–149.

[152] K. JOHNSON and N. R. WALLACH, Composition series and intertwining operators for the spherical principal series, *Bull. Amer. Math. Soc.* **78** (1972) 1058–1059.

[153] A. JOSEPH, Commuting polynomials in quantum canonical operators and realizations of Lie algebras, *J. Math. Phys.* **13** (1972) 351–357.

[154] A. JOSEPH, Gel'fand–Kirillov dimension for algebras associated with the Weyl algebra, *Ann. Inst. H. Poincaré* (A) **17** (1972) 325–336.

[155] A. JOSEPH, A characterization theorem for realizations of $\mathfrak{sl}(2)$, *Proc. Cambridge Phil. Soc.* **75** (1974) 119–131.

[156] A. JOSEPH, Minimal realizations and spectrum generating algebras, *Commun. Math. Phys.* **36** (1974) 325–338.

[157] A. JOSEPH, Symplectic structure in the enveloping algebra of a Lie algebra, *Bull. Soc. Math. France* **102** (1974) 75–83.

[158] A. JOSEPH, The Weyl algebra – Semisimple and nilpotent elements, *Amer. J. Math.* **97** (1975) 597–615.

[159] A. JOSEPH, The minimal orbit in a simple Lie algebra and its associated maximal ideal, *Ann. Sci. Ecole Norm. Sup.* **9** (1976) 1–29.

[160] A. JOSEPH, Second commutant theorems in enveloping algebras, to appear.

[161] A. JOSEPH, A wild automorphism of $U\mathfrak{sl}(2)$, to appear.

[162] A. JOSEPH, Sur les algèbres de Weyl, Lecture Notes.

[163] Y. KOSMANN and S. STERNBERG, Conjugaison des sous-algèbres d'isotropie, *C. R. Acad. Sci. Paris* (A) **279** (1974) 777–779.

[164] G. LEGER and E. LUKS, Sur les fonctions polynômes invariantes sur les algèbres de Lie, *C. R. Acad. Sci. Paris* (A) **280** (1975) 1177–1179.

[165] J. LEPOWSKY, The subquotient theorem and some applications, in: *Harmonic Analysis on Homogeneous Spaces*, Proc. Symp. Math., (Am. Math. Soc., Providence, R.I., 1973).

[166] J. LEPOWSKY, Conical vectors in induced modules, *Trans Amer. Math. Soc.* **208** (1975) 219–272.

[167] J. LEPOWSKY, Existence of conical vectors in induced modules, *Ann. Math.* **102** (1975) 17–40.

[168] J. LEPOWSKY, Uniqueness of imbedding of certain induced modules, to appear.

[169] J. LEPOWSKY, On the uniqueness of conical vectors, to appear.

[170] J. LEPOWSKY, A generalization of H. Weyl's "unitary trick", *Trans. Amer. Math. Soc.* **216** (1976) 229–236.

[171] J. LEPOWSKY and G. W. MCCOLLUM, Cartan subspaces of symmetric Lie algebras, *Trans. Amer. Math. Soc.* **216** (1976) 217–228.

[172] M.-P. MALLIAVIN-BRAMERET, Sur les anneaux d'invariants de groupes nilpotents, *C. R. Acad. Sci. Paris* (A) **280** (1975) 1173–1175.

[173] J. C. MCCONNELL and J. C. ROBSON, Homomorphisms and extensions of modules over certain differential polynomial rings, *J. Algebra* **26** (1973) 319–342.

[174] J. MICKELSON, Step algebras of semisimple subalgebras of Lie algebras, *Rept. Math. Physics* **4** (1973) 307–318.

[175] J. MICKELSON, On irreducible modules of a Lie algebra which are composed from finite dimensional modules of a subalgebra, *Ann. Acad. Sci. Fenn.* (A) **598** (1975) 1–16.

[176] J. MICKELSON, On certain irreducible modules of the Lie algebra $\mathfrak{gl}(4, \mathbf{C})$, *Ann. Acad. Sci. Fenn.* (A) **1** (1975) 285–296.

[177] A. I. OOMS, On Lie algebras having a primitive universal enveloping algebra, *J. Algebra* **32** (1974) 488–500.

[178] H. OZEKI and M. WAKIMOTO, On polarizations of certain homogeneous spaces, *Hiroshima Math. J.* **2** (1972) 445–482.

[179] A. M. PERELOMOV and V. S. POPOV, Casimir operators for semisimple Lie groups (in Russian), *Izv. Akad. Nauk SSSR*, Ser. Mat. **32** (1968) (English Transl.: *Math. URSS Izv.* **2** (1968) 1313–1335.

[180] N. N. SHAPOVALOV, On a conjecture of Gelfand–Kirillov (in Russian) *Funkcion. Anal. Prilož.* **7** (1973) 165–166.

[181] N. N. SHAPOVALOV, Structure of the algebra of regular differential operators (in Russian) *Funkcion. Anal. Prilož.* **8** (1974) 43–54.

[182] S. J. TAKIFF, Invariant polynomials on Lie algebras of inhomogeneous unitary and special orthogonal groups, *Trans. Amer. Math. Soc.* **170** (1972) 221–230.

[183] P. TAUVEL, Sur les représentations des algèbres de Lie nilpotentes, *C. R. Acad. Sci. Paris* (A) **278** (1974) 977–979.

[184] A. VAN DEN HOMBERGH, Sur les suites de racines dont la somme des termes est nulle, *Bull. Soc. Math. France* **102** (1974) 353–364.

[185] A. VAN DEN HOMBERGH, A note on Mickelson's step algebra, *Indag. Math.* **37** (1975) 42–47.

[186] A. VAN DEN HOMBERGH, Harish-Chandra modules and representations of step algebras, Thesis, Nijmegen University (1976).

[187] A. VAN DEN HOMBERGH, Note on a paper by Bernhstein, Gel'fand and Gel'fand on Verma modules, *Indag. Math.* **36** (1974) 352–356.

[188] M. WAKIMOTO, Polarizations of certain homogeneous spaces and most continous principal series, *Hiroshima Math. J.* **2** (1972) 483–533.

[189] M. WAKIMOTO, On the irreducibility of induced representations of $SU(2, 1)$, *Hiroshima Math. J.* **3** (1973) 391–406.

[190] N. WALLACH, On the Enright-Varadarajan modules: a construction of the discrete series, *Ann. Sci. Ecole Norm. Sup.* **9** (1976) 81–101.

[191] F. L. WILLIAMS, The cohomology of semi-simple Lie algebras with coefficients in a Verma module, to appear.

[192] M. K. F. WONG and HSIN-YANG YEH, Eigenvalues of the invariant operators of the orthogonal and symplectic groups, *J. Math. Phys.* **16** (1975) 1239–1243.

[193] L. ABELLANAS and L. MARTINEZ ALONSO, A general setting for Casimir invariants, *J. Math. Phys.* **16** (1975) 1580–1584.

[194] J. ALEV, Un théorème d'Eisenbud–Evans dans les algèbres enveloppantes, *C. R. Acad. Sci. Paris* (A) **282** (1976) 763–765.

[195] W. BORHO, Berechnung der Gelfand–Kirillov Dimension bei induzierten Darstellungen, to appear.

[196] W. BORHO and J. C. JANTZEN, Über primitive Ideale in der Einhüllenden einer halbeinfachen Lie-Algebra, to appear.

[197] W. CASSELMAN and M. S. OSBORNE, The restriction of admissible representations to $\mathfrak{n}$, to appear.

[198] N. CONZE and M. DUFLO, Représentations induites des groupes semi-simples complexes, to appear.

[199] J. DIXMIER, Sur les algèbres enveloppantes de $\mathfrak{sl}(n, \mathbf{C})$ et $\mathfrak{af}(n, \mathbf{C})$, *Bull. Sci. Math.* **100** (1976) 57–95.

[200] J. Dixmier, Polarisations dans les algèbres de Lie, II, to appear.

[201] J. Dixmier, Idéaux primitifs dans les algèbres enveloppantes, to appear.

[202] M. Duflo, Représentations irréductibles des groupes semisimples complexes, dans: *Analyse harmonique sur les groupes de Lie*, Lecture Notes in Math. **497** (Springer, Berlin, 1975) 26–88.

[203] M. Duflo, Sur la classification des idéaux primitifs dans l'algèbre enveloppante d'une algèbre de Lie semi-simple, to appear.

[204] T. J. Enright and V. S. Varadarajan, On an infinitesimal characterization of the discrete series, *Ann. Math.* **102** (1975) 1–15.

[205] H. Garland and J. Lepowsky, Lie algebra homology and the Macdonald–Kac formulas, to appear.

[206] A. M. Gavrilik and A. U. Klimyk, Irreducible and indecomposable representations of the $\mathfrak{so}(n, 1)$ and $\mathfrak{iso}(n, 1)$ algebras, to appear.

[207] J. C. Jantzen, Darstellungen halbeinfacher algebraischer Gruppen und zugeordnete kontravariante Formen, Thesis, Bonn (1973).

[208] J. C. Jantzen, Kontravariante Formen auf induzierten Darstellungen halbeinfacher Lie-Algebren, to appear.

[209] A. Joseph, Sur les vecteurs de plus haut poids dans l'algèbre enveloppante d'une algèbre de Lie semi-simple complexe, *C. R. Acad. Sci. Paris* (A) **281** (1975) 835–837.

[210] A. Joseph, A preparation theorem for the prime spectrum of a semisimple Lie algebra, to appear.

[211] A. Joseph, Primitive ideals in the enveloping algebras of $\mathfrak{sl}(3)$ and $\mathfrak{sp}(4)$, to appear.

[212] A. Joseph, A generalization of Quillen's lemma and its application to the Weyl algebras, to appear.

[213] B. Kostant, On the tensor product of a finite and infinite dimensional representation, *J. Functional Anal.* **20** (1975) 257–285.

[214] C. Y. Lee, Invariant polynomials of Weyl groups and applications to the centres of universal enveloping algebras, *Can. J. Math.* **26** (1974) 583–592.

[215] J. Lepowsky, Linear factorization of conical polynomials over certain nonassociative algebras, *Trans. Amer. Math. Soc.* **216** (1976) 237–248.

[216] J. Lepowsky, A generalization of the Bernstein–Gelfand–Gelfand resolution, to appear.

[217] J. Lepowsky, Generalized Verma modules, the Cartan–Helgason theorem and the Harish-Chandra homomorphism, to appear.

[218] A. V. Lutsjuk, Homomorphisms of the modules M_χ (in Russian), *Funkcion. Anal. Prilož.* **8** (1974) 91–92.

[219] C. Martin, Sur certaines représentations de l'algèbre de Lie $\mathfrak{so}(4, 1)$ et de l'algèbre de Lie du groupe de Poincaré, *Ann. Inst. H. Poincaré* (A) **20**) (1974) 373–402.

[220] J. C. McConnell, Representations of solvable Lie algebras, III, Cancellation theorems, to appear.

[221] C. Moeglin, Factorialité dans les algèbres enveloppantes, *C. R. Acad. Sci. Paris* (A) **282** (1976).

[222] H. Moscovici and A. Verona, Remarques sur les idéaux premiers des algèbres enveloppantes, *Rev. Roumaine Math. Pures Appl.* **20** (1975) 423–428.

[223] H. Moscovici and A. Verona, Sur le spectre des anneaux de polynômes tordus *Rev. Roumaine Math. Pures Appl.* **21** (1976) 531–538.

[224] Nghiem Xuan Hai, Sur certaines représentations d'une algèbre de Lie résoluble complexe, to appear.

[225] A. I. Ooms, On Lie algebras with primitive envelopes, Supplements, to appear.

[226] R. Rentschler, Comportement de l'application de Dixmier par rapport à l'anti-automorphisme principal pour des algèbres de Lie résolubles, *C. R. Acad. Sci. Paris* (A) **282** (1976) 555–557.

[227] R. Rentschler, Idéaux complètement premiers de l'algèbre enveloppante de $\mathfrak{g} = \mathfrak{so}(5, \mathbf{C})$ et l'espace des orbites de $\mathfrak{g}^*$, to appear.

[228] S. Sternberg, Symplectic homogeneous space, *Trans. Amer. Math. Soc.* **212** (1975) 113–130.

[229] P. Tauvel, Polarisations et représentations induites des algèbres de Lie résolubles, *Bull. Sci. Math.* **100** (1976) 33–44.

[230] D. P. Zelobenko, Cyclic modules for a semisimple complex Lie group (in Russian), *Izv. Akad. Nauk SSSR, Ser. Mat.* **37** (1973) 502–515.

SUBJECT INDEX